PROCEEDINGS OF THE SEVENTH MULTIDISCIPL[I]
AND THE ENGINEERING AND ENVIRON[M]
HARRISBURG-HERSHEY/PENNSYLVANIA

HYDROGEOLOGY AND ENGINEERING GEOLOGY OF SINKHOLES AND KARST – 1999

Edited by

BARRY F. BECK, ARTHUR J. PETTIT & J. GAYLE HERRING
P.E. LaMoreaux & Associates, Inc., Oak Ridge, Tennessee

Sponsored by
P.E. LaMoreaux & Associates, Inc.
and Co-Sponsored by
The Federal Highway Administration, U.S. Department of Transportation
The Association of Ground Water Scientists and Engineers (NGWA)
The Geo-Institute of the American Society of Civil Engineers
The Virginia Water Resources Research Center
The Association of Engineering Geologists
The U.S. Environmental Protection Agency
The Pennsylvania Geological Survey
The Mid-Atlantic Karst Consortium
and The Karst Waters Institute

A.A. BALKEMA / ROTTERDAM / BROOKFIELD / 1999

Cover photo: Trick-or-Treat! This large cover-collapse sinkhole, approximately 60 feet by 30 feet and 10 to 20 feet deep, developed in the early morning hours of Halloween, 1995, beneath South Street in Frederick, Maryland, USA. The motorist was injured but was able to climb out of the hole unaided. The overpass in the background is Interstate-70. The sinkhole developed in sandy residual soils over the Grove Limestone.
Photo by A. David Martin.

The papers herein are the work and responsibility of the individual authors. Although every effort possible has been made to insure the professional quality of this volume, the editors and P.E. LaMoreaux & Associates, Inc., can make no warranty as to the validity of any of the data or recommendations discussed herein.

Authorization to photocopy items for internal or personal use, or the internal or personal use of specific clients, is granted by A.A. Balkema, Rotterdam, provided that the base fee of US$1.50 per copy, plus US$0.10 per page is paid directly to Copyright Clearance Center, 222 Rosewood Drive, Danvers, MA 01923. For those organizations that have been granted a photocopy license by CCC, a separate system of payment has been arranged. The fee code for users of the Transactional Reporting Service is: 90 5809 046 9/99 US$1.50 + US$0.10.

Published by
A.A. Balkema, P.O. Box 1675, 3000 BR Rotterdam, Netherlands
Fax: +31.10.413.5947; E-mail: balkema@balkema.nl; Internet site: www.balkema.nl
A.A. Balkema Publishers, Old Post Road, Brookfield, VT 05036-9704, USA
Fax: 802.276.3837; E-mail: info@ashgate.com

ISBN 90 5809 046 9

© 1999 A.A. Balkema, Rotterdam
Printed in the Netherlands

Table of contents

1 Keynote paper

Karst hydrology: Recent developments and open questions 3
W.B.White

2 Development and distribution of sinkholes and karst

Measuring strain and stress from sinkhole distribution: Example of the Velebit mountain range, Dinarides, Croatia 25
S.Faivre & Ph.Reiffsteck

Shallow karst in a breached anticline 31
R.E.Gray & R.J.Turka

Spatial relationships between new and old sinkholes in covered karst, Albany, Georgia, USA 37
J.A.Hyatt, H.P.Wilkes & P.M.Jacobs

An investigation into the factors causing sinkhole development at a site in Northampton County, Pennsylvania 45
S.A.Jenkins & J.E.Nyquist

Deep karst conduits, flooding and sinkholes: Lessons for the aggregates industry 51
J.L.Lolcama, H.A.Cohen & M.J.Tonkin

A geologic framework in karst: US Geological Survey contributions to the hydrogeology of the Ozarks of Missouri 57
R.C.Orndorff, S.Šebela, D.J.Weary, R.C.McDowell, R.W.Harrison & R.E.Weems

Topographic and hydrogeologic controls on sinkhole formation associated with quarry dewatering 63
S.Strum

Applications of GIS technology to the triggering phenomena of sinkholes in Central Florida 67
D.Whitman & T.Gubbels

3 Prediction of sinkhole development and planning studies in karst

Casual hydrofracturing theory and its application for sinkhole development prediction in the area of Novovoronezh Nuclear Power House-2 (NV NPH-2), Russia
A.V.Anikeev — 77

Karst atlas for Kentucky
J.C.Currens & J.A.Ray — 85

Geohazard map of cover-collapse sinkholes in the 'Tournaisis' area, southern Belgium
O.Kaufmann & Y.Quinif — 91

Predicting the distribution of dissolution pipes in the chalk of southern England using high resolution stratigraphy and geomorphological domain characterisation
J.Lamont-Black & R.Mortimore — 97

Is human activity accelerating karst evolution?
B.Paukštys — 103

Site selection and design considerations for construction in karst terrain/sinkhole-prone areas
B.A.Memon, A.F.Patton, L.D.George & Th.S.Green — 107

Not your typical karst: Characteristics of the Knox Group, southeastern US
J.C.Redwine — 111

4 Engineering and remedial treatment of sinkholes and other karst features

Detection and treatment of karst cavities in Kuwait
W.A.Abdullah & M.A.Mollah — 123

Engineering approaches to conditions created by a combination of karst and faulting at a hospital in Birmingham, Alabama
T.Cooley — 129

Remedial measures for residential structures damaged by sinkhole activity
R.C.Kannan & N.S.Nettles — 135

Hydrogeological monitoring strategies for investigating subsidence problems potentially attributable to gypsum karstification
J.Lamont-Black, P.L.Younger, R.A.Forth, A.H.Cooper & J.P.Bonniface — 141

Engineers challenged by Mother Nature's twist of geology
C.M.Reith, A.W.Cadden & C.J.Naples III — 149

Compaction grouting versus cap grouting for sinkhole remediation in east Tennessee
T.C.Siegel, J.J.Belgeri & M.W.Terry — 157

The role of the geological engineering consultant in residential sinkhole damage investigations
T.J.Smith & W.A.Ericson — 165

Karst and engineering practice V.V.Tolmachev	171
Replacing the Beards Creek bridge S.E.Walker & S.L.Matzat	179

5 Applications of geophysics to karst investigations

Application of electrical resistivity tomography and natural-potential technology to delineate potential sinkhole collapse areas in a covered karst terrane W.Zhou, B.F.Beck & J.B.Stephenson	187
Characterizing karst conditions at a low-level radioactive DOE site R.C.Benson & L.Yuhr	195
Microgravity techniques for subsurface investigations of sinkhole collapses and for detection of groundwater flow paths through karst aquifers N.C.Crawford, M.A.Lewis, S.A.Winter & J.A.Webster	203
Two-dimensional resistivity profiling; geophysical weapon of choice in karst terrain for engineering applications M.H.Dunscomb & E.Rehwoldt	219
Karst system characterization utilizing surface geophysical, borehole geophysical and dye tracing techniques S.George, T.Aley & A.Lange	225
Integrated geophysical surveys applied to karstic studies R.McDonald, N.Russill & R.Davies	243
A case study of the reliability of multi-electrode earth resistivity testing for geotechnical investigations in karst terrains M.J.S.Roth, J.R.Mackey, C.Mackey & J.E.Nyquist	247
Remediation of leakage of an earth-filled dam and reservoir by geophysically-directed grouting, Washington, Missouri D.G.Taylor & A.L.Lange	253

6 Governmental role in karst areas: Regulations and education

Education about and management of sinkholes in karst areas: Initial efforts in Lebanon County, Pennsylvania E.D.Buskirk Jr, M.D.Pavelek II & R.Strasz	263
Irish methodologies for karst aquifer protection D.Daly & D.Drew	267
Maryland's zone of dewatering influence law for limestone quarries M.K.Gary	273

Regulating construction of manure storage systems in sinkhole prone areas of Minnesota   279
D.B.Wall

7 Dye tracing and the delineation of karst groundwater basins

Karst groundwater basin delineation, Fort Knox, Kentucky   287
D.P.Connair, S.A.Engel & B.S.Murray

Dye study tracks historical pathway of VOC-bearing industrial waste water from failed pond at metals coating facility   293
L.D.George & G.M.Ponta

A method for correction of variable background fluorescence in filter fluorometry   301
S.R.Lane & C.C.Smart

Non-linear curve-fitting analysis as a tool for identifying and quantifying multiple fluorescence tracer dyes present in samples analyzed with a spectrofluorophotometer and collected as part of a dye tracer study of groundwater flow   307
R.B.Tucker & N.C.Crawford

8 Environmental hydrogeology in karst terrane

Shallow lateral DNAPL migration within slightly dipping limestone, southwestern Kentucky   315
M.Jancin & W.F.Ebaugh

Environmental characterization of karstic terrains: A case study for the practical application of stable isotope ratios and anion/cation analysis of ground water   323
J.M.Mason, W.J.Gabriel & D.I.Siegel

The protection of a karst water resource from the example of the Larzac karst plateau (south of France). A matter of regulations or a matter of process knowledge?   331
V.Plagnes & M.Bakalowicz

Resources and quality of waters in limestone areas of Peddavanka watershed, A.P., India   339
A.S.Sudheer & S.Srinivasa Gowd

Transport and variability of trace metals in a karst aquifer based on spring chemistry   345
D.J.Vesper

Use of GORE-SORBER® modules to screen for organic contaminants in karst springs   351
D.J.Vesper, J.E.Rice & R.F.Fenstermacher

9 Waste disposal and storage in karst terrane

Design of geotechnical fabrics for septic systems in karst   359
J.A.Fischer, J.J.Fischer & R.S.Ottoson

The environmental hazards of locating wastewater impoundments in karst terrain   365
B.A.Memon, M.Mumtaz Azmeh & M.Wallace Pitts

Stability evaluation for the siting of municipal landfills in karst 373
M.Z.Yang & E.C.Drumm

10 Stormwater management and flood hazards in karst terrane

Simulating time-varying cave flow and water levels using the Storm Water Management Model (SWMM) 383
C.W.Campbell & S.M.Sullivan

Stormwater management design in karst terrane adjusting hydrology models and using karstic features 389
J.C.Laughland

A review of stormwater best management practices for karst areas 395
M.S.McCann & J.L.Smoot

11 Special session on highways in karst – Design, construction and repair

Highway engineering geology of karst collapse features in the Sacramento Mountains, Otero and Lincoln Counties, New Mexico 401
R.M.Colpitts & W.R.Hahman

Road and bridge construction across gypsum karst in England 407
A.H.Cooper & J.M.Saunders

Remediation of sinkholes along Virginia's highways 413
D.A.Hubbard Jr

The vulnerability map of karst along highways in Slovenia 419
S.Šebela, A.Mihevc & T.Slabe

The system of antikarst protection on railways of Russia 423
V.V.Tolmachev, S.E.Pidyashenko & T.A.Balashova

Geotechnical engineering and geology for a highway through cone karst in Puerto Rico 431
L.Vazquez Castillo & C.Rodriguez Molina

Fieldtrip guidebook

Pre-conference field trip: Karst geology and hydrogeology of the Ridge and Valley and Piedmont Provinces of south-central Pennsylvania 449
W.E.Kochanov, H.L.Delano, C.deWet, S.B.Gaswirth, D.A.Hopkins, J.L.Lieberfinger, H.L.Reccelli-Snyder, R.A.Hoover & A.E.Becher

Email addresses 475

Author index 477

1 Keynote paper

Karst hydrology: Recent developments and open questions

WILLIAM B.WHITE Department of Geosciences and Materials Research Laboratory, The Pennsylvania State University, University Park, Pa., USA

ABSTRACT

Karst aquifers are those that contain dissolution-generated conduits that permit rapid transport of ground water, often in turbulent flow. The conduit system receives localized inputs from sinking surface streams and as storm runoff through sinkholes. The conduit system interconnects with ground water stored in fractures and in the granular permeability of the bedrock. As a conceptual framework, the basic components of karstic aquifers seem to be generally accepted. Progress in the decade of the 1990's has focused mainly on quantifying the conceptual model. The equilibrium chemistry of limestone and dolomite dissolution has been reliably established and there are formal models for the kinetics of dissolution. Kinetic models have been used to calculate both fracture enlargement to protoconduits (0.01 m aperture) and the enlargement of protoconduits to the size of typical cave passages. Modeling of ground water flow in karstic aquifers has been less successful. Progress has been made in the use of water budgets, tracer studies, hydrograph analysis, and chemograph analysis for the characterization of karstic aquifers. Topics on which progress is needed include (a) the construction of models that describe the complete aquifer including the interactions of all components, (b) models for clastic sediment transport within the aquifer, and (c) working out processes and mechanisms for contaminant transport in karst aquifers. An optimistic assessment at the end of the millennium is that a complete model for karstic aquifers is visible on the horizon.

INTRODUCTION

Understanding properties, characterization and evolution of karstic carbonate aquifers has improved substantially over what it was only a few years ago. Concepts of karst hydrology written only a decade ago (White, 1988; Dreybrodt, 1988, Ford and Williams, 1989) have some need of updating and important new results need to be added to the conceptual framework. Some updating has been outlined in two previous reviews (White, 1993, 1998). The understanding of karst hydrology has now progressed to a point where a synthesis of the entire subject can be visualized, at least in broad outline, although much detail needs to be filled in. It is the purpose of this review to outline this synthesis, to summarize those parts where substantial new detail has been established, and to indicate the blank spaces where additional research is necessary. In a sense, what is being done is to organize and systematize a lot of answers and then go back and say, "what was the question?"

The characterizing features of karst aquifers are the open conduits which provide low resistance pathways for ground water flow and which often short circuit the granular or fracture permeability of the aquifer. Conduit flow often has more in common with surface water than with ground water. Karst hydrology requires a mix of surface water concepts and ground water concepts.

The attitude of the hydrogeological community toward conduit permeability has come almost full circle. In the early days - say the 1960's - the general opinion was that conduits had nothing to do with ground water hydrology. Certainly, they existed. No one denied the existence of caves. But it was felt that in the active ground water system, conduits were simply water-filled cavities that did not disrupt the flow field. Thrailkill (1968) captured the essence of this thinking in a drawing reproduced in Figure 1. As the conduit system became understood as the main flow path from recharge source to spring discharge in many aquifers, attention became focused on the conduits. Great advances in cave exploration and survey, systematic tracer studies delineating ground water basins, and physical and chemical studies of karst springs in the decades of the 1970's and 1980's gave the impression that karst hydrology was the hydrologic properties of conduit systems. The new thinking of the decade of the 1990's has been to put all of the components back together again so that investigators are (finally!) talking about conduit flow, fracture flow, matrix flow, and the coupling between them.

All aquifers evolve, at least on a scale of tens or hundreds of thousands of years, as rivers downcut to lower base levels, tectonic forces shift elevations of recharge areas, and soils thicken or are eroded away. Within silicate rock aquifers, however, permeability changes only slowly with dissolution or precipitation of minerals within the pore spaces or along fractures. In contrast, the permeability of karst aquifers arises primarily from the enlargement of joints and bedding plane partings as circulating groundwater removes the carbonate bedrock. On a geologic time scale, dissolution is a very rapid process with significant modifications of flow

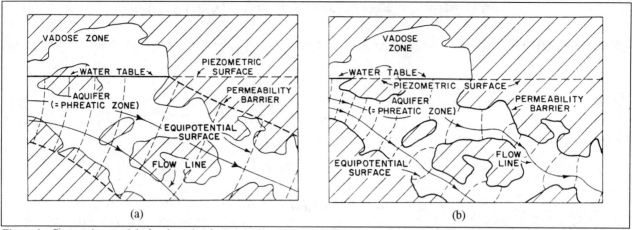

Figure 1. Contrasting models for the role of conduits in carbonate aquifers. (a) Water-filled cave in continuous flow field. (b) Cave = aquifer. Adapted from Thrailkill (1968).

paths and the development of new flow paths requiring only thousands of years. As a result, karst aquifers must be regarded as works in progress. What is observed is only a snapshot of a system that is rapidly changing. A great deal of effort has gone into modeling the evolution of karst aquifers as the dissolution process progresses. This is an area where great progress has been made in the past decade. From the point of view of practical hydrology, water resource development, resource protection, and contaminant transport of the aquifer must be taken as it exists at the present moment of time. The snapshot is what must be analyzed.

THE CONCEPTUAL FRAMEWORK

A shift in the conceptualization of karst aquifers that took place gradually, but only in the past two decades has been specifically articulated, is a recognition of the close relationship between surface water and ground water in karstic regions.

Surface water is discussed in terms of the drainage basin. Upstream from some specified gauge point is a pattern of surface drainage channels with streams that are fed by ground water discharge and by overland storm flow from the land surface that slopes down into the channels. It is possible to draw a drainage divide that encircles the surface water basin so that all precipitation that falls within the divide passes the gauge point by way of the main drainage channel. In karst regions, surface water becomes ground water when it sinks into the stream bed or into swallets. Karst ground water becomes surface water when it emerges from springs. These springs are often of high discharge and form the headwaters of sizable surface streams.

The ground water that discharges from karst springs is usually collected from a limited volume of the carbonate rocks that make up the aquifer. Tracer tests on sinking streams and cave streams, surveyed cave passage, and considerations of structural and lithologic boundaries often permit the delineation of a ground water basin for each spring (or group of springs). Identification of groundwater basins allows the writing of water budgets for karst springs just as they are written for surface water basins, a technique that cannot be easily applied to other types of aquifers. Ground water basin divides, however, are not firmly fixed as surface water divides are fixed by topographic highs. Ground water divides may shift depending on rates of recharge. Underground piracy routes and ground water spillover routes into adjacent ground water basins are common.

Most investigators are in essential agreement on the components of the karst surface water/ground water system in regions of fluviokarst and doline karst. The majority of the karst regions of central and eastern United States are fluviokarst with some limited areas of doline karst. The conceptual model can be drawn in various ways but the essential features are those shown in Figure 2. This model provides the basis for the discussion that follows.

Recharge

Precipitation falling into a karstic drainage basin can be divided into the following components:
  a. Allogenic recharge: Surface streams draining from non-carbonate portions of the basin and entering the carbonate aquifer through swallets.
  b. Diffuse (or dispersed) infiltration: Rainfall directly onto the karst surface and from there entering the aquifer as infiltration through the soil and the fracture and matrix permeability of the underlying carbonate bedrock.
  c. Internal runoff: Storm flow into closed depressions where the storm water enters the aquifer quickly through sinkhole drains.
  d. Overflow from caprock or perched aquifers: Precipitation collected on clastic plateau caprock or other perched aquifers above the carbonate aquifer. Overflow water enters the aquifer through vertical shafts or solutionally widened fractures in the vadose zone.

Allogenic recharge can be measured by stream gauges placed just upstream from the swallets. Placement of the gauges requires care because at many locations water loss will occur at a sequence of sink points distributed along the stream bed. In broad brush terms, however, allogenic recharge is determined by the areas of the allogenic catchments and the annual precipitation at the location.

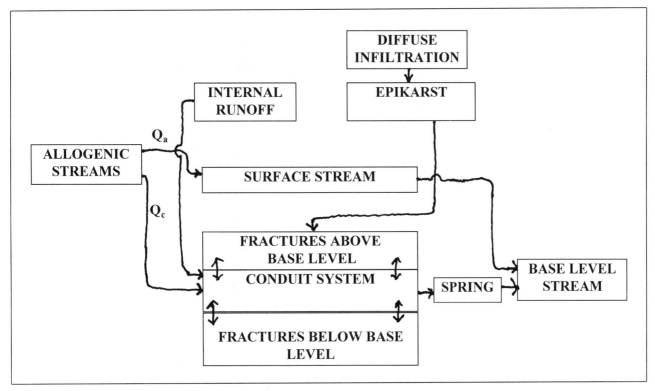

Figure 2. Conceptual model for carbonate aquifer.

Allogenic recharge can be measured by stream gauges placed just upstream from the swallets. Placement of the gauges requires care because at many locations water loss will occur at a sequence of sink points distributed along the stream bed. In broad brush terms, however, allogenic recharge is determined by the areas of the allogenic catchments and the annual precipitation at the location.

The underground drainage system is not equally well developed in all regions. We can define a "carrying capacity" for the conduit drainage system, Q_c. The carrying capacity represents the entire underground drainage route, from swallet to spring, including obstructions in the conduit caused by breakdown or silt infilling, and it also includes the increase in flow volume caused by rising heads during storm events. By comparing the carrying capacity of the underground system with the available allogenic recharge, Q_a, we can distinguish three possibilities.

a. $Q_c > Q_a$ (max)
b. Q_a (max) $> Q_c > Q_a$ (base)
c. $Q_c > Q_a$ (min)

In case (a) the underground system is capable of carrying the allogenic runoff from even the most extreme storms. All allogenic runoff will become part of the allogenic recharge and there will be no surface flow across the karst (and likely no surface stream channel either). In case (b), base flows will be lost to the subsurface but storm flow will overtop the swallets when the runoff, Q_a, becomes equal to Q_c. Excess storm runoff from the allogenic basins will spill over into the surface channel and remain as surface water. In case (c) the carrying capacity is not sufficient to transmit even the base flow of the allogenic basin. There will be a perennial surface stream flowing through the karst although the flow volume would likely be less that expected from the overall area of the basin.

Diffuse infiltration in karst drainage basins is not intrinsically different from infiltration, as described in any hydrogeology textbook, in any other type of basin. Initial precipitation is taken up as soil moisture and when the soils are saturated, water begins to percolate downward through the matrix porosity or along fractures in the bedrock, ultimately to reach the water table and storage in the phreatic zone of the aquifer. In karst drainage basins, diffuse infiltration recharges mainly the fracture (and possibly matrix) porosity of the aquifer. What is somewhat different in karst terrain is the presence of the epikarst (or subcutaneous zone). The soil/carbonate rock interface is usually sharp with the weathered rock zone or C-horizon often missing. Karst soils tend to be well leached with very little carbonate material surviving. The bedrock surface, however, is often highly irregular with deep crevices dissolved along fractures (known as "grikes" or "cutters"). These are soil filled and serve as reservoirs for infiltrating water. Water may move laterally along the crevices for considerable distances before finding an open fracture along which it can descend into the bedrock. Because of temporary storage in the epikarst, diffuse infiltration water may require days to weeks to reach the water table. Where soils are thin and the bedrock highly fractured, diffuse infiltration reaches the subsurface rapidly giving sharp peaked hydrographs that lag only slightly behind the precipitation. This is illustrated by excellent data from the Planina Cave in Slovenia (Kogovsek, 1981) (Figure 3). Paul Williams (1983, 1985) called attention to the importance of this component of the hydrologic

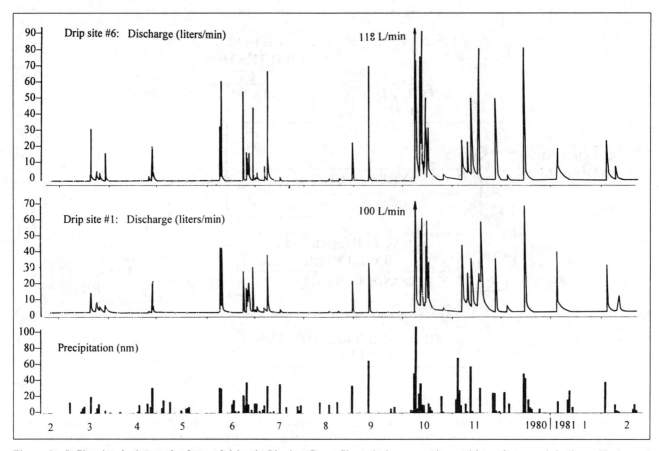

Figure 3. Infiltration hydrographs for roof drips in Planina Cave, Slovenia in comparison with surface precipitation. Horizontal scale labels: months in 1980/1981. From Kogovsek (1981).

system and called it the "subcutaneous zone". French researchers were already using the term "epikarst" for the same component of the karst drainage system and that name seems to have come into common usage.

Precipitation that enters the karst aquifer from the overlying soil is known in some literature as autogenic recharge. It seems worthwhile, as suggested by Gunn (1983), to separate the autogenic recharge into two components: the diffuse infiltration described above and what is here called "internal runoff". Internal runoff is storm flow and is exactly equivalent to overland flow in the usual description of the hydrologic cycle. However, in many karst regions, overland flow discharges into sinkholes rather than surface streams. Some sinkholes have open drains that connect directly to the underlying conduit system. Others are soil filled but with frequent soil piping failures which also indicate connections to an open conduit system. Sinkhole drains are only rarely large enough to permit human exploration. They can sometimes been seen in profile in quarries and roadcuts. In central Pennsylvania they appear as jagged, irregular, vertical openings on the order of 5 - 20 cm in cross-section. The typical openings are too small to function as caves but they are certainly large enough to function as storm drains.

Precipitation that enters the karst aquifer from the overlying soil is known in some literature as autogenic recharge. It seems worthwhile, as suggested by Gunn (1983), to separate the autogenic recharge into two components: the diffuse infiltration described above and what is here called "internal runoff". Internal runoff is storm flow and is exactly equivalent to overland flow in the usual description of the hydrologic cycle. However, in many karst regions, overland flow discharges into sinkholes rather than surface streams. Some sinkholes have open drains that connect directly to the underlying conduit system. Others are soil filled but with frequent soil piping failures which also indicate connections to an open conduit system. Sinkhole drains are only rarely large enough to permit human exploration. They can sometimes been seen in profile in quarries and roadcuts. In central Pennsylvania they appear as jagged, irregular, vertical openings on the order of 5 - 20 cm in cross-section. The typical openings are too small to function as caves but they are certainly large enough to function as storm drains.

Many carbonate aquifers in eastern United States occur in low-dip Mississippian limestones. The limestones are exposed on valley walls and beneath valley floors in a dissected plateau topographic setting. Examples are the Cumberland Plateau of Tennessee and Alabama, the Mammoth Cave Plateau of southcentral Kentucky, and many other examples. In this setting, large areas of limestone are capped by shales and sandstones which are the rocks exposed at the surface over much of the plateaus. Runoff from the clastic caprock or discharge from perched aquifers within the caprock sequence drain from the edge of the plateaus and penetrate the vadose zone of the underlying limestone aquifer through solutionally widened fractures or vertical shafts. In locations where the caprock itself is permeable, water infiltrates over the entire caprock area as a form of diffuse infiltration.

Flow Systems

Water recharged into karst aquifers moves down gradient through them using some combination of highly anisotropic pathways. Karst aquifers are usually discussed in terms of a triple porosity model or triple permeability model.
  a. Matrix permeability: The intergranular permeability of unfractured bedrock.
  b. Fracture permeability: Mechanical joints, joint swarms, and bedding plane partings, all of these possibly enlarged by solution.
  c. Conduit permeability: pipe-like openings with apertures ranging from 1 cm to a few tens of meters.

The Paleozoic carbonate rocks of Eastern United States tend to have a very low matrix permeability. The matrix permeability is sufficiently small that matrix flow has often been neglected. The Tertiary limestones that make up the Floridan aquifer or the carbonate aquifers of Puerto Rico and other Caribbean Islands have much higher matrix permeabilities and matrix flow is an important part of the overall ground water flow system.

The term "fracture" is used to encompass single joints, joint swarms, and bedding plane partings. In the pristine state these have apertures in the range of 50 - 500 micrometers. Fractures in carbonate aquifers are rarely in a pristine state. They are enlarged by dissolution and eventually reach a width (about one centimeter) where they would be called "conduits". Essential parameters are effective fracture aperture and fracture spacing.

Conduits are solutionally enlarged pathways through the carbonate aquifer. Ground water flow in conduits is localized, and under normal gradients flows in a turbulent regime. The onset of non-Darcian behavior occurs when the aperture exceeds about one centimeter. Caves are conduit fragments large enough (and provided with entrances) for human exploration. Within any given karst drainage basin, the accessible caves usually tally up to much less (often less than 1%) of the conduit length that swallet-spring distances and tracer tests demand must be present. There is the unavoidable problem that the minimum size for a conduit is 0.01 meter while the minimum size for direct exploration and survey is about 0.5 meter. One suspects that there is a very large component of conduit porosity with a size scale in the 0.01 to 0.5 meter range about which we know almost nothing.

Even within the context of the generally accepted conceptual model for karst aquifers, it is not precisely known whether the openings that provide the porosity really form three distinct groups. In terms of effective aperture, the aperture of the smaller joints is comparable to the intergranular pathways in the more porous young limestones and indeed matrix flow may dominate over fracture flow in some of these rocks. Fractures are modified by dissolution. Fracture swarms form zones of high permeability and individual fractures are observed to have apertures in excess of 0.01 meter although they may not be continuous at this aperture.

Permeability is generally extracted from hydraulic conductivity which is turn is determined by laboratory measurements on rock cores or by pumping, slug, or packer tests on wells drilled into the aquifer. Intergranular or matrix permeability can be measured on core samples (given enough cores to account for facies changes laterally along the beds and lithologic changes in the stratigraphic section). What is measured with pumping or other tests on wells is highly sensitive to exactly where the wells are placed. Mostly, the hydraulic conductivity extracted from pumping tests is a measure of the fracture permeability in the vicinity of the well bore and again, multiple sampling is necessary to assure that the results are representative of the aquifer as a whole. Some guidance for well placement in aquifers with fracture permeability dominated by near vertical joints and joint swarms is provided by fracture trace mapping from aerial photographs (Lattman and Parizek, 1964; Parizek, 1976). It is much more difficult to locate wells in aquifers with fracture permeability dominated by bedding plane partings although the systematic use of packer tests can be helpful. In general, conduit permeability cannot be defined, let alone measured. Simple observation of roadcuts and quarries shows that fully developed conduits are sparse, making up only a minute percentage of the aquifer cross-section. Yet conduits completely dominate the flow behavior of the aquifer when they are present. The net result is that the effective permeability of a karst aquifer is highly scale dependent as described by Teutsch and Sauter (1991) and Sauter (1992) (Figure 4).

Discharge

Ground water from karst aquifers is usually discharged back to surface routes through large springs. Spring discharges account for the runoff from the entire karst ground water basin including any allogenic inputs. Some identified spring types are:
  a. Open conduit gravity springs where the water emerges from an open cave mouth, sometimes accessible to direct exploration. More often the conduit is blocked by hillside talus so that the spring discharge issues from the rubble pile.
  b. Alluviated conduit springs where water emerges from a rise pool formed by glacial material or alluvial sediment acting as a dam for the conduit.
  c. Rise pools discharging water from shallow flooded conduits (shallow = a few meters to a few tens of meters)
  d. Artesian springs where water rises under pressure from considerable depths (considerable = 100's of meters)
  e. Springs that discharge from solutionally widened fracture swarms.

Karst springs exhibit a tremendous variety of physical forms and rates of discharge. Structural or stratigraphic constraints sometimes result in spring clusters where the discharges from multiple drainage basins resurface in the same general area. Distributary systems are common so that the same ground water basin may discharge from more than one spring. Rates of discharge vary by many orders of magnitude.

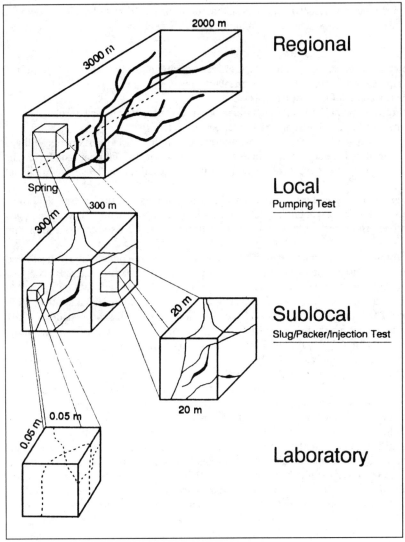

Figure 4. Scale dependence of carbonate aquifer permeability. From Sauter (1992)

Karst springs exhibit a tremendous variety of physical forms and rates of discharge. Structural or stratigraphic constraints sometimes result in spring clusters where the discharges from multiple drainage basins resurface in the same general area. Distributary systems are common so that the same ground water basin may discharge from more than one spring. Rates of discharge vary by many orders of magnitude.

Very useful information on spring discharge systems has been provided by cave divers over the past decade although little of the information is formally published in the scientific literature. In terrain of moderate relief many active conduits under low flow conditions are found to consist of segments of stream passage terminated by sumps. Sump diving has shown many of the water-filled segments to be relatively shallow and to connect with further segments of air-filled conduit. Diving at sumped spring mouths has also led to air-filled cave, justifying category (c). The large springs of Florida and the Bahamas with their associated conduit systems are flooded by post-Pleistocene sea level rise and are a more extreme example of category (c). A very few deep dives, taken to the very limit of SCUBA techniques, provides some information on the internal structure of (d) artesian springs. Some maximum depths that have been reached are: 205 m in fountaine de Vaucluse (France), 238 m in nacimiento del Rio Mante (Mexico) (Courbon et al., 1989), and 282 m in cenote Zacaton (Mexico) (Kristovich and Bowden, 1995). The fountaine de Vaucluse has been plumbed by instruments to a bottom depth of 308 m.

With respect to category (e), there is another bit of nomenclature that needs to be discussed. In the early days of karst hydrology, Shuster and White (1971) examined the chemical properties and physical setting of 14 springs in the folded Ordovician carbonates of central Pennsylvania. They found two quite distinct populations. One group of springs discharged from fractured dolomites, displayed nearly constant temperature and chemical composition without regard to seasonal cycles or storm inputs, and was recharged mainly from diffuse infiltration through thick soils of the valley floor. The second group discharged from highly karstic limestones, sometimes from open cave systems, displayed temperatures that varied seasonally and chemical compositions that responded rapidly to storm pulses passing through the system, and was recharged both from internal runoff into sinkholes and from a large component of mountain runoff sinking at the limestone/shale contact. They labeled the first set "diffuse flow springs" and the second group "conduit flow springs" and these terms have become widely used. Entirely by chance, this early investigation was made in a hydrologic setting where the response functions of the two sets of springs were near the extreme opposite ends of the range. Further studies by many investigators in many different regions and geological settings suggest that the real world is more complicated. Some have made the flat statement that all carbonate springs are conduit springs and that the term "diffuse flow" spring should be discontinued (e.g. Worthington, 1991; Schindel et al., 1996). The argument is that because the critical aperture at which non-linear flow regimes appear is about 0.01 meter, even springs fed by clusters of solutionally widened fractures will exceed this limit. A pertinent point. However, from the point of view of springs as water supplies, there is value in distinguishing between springs fed mainly by fracture systems and springs fed by natural storm drains although perhaps they should not be called "diffuse flow" springs.

Recent literature has emphasized a distinction proposed by Worthington (1991) between "underflow" springs that carry the base flow from ground water basins and "overflow" springs that carry only part of the discharge or become active only during storm flow. What is important from the viewpoint of water quality monitoring and contaminant transport is that the underflow springs are often hidden, for example by rising in the bed of the surface stream. Failure to locate all spring orifices means that water budgets will not balance, tracer tests may be misinterpreted, and contaminants may escape unnoticed.

Springs enter prominently into the discussion of karst aquifers because their flow behavior, turbidity, and chemistry reflect a composite of everything that has happened upstream. Springs are, therefore, the appropriate gauging points, sampling points, and monitoring points for karst aquifers.

CHEMISTRY
Equilibrium

One triumph is the complete delineation of the chemical reactions for the dissolution of calcite and dolomite. The basic chemistry reached a mature formulation in the early 1980's with publication of reliable values for equilibrium constants (Plummer and Busenberg, 1982), identification of pertinent complexes and their role in the chemistry, and the establishment of useful interpretative parameters, hardness, saturation index, and calculated carbon dioxide partial pressure. This aspect of karst hydrology has reached the form of standard textbook presentation (White, 1988; Langmuir, 1996; Drever, 1997). Chemical calculations can be reliably performed in several speciation and reaction path computer programs such as WATEQ4F, MINTEQA2 and PHREEQE. See Langmuir (1996), in particular, for discussion of the range of application and the limitations of the various programs.

Kinetics

The development of karst aquifers is mostly a matter of the kinetics of the dissolution reactions rather than the final equilibrium state of these reactions. A good starting point is the venerable Plummer - Wigley - Parkhurst equation (Plummer et al., 1978) which attempted to sort out different chemical reactions, the individual rates of which summed to provide an overall dissolution rate for calcite. The reactions are

$$CaCO_3 + H^+ \Rightarrow Ca^{2+} + HCO_3^- \quad (1)$$

$$CaCO_3 + H_2CO_3 \Rightarrow Ca^{2+} + 2HCO_3^- \quad (2)$$

$$CaCO_3 + H_2O \Rightarrow Ca^{2+} + HCO_3^- + OH^- \quad (3)$$

Each reaction is described for a forward reaction term in the rate equation.

$$\text{Rate} = k_1 a_{H^+} + k_2 a_{H_2CO_3} + k_3 a_{H_2O} - k_4 a_{Ca^{2+}} a_{HCO_3^-} \quad (4)$$

The first term is mass transfer controlled but the second and third terms are reaction rate controlled so that in the pH range of karst ground waters the dissolution rate is only slightly dependent on flow regime.

Considerable additional progress has been made toward understanding carbonate dissolution kinetics. Further investigations of reaction rates (Palmer, 1991; Dreybrodt and Buhmann, 1991; have made use of a generic rate equation

$$\text{Rate} = \frac{A}{V}\frac{dC}{dt} = k\left(1 - \frac{C}{C_s}\right)^n \quad (5)$$

where A = area, V = volume of solution, k = reaction rate constant, C = concentration of dissolved carbonate, C_s = equilibrium saturation concentration for dissolved carbonate, and n = reaction order. By treating the reaction order as an empirical parameter, it was possible to obtain very good fits to experimental data (e.g. Figure 5). What shows up clearly in this analysis of the data is the break in reaction rate at about 85 percent saturation. When the water is undersaturated, the reaction kinetics are roughly first order; when the water is approaching saturation, the reaction order increases to near fourth order. This break is of profound importance in the development of conduit permeability.

Further investigations of dissolution kinetics under conditions of turbulent flow and under conditions of near equilibrium (Dreybrodt and Buhmann, 1991; Svensson and Dreybrodt, 1992; Dreybrodt et al., 1996; Liu and Dreybrodt, 1997) reveals additional controls.
   a. The hydration reaction of aqueous CO_2 to H_2CO_3, which has a time constant of about 30 seconds, is shown to be rate controlling when A/V ratios are high and under some conditions of turbulent flow.
   b. Adsorption of ions on the reactive surface becomes rate-control-ling under near-saturation conditions.
   c. A diffusion boundary layer becomes important under conditions of turbulent flow.

EVOLUTION OF CONDUIT PERMEABILITY
Critical Thresholds

Fracture aquifers are found in rocks other than carbonates. There are fractured quartzites, fractured granites and fractured basalts all of which can be productive aquifers under proper circumstances. Some moderate success has been achieved by modeling fracture aquifers using the same approach used in porous media aquifers, the only requirement being to keep the scale sufficiently

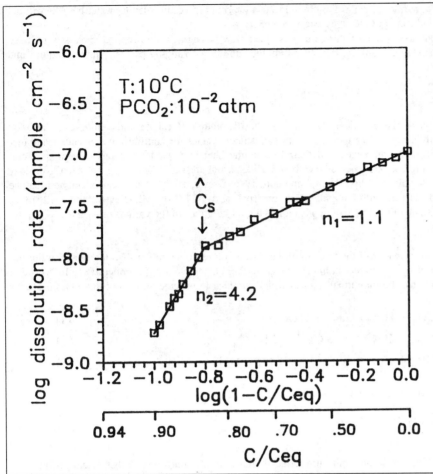

Figure 5. Typical dissolution rate data for calcite in a closed system with P_{CO_2} = 0.01 atm. From Dreybrodt (1998)

large (or the fracture spacing sufficiently small) that the discontinuity between fractured and unfractured rock is blurred out. These models demand that fracture flow be laminar and that the flow field follow Darcy's law, at least on the average. What is different about fractured carbonates is that the fractures enlarge with time and continued ground water circulation.

The evolution from fractures to conduits is discontinuous. There are three thresholds that appear when the aperture reaches roughly 0.01 meter.

a. Hydrodynamic threshold: The onset of turbulent flow and resulting loss of Darcian behavior.
b. Kinetic threshold: A shift in dissolution rate from 4th order kinetics to linear kinetics.
c. Transport threshold: Apertures and flow velocities become sufficient for the transport of clastic material.

The hydrodynamic threshold is (at least in principle) quite straightforward. Flow of water in small fractures is laminar and is described by the version of Darcy's law known as the Hagen-Poiseuille equation

$$h_f = \frac{12 \eta \bar{v} L}{\rho g B^2}$$

where h_f = head loss, η = viscosity of fluid, $\bar{v}$ = velocity, L = path length along fracture, ρ = density of fluid, g = acceleration due to gravity, B = full aperture (not half-aperture as is often used in fracture flow calculations). For a choice of hydraulic gradient (head/path length), mean flow velocity can be calculated as a function of aperture. These values in turn can be used to calculate a Reynolds number as a function of aperture. When the Reynolds number reaches the range of 500, there will be an onset of turbulence. For the hydraulic gradients typically observed moderate relief karst aquifers, this occurs when the aperture reaches 0.01 meter.

The kinetic threshold is more difficult and also dependent on the choice of rate equation. When unsaturated water enters a fracture, there is a rapid reaction and uptake of dissolved carbonate. If the kinetics were strictly linear, goes the argument, the water would quickly become saturated, dissolution would cease, and there would be no dissolutional enlargement of fractures deep within the aquifer. There would be surface karst but no caves. However, if the rate shifts from first order to fourth order when the water is only 85 percent saturated, the rate slows down by some orders of magnitude, thus permitting slow, laminar flow to carry the slightly undersaturated water deep into the aquifer. Dissolution continues along the entire flow path but at a slow rate. The significant concept is the "breakthrough". Of all possible pathways through the aquifer along joints, joint swarms, and bedding plane partings, one pathway will, only by chance, have a slightly larger average initial aperture than the others. This pathway will gradually enlarge until it's aperture is such that water with less than the 85 percent saturation can penetrate the entire aquifer. When this happens, the kinetics shift from fourth order to first order and the rate of dissolution dramatically increases. In effect, the shift in kinetics triggers a runaway process which allows the chosen pathway to enlarge rapidly. This is the kinetic threshold or kinetic trigger that explains why the number of fully developed conduits is usually (but not always) much smaller than the number of fractures.

Transport of clastic particles ranging in size from colloids to boulders either as bedload or as suspended load is a complex process that has not been completely worked out for karstic conduits. However, standard diagrams showing the critical velocity for sediment movement as a function of particle size show that the critical velocities are indeed very similar to the velocities for the onset of turbulence. Thus the threshold for sediment transport occurs at about the same aperture as the other thresholds.

The coincidence of all three critical thresholds at an aperture of 0.01 meter provides a natural dividing line between fracture aquifers and conduit aquifers. It also separates the process of conduit development into an initiation phase as the mechanical fracture enlarges to the critical thresholds and an enlargement phase as the protoconduit enlarges to a fully developed conduit of the size of typical cave passages.

Initiation

The basic concept for initiation of conduit permeability is that of an initial pathway consisting of a sequence of joints, joint swarms, and bedding plane partings connecting the recharge area to the (future) discharge area. Water is driven through the sequence of fractures by the head difference between inlet and outlet. Because the hydraulic gradient and the dominant fracture orientations are unlikely to be parallel, the pathway will have a certain tortuosity in both vertical and horizontal planes. The total path length can range from tens of meters to tens of kilometers. In the initiation phase, the aperture increases slowly under fourth order kinetics until it becomes sufficiently wide to permit undersaturated water to penetrate the entire distance from inlet to exit.

Gradual enlargement of single fractures lends itself to accurate geochemical modeling. Because fracture flow is laminar in the initiation phase, flow velocities can be described by the Hagen-Poiseuille equation. Using fourth order kinetics, Dreybrodt (1990) constructed a model by which he was able to show that the sensitive parameters are the initial aperture, the path length, and the hydraulic gradient. The model is substantially less sensitive to rate constants which themselves are determined by details of bedrock lithology. Early calculations of time to breakthrough (White, 1988) suggested that the critical thresholds could be reached in as little as 3000 - 5000 years. Dreybrodt's new calculations have extended this time to values in the range of 10,000 years for wide fractures and short travel distances under the hydraulic gradients commonly found in nature and much longer times, up to millions of years, for the least favorable conditions. The sensitivity of breakthrough time to initial aperture led Groves and Howard (1994-a) to the conclusion that there is a minimum aperture, depending on hydraulic gradient, of a few hundred micrometers below which breakthrough would not occur and conduits would not form.

An important practical result of modeling the initiation phase of conduit development is that in the special case of wide initial fractures, short pathlengths, and high gradients, breakthrough could occur is as little as 100 years, a number within the design lifetimes of dams and related hydraulic structures (Dreybrodt, 1992, 1996). Dreybrodt calculates that the breakthrough event will occur in less than 100 years if

$$\frac{i}{\lambda} > \left(5.3 \times 10^{-8}\right) a_0^{-2.63} P_{CO_2}^{-0.77} \qquad (7)$$

where i = hydraulic gradient, λ = path length in m, a_0 = initial fracture aperture in cm, and P_{CO_2} = carbon dioxide partial pressure of the water entering the fracture in atmospheres. This is a very important result because it is widely assumed that dissolution rates are so slow, that modification of the bedrock beneath or around a dam would not occur over the lifetime of the structure. Dreybrodt's calculations imply that leaks due to dissolutional enlargement are possible. Certainly, such leaks around dams on gypsum terrain have been documented.

The next step is to extend the calculations from single fractures to the assemblage of fractures that make up the actual flow path in a real aquifer. Groves and Howard (1994-b) made one attempt on a grid of fractures with a relatively wide fracture spacing. They concluded that pathway selection takes place very early in the process and that the preferred pathway has been identified before the critical thresholds are reached. A model with a smaller fracture spacing was developed by Siemers and Dreybrodt (1998). They introduced a probability that any given fracture segment on the grid was connected to other fractures. If the fracture was connected, it could contribute to the flow path. If it wasn't, it couldn't. Their results (Figure 6) show a remarkable similarity to the initiation paths described by Ewers and Quinlan (1981) based on intuition and the results of laboratory dissolution experiments.

The fluid flow portion of modeling the initiation phase of conduit development also applies to the overall flow regime in the fracture permeability portion of the aquifer. The simplest geometry is a plane parallel walled fracture of fixed aperture and uniform cross-section. Real fractures do not have this geometry. Within this approximation, the flow through the fracture is proportional to the cube of the aperture - the cube law of fracture flow. Introducing rough walls, variable aperture, and tortuosity to the fracture requires a correction to the cube law (Oron and Berkowitz, 1998). A new attempt (Hanna and Rajaram, 1998) to model dissolutional widening of rough and irregular fractures predicts breakthrough times (defined by these authors as the onset of turbulence) about half of what is predicted for smooth fractures. One of the curious results of this study is that it predicts breakthrough and rapid conduit growth can occur in irregular channels even in a regime of linear kinetics whereas water moving in smooth-walled fractures becomes rapidly saturated in the linear regime.

Conduit Enlargement and Integration

Once the sequence of fractures that mark out the optimum flow path have been chosen and the effective aperture reaches the critical value of about 0.01 meter, undersaturated water traverses the entire width of the aquifer. Dissolution kinetics shift into the near-linear regime. Using the generic rate equation, Palmer (1991) was able to show that, perhaps somewhat surprisingly, the rate of retreat of the conduit wall depended only on the degree of undersaturation and was described by the equation

$$S = \frac{31.56 \, k_1 \left(1 - \frac{C}{C_s}\right)^n}{\rho_R} \qquad (8)$$

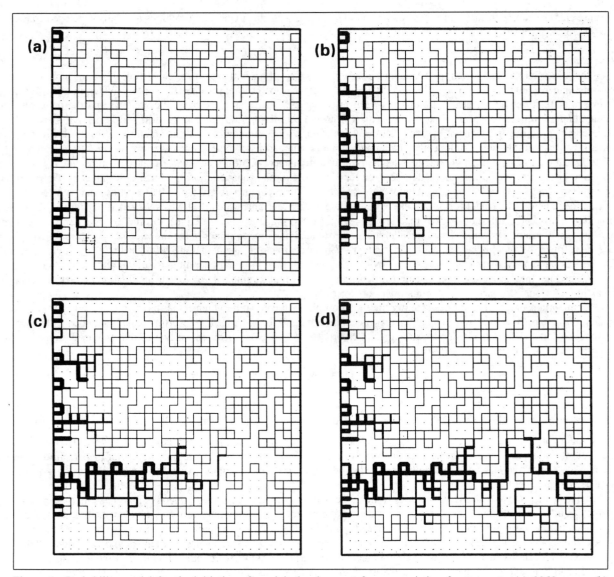

Figure 6. Probability model for the initiation of conduit development from preexisting fracture set. (a) 9960 years, (b) 13,657 years, (c) 17,367 years, (d) breakthrough achieved after 17,762 years. Probability of any given fracture being continuous = 0.7. From Siemers and Dreybrodt (1998).

where S = rate of wall retreat in cm/year, k_1 is the first order reaction rate constant, C = concentration of dissolved carbonate, C_S = saturation concentration of dissolved carbonate, n = reaction order, and ρ_R = density of bedrock. Using experimentally determined values of k_1 and n, Palmer found that the maximum rate of wall retreat to be 0.01 to 0.1 cm/year. Dreybrodt and his colleagues arrived at similar values using his kinetic models. At this rate of dissolution, an active conduit can enlarge from the threshold value of 0.01 meters to a diameter of a meter or more in times on the order of a few thousand years.

The final step in the geochemical modeling of the development of conduit permeability would be to predict the actual pattern of the conduit system. Thus far, this analysis has not met with much success. Howard and Groves (1995) made some calculations for conduit development under conditions of turbulent flow. Unlike the highly selective passage enlargement that takes place during the initiation phase, the enlargement phase is marked by more uniform passage growth. The distinction between single conduit caves, branchwork caves, network maze caves, and anastomotic maze caves must be made on the basis of geological constraints, and arrangement of recharge and discharge areas. Theoretical calculations based on fluid mechanics and chemical kinetics have not been much help.

CHARACTERIZATION OF KARST AQUIFERS

Although karst aquifers are exceedingly difficult to model because of their exceptionally heterogeneous permeability distribution and equally difficult to engineer as water supplies or to protect from contaminants, they do have some positive attributes. One important attribute is the number of characterization techniques that can be applied. Most provide information on the entire groundwater basin. Perhaps most important of the attributes of karst aquifer is the fact that most of the groundwater resurfaces at a single spring or limited group of springs. The karst spring is like a perfectly placed well in other aquifers. Water discharging from a karst spring carries an imprint of everything upstream in the aquifer. The problem is to read the imprint. A third important attribute is the diversity of possible sampling. Flow measurements and samples for analysis can be collected from the various inputs, flow paths, and outputs outlined in Figure 2.

Water Budgets

The notion of balancing inputs and outputs is a very powerful tool because it depends only on mass conservation of the water moving through the aquifer. However, water budgets do require continuous flow monitoring, thus dedicated data loggers and related equipment. Perhaps because of time and equipment requirements, relatively few water budget studies have been undertaken.

A useful concept borrowed from surface water hydrology is normalized base flow. It was long ago recognized that karst aquifers with well developed conduit system had low storage capacity because of rapid draining through the conduits (E.L. White, 1977). Base flows from springs draining these aquifers were exceptionally low. Normalized base flow is simply the mean spring base flow divided by the area of the catchment

$$Q_N = \frac{Q_B}{A} \qquad (9)$$

Quinlan and Ray (1995) tested the basic concept on a large number of drainage basins (Figure 7). If the normalized base flow can be calibrated to the specific climatic and hydrogeologic regime, measurements of base flow can provide reasonably accurate estimates of basin area. These can be used to confirm basin boundaries drawn by means of other techniques.

Tracer Tests

Tracing the path of karst ground water by means of added substances that can be detected when they reappear at the outlet point has been a standard procedure for more than 100 years. What has happened in the past decade or two has been the introduction of many new tracer techniques with resulting improvement both in the sensitivity of the tests and in their reliability. Among the many tracers that have been used, fluorescent dyes have emerged as the tracer of choice by most investigators.

The evolutionary history of tracing with fluorescent dyes is:

    a. Visual observation of dye

    b. Use of dye receptors (charcoal packets) to capture dye followed by elutriation and visual evaluation of test.

    c. Use of dye receptor followed by confirmation of dye by measurement of the characteristic fluorescence spectrum of the elutriate.

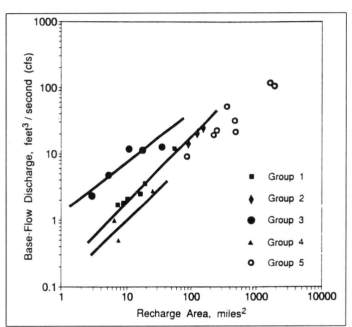

Figure 7. Relationship between mean base flow and ground water basin area for five groups of karst springs. The springs were grouped in terms of the hydrogeology of the basin. From Quinlan and Ray (1995).

    d. Quantitative tracer tests using automatic water samplers combined with spectrofluorometric analysis to determine dye concentrations directly in water samples.

    e. Quantitative tracer tests using continuous in-situ water analysis with an on-site spectrofluorophotometer.

As the sophistocation of the instrumentation has advanced, so also have the technicalities of dye injection, precautions against cross-contamination of tests, record keeping, chain-of-custody, and Q_A/Q_C considerations. Tracer testing has passed into the hands of specialists who have the necessary laboratory equipment and have established acceptable procedures to meet legal and regulatory standards.

Dye analysis by fluorescence spectroscopy permits quantitative analysis of dye concentrations using fluorescence peak intensities calibrated by suitable standards. Deconvolution of dye spectra allows the separation of overlapping fluorescence bands and thus permits the simultaneous injection of multiple dyes (Table 1). Automatic samplers or in-situ spectrofluorophotometers allows the determination of dye breakthrough curves and thus precise time-of-travel from dye injection point to spring. The shape of the dye release curve also gives information that can be interpreted in terms of the flow path through the aquifer. Careful attention to possible contamination in combination with more sensitive methods of instrumental analysis has reduced the detection limit for dye from the part-per-billion range to the part-per-trillion range. Extremely low detection limits means that it is now possible to attach significance to negative tests whereas with more qualitative tracing, only a positive test could be used; a negative result was merely indeterminant.

Table 1. Fluorescent dyes suitable for tracer studies.

Common Name	Color Index	Fluorescence Elutriate	Wavelength (nm) Water
Sodium fluorescein	Acid yellow 73	515.5	508.
Eosine	Acid red 87	542.	535.
Rhodamine WT	Acid red	568.5	576.
Sulphorhodamine B	Acid red 52	576.5	585.
Optical brightner	Tinopal CBS-X	398.0	397.

Data courtesy of Crawford & Associates, Bowling Green, KY

Spring Hydrographs

Karst drainage basins respond to storm inputs, primarily from swallets and internal runoff. The storm hydrograph at the swallet can be thought of as an input function with the aquifer acting as a transfer function. The two together form an output function which is the hydrograph observed at the spring. Three cases can be distinguished (Figure 8-a). If the response time of the aquifer is fast with respect to the mean spacing between storms, the individual storm pulses will appear in the spring hydrograph. These have the same shape as surface stream hydrographs with a steep rising limb, a crest, and a slower recession limb. In the intermediate case the response time of the aquifer is comparable to the mean spacing between storms. The hydrograph has a certain amount of structure but the individual storm pulses are smeared out. Finally, the hydrograph for a slow response aquifer shows only the effects of seasonal wet and dry periods. Individual storm events are area completely smeared out.

The response time of karst aquifers depends on at least these factors: (a) the contribution of allogenic recharge and internal runoff, (b) the carrying capacity and internal structure of the conduit system, and (c) the area of the ground water basin. The lack of individual storm pulses in the hydrograph does not, in itself, imply absence of conduits or a lack of karstic behavior to the ground water basin. A smoother, less structured hydrograph may indicate an aquifer mainly recharged by diffuse infiltration rather than by sinking streams or internal runoff. Quinlan *et al.* (1991) offer the additional caveat that a sinking stream located close to the spring can produce a fast response hydrograph in a spring which otherwise is a slow response basin.

The recession limbs of individual storm hydrographs are expected to be exponential

$$Q = Q_0 e^{-t/\tau} \tag{10}$$

where τ is the aquifer response time. Semilog plots of recession limbs, however, frequently produce two or more straight line

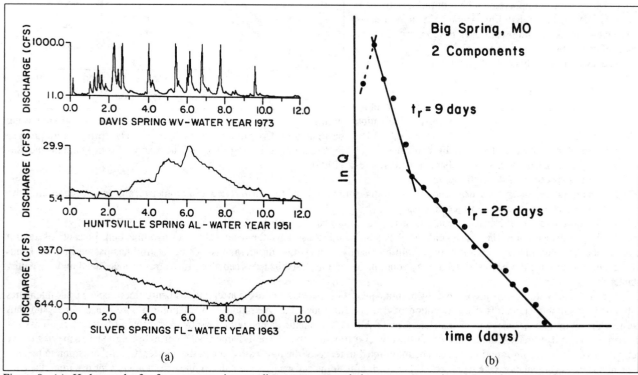

Figure 8. (a) Hydrographs for fast response, intermediate response, and slow response springs. Data from U.S. Geological Survey gauging records. (b) Semi-log plot of regression limb of a storm hydrograph from Big Spring, Missouri showing quickflow and slowflow components. From White (1993).

segments (Figure 8-b). There is a portion of the aquifer porosity that drains quickly, with time constants on the order of a few days, and a second portion that drains slowly with time constants on the order of weeks. Following Atkinson (1977), the fast response portion is termed "quickflow" and the slow response portion is termed "slowflow". The noncommittal labeling reflects ignorance of what is happening inside the aquifer. It may be that quickflow represents the draining of the conduits while the slowflow portion represents draining the fractures into the conduits. If so, the implication is that there are two quite distinct types of permeability, at least in the fast response aquifers.

Hydrograph analysis has a long history in karst hydrology going back to the pioneering work of David Burdon (Burdon and Papakis, 1963), to the work of Yugoslav karst hydrogeologists (Milanovic, 1981; original in Serbo-Croatian, 1979), and to work in the Mendip Hills of southwest England (Atkinson, 1977). Atkinson concluded from an analysis of spring hydrographs that 50 percent of the spring discharge was by quickflow and 50 percent by slowflow (or diffuse flow). Atkinson also concluded that 92 percent of the storage was in the fracture permeability.

Spring Water Chemistry

The hardness of karst spring waters often fluctuates with the season of the year and in some springs the hardness also responds to storm inputs to the ground water basin. By monitoring the water chemistry of a suite of springs on a two-week interval over an entire water year, Shuster and White (1971) showed that the fluctuations in hardness, expressed as the coefficient of variation of the measurements on each spring (coefficient of variation = standard deviation/mean), was in the range of 1 - 2 percent for one set of springs and in the range of 20 - 25 percent for a second set of springs. The question is: what is the source of the hardness fluctuations? Shuster and White claimed that a high coefficient of variation was an indication that the spring was draining an open conduit system. Ternan (1972) made similar measurements on springs in the English Pennines and claimed that the key variable was flow-through time (Figure 9-a). Worthington *et al.* (1992) proposed that the key variable was the fraction of allogenic recharge (Figure 9-b). Ternan's regression line has an r^2 of 0.66; Worthington *et al.* of 0.77. Who was correct? To a certain extent,

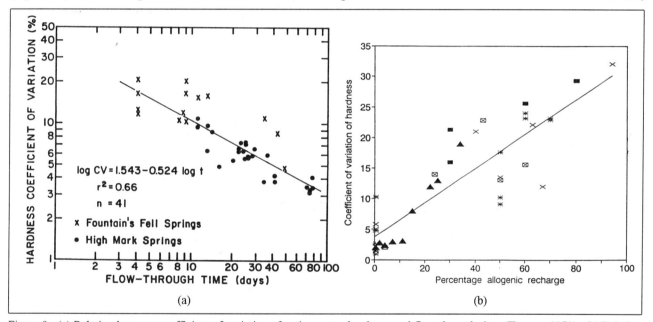

Figure 9. (a) Relation between coefficient of variation of spring water hardness and flow-through time (Ternan, 1972) (b) Relation between coefficient of variation of spring water hardness and fraction of ground water basin contributing allogenic recharge. (Worthington *et al.*, 1972).

everyone. Springs with high coefficients of variation in hardness must be fed by conduits to achieve a high throughput time. The sudden drops in hardness are due to injections of low-hardness storm water and this requires sinking stream recharge. For the converse, however, Worthington *et al.* make an important correction. Low coefficient of variation does not imply an absence of conduits; only that the recharge is mainly from diffuse infiltration rather than allogenic recharge or internal runoff. Hardness fluctuations can be easily measured by monitoring specific conductance which scales with hardness in carbonate ground waters. Variability in the specific conductance of a spring is a good indication of the proportion of allogenic and internal runoff water compared with the proportion of diffuse infiltrations.

The observation of chemical variability resulted from analyses based on a sampling interval of two weeks, an interval originally selected for convenience only. A continuous record of concentrations of dissolved ions (or specific conductance as a useful surrogate) shows sharp dips in concentration when a storm pulse is moving through the system followed by a slow recovery to base flow chemistry. A set of these records, termed "chemographs", obtained from the Maramec Spring in Missouri (Dreiss, 1989) shows that Ca^{2+}, Mg^{2+}, and HCO_3^- give identical records further supporting the idea using specific conductance to track carbonate hardness.

The chemical variability obtained with weekly or biweekly sampling intervals is only the result of picking random points from the much more detailed record of the chemograph.

Spring records obtained by superimposing various types of chemographs on the hydrograph and the precipitation is an extremely powerful way of learning about a karst aquifer. Figure 10 shows one of the results of a beautiful study by Ryan and Meiman on the Big Spring basin in the south-central Kentucky karst. These are records spanning only two and a half days during a series of storms in September, 1992. First there is the hourly precipitation. There were three distinct rain events during the period of observation. Second, there is the discharge, first rather low, and then rising quickly in response to the first and most intense rain event. The discharge remains high through the three rain events and then begins to decrease. Third is the specific conductance, initially high, characteristic of water that had had sufficient time to react with the limestone wall rock. There is no change in specific conductance at the rising limb of the hydrograph. Only after a full day does the specific conductances begin to decrease and low hardness storm water reaches the spring. Finally there are shown data on turbidity and fecal colliform counts. These also peak, but only after the hydrograph as begun falling and after the conductance has also decreased. An exceedingly clever part of the Ryan and Meiman study was the introduction of a dye trace in the swallet at the extreme distal end of the basin. Since they could not anticipate the storm, it was necessary to shift the time of dye injection to coincide with the onset of the first storm, and then to offset the time of dye breakthrough by the same amount. When this correction is made, the peak in the dye pulse exactly coincides with the peak in the turbidity and fecal colliform curves.

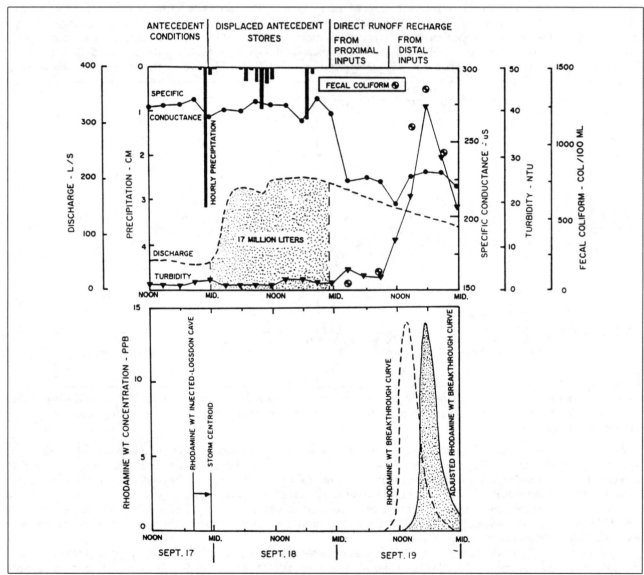

Figure 10. Hydrographs and chemographs for Big Spring, south-central Kentucky for a storm event of September, 1992. From Ryan and Meiman (1996).

There is a rich reward in hydrologic interpretation to be found in the data of Figure 10. There is almost no lag between the onset of the rain event and the rising limb of the hydrograph. Storm runoff into sinkholes and swallets upstream in the basin raise the hydraulic head and force water out of storage in flooded sections of conduit. This water has the same chemistry as the water discharging from the spring before the storm and thus there is no change in the specific conductance. The drop in specific conductance coincides with the arrival at the spring of swallet and sinkhole water from the middle part of the basin (confirmed in a second set of measurements with a different dye injected near the middle of the basin). Water from the distal portions of the basin do not arrive until later as indicated by the dye pulse and it is this portion of the recharge that carries the turbidity and colliform pollution. It was possible to integrate the hydrograph and calculate the quantity of water in storage in the flooded conduit portion of the system.

The results shown in figure 10 have very important implications for the monitoring of ground water basins in karst. Pollutants came through in a pulse but a pulse that greatly lagged the rising limb of the hydrograph. Reliable sampling requires precise timing and a preknowledge of the hydrology of the basin.

Wise Use of Test Wells in Karst

Aquifer characterization by drilling test wells is the standard approach for most aquifers but is something of a bad odor for karst aquifers. And indeed, if one drills the well where it is most convenient to park the drill rig, one is likely to learn very little about the karst aquifer. A realization of the 90's is that test wells in karst can be useful if one locates them properly and if one asks them the right sort of questions.

Wells rarely penetrate the main conduit of a karst ground water basin unless by extreme good luck or by deliberately targeting the conduit by sending cave explorers into it with a magnetic induction ("cave radio") location device. Wells that penetrate the conduit can be used to install instrument packages for continuous monitoring of temperature, specific conductance, and other variables. A transducer can permit the well to be used as a piezometer to measure pressure head in the conduit under flood flow conditions. Groves and Meiman (personal communication) have had great success in using wells drilled into the Hawkins - Logsdon River master conduit in Mammoth Cave to measure hydrographs in pressure head.

Downhole video cameras can be used to locate open conduits, fracture zones and other features in wells drilled in karst. A useful technique, where it can be applied, is to pump the well dry, then use the video camera to note zones where water is pouring into the well bore. This in turn identifies zones where packers can be used to characterize productive zones in the aquifer.

Quinlan *et al.* (1991) have recommended measuring the coefficient of variation of specific conductance of well water as a means of identifying those wells that are sensitive to contaminant movement. This technique has the advantage that it can be applied to domestic wells with pumps already installed without any need to disturb the well or its equipment.

Wells drilled on fractures and fracture intersections can be pump tested to provide information on the hydraulic conductivity of the fracture system. As always, such data must be interpreted with caution. Packer and slug tests provide additional information. Worthington and Ford (1997) offer some useful guidelines for borehole tests in carbonate aquifers.

Modeling Karst Aquifers

This review ends with a short mention of an area of intensive effort in karst hydrology - that of constructing realistic and useful computer models of karst aquifers. The fundamental problem facing the modellers is how to deal with the extreme range in aquifer permeability and its spatial distribution.

Careful water table mapping by Quinlan and Ray (1981) in the Mammoth Cave Area shows that large conduits, at least, act as ground water drains. Their location is marked by troughs in the water table much in the same way as surface streams mark troughs in the water table. Hydraulic gradients and the associated flow field within the matrix and fracture system are directed toward the conduits rather than toward some surface region of ground water discharge. The conduit, in turn, discharges its water at a spring. If the conduit system receives water from allogenic surface basins, it will be subject to flooding. Because the conduit lies in a ground water trough, only a volume of water sufficient to fill the trough is required. Indeed, the trough can be overfilled with the development of a groundwater mound above the conduit. Groves and Meiman (personal communication) have observed heads of 30 meters above the Logsdon-Hawkins River master conduit in Mammoth Cave during flood flow. The matrix and fracture system drains into the conduit during base flow but can be recharged from the conduit during flood flow. There must be a coupling function that interconnects the two flow regimes.

Early attempts to model karst aquifer assumed Darcy flow and also assumed that fracture permeability could be described as an average if viewed on a sufficiently large scale. Conduits were ignored. These models gave way to double continuum models with a fracture permeability and a conduit permeability (matrix permeability is concealed in the fracture permeability) which work better. The current generation of models are called discrete models and contain an explicit conduit system put in "by hand" either based on actual exploration or by best guess as to where the conduits should be (Mohrlok and Sauter, 1997; Mohrlok et al, 1997). Both double continuum and discrete models do a reasonable job of predicting the spring discharge hydrographs including the peaks and recessions due to storm flow. A reasonably accurate quantitative description of karst aquifers seems possible and may be achieved early in the new millennium.

OPEN QUESTIONS

The author of this review has been frequently accused of ending his papers with questions rather than answers. Maybe this is because the questions are often more interesting than the answers. There seems no reason to make an exception in the present report. Here are some questions that require further investigation.

a. Using rate equations with reaction orders as empirical parameters to be determined experimentally has been very helpful in modeling the evolution of conduit permeability. However, fractional orders imply that more than one process is rate-controlling at the atomic scale. In spite of the mountain of literature concerning the dissolution kinetics of carbonate minerals and carbonate rocks, the end is not in sight.

b. The current generation of geochemical models are homing in on the initiation step in conduit development and on the enlargement phase of uniform single conduits. An obvious next step is to model the entire karst drainage system - to reduce Figure 2 to a realistic set of calculations. Success would bring the subject very close to a final theory for the origin and evolution of carbonate aquifers.

c. Water budget calculations are a powerful tool for the discrimination of the various flow paths in the conceptual karstic aquifer. It would be of great interest to attempt a complete budget - with all of the instrumentation that would require - on a carefully selected aquifer.

d. Detailed measurements of spring hydrographs, chemographs, and other parameters in the style used by Ryan and Meiman (1996) in the Big Spring, Kentucky ground water basin should be repeated for other ground water basins. Ryan and Meiman's results contains a most interesting set of leads and lags which may appear differently in other basins and which collectively give more insight into the internal structure of the aquifer than any other set of measurements so far undertaken.

e. Modeling efforts are making progress. An accurate description of aquifer permeability and a means of estimating permeability distribution is needed as well as an improved calculation for describing the coupling between the different parts of the aquifer.

f. Often overlooked in discussions of carbonate aquifer is the flux of clastic sediments which is also transported. Especially in those ground water basins for which all drainage is underground, clay, silt, sand, and cobbles washed into swallets from allogenic drainage basins much be eventually transported through the aquifer and flushed out the spring. Otherwise, the underground system would eventually fill with sediment and allogenic runoff would be forced to return to surface routes. Likewise soil and other weathering debris from the karst surface that transported underground by soil piping must eventually be transported through the system. Although it seems clear that much sediment transport is episodic with sediment movement only during storm flows, there does not exist even a preliminary model for the entire process.

g. Much research, particularly applied research, has been directed to the problem of contaminant transport in karst aquifers. There have been many investigations of spill sites, dump sites, and Superfund sites which demonstrate all too clearly the rapid movement of contaminants in open conduit aquifers. Many further investigations address the question of proper monitoring of landfills and waste sites in karst. What is needed as backup research for practical questions are investigations of transport mechanisms of heavy metals, LNAPLs and DNAPLs in karst aquifers. Other than a general recognition that LNAPLs float on underground streams and pond behind obstructions, most of the transport questions have not even been properly framed.

ACKNOWLEDGEMENTS

The objective was to review some of the accomplishments in karst hydrology over roughly the past decade. I thank those whose work was cited for sharing their work with reprints and preprints. To those whose equally good work was not cited, I can only apologize. I freely admit to only skimming some of the highlights and in so doing omitting much that would have been covered in a truly comprehensive review. I thank Elizabeth White for her assistance in the preparation of this manuscript.

REFERENCES

Atkinson, T.C., 1977, Diffuse flow and conduit flow in limestone terrain in the Mendip Hills, Somerset (Great Britain): Journal of Hydrology, v. 35, p. 93-110.

Burdon, D.J. and Papakis, N., 1963, Handbook of Karst Hydrogeology. Athens: United Nations Special Fund, 276 p.

Courbon, P., Chabert, C., Bosted, P., and Lindsley, K., 1989, Great Caves of the World. St Louis, MO: Cave Books, p. 369.

Dreiss, S.J., 1989, Regional scale transport in a karst aquifer. 1. Component separation of spring flow hydrographs: Water Resources Research, v. 25, p. 117-125.

Drever, J.I., 1997, The Geochemistry of Natural Waters. Upper Saddle River, NJ: Prentice Hall, 436 p.

Dreybrodt, W., 1988, Processes in Karst Systems. Berlin: Springer-Verlag, 288 p.

Dreybrodt, W., 1990, The role of dissolution kinetics in the development of karst aquifers in limestone: A model simulation of karst evolution: The Journal of Geology, v. 98, p. 639-655.

Dreybrodt, W. and Buhmann, D., 1991, A mass transfer model for dissolution and precipitation of calcite from solutions in turbulent motion: Chemical Geology, v. 90, p. 107-122.

Dreybrodt, W., 1992, Dynamics of karstification: A model applied to hydraulic structures in karst terranes: Applied Hydrogeology, v. 3, p. 20-32.

Dreybrodt, W., 1996, Principles of early development of karst conduits under natural and man-made conditions revealed by mathematical analysis of numerical models: Water Resources Research, v. 32, p. 2923-2935.

Dreybrodt, W., Lauckner, J., Liu, Z., Svensson, U. and Buhmann, D., 1996, The kinetics of the reaction $CO_2 + H_2O \longrightarrow H^+ + HCO_3^-$ as one of the rate limiting steps for the dissolution of calcite in the system $H_2O-CO_2-CaCO_3$: Geochimica et Cosmochimica Acta, v. 60, p. 3375-3381.

Dreybrodt, W., 1998, Principles of karst evolution from initiation to maturity and their relation to physics and chemistry: in Yuan, D.X. and Liu, Z.H., Eds., Global Karst Correlation, Beijing, Science Press, p. 33-49.

Ewers, R.O. and Quinlan, J.F., 1981, Cavern porosity development in limestone: A low dip model from Mammoth Cave, Kentucky: Proceedings of the Eighth International Congress of Speleology, p. 727-731.

Ford, D.C. and Williams, P.W., 1989, Karst Geomorphology and Hydrology. London: Unwin Hyman, 601 p.

Groves, C.G. and Howard, A.D., 1994-a, Minimum hydrochemical conditions allowing limestone cave development: Water Resources Research, v. 30, p. 607-615.

Groves, C.G. and Howard, A.D., 1994-b, Early development of karst systems. 1. Preferential flow path enlargement under laminar flow: Water Resources Research, v. 30, p. 2837-2846.

Gunn, J., 1983, Point-recharge of limestone aquifers - A model from New Zealand karst: Journal of Hydrology, v. 61, p. 19-29.

Hanna, R.B. and Rajaram, H., 1998, Influence of aperture variability on dissolutional growth of fissures in karst formations: Water Resources Research, v. 34, p. 2843-2853.

Howard, A.D., and Groves, C.G., 1995, Early development of karst systems 2. Turbulent flow: Water Resources Research, v. 31, p. 19-26.

Kogovsek, J., 1981, Vertical percolation in Planina Cave in the period 1980/81: Acta Carsologica, v. 10, p.111-125.

Kristovich, A. and Bowden, J., 1995, Zacaton: AMCS Activities Letter, No. 21, p. 38-43.

Lattman, L.H. and Parizek, R.R., Relationship between fracture traces and the occurrence of ground-water in carbonate rocks: Journal of Hydrology, v. 2, p. 73-91.

Liu, Z. and Dreybrodt, W., 1997, Dissolution kinetics of calcium carbonate minerals in H_2O-CO_2 solutions in turbulent flow: The role of the diffusion boundary layer and the slow reaction $H_2O + CO_2 \longrightarrow H^+ + HCO_3^-$: Geochimica et Cosmochimica Acta, v. 61, p. 2879-2889.

Milanovic, P.T., 1981, Karst Hydrogeology. Littleton, CO: Water Resources Publications, 434 p.

Mohrlok, U. and Sauter, M., 1997, Modelling groundwater flow in a karst terrane using discrete and double-continuum approaches - importance of spatial and temporal distribution of recharge: Proc. 6th Conference on Limestone Hydrology and Fissured Media, La Chaux-de-Fonds, Switzerland, p. 167-170.

Mohrlok, U., Kienle, J. and Teutsch, G., 1997, Parameter identification in double-continuum models applied to karst aquifers: Proc. 6th Conference on Limestone Hydrology and Fissured Media, La Chaux-de-Fonds, Switzerland, p. 163-166.

Oron, A.P. and Berkowitz, B., 1998, Flow in rock fractures: The local cubic law assumption reexamined: Water Resources Research, v. 34, p. 2811-2825.

Palmer, A.N., 1991, Origin and morphology of limestone caves: Geological Society of America Bulletin, v. 103, p. 1-21.

Parizek, R.R., 1975, On the nature and significance of fracture traces and lineaments in carbonate and other terranes. in V. Yevjevich, Ed., Karst Hydrology and Water Resources, Fort Collins, CO: Water Resources Publications, p. 47-108.

Plummer, L.N., Wigley, T.M.L., and Parkhurst, D.L., 1978, The kinetics of calcite dissolution in CO_2-water systems at $5°$ to $60°$ C and 0.0 to 1.0 atm CO_2: American Journal of Science, v. 278, p. 179-216.

Plummer, L.N. and Busenberg, E., 1982, The solubilities of calcite, aragonite and vaterite in CO_2-H_2O solutions between 0 and 90 $°C$, and an evaluation of the aqueous model for the system $CaCO_3$-CO_2-H_2O: Geochimica et Cosmochimica Acta, v. 46, p. 1011-1040.

Quinlan, J.F., Smart, P.L., Schindel, G.M., Alexander, Jr., E.C., Edwards, A.J., and Smith, A.R., 1991, Recommended administrative/regulatory definition of karst aquifer, principles of classification of carbonate aquifers, practical evaluation of vulnerability of karst aquifers, and determination of optimum sampling frequency at springs: Proceedings of the Third Conference on Hydrogeology, Ecology, Monitoring, and Management of Ground Water in Karst Terranes, p. 573-635.

Quinlan, J.F. and Ray, J.A., 1995, Normalized base-flow discharge of ground water basins: A useful parameter for estimating recharge area of springs and recognizing drainage anomalies in karst terranes. in B.F. Beck, Ed., Karst Geohazards, Rotterdam, Balkema, p. 149-164.

Ryan, M. and Meiman, J., 1996, An examination of short-term variations in water quality at a karst spring in Kentucky: Ground Water, v. 34, p. 23-30.

Sauter, M., 1992, Quantification and forecasting of regional groundwater flow and transport in a karst aquifer (Gallusquelle, Malm, SW Germany): Tubinger Geowissenschaftliche Arbeiten, Part C, No. 13, p. 151.

Schindel, G.M., Quinlan, J.F., Davies, G. and Ray, J.A., 1996, Guidelines for wellhead and springhead protection area delineation in carbonate rocks: U.S. Environmental Protection Agency report EPA 904-B-97-003, p. 126.

Shuster, E.T. and White, W.B., 1971, Seasonal fluctuations in the chemistry of limestone springs: A possible means for characterizing carbonate aquifers: Journal of Hydrology, v. 14, p. 93-128.

Siemers, J. and Dreybrodt, W., 1998, Early development of karst aquifers on percolation networks of fractures in limestone: Water Resources Research, v. a34, p. 409-419.

Svensson, U. and Dreybrodt, W., 1992, Dissolution kinetics of natural calcite minerals in CO2-water systems approaching calcite equilibrium: Chemical Geology, v. 100, p. 129-145.

Ternan, J.L., 1972, Comments on the use of a calcium hardness variability index in the study of carbonate aquifers: with reference to the Central Pennines, England: Journal of Hydrology, v. 16, p. 317-321.

Teutsch, G. and Sauter, M., 1991, Groundwater modeling in karst terranes: Scale effects, data acquisition and field validation: Proceedings of the Third Conference on Hydrogeology, Ecology, Monitoring, and Management of Ground Water in Karst Terranes, Nashville, TN, p. 17-35.

Thrailkill, J., 1968, Chemical and hydrologic factors in the excavation of limestone caves: Geological Society of America Bulletin, v. 79, p.19-46.

White, E.L., 1977, Sustained flow in small Appalachian watersheds underlain by carbonate rocks: Journal of Hydrology, v. 32, p. 71-86.

White, W.B., 1988, Geomorphology and Hydrology of Karst Terrains. New York: Oxford University Press, 464 p.

White, W.B., 1993, Analysis of karst aquifers: in Alley, W.M., Ed., Regional Ground-Water Quality, New York: Van Nostrand Reinhold, p. 471-489.

White, W.B., 1998, Groundwater flow in karstic aquifers: in Delleur, J.W., Ed., The Handbook of Groundwater Engineering, Boca Raton, FL: CRC Press, p. 18-1 - 18-36,

Williams, P.W., 1983, The role of the subcutaneous zone in karst hydrology: Journal of Hydrology, v. 61, p. 45-67.

Williams, P.W., 1985, Subcutaneous hydrology and the development of doline and cockpit karst: Zeitschrift fur Geomorphologie, v. 29, p. 463-482.

Worthington, S.R.H., 1991, Karst Hydrogeology of the Canadian Rocky Mountains: Ph.D. thesis, Department of Geography, McMaster University, Hamilton, Ontario, 227 p.

Worthington, S.R.H., Davies, G.J., and Quinlan, J.F., 1992, Geochemistry of springs in temperate carbonate aquifers: Recharge type explains most of the variation: Proceedings 5th Conference on Limestone Hydrology and Fissured Media, Neuchatel, Switzerland, p. 341-347.

Worthington, S.R.H., and Ford, D.C., 1997, Borehole tests for megascale channeling in carbonate aquifers: Proc. 6th Conference on Limestone Hydrology and Fissured Media, La Chaux-de-Fonds, Switzerland, p. 195-198.

2 Development and distribution of sinkholes and karst

Measuring strain and stress from sinkhole distribution: Example of the Velebit mountain range, Dinarides, Croatia

S. FAIVRE Department of Geography, Zagreb, Croatia, and Laboratoire de Géographie Physique 'Géodynamique des Milieux Naturelles et Antropisés' UPRES-A 6042-CNRS, Clermont-Ferrand, France

PH. REIFFSTECK Laboratoire Central des Ponts et Chaussées, Paris, France

ABSTRACT

The sinkholes appear to be a sensitive indicator of the recent tectonic activity. Their spatial distribution allows us the measurement of strain and stress to which the studied area has been submitted during recent periods. Our analysis is based on the center to center method (Ramsy, 1967 ; Fry, 1979), dealing with the determination of strain ellipse. This method is applied to the analysis of topographical changes in distribution of sinkholes. As a result we have obtained a considerable number of local results. To pass from a local to a regional level we have used Panozzo's projection method (Panozzo, 1984). Both of these methods were adapted to the study of sinkholes. The results represent a very good base for the interpretation of recent local and regional deformation of the studied zone.

INTRODUCTION

The character of orogenic belts varies with paleogeography and tectonics according to the pre-existing structure and relief as well as to the stress conditions during orogenesis. Crustal stress has also caused the development of faults which results in the fault generated relief forms. At a more local scale, fracture patterns related to the crustal stress constitute perhaps the most significant single structural control of landform development (e.g., Birot, 1952). Stress, either residual or applied, has been invoked in explanation of a number of minor forms, developed mainly in granite (e.g. Jennings and Twidale, 1971), but also in other lithological settings (Twidale & Lageat, 1994). From this, two questions arise : Can the stress and strain be measured from relief ? Can the deformation rate be measured from relief ?

In order to test these hypothesis one has to apply the analyses to a specific study area. Although there are similarities between tectonic regions, each has a unique tectonic history and as a consequence possesses unique landform sequences. One such unique structural region is the Velebit mountain range (Fig.1) with the typical karstic landforms. It is a part of the Outer Dinarides, which is in turn a part of the Adriatic - Eurasia collision zone. Its numerous specific relief forms and the recent tectonic activity make the Velebit mountain range pertinent for such analyses.

Searching for a landform which could be the sensitive indicator of tectonic activity, we look for help within the most widespread karst forms which bear a close relation to tectonics. Since the latest postulates about sinkhole morphogenesis emphasize the major influence of faults and fractures (Aubert, 1966 ; Beck, 1984 ; Day, 1984 ; Kemmerly, 1982 ; Mihljević, 1994 ; Faivre, 1992) we have chosen the sinkholes as geomorphological markers. The specific goal of this study is the measurement of strain and stress from the sinkhole distribution.

STUDY AREA

The Velebit mountain range is the biggest elevation in the Republic of Croatia. With its surface

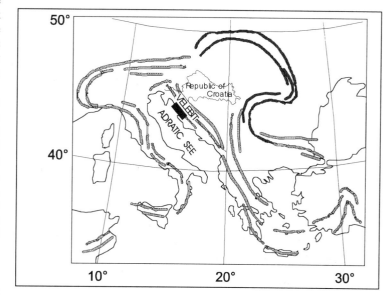

Fig.1. Location of the Velebit mountain range.

of 2 274 km² and the length of 145 km, it is also the longest mountain range of the Dinarides mountainous system. It extends along the Adriatic coast in a dinaric NW-SE direction. The studied area consists of three main orographical units : north Velebit, central and south Velebit.

The distinct domination of carbonate rocks (limestones and lesser dolomites), specific climate properties, together with the tectonic states, play a major role in the morphogenesis of the karst relief. As a typical karst domain, the Velebit mountain range offers plenty of karstic relief forms. Its strongly fractured carbonate sediments provide a great predisposition for the development of sinkhole. The great frequency of these forms allows us to undertake the proposed study.

TECTONIC SETTING

The structures of Dinaric and Adriatic areas are characterised by inverse and thrust relations. They are the consequence of the gradual movements of the Adriatic microplate and the deformation of the Earth's crust in the narrow zone of the European plate. Major features can be coherently interpreted as consequences of a succession of shortening mechanisms and block readjustments, driven by convergence between Africa and Eurasia. It is known that the regional stress approximately SW-NE in orientation, formed the structures of Velebit. The recent Quaternary period is characterized by reorganization of the structure of Velebit by refolding and new faulting with a stress approximately N-S (Prelogović, 1995). This change in the stress orientation is the result of the change in motion of the Adriatic plate from a NE to a NW direction. The consequence is the accentuated compression or the rotation of structures which are unfavorably oriented to the new stress.

The Velebit mountain range is characterized by anticline folding and high degree of fracturing. The faults represent a major characteristic of the structural unit. The most important one is the fault of Velebit (1 at Fig. 6). It is an overthrusted reverse fault with an overturn to the SW and S. The whole Velebit fault zone situated on the littoral side of the mountain is characterized by faults of the same type. Thrusting occurred on those shallow landward-dipping faults (Anderson, Jackson, 1987). The inland faults are of relatively strong inclination with frequent variation. Recently the regional faults have been successively reactivated as right lateral strike-slip faults. The most important faults are those which determine the three major parts of the Velebit mountain range.

THE APPLICATION OF THE CENTER TO CENTER TECHNIQUE TO THE SINKHOLE ANALYSIS

First determinations of finite strain in the plane were effectuated by direct observations of deformed shapes of objects that are known to have been spherical before deformation (ex. ooids). Later, instead of looking at shape deformations of objects, Ramsay (1967) took into consideration the variation in minimal distance between their centers. Thus, different kinds of materials which have undergone homogeneous strain can be used. This makes the wider use of the method possible. The initial distance between centers of two markers (nearest neighbors) of one population with random distribution in undeformed material is statistically the same in any direction of measurement. But after deformation, these distances will increase following λ_1 and reduce following λ_3 depending on the type of finite strain ellipsoid.

A graphical technique called the Fry method (Fry, 1979) is useful for determining the strain ellipse from a large number of points. In essence, it involves plotting the length and orientation of a large number of center - to - center lines relative to a single reference point. The method consists of determining a field with random and homogeneous distribution of objects, assigning its central point and superimposing the field as many times as there are objects taking for the center each point successively. Thus, we obtain a field of points which we call a « cumulated field ». A circular zone appears in the central part of the field. If the field is subjected to a deformation the central zone becomes elliptical because objects under deformation draw apart in the direction of the highest extension and they draw nearer in the direction of the highest compression.

The application of the Fry method, aiming to determine general deformation of rock complex, is applied to the analysis of topographical changes in sinkhole distribution, as proposed by Mihljević (1994). On the Velebit mountain range, with its 2 274 km², we find approximately 40 000 sinkholes. The sinkhole density is especially high in its northern and central part where we find as many as 121 sinkholes per square kilometer.

After many different experiments we find that 50-100 is the optimal number of objects because sinkholes occupy much more space than e.g. ooids in oolitic limestone. Therefore we form 623 fields across the whole Velebit mountain range with the variation from 20-120 sinkholes. The homogeneity of distribution seems to be much more important than the number of objects in the field. As our field of origin is not completely random, since sinkhole successions reveal joint and fracture systems, we create the fields within the major fault zones to approach the homogeneity of the medium, so that fields are not intersected by major faults.

All the zones have been digitised by ARC INFO from 1: 25 000 maps and then transferred to program ELLIPSE, written in Pascal language, which assures the determination of ellipses using statistics laws. It is of utmost importance to perform as many translations as there are points. The cumulated field has been cut in 20° sectors. Every point as a part of one sector is projected on its bisector (Fig. 2). Thus the distribution in classes with minimum of 6 points is realised. Then we verified if sinkhole occurrence along the bisector conformed to the law of Poisson distribution with a chi-square test. We assume a theoretical distribution. We suppose that the random variable is according to Poisson's law (Fry, 1979). See Press (1992) for development of the equations.

Furthermore, to check the Poisson random variable, the chi-square test is used for all 18 sectors. Once the calculation is completed we estimate the significance of the chi-square test using the chi-square probability function. The value of the chi-square test will be used to establish theoretical ellipse. The ellipse has been determined as the set of points that fit the Poisons law at 95 % for each sector. Consequently, we determine the long and short radiuses of the ellipses with the Gauss Newton method of optimisation taking all the points into account.

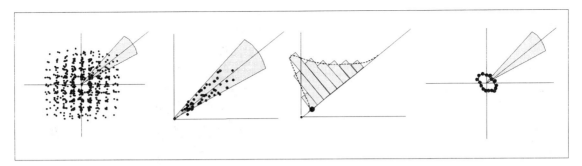

Fig. 2. Different phases of the method

The next phase in determination of ellipse orientations is realised in the iterative manner. The theoretical ellipse rotates through the sectors to allow the calculation of the correlation between the real and the theoretical ellipse. The correlation is based on the observed points weighed in terms of the chi-square law, for any angle. Finally, we obtain the ellipses with their angle, the correlation coefficient between the real and the theoretical ellipse and the ellipticity coefficient. The correlation between the real and the theoretical ellipse gives us another possibility to test the quality of the results, that is, the quality of the fields delineation. Different phases of the method applied to one field of sinkholes from the central Velebit are represented in the Fig. 3. In this way we have obtained 623 ellipses on the Velebit mountain range.

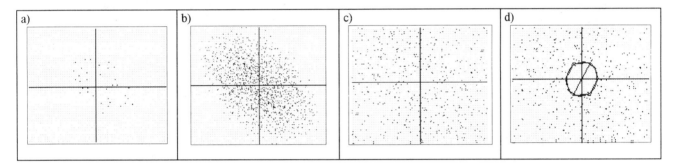

Fig. 3. Centre to centre method
a) Centres of sinkholes digitised by ARC/Info ; b) Cumulated field ; c) Zoom of the central zone ; d) Ellipsoid of deformation

It must be emphasised that the average stress field pattern may vary from one scale to another (Rebaï et al., 1992). Nevertheless, variations of stress directions at a given scale are consistent with the kinematics of faults of the same scale. Using these ellipses we may analyse field by field at a local scale comparing the result of each one with the relief assemblage. Then, the need for summing up these results occurs. To pass from a local to a regional scale, we must find another method.

PANOZZO'S PROJECTION METHOD

In order to quantify the principal orientation from the whole set of ellipses previously obtained in the Velebit zone, we use the projection method (Panozzo, 1984). The aim of this method is to analyse a two-dimensional strain presented as the preferred orientation. The basic operation of the proposed strain analysis is to project sets of straight lines or line segments on the x-axis. Then the sets of projected lines are rotated through angles from 0° to 180°. The projection of a single line of length l is given by $P(\alpha) = l \cos(\alpha_i + \alpha)$. $P(\alpha)$ is computed for each slope and summed to obtain $A(\alpha)$ for m lines of length l and initial orientation α_i. $A(\alpha)$ is independent of the position of line in the plane. Thereafter the preferred orientation is obtained from the minimum and maximum value of $A(\alpha)$. It gives $\alpha = 90° - \alpha_{min}$ or $180° - \alpha_{max}$. The population studied here is composed of ellipses obtained with the Fry method analysis, therefore, the ellipses will be projected on an axis taking into account their ellipticity coefficient. This problem is solved using projection of each ellipse approximated by a set of short straight line segments. The histogram of total length of projection of angle of rotation using 10° amplitude classes can be drawn. In this way we can also determine the global slenderness.

This method has been applied to the different zones of the Velebit mountain range following three main morphostructural zones : north Velebit, central Velebit and south Velebit. Each zone is decomposed into two blocks, taking the principal fault as a border. The results are synthesised in the Fig. 4. We have represented the obtained preferred orientations with principal faults used for zoning. The results are expressed in the trigonometrical direction of measurement. These ellipsoids represent the orientation of principal deformation. We can clearly see the regular change in orientation following the Velebit curvature from the north to the south.

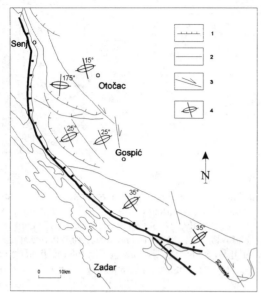

Fig. 4. Strain ellipses from sinkholes analysis on six main blocks of the Velebit mountain range
Legend: 1 - reverse fault, 2 - normal fault, 3 - strike-slip fault, 4 - strain ellipsoid.

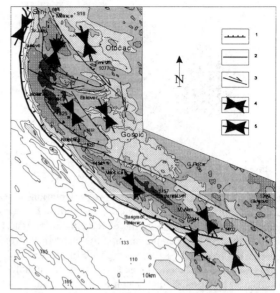

Fig. 5. Comparison between maximum stress orientations of the field (Preologović et al., 1998) and sinkhole measurements
Legend: 1 - inverse fault, 2 - normal fault, 3 - strike-slip fault, 4 - stress direction obtained from sinkholes measurement, 5 - stress directions obtained from the field measurement

DEFORMATION RATE

It is important to point out that such strain determinations do not only provide the orientation of the two principal strain axes. They also indicate the strain rate. Trying to define the zones of increased strain intensity the coefficient of ellipticity has been used. The more the ellipse is flattened, the stronger is the strain. The coefficients of ellipticity have been obtained with the Fry method for the the 623 points of measurements. They are further interpolated to the whole studied zone using the program written by Petar Bašić in the MATLAB computer software.

Observing the map (Fig. 6) the first striking thing is a large zone of a relatively high strain intensity along the coast. This indicates a close contact with the Adriatic microplate which is characterised in this very area by one of the highest compressions of all (Cigrovski-Detelić, 1998). The black zones coincide with the large Velebit fault zone.

The maximal deformation rate is marked on the central part of the Velebit mountain range which also coincides with its maximal curvature. This may be correlated to the sinkhole density map because it seems that the deformation rate is inversely proportional to the sinkhole density. This confirms the hypothesis that the sinkhole density decreases near active faults.

DISCUSSION

Measuring the strain from deformed relief permits the deduction of the stress orientation. Since the dynamic of the Adriatic boundary is of a complex nature it leads to a variety of stress states. This study confirms that the change of stress orientation depends on the scale of the analyse, therefore, it is essential to clearly specify the scale when discussing stress orientations. Application of the Fry method allows the determination of the local deformation field for each of 623 studied areas. Those results can be interpreted only in comparison with the same scale tectonic setting. They can be used in a detailed analysis of the expression of deformation in the recent relief. Many of those local results coincide very well with the results of the in situ measurement (Prelogović et al. 1998).

The second method of projection gives us the opportunity to change the scale of the study and to verify our primary local results with those

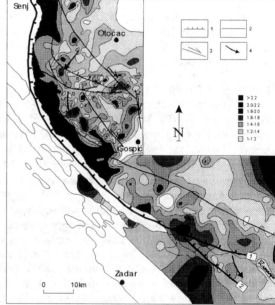

Fig. 6. Deformation rate measured from sinkhole distribution
Legend: 1 - reverse fault, 2 - normal fault, 3 - strike-slip fault, 4 - relative motion.

previously confirmed by different authors. This method allows us to define the deformation field by main blocks and three main parts of Velebit. Our results obtained for six main blocks of the Velebit mountain range correspond clearly with a recent NNW motion of the Adriatic microplate (e.g., Del Ben et al., 1991 ; Slejko, 1993). Zoback et al. (1989) have concluded that in several plates the maximum horizontal stress is subparallel to the direction of the absolute plate motion. In Fig. 4 it can be observed that the stress direction of five out of six main analysed blocks coincide with the direction of motion of the Adriatic microplate. The west and east block of the north Velebit show different maximum stress direction due to the different stress regime related to transtensional processes.

Speaking of the main Velebit morphostructures, the gradual change in stress direction can be observed following the Velebit curvature from N355°, N330° to N325°. The mean value of these three parts gives the regional stress value of N350° for the whole Velebit mountain range.

The recent stress field of Central Europe and its adjoining area is presented by Grüntal & Stromayer (1992) in terms of directions of maximum horizontal stress. Both of their models show the same (NW-SE ; NNW-SSE) stress direction for the north part of the Outer Dinarides which, of course, include the Velebit mountain range. These results are also in accordance with the motion of the Adriatic plate, as well as with results acquired from sinkhole analysis.

Prelogović et al. (1998) in their study of the structural framework and seismotectonic activity of the northern Velebit give the following results of the stress measurement in the field on the outcrops of main faults : the western block 20/25°N-200/205°N and 160°-340°N for the eastern block. These results perfectly coincide with the ones obtained from the sinkholes (Fig. 5). This is one an additional confirmation for the presented method and the obtained results.

CONCLUSIONS
- The tectonic regime influences the development of relief in such a way that it even permits the measurement of the strain and stress, from relief (case of sinkholes).
- These results open a great number of possibilities for further analysis such as : the study of the relative tectonic activity ; the study of block movements and movements on faults and drawing of the strain and stress trajectories
- This study also contributes to the improvement of knowledge about sinkhole development. Generally the location of sinkholes is determined by the nature of the rock and climate conditions, but their precise location is determined by the tectonic regime.
- This analysis contributes to the study of the morphostructural conditions and recent kinematics of the Velebit mountain range on the local and regional level. It is discovered that sinkholes reflect the recent deformations of the area as well.
- This study confirms that geomorphological analysis can contribute to the study of nature and magnitude of recent and contemporary tectonic activity. Moreover, this kind of study represents a very good base for further complex explanations of the relief development.

ACKNOWLEDGEMENTS

This study is the result of co-operation between the Zagreb University, Croatia and the University Blaise Pascal, Clermont-Ferrand, France, facilitated by the French government. S.F. particularly wishes to thank Professor Yannick Lageat, Professor René Neboit-Guilhot and Professor Andrija Bognar, as well as Marie-Françoise Andre, the director of the « Laboratoire de Géographie Physique ». S.F. is grateful to Eduard Prelogović, Professor at the Zagreb University, for helpful suggestions, and to Petar Bašić for help in writing the interpolation program.

REFERENCES

Anderson, H., and Jackson J., 1987, Active tectonics of the Adriatic region : Geophys. J. R. astr. Soc. 91, 937-983 p.

Aubert, D., 1966, Structure, activité et évolution d'une doline : Bull. de la Soc. Neuchateloise des Sciences Nat., 89, 113-120 p.

Beck, F.B., (ed.) 1984, Sinkholes : Their geology, engineering and environmental impact : Florida Sinkhole Research Institute. Balkema, Rotterdam, 429 p.

Birot, P., 1952, Le relief granitique dans le nord-ouest de la Péninsule Ibérique : Proceedings, 17th International Geographical Union (Washington, DC), 301-303 p.

Cigrovski-Detelić, B., 1998, The use of GPS measurements and geotectonic information in the analysis of the CRODYN geodynamic network : Unpublished PhD. Thesis. University of Zagreb 145 p., (in Croatian).

Jennings, J.N., and Twidale, C.R., 1971, Origin and implications of the A-tent, a minor granite landform : Australian Geographical Studies 9, 41-53 p.

Twidale, C.R., and Lageat, Y., 1994, Climatic geomorphology : a critique : Prog. Phys. Geogr., 18, 319-334 p.

Day, M.J., 1984, Predicting the location of surface collapse within karst depression : a Jamaican example : In : Beck, F.B. (ed.) Sinkholes : Their geology, engineering and environmental impact. Florida Sinkhole Research Institute. Balkema, Rotterdam, 429 p.

Del Ben, A., Finetti, I., Rebez, A., Slejko, D., 1991, Seismicity and seismotectonics at the Alps-Dinarides contact : Bol. di Geofisica Teorica ed Applicata, Vol 33, No 130-131, 155-176 p.

Faivre, S., 1992, The analysis of doline density on the north Velebit and Senjsko bilo : Proced. of the Internat. Symp. « Geomorphology and sea », Faculty of Natural Science and Mathematics, Department of Geography, Zagreb, 13-24 p.

Fry, N., 1979, Random point distributions and strain measurement in rocks : Tectonophysics, 60, 89-105 p.

Grüntal, G., Stromeyer, D., 1992, The recent crustal stress field in central Europe : trajectories and finite element modelling : J. Geophys. Research, Vol. 97. No. B8, 11805-11820 p.

Kemmerly, P.R., 1982, Spatial analysis of a karst depression population : clues to genesis : Bull. Geol. Soc. Am. 93, 1078-1086 p.

Mihljević, D., 1994, Analysis of spatial characteristics in distribution of sink-holes, as an geomorphological indicator of recent deformations of geological structures : Acta Geogr. Croatica, Vol.29, 29-36 p.

Panozzo, R., 1984, Two-dimensional strain from the orientation of lines in plane : Journal of Structural Geology, Vol.6, No.1/2, 215-221 p.

Prelogović, E., 1995, Geological structure of Velebit mountain range : Paklenicki zbornik Vol.1, Simpozij povodom 45. Godisnjice NP « Paklenica », Starigrad-Paklenica, 49-54 p., (in Croatian).

Prelogović, E., Kuk, V., Buljan, R., 1998, The structural fabric and seismotectonic activity of northern Velebit : some new observations : Zbornik RGN fakulteta, Vol.10., Zagreb, in press.

Press, W.H., Teukolsky, S.A., Vetterling, W.T., Flannery, B.P., 1992, Numerical recipes in Fortran : 2nd edit., Cambridge University Press, 614. p.

Ramsay, J.G., 1967, Folding and Fracturing of Rocks : McGraw-Hill, New York, 586 p.

Rebaï, S., Philip, H., and Taboada, A., 1991, Modern tectonic stress field in the Mediterranean region : evidence for variation in stress directions at different scales : Geophys.J. Int.,110,.106-140 p.

Slejko, D., 1993, A review of the Eastern Alps - Northern Dinarides seismotectonics : In : Boschi, E. Mantovani, E., and Morelli, A., Eds., : Recent Evolution and Seismicity of the Mediterranean Region, NATO ASI Series, Kluwer Academic Publishers, 251-260 p.

Zoback, M. L., Zoback, M.D., Adams, J., Assumpçao, M., Bell, S., Bergman, E.A., Blümling, P., Brereton, N.R., Denham, D., Ding, J., Fuchs, K., Gay N., Gregerson, S., Gupta, H.K., Gvishiani, A., Jacob, K., Klein, R., Knoll, P., Magee, M., Mercier, J.L., Müller, B., Paquin, C., Rajendran, K., Stephansson, O., Suarez, G., Suter, M., Udias, A., Xu, Z.H. & Zhizhin, M., 1989, Global patterns of tectonic stress : Nature, 341 (6240), 291-298 p.

Shallow karst in a breached anticline

RICHARD E.GRAY & ROBERT J.TURKA   GAI Consultants, Inc., Monroeville, Pa., USA

ABSTRACT

The Canaan Valley in Tucker County, West Virginia, formed as a result of erosion (breaching) of the doubly plunging Blackwater Anticline. The valley is largely underlain by limestone. Due to the dip of the rock strata away from the crest of the anticline, the limestone outcrops in adjacent watersheds at elevations significantly lower than the valley floor. These conditions might be expected to result in extensive karst development under the ridges surrounding the valley and leakage from the valley. A detailed geologic study to determine the extent of leakage from the valley through the limestone utilized a variety of evaluation methods including remote sensing/thermal imagery, water balances, piezometric levels, drainage basin analysis and ground water tracing using dyed lycopodium spores. The results determined that the valley contained only shallow karst development with no leakage into adjacent watersheds. Eliminating the concern of leakage was a major factor in obtaining approval for a large pump storage project. However, environmental issues eventually stopped the project.

INTRODUCTION

The Canaan Valley in Tucker County, West Virginia, is approximately 3.5 miles wide and 13 miles long at an average elevation of 3250 feet. The surrounding ridges rise over elevation 4000 feet. The valley formed as a result of erosion (breaching) of the northeast-southwest trending, doubly plunging Blackwater Anticline. The Blackwater River forms in the valley, flows through extensive wetlands on the flat valley floor and it exits the valley through a water gap on the western side. The valley has been considered as a possible reservoir site for many years and has been studied extensively since it has apparently ideal geologic conditions for leakage. These conditions are:
1. The presence of a soluble limestone beneath a large portion of the valley floor.
2. The dip of the rock strata away from the valley.
3. Outcropping of the limestone in adjacent watersheds at elevations significantly lower than the valley floor.

Interest in utilizing the valley for a pump storage reservoir prompted detailed geologic studies in the early 1970's to determine if there was leakage from the valley through the limestone.

GEOLOGY

Figure 1 is a geologic map of the Canaan Valley area. As a result of erosion and geologic structure the central valley floor is underlain by the Pocono Formation (Mp), which is surrounded by belts of the overlying Greenbrier Limestone (Mg) and Mauch Chunk Formation (Mmc). The ridges surrounding the valley are capped with the resistant sandstones of the Pottsville Formation (Ppv). The following is a general description of these formations from oldest to youngest.

The Pocono Formation is composed mainly of sandstone with occasional shale seams. The outcrop of the Pocono Formation which occupies the approximate center of the Canaan Valley, is about seven miles long and one to two miles wide. The sandstone members of the Pocono Formation are resistant rocks.

The Greenbrier Limestone, which underlies a major portion of the valley, is composed of thick beds of limestone separated by shales. The Greenbrier Limestone increases in thickness to the southwest. Near the mouth of the Blackwater River, about nine miles west of the valley, it is less than 300 feet thick, while at the south end of the valley it is over 400 feet thick.

The Mauch Chunk Formation, which comprises a portion of the valley floor and the lower portions of the valley walls, is 400 to 600 feet thick. It is primarily composed of nonresistant red shale and claystone with occasional lenses of fine grained sandstone. Generally, the Mauch Chunk is covered by a soil mantle and very few outcrops of the formation are present in the valley.

The steep walls of the Canaan Valley are formed by resistant sandstones of the overlying Pottsville Formation which is 500 to 600 feet thick. It is characterized by hard, resistant sandstones separated by sandy shales. Several thin coal seams are present, but they are too thin to mine.

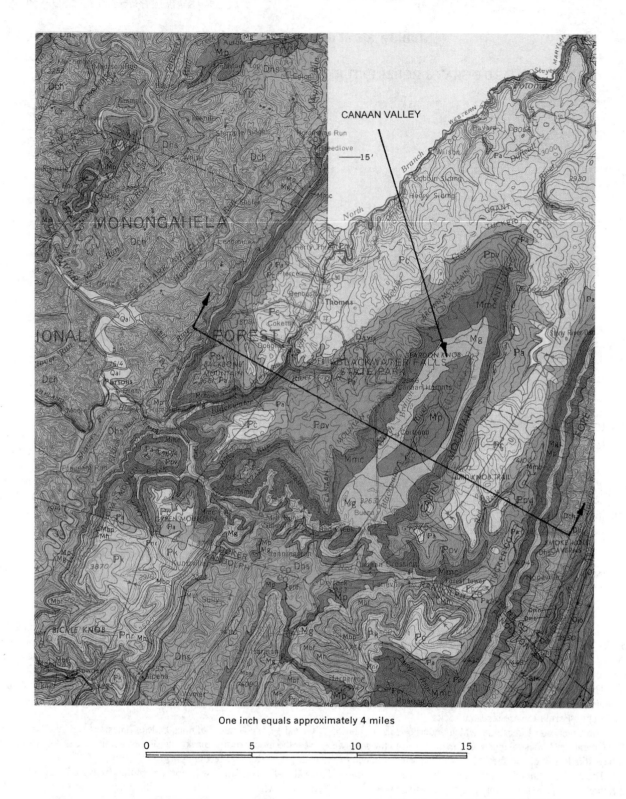

Figure 1. Geology of Canaan Valley area

Figure 2, an east-west section, shows the dip of the rock strata downward from the axis of the Blackwater Anticline and from the Canaan Valley into the adjacent Stony River Syncline to the east and the North Potomac Syncline to the west. Since the Blackwater Anticline is basically an elliptical dome, the rock strata also dip to the north and south. The Greenbrier Limestone is deeply buried along the synclinal axis but outcrops further to the west in the North Potomac Syncline, approximately 600 feet lower than and about nine miles away from the Canaan Valley. The structural conditions are similar on the east side (the Allegheny Front) of the Stony River Syncline but the change in elevation (approximately 400 feet) and distances (approximately five miles) are less. As the Greenbrier Limestone is deeply buried in both synclines, dissolution is very unlikely because the limestones in both synclines have always been far below the water table. W. E. Davies (1960) found "...features of caves along the Allegheny Front and along major valleys in the eastern part of the Allegheny Plateaus show that solution is confined to a zone along the edge of the limestone outcrop. The termination of passages a short distance back from the plateau front and their slope upwards to the top of the limestone result from the intersection of an active zone of movement in the ground water in the limestone formation. The apparent absence of large solution openings in limestone beneath the Allegheny Plateau suggests that integrated solution development is absent deep in the ground water..." All geologic work on Canaan Valley done prior to the 1970's resulted in conclusions similar to those of Davies.

The plunge of the Blackwater Anticline to the north deeply buries the Greenbrier Limestone so it does not outcrop for a considerable distance (in excess of 12 miles). South is the only potential direction for leakage since there is no leakage from the valley to the north, east and west where the Greenbrier Limestone dips away from the valley and becomes deeply buried. To the south, headward erosion of Elk Lick Run to the southwest and a tributary of Red Creek to the southeast into the anticlinal structure along with localized structural changes have exposed the Greenbrier Limestone (see Figure 1).

FIELD STUDIES

Prior studies of the leakage potential of the valley were conducted in 1913 (Fuller), 1923 (Scheidenhelm), 1929 (Reger), and 1961 (Charles T. Main, Inc.). In spite of the obvious geological conditions favoring leakage, all of these investigations concluded that drainage was internal to the valley and that leakage was not a problem. From 1970 to 1974 extensive geologic studies were undertaken which in addition to typical geologic evaluations, included special studies such as:

Remote Sensing/Thermal Imagery
Water Balances
Piezometric Levels
Drainage Basin Analysis
Ground Water Tracing

All of this work confirmed the findings of the earlier studies. Springs outside the valley on the slopes to the south and southwest were all determined to be derived from local sources of surface and groundwater and to have no connection to the Canaan Valley. Sinkholes and caves within the Canaan Valley were all found to drain into the Blackwater River System within the valley. For brevity, only the ground water tracing program is described.

DYED SPORE TRACING

A subsurface stream tracing program was undertaken to determine the character and direction of subsurface flow in the south end of the valley, the area thought to have the highest potential for leakage. The stream tracing involved injection of dyed lycopodium (club moss) spores into surface flows that disappeared into sinkholes, and trapping the spores in springs and streams with strategically placed nanoplankton nets. Nets were placed so that every possible direction of subsurface flow was covered. Details on the use of dyed spores for ground water tracing are presented by Atkinson (1973). Gardner and Gray (1976) report on use of this technique in Canaan Valley.

The tracing study was conducted primarily to characterize the subsurface flows at the south end of the valley. Within the valley, all of the subsurface flows are to the north, congruous with the surface drainage. The results of the study are presented in Figure 3 where the flow directions and distances are shown by the arrows.

The results of the spore tracing indicate that the subsurface flow in the Greenbrier Limestone is shallow and fairly contiguous with the surface drainage, and that subsurface pirating of flow from one drainage basin to another occurs only locally beneath small drainage divides, but does not occur beneath the larger ridges, particularly those ridges capped by thick sequences of non-soluble rocks.

The tracing study showed, in a few cases, that surface streams may flow into sinkholes and bypass their former surface flow valleys by flowing under low ridges and reappearing as springs further downstream. In all cases, these ridges are within the drainage basin of the main stream and the ridges are composed entirely of Greenbrier Limestone. For example, the injection point of the spores recovered in the Red Creek Basin was located in the gap forming the drainage divide between Canaan Valley and Red Creek drainage basin, with the injection point just inside the Red Creek Basin. The Greenbrier Limestone is relatively high and well exposed in this area, maximizing the possibility of solution development. Dyed spore tracing revealed that subsurface flow stays within the major drainage basin and does not cross the major drainage divide. The same situation was observed where the entire flow of Elk Lick Run disappeared into a sinkhole, and reappeared a short distance downstream without crossing any major divides.

The fact that no spores were detected in flows from the Greenbrier Limestone outside the valley precludes the possibility of leakage from the south end of the valley. This is reinforced by the fact that no major drainage divides were crossed in areas where pirating is even more likely, such as the drainage basins of Red Creek and Elk Lick Run where hydraulic gradients are higher, and the Greenbrier Limestone is not covered by nonsoluble, low permeability rocks.

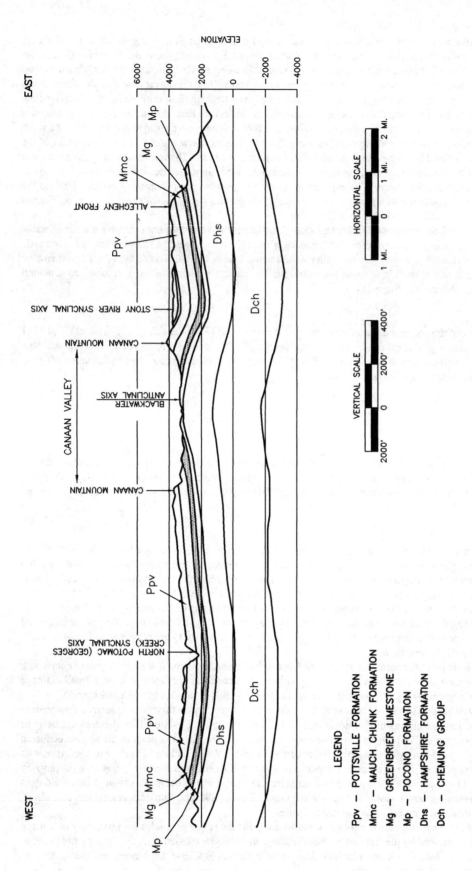

Figure 2. East-west cross-section through the Canaan Valley area

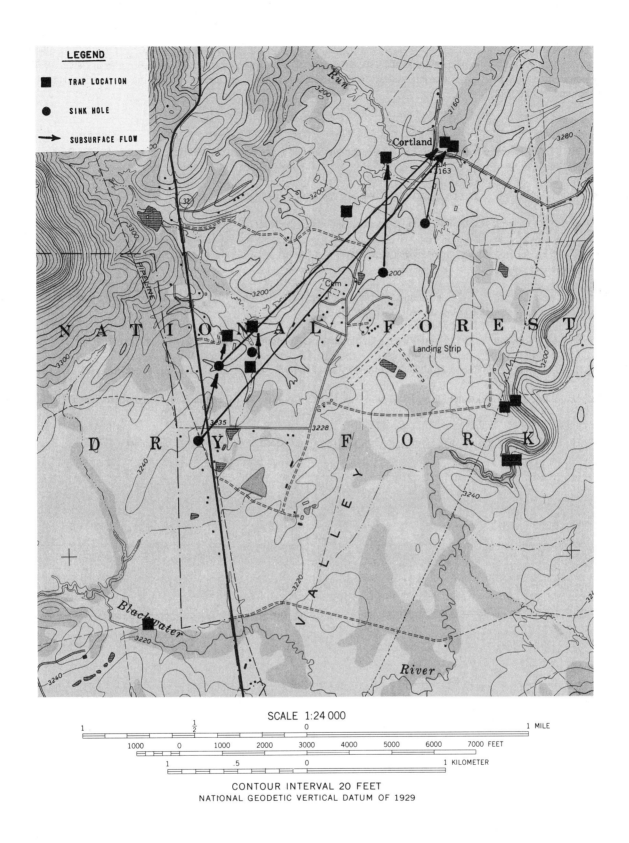

Figure 3. Results of ground water tracing in the southern end of the Canaan Valley

CONCLUSIONS

All evidence provided by the field observations, tests, data and analyses indicates that there is no leakage from Canaan Valley. The shallow solution development in the Canaan Valley, with no solution conduits occurring beneath the surrounding ridges is of the fluviokarst type (White, 1988).

REFERENCES

Atkinson, T.C., 1968, Tracing Swallet Waters Using Lycopodium Spores: Transactions Cave Research Group of Great Britain, V. 10, N. 2, p. 99-106.

Charles T. Main, Inc., 1961, Report on Blackwater River Storage Dam and Reservoir.

Davies, W.E., 1960, Origin of Caves in Folded Limestone, Bulletin of the National Speleogical Society, Bulletin Vol. 22, Part 1.

Fuller, M.L., 1913, Geological Report on Blackwater River Projects of West Virginia, Associated Geological Engineers, Boston, Massachusetts.

Gardner, G.D., and Gray, R.E., 1976, Tracing Subsurface Flow in Karst Regions Using Artificially Colored Spores, Bulletin of the Association of Engineering Geologists, Vol. XIII, No. 3.

Geologic Map of West Virginia, 1968, West Virginia Geological and Economic Survey.

Reger, D.B., 1929, Geologist's Report on Water Tightness of Blackwater Reservoir, Tucker County, West Virginia.

Scheidenhelm, F.W., 1923, Blackwater Storage Reservoir Site, Tucker County, West Virginia, with Reference to Water-Tightness, New York.

White, W.B., 1988, Geomorphology and Hydrology of Karst Terrains, Oxford University Press, New York.

Spatial relationships between new and old sinkholes in covered karst, Albany, Georgia, USA

JAMES A. HYATT & HOLLY P. WILKES *Department of Physics, Astronomy and Geosciences, Valdosta State University, Ga., USA*

PETER M. JACOBS *Department of Geography, University of Wisconsin, Whitewater, Wis., USA*

ABSTRACT

In Dougherty County, Georgia, thousands of cover subsidence and cover collapse sinkholes have formed by the piping of overburden into cavities in the Ocala Limestone. Although sinkholes may form slowly, sudden collapse is often triggered by heavy rainfall and/or flooding that saturates and weakens soil arches that overly cavities. A spectacular example of this occurred in July 1994 when flooding from Tropical Storm Alberto (TSA) triggered the collapse of >300 sinkholes (Hyatt and Jacobs, Geomorphology, 17:305-316). Here we examine relationships between the July, 1994 (new) sinkholes and pre-existing (old) sinkholes within 71 km^2 of suburban Albany. Our purposes are to contrast the dimensions, spatial distributions, and overburden characteristics for these two populations, and to test whether elevation, and the locations of old sinkholes have influenced the locations of new sinkholes.

The dimensions (length, width, depth, perimeter) and locations of 311 new sinkholes were determined using survey equipment, a global positioning system, and reports from local residents. Similar measures for 329 old sinkholes were derived from digitized USGS 1:24000 map sheets. New sinkholes cluster within TSA flood limits, and are found preferentially in sandy alluvial lowlands as identified by digital elevation modeling. Old sinkholes are significantly larger, more pan-like in cross-sectional form, and group dimensionally into three statistically significant elevation classes. These classes are more widespread topographically than are new sinkholes, probably reflecting the influence of frequent flooding in lowlands, and enhanced piping in uplands. Clustering of new sinkholes within old sinkholes does occur for a few selected sites. Nearest neighbor analysis, however, indicates that statistically new sinkholes have not clustered near old sinkholes. This implies that the locations of old sinkholes by themselves have limited predictive utility to identifying sites for new sinkholes.

INTRODUCTION

In July 1994 Tropical Storm Alberto (TSA) released >530 mm (21 in) precipitation in 19 h causing extensive flooding of the Flint River in southwest Georgia. In addition to flooding, runoff from TSA triggered the collapse of >300 sinkholes within a 71 km^2 region of covered karst in suburban Albany, Georgia (Figure 1). These new sinkholes are unusual for two reasons. First, they formed in response to a single hydrologic event and therefore represent, with certainty, a single sinkhole population that differs in age from the 329 pre-existing sinkholes within the same 71 km^2 (Figure 1). As a consequence, this data set is ideal for testing whether the location of new sinkholes has been influenced by the location of pre-existing or old sinkholes (cf. Drake and Ford, 1972; Kemmerly, 1982). Secondly, the Albany data set is unusual because of the role that flooding played in triggering collapse. Hyatt and Jacobs (1996) argue, based on empirical denudation rates, that cavity production by bedrock dissolution outpaces cavity infilling associated with the development of sinkholes. Consequently, for many locations the formation of sinkholes likely has been limited by the frequency of triggering events rather than the rate at which subsurface cavities have developed. Although analysis of the new sinkholes provides insight into spatial trends for a single collapse event (Hyatt and Jacobs, 1996), spatial characteristics of the old sinkholes provides a time-integrated view of the development of both cover collapse and cover subsidence sinkholes.

In this paper we further examine sinkholes in suburban Albany in order to consider the influence of surface elevation, flood frequency, and pre-existing sinkholes on the development of new sinkholes. Our specific objectives are to: *(1)* contrast dimensions, distributions, and overburden characteristics for populations of new and old sinkhole; *(2)* test whether elevation has influenced the locations and dimensions of sinkholes; and *(3)* determine whether the presence of pre-existing sinkholes influences the development of new sinkholes at Albany.

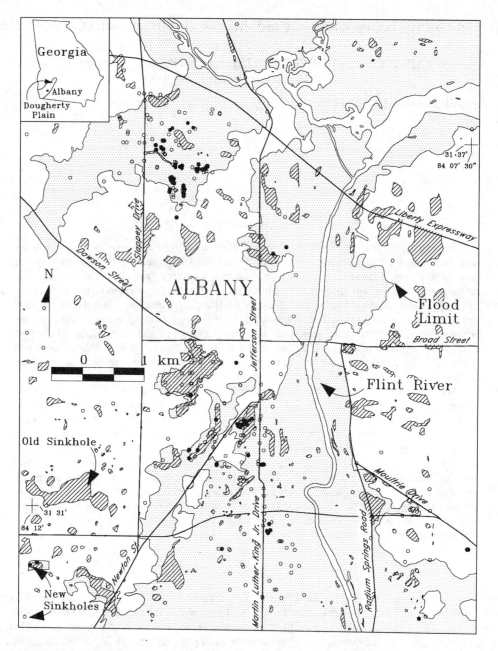

Figure 1. Location of Dougherty Plain and Albany Georgia (inset) and 71 km² study site showing locations of new sinkholes triggered by TSA, and pre-existing or old sinkholes. Filled circles (•) show new sinkhole locations determined by global positioning system, whereas open circles identify street address locations for new sinkholes. Stippled region identifies the limits of TSA flooding.

STUDY SITE

New and old sinkholes were examined within a 71 km² area encompassing most of suburban Albany, and portions of the surrounding Dougherty Plain (Figure 1). The Dougherty Plain is a mantled karst surface situated between dissected Fall Line Hills to the northwest and a fluvially dominated Tifton Upland surface to the southeast (Clark and Zisa, 1976). The Plain is underlain by > 1500 m of Cretaceous and younger Gulf Coastal Plain carbonate and clastic sedimentary rocks. This sequence is capped by 3 to 36 m of alluvium and residuum that directly overlies late Eocene limestone of the Ocala formation (Hicks et al., 1987).

Previous studies of the Dougherty Plain indicate the presence of thousands of suffosion sinkholes (Herrick and Legrand, 1964; Brook and Allison, 1983; Beck and Arden, 1984), or ravelling sinks as they are referred to in the engineering literature (Sowers, 1975). These sinkholes form by cover collapse and cover subsidence mechanisms (Beck and Sinclair, 1986). Cover collapse occurs suddenly when soil arches that overly subsurface cavities fail. Often collapse occurs in response to a triggering event such as an intense rainstorm, or in the case of the new sinkholes at Albany, in response to flooding. In contrast, cover subsidence sinkholes form more gradually as overburden is semi-continuously piped downward into openings in the underlying bedrock (Wilson et al. 1987). Although considerable variation in cross-sectional form may occur for both types of sinkholes, the rapidly formed cover collapse sinks initially have bowl, undercut, or shaft forms with relatively small diameter to depth ratios (Hyatt and Jacobs, 1996). Cover subsidence sinkholes, however, on average are shallow, pan-shaped depressions with larger diameter-to-depth ratios.

METHODS

Sinkhole dimensions and locations were obtained by field surveys and from 1:24000 USGS quadrangle map sheets. New sinkholes were examined in the field less than two months after TSA using techniques described by Hyatt and Jacobs (1996) and summarized briefly here. Seventy-seven sinkholes were surveyed to determine sinkhole length, width, perimeter, and depth. Twenty-four of these sinkholes had depth obstructions, and dimensions reported in this paper are derived from the remaining 53 sinkholes. Sinkhole center locations were obtained to within 3 m using a Trimble Global Positioning System and post-processed base station files. These locations were merged with an additional 34 sinkhole locations obtained from the Albany-Dougherty Planning Commission, and 201 sinkhole locations reported by citizens to the Albany City Engineers Office. In contrast, all dimensions and locations for old sinkholes were derived from 1:24000, 5 ft contour interval USGS map sheets. For this study, a sinkhole was defined by a minimum of one closed hatchured contour line. All sinkhole dimensions, including elevation, were derived manually by digitizing contours from the USGS map sheets. Once digitized, these data were converted to NAD83 State Plane Space in ArcInfo and imported into AutoCad to determine length, width, perimeter, area, and centroid coordinates. Sinkhole density counts were derived from these files by programming a 1 km^2 search window to step across the study site along a 250 m node spacing. In addition to data derived from the USGS map sheets, several digital elevation model (DEM) files were downloaded and used to examine topography in relation to sinkhole locations. Sinkhole dimensions were analyzed statistically to determine measures of central tendency, and a variety of inferential statistical tests were performed to identify subgroups within the data set.

RESULTS AND ANALYSIS

Comparing dimensions and overburden characteristics for old and new sinkholes

Statistical analysis of the dimensions of new and old sinkholes indicates that old sinkholes are larger (Table 1), but that variance for most dimensions does not differ between groups. Median length, width, perimeter, and diameter-to-depth ratio (L,W,D,DD) for old sinkholes are 24 to 48 times larger than they are for new sinkholes, although median depth and asymmetry values are only marginally larger for old sinkholes. Statistical comparison of mean log transformed sinkhole dimensions unequivocally indicate that old sinkholes are larger ($p<0.001$). In contrast, f-test comparisons show that, excluding asymmetry and DD, the variance of transformed L, W, and P dimensions do not differ between new and old sinkholes.

Overburden type and thickness is known to influence the development, location, and size of sinkholes in some locations (Palmquist, 1979). Hyatt and Jacobs (1996) compared the locations of new sinkholes with the thickness and type of overburden as derived from published maps (Hicks et al. 1987). They conclude that, while new sinkholes are more frequent in thin overburden (< 12 m thick), they are not restricted to those classes. Similar trends occur for old sinkholes, with frequencies in each overburden class not differing between new and old groups by more than 7%. Thus, unlike many other locations described in the literature, overburden thickness does not appear to have a dominating influence on sinkhole development at Albany. The type of overburden appears to have a somewhat stronger influence on the location of new sinkholes. Nearly 60% of new sinkholes formed in predominantly sandy overburden (Hyatt and Jacobs, 1996), while old sinkholes are more evenly distributed, with 43% in sand, 19% in sand and clay, and 38% in clayey overburden. Variance tests indicate no difference between the transformed dimensions of sinkholes located in different overburden types. Similarly, no significant difference exists between mean dimensions for sand and sand-clay classes, and for sand and clay classes. However, except for width, all dimensions do differ significantly ($p<0.05$) for sinkholes that formed in sand and clay as compared with sinkholes that formed in clay.

Distribution of new and old sinkholes by elevation class

Sinkholes occur at elevations ranging from 160 to 245 ft asl (Figure 2). New sinkholes are more tightly grouped than are old sinkholes, with nearly 90% of new sinkholes occurring between 180 and 200 ft asl. This distribution clearly reflects the effect of TSA flooding which peaked at 193 ft asl., triggering the collapse of many sinkholes within the limits of flooding (Hyatt and Jacobs, 1996).

The importance of elevation is more evident when comparing sinkhole density with digital topography (Figure 3). Sinkhole

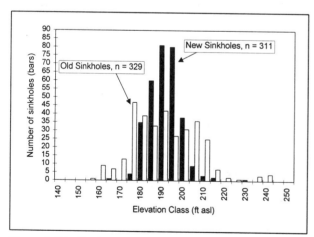

Figure 2. Distribution of new and old sinkholes by 5 foot elevation interval.

Table 1. Descriptive statistics for new and old sinkholes. Note the larger dimensions, excluding depth, for old sinkholes.

Dimension	Median	Mean	Std	Range		
New Sinkholes (reported by Hyatt and Jacobs, 1996), n=53						
Length (m)	1.9	4.7	9.2	0.3	to	44.0
Width (m)	1.6	3.1	12.3	0.3	to	20.9
Perimeter (m)	6	16	294	1	to	256
Depth (m)	0.7	0.9	0.8	0.1	to	4.3
Area (m^2)	na	na	na	na	to	na
Diameter:Depth	2.8	3.3	2.0	0.6	to	10.8
Asymmetry (L:W)	1.2	1.3	0.3	1.0	to	2.1
Old Sinkholes (this study see Figure 1), n = 329						
Length	92	160	190	15	to	1384
Width	47	78	88	8	to	892
Perimeter	236	443	595	45	to	5774
Depth[a]	0.8	1.3	1.1	0.8	to	6.9
Area	2937	16049	43065	126	to	462830
Diameter:Depth	69	109	110	7	to	791
Asymmetry	1.7	2.3	2.1	1.0	to	25.5

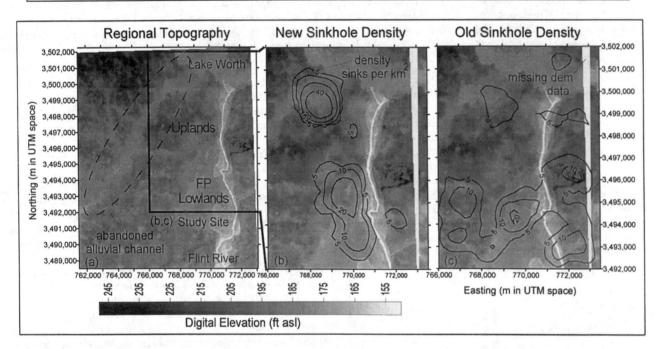

Figure 3. Digital elevation model for *(a)* Albany West USGS 1:24,000 quadrangle map sheet. Note location of study site, and alluvial lowlands, abandoned alluvial channel, and upland topographic units. Overlay contours show *(b)* the density of new sinkholes and *(c)* the density of old sinkholes in relation to digital elevation. See text for discussion.

locations are described in relation to three topographic units that are visible in Figure 3*a*. The first, a low-lying flood plain and river terrace complex, parallels the Flint River and broadens to nearly 4 km in width at the southern end of the study site. Secondly, an elongate elevation trough extends from Lake Worth southeast across the map. Although currently unoccupied by a surface stream, the pattern of numerous tributary topographic lows that join this trough imply that it may be an abandoned alluvial channel. Finally, irregular upland surfaces occur at elevations higher than the alluvial plain and abandoned channel. New sinkholes have clustered almost exclusively in the alluvial lowlands and adjacent to the abandoned alluvial channel, where TSA floodwaters were able to inundate the landscape. In contrast, old sinkholes, which are of varying ages and likely have formed by both cover subsidence and cover collapse mechanisms, are more widely dispersed with density point maxima occurring in all three topographic units (Figure 2,3*c*).

Given the relationships between sinkhole density and topography described above, we tested whether the dimensions of old sinkholes change with elevation, and how such change compares to the topographic units identified in Figure 3. To address these questions, old sinkhole were grouped into 5 ft elevation classes similar to those depicted in Figure 2. Multiple one way analysis of variance (ANOVA) was performed on log-transformed length, width, area, and perimeter dimensions to determine whether significant differences exist between groups. Adjacent elevation classes that did not differ significantly ($p<0.05$) were grouped and ANOVA tests were repeated until no new groupings were made. Three elevation classes emerged from this grouping procedure (Table 2,3). Class one includes sinkholes at elevations below 175 ft asl, which closely corresponds to the alluvial lowland topographic unit in Fig 3. Class two includes sinkholes at elevations between 175 and 195 ft asl, which closely corresponds to elevations within the abandoned alluvial channel. Finally, elevation class three, corresponding to the upland topographic unit, includes sinkholes at elevations greater than 195 ft asl.

Descriptive statistics for elevation classes indicate that sinkholes become progressively larger with increasing elevation (Table 2). Median L, W, P, and area dimensions increase by factors of 1.7 to 2.6 between class 1 and class 2, and by factors of 1.6 to 3.3 between class 2 and class 3. Mean dimensions also increase with each class. Furthermore, although variance does not differ significantly for most dimensions, differences between elevation class means are significant at the $p<0.05$ level (Table 3). Only depth and asymmetry show little change between groups.

Table 2. Descriptive statistics for sinkhole dimensions for sinkholes grouped by elevation class. Note, for all dimensions excluding depth and asymmetry, median and mean value increases with increasing elevation.

Dimension	Median	Mean	Std	Range		
Class 1 (elevations less than 175 ft asl), n=77						
Length (m)	48	98	145	15	to	947
Width (m)	28	43	57	8	to	440
Perimeter (m)	132	257	388	49	to	2573
Area (m^2)	1032	6058	21697	131	to	144269
Depth (m)[a]	0.8	1.5	1.5	0.8	to	6.9
Diameter:Depth	35	58	53	7	to	241
Asymmetry (L:W)	1.7	2.5	2.7	1.0	to	17.7
Class 2 (elevations 175 to less than 195 ft asl), n=141						
Length	85	159	218	15	to	1383
Width	48	81	98	11	to	692
Perimeter	218	461	726	45	to	5774
Area	2836	17474	54120	134	to	462830
Depth	0.8	1.2	1.0	0.8	to	5
Diameter:Depth	66	117	127	14	to	793
Asymmetry	1.6	2.1	2.3	1.0	to	26.0
Class 3 (elevations greater than 195 ft), n=111						
Length	165	201	166	17	to	932
Width	77	103	94	9	to	508
Perimeter	404	550	498	48	to	3044.0
Area	9344	21271	37394	126	to	267011
Depth	0.8	1.3	1.0	0.8	to	5
Diameter:Depth	121	152	125	13	to	633
Asymmetry	1.8	2.2	1.3	1.0	to	11

Nearest neighbor analysis of the distribution of new and old sinkholes

Nearest neighbor analysis (NNA) is often used to identify clustering within sinkhole populations and to test for associations between different populations (Williams, 1972; Drake and Ford, 1972; Kemmerly 1982). NNA makes use of a Clark-Evans index (R) that compares mean nearest neighbor distance with an expected distance for a random distribution having the same sinkhole density as the observed distribution. R varies from 0 for perfectly clustered distributions through 1.0 for random distributions to a limit of 2.15 for uniform distributions. New sinkhole nearest neighbor distances are moderately clustered (R = 0.55) (Hyatt and Jacobs, 1996), whereas old sinkholes are more randomly distributed (R = 0.90). These statistics are consistent with density distribution maps (Figure 3b,c).

NNA may also be applied to the distance between new and old sinkholes in order to gain insight into the degree to which one population clusters around another. Clustering is expected if a primary (older) population of sinkholes influences the location of a secondary (newer) population of sinkholes. As explained by Kemmerly (1982), this multigenerational development may occur when static water-level drawdown associated with the enlargement of the primary sinkhole generates locally steeper hydraulic gradients, which facilitate the development of new sinkholes nearby. In the covered karst of Albany, this mechanism may result in a preferential grouping of new suffosion sinkholes around old sinkholes. NNA provides a means for testing this hypothesis. The R value for new-to-old sinkhole distances is clearly random (R = 0.96), whereas the R value for old-to-new sinkhole distances of 0.71 is not. The latter differs significantly from random (p=0.001), and suggests weak clustering, with old sinkholes spaced 29% closer to new sinkholes than would be expected for a random distribution. These findings are somewhat contradictory. A more clear indication that new have not preferentially formed near old sinkholes is made by comparing nearest neighbor distances for new-to-old sinkholes with nearest neighbor distances for old-to-new sinkholes. These comparisons are made for 1st order (nearest) through 12th order (12th nearest neighbor) sets of nearest neighbor distances (Figure 4). If one population clusters about the other, nearest neighbor distances should

Table 3. Comparisons of variances and means for sinkholes grouped by elevation class.

Null Hypothesis [a,b]	Tests for differences in variances[c]			Tests for differences in means[c]		
	F	p	outcome	T	p	outcome
LogLenCl1=LogLenCl2	1.166	0.046	no difference	16.960	0.001	significant difference
LogLenCl1=LogLenCl3	1.083	0.714	no difference	-6.394	<0.001	significant difference
LogLenCl2=LogLenCl3	1.076	0.691	no difference	-3.338	0.001	significant difference
LogWidCl1=LogWidCl2	1.492	0.054	no difference	-5.098	<0.001	significant difference
LogWidCl1=LogWidCl3	1.851	0.152	no difference	13.210	<0.001	significant difference
LogWidCl2=LogWidCl3	1.051	0.776	no difference	17.860	0.009	significant difference
LogAreaCl1=LogAreaCl2	1.365	0.133	no difference	-4.401	<0.001	significant difference
LogAreaCl1=LogAreaCl3	1.437	0.092	no difference	13.390	<0.001	significant difference
LogAreaCl2=LogAreaCl3	1.052	0.773	no difference	17.315	0.002	significant difference
LogPerCl1=LogPerCl2	1.299	0.207	no difference	16.580	<0.001	significant difference
LogPerCl1=LogPerCl3	1.232	0.331	no difference	-6.468	<0.001	significant difference
LogPerCl2=LogPerCl3	1.054	0.777	no difference	-2.947	0.004	significant difference
LogDiaDepCl1=LogDiaDepCl2	1.328	0.173	no difference	-4.820	<0.001	significant difference
LogDiaDepCl1=LogDiaDepCl3	1.420	0.855	no difference	-7.550	<0.001	significant difference
LogDiaDepCl2=LogDiaDepCl3	1.274	0.185	no difference	-2.730	0.007	significant difference
LogAsyCl1=LogAsyCl2	1.354	0.122	no difference	1.028	0.305	no difference
LogAsyCl1=LogAsyCl3	1.923	0.002	significant difference	-0.485	0.628	no difference
LogAsyCl2=LogAsyCl3	1.420	0.055	no difference	-1.932	0.055	no difference

a – Cl refers to elevation class: 1 sinkholes elevations <175 ft asl; 2 sinkholes elevations 175 ft to < 195 ft asl; 3 sinkhole elevations >195 ft asl.
b - All null hypotheses are tested with log transformed data; for asymmetry it was first necessary to translate values away from an asymmetry of 1 (by subtracting 0.9) before the log transform was applied. The normality of z-scores for all transformed dimensions, except depth, were confirmed by a Kolmogorov-Smirnov non-parametric statistical test. T-tests and F-tests were not performed for depth because log normality could not be confirmed.
c – Significant difference is defined as p<0.05. Asymmetry class 1 and 3 comparisons use an unpaired t-test, which assumes unequal variance. All other comparisons of means were by general t-tests, which assume similar variance.

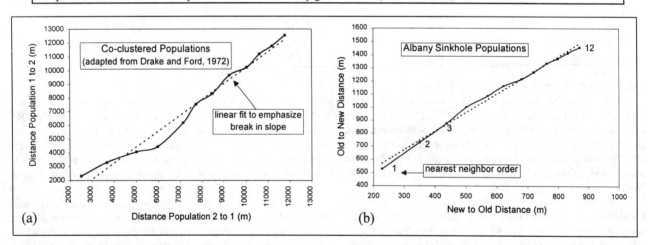

Figure 4. Nearest neighbor plots showing (a) an example of co-clustered sinkhole populations (adapted from Drake and Ford, 1972), and (b) the Albany data. Note that Albany data do not deviate from linearity indicating an absence of co-clustering.

increase markedly once the nearest neighbor order exceeds the average number of secondary sinkholes that cluster around the primary sinkhole. This will result in a distinct break in slope like the example given in Figure 4a that is adapted from co-clustered sinkhole populations in the Mendip Hills, England (Drake and Ford, 1972). Although the Albany data deviate slightly from a straight line (Figure 4b), they do not show any marked break in slope as would be expected if new sinkholes clustered preferentially around old sinkholes, leading us to conclude that old sinkholes had little influence on the development of new sinkholes.

DISCUSSION AND CONCLUSIONS

Striking differences exist between the size, form, and spatial distribution of new and old sinkholes at Albany. New sinkholes are small and have small diameter-to-depth ratios (Figure 5); they have cone-like cross-sectional forms (Hyatt and Jacobs, 1996); they cluster tightly within narrowly constrained elevation classes (Figures 2,3); and they are all the same age. These characteristics argue very strongly for a cover collapse origin, with flooding being the primary triggering agent for collapse. In contrast, old sinkholes are significantly larger (Table 1), and they are more pan-like in cross-sectional form, as is indicated by large diameter-to-depth ratios (Figure 5). Also, they occur over a wider range of elevations and are of varying, albeit, unknown ages. In all likelihood, old sinkholes have formed both by cover collapse and by cover subsidence mechanisms.

Elevation has contributed to differences in the dimensions of old sinkholes. They clearly group into 3 significantly different elevation classes (Table 3, Figure 5). Furthermore, the elevation for these classes closely corresponds to topographic regions identified in DEM maps of the study site (Figure 3). Elevation likely is important for two interrelated reasons. First, lowlands are more frequently flooded than are uplands. Sinkholes in elevation class 1 occur within the limits of the 3-year recurrence interval (RI) flood event as calculated from Flint River stage records. Elevation class 2 is subject to less frequent flooding, spanning the 3-year to 500-year RI flood limits. This upper value is based on estimates of the RI for TSA (Hyatt and Jacobs, 1996), which closely corresponds to the upper limit of this elevation class. Class 3 likely has not been flooded by the Flint River. Thus, old sinkholes within the floodplain and river terrace lowlands have experienced more flood-induced triggering events than have sinkholes in class 2, the abandoned alluvial channel (Figure 3). Although more difficult to confirm, elevation may also be important because of its influence on piping within overburden. In uplands, the depth to the water table, on average, will be greater than in lowlands, resulting in locally steeper hydraulic gradients within the overburden, favoring piping. In summary, the observed trend of increasing sinkhole size with elevation class (Table 2, Figure 5) may reflect these controls. Class 1 sinkholes are small, perhaps because frequent triggering events do not permit large cavities to form in the overburden, and overbank deposition and slope wash further reduces the size of sinkholes. Class 2 sinkholes are larger in part because of less frequent triggering events for collapse, which provides more time for cavity growth, and because of enhanced piping. Class 3 sinkholes, are not subject to flood-induced triggering events, and have the steepest hydraulic gradients, contributing to the growth of large sinkholes.

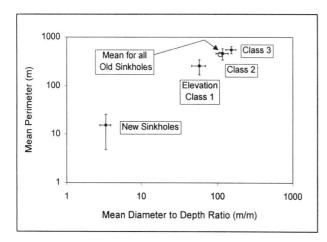

Figure 5. Comparison of mean diameter-to-depth ratios and mean perimeter for all sinkholes. Error bars denote 95% confidence intervals about the means. Note the clear separation between new and old sinkhole groups, and for different elevation classes.

Finally, although some old sinkholes do contain several new sinkholes (Wilkes and Hyatt, 1998), nearest neighbor analysis of new and old sinkhole locations (Figure 4) clearly indicates that the vast majority of new sinkholes have not clustered around old sinkholes. Thus, a multigenerational model (cf. Kemmerly, 1982) for the development of sinkholes does not appear to apply to the Dougherty Plain at Albany.

ACKNOWLEDGMENTS

We thank Randy Wethersby (Albany Planning and Development, GIS Department) for providing access to spatial data files, Chris Strom (South Georgia Regional Development Center) for assistance in converting between State Plane and UTM space. Also, David Gibson (Valdosta State University) provided valuable advice concerning statistical analyses. Valdosta State University supported the initial impetus for this study through a faculty development grant to JAH.

REFERENCES

Beck, B. F. and Arden, D. D., 1984. Karst hydrogeology and geomorphology of the Dougherty Plain, Southwest Georgia. Southeastern Geological Society Guidebook No. 26, Southeastern Geologic Society, Tallahassee, 59 pp.

Beck, B. F., and Sinclair, W. C., 1986. Sinkholes in Florida, an introduction. Florida Sinkhole Research Institute, University of Central Florida. Report 85-86-4, 16 pp.

Brook, G. A., and Allison, T. L., 1983. Fracture mapping and ground subsidence susceptibility modelling in covered karst terrain: Dougherty County, Georgia. *In:* P. H. Dougherty (editor), Environmental Karst, GeoSpeleo Publications, Cincinnati, pp. 91-108.

Clark, W. Z., and Zisa, A. C., 1976. Physiographic map of Georgia. Georgia Geologic Survey, Statewide Map 4.

Drake, J. J., and Ford, D. C., 1972. The analysis of growth patterns of two-generation populations: the example of karst sinkholes. Canadian Geographer XVI:381-384.

Herrick, S. M., and LeGrande, H. E., 1964. Solution subsidence of a limestone terrane in southwest Georgia. International Association of Scientific Hydrology, Bulletin 9: 25-36.

Hicks, D. W., Krause, R. E., and Clarke, J. S., 1981. Geohydrology of the Albany Area, Georgia. Department of Natural Resources, Environmental Protection Division and Georgia Geologic Survey, 31 pp.

Hicks, D. W., Gill, G. E., and Longsworth, S. A., 1987. Hydrogeology, chemical quality, and availability of ground water in the upper Floridan Aquifer, Albany area, Georgia. U. S. Geological Survey Water-Resources Investigations Report 87-4145, 52 pp.

Hyatt, J. A., and Jacobs, P. M., 1996. Distribution and morphology of sinkholes triggered by flooding following Tropical Storm Alberto at Albany, Georgia, USA. Geomorphology 17:305-316.

Kemmerly, P. R., 1982. Spatial analysis of a karst depression population: clues to genesis. Geological Society of America Bulletin 93:1078-1086.

Palmquist, R., 1979. Geologic controls on doline characteristics in mantled karst. Z. Geomorph. N.F. Suppl. Bd 32, pp. 90-106.

Sowers, G. F., 1975. Failures in limestone in humid subtropics. Journal of the Geotechnical Engineering Division, Proceedings ASCE, GT-8, pp. 771-787.

Wilkes, H. P., and Hyatt, J. A. 1998. Distribution and analysis of sinkholes in Albany, Georgia prior to Tropical Storm Alberto. Georgia Journal of Science, 56: p. 54.

Williams, P. W., 1972*b*. The analysis of spatial characteristics of karst terrains. *In:* R. J. Chorley, (editor), Spatial Analysis in Geomorphology. Harper & Row, Publishers, New York, pp. 135-163.

Wilson, W. L., McDonald, K. M., Barfus, B. L., and Beck, B. F., 1987, Hydrogeologic factors associated with recent sinkhole development in the Orlando area, Florida, Florida Sinkhole Research Institute, University of Central Florida, Report No. 87-88-4, 109 pp.

An investigation into the factors causing sinkhole development at a site in Northampton County, Pennsylvania

STEPHEN A. JENKINS & JONATHAN E. NYQUIST   Temple University, Philadelphia, Pa., USA

ABSTRACT

Stratigraphic and structural analyses combined with electrical resistivity soundings and drilling were used to investigate factors causing the development of sinkholes at a site in Northampton County, Northeastern Pennsylvania. By compiling and contrasting data obtained from these investigative techniques it was possible to hypothesize the manner in which local carbonate geology determines the development of sinkholes in the area. Bedrock at the site is comprised mostly of the Epler Formation, which contains both dolomite and limestone. It seems, however, that sinkholes forming in the area relate to voids that occur only in the limestone units. Therefore in areas underlain primarily by dolomite the development of sinkholes is much less likely. Dolomite bedrock topography can be quite different from limestone bedrock topography. Limestone often forms bedrock pinnacles in the subsurface, whereas dolomite, being less soluble, tends to have a more uniform subsurface topography. We found that although resistivity could not distinguish limestone from dolomite, the method could distinguish the two types of subsurface bedrock topography, and although resistivity was not able to conclusively detect air-filled voids directly, it was able to detect clay and water filled features within the bedrock. We conclude that resistivity can be used to ascertain whether the subsurface geology is characteristic of bedrock prone to the development of voids, and to decide where to drill when searching for direct evidence of voids, but only with a thorough understanding of the local geology.

INTRODUCTION

Metzgar Athletic Field and Easton Airport are adjacent sites located on the outskirts of Easton, PA about 100 miles north of Philadelphia. Sinkholes have been a recurring problem at both the field, and to a lesser extent the airport, since the sites were developed in the 1960's (Chen an Roth, 1997). To investigate the factors responsible for the development of these sinkholes, we used aspects of geophysics, structural geology, and stratigraphy. The geophysical techniques tested were 2-D electrical resistivity soundings, and ground penetrating radar. It quickly became apparent that radar would not work because the local clay soil quenched the radar signal. To better interpret the resistivity data we conducted a detailed study of the stratigraphy and structural geology at nearby outcrops, which were directly correlative to the geology at the airport. In this manner we were able to interpret the causes of sinkhole development and relate them directly to the geology.

STRATIGRAPHY

We collected stratigraphic data from two locations: a road-cut located approximately 500 m (1635 ft) west of Metzgar Field, along Bushkill Road in the town of Tatamy, and an outcrop in the Bushkill Creek, 40 m (130 ft) west of the road-cut (Figure 1). The strata were all carbonate rock of the Beekmantown Group (Drake, 1967), which strikes at 60° and dips consistently between 42° and 47° to the south. The section presented in Figure 2 constitutes a 67 m (219 ft) section of the 110m (350 ft) long road-cut starting from the northern end. There are two distinct sinkholes visible in this section of the road-cut and we wanted to identify any lithologic criteria relevant to the development of these sinkholes.

We identified three general lithologies: a relatively soft limestone-rich unit, a very hard chert-carbonate, and a very hard crystalline dolomitized-carbonate. Four distinct faults dipping at 45° were also identified. X-ray diffraction (XRD) analyses of rock samples taken from the outcrop confirmed that the hard carbonate is dolomite.

The outcrop consists primarily of dolomite and chert; limestone units only begin to come in towards the southern end of the outcrop (Figure 2). This pattern makes sense because the highest part of the outcrop is comprised of resistant dolomite and chert. The outcrop gets lower and eventually fades out at the southern end corresponding to an increase in limestone. Also, as can be seen in Figure 2, the only two sinkholes exposed in the outcrop occur in limestone units and not in the dolomite, which reflects the higher solubility of limestone. When the road-cut is traced along strike to the river the same pattern is observed. In the northern part of the river, which correlates to the northern part of the road-cut, resistant dolomite and chert units jut out of the river. While in the southern part, which correlates to the southern part of the road-cut, dolomitic beds do not exist and no units are exposed above the water surface.

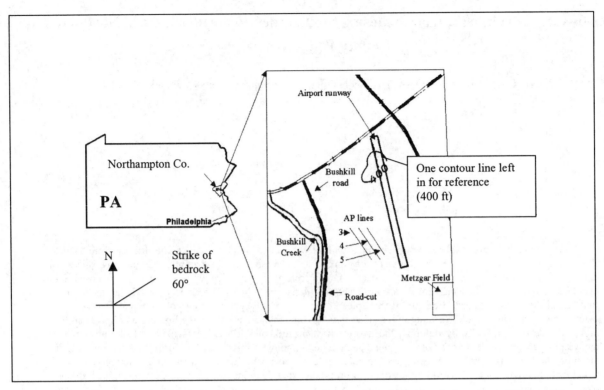

Figure 1. Map of Pennsylvania with the location of Northampton County. Also shown are the locations of Easton Airport, Metzgar Field, the Bushkill Creek, the Tatamy road-cut, and the AP series of resistivity lines.

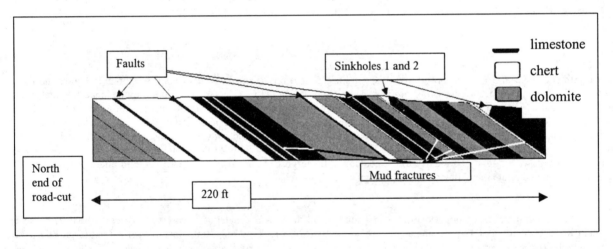

Figure 2. Stratigraphic cross-section of the Tatamy road-cut.

GEOPHYSICAL METHODS

To investigate the subsurface geology we used the Sting/Swift multi-electrode resistivity system (Advanced Geosciences Incorporated). All materials have a relative resistance to electrical current. Knowing the range of resistivity values for specific materials allows for identification of a material based on its measured resistivity (Ward, 1990). The resistivity system uses a pair of electrodes to send electrical current into the ground and measures the potential drop between another pair of electrodes. It then calculates and stores an apparent resistivity value for a specific spot in the subsurface. The Sting/Swift is a multi-electrode system (our system has 28 electrodes) and therefore can vary the separation of both the source and sink of the current, and the electrodes measuring the potential drop. Hundreds of measurements with varying separations are conducted in a matter of minutes (up to 60 minutes when using a 28 electrode system). In this manner apparent resistivity values are obtained for locations at varying vertical and horizontal locations in the subsurface. The system uses an inversion program to produce a 'best fit' model in the form of a resistivity cross-section from which it is possible to interpret the subsurface geology based on the ranges of resistivity (deGroot-Hedlin, and Constable, 1990, Griffiths and Barker, 1993, Loke, 1996).

Instruments such as the Sting/Swift system are creating a renewed interest in resistivity methods. Applications for resistivity include, mapping bedrock topography, locating subsurface voids and chemical plumes, finding archeological sites and detecting leaks in dams. Instruments such as the Sting/Swift have brought about this boom in resistivity applications because multi-electrode systems are capable of making hundreds of measurements in a short period of time. Also the explosion in computer processing speed has made it possible for average computer to process the data in the field.

In this study resistivity sounding was used in an attempt to map the subsurface geology and locate any voids in the near-surface bedrock. To see deeper into the earth larger electrode spacing is required, however, as electrode spacing is increased near-surface detail is lost. For this reason we chose an electrode spacing of between one and five meters. During the fieldwork we collaborated with a team of engineers from Lafayette College lead by Dr. Mary Roth. They were studying foundation stability in Metzgar Field using an identical Sting/Swift system (see Roth, et al., this volume). Together we collected more than 70 resistivity soundings, both in Metzgar Field and in the surrounding area.

RESULTS AND DISCUSSION OF GEOPHYSICAL DATA

Lines M-22, M-26, and M-61 (Figure 3) are examples of resistivity cross-sections constructed from resistivity data collected in Metzgar Field. The lines used to collect this data all ran perpendicular to strike and used a dipole-dipole array with an electrode spacing of between 1 and 5m. The black and white scale under each of the cross-sections represents resistivity values in ohm-meters. The more electrically conductive a material is the less resistant it will be to the flow of electrical current. Notice the low resistivity layer near the surface in most of the sections; this corresponds to a clay soil layer. Carbonate bedrock is usually much less conductive than soil. This phenomenon can be used to interpret the soil-rock interface in the cross-sections. For example in M-61 we believe this interface lies in the zone where the resistivity values move from 200 ohm-m to 700 ohm-m. Air is extremely resistive; therefore air-filled voids should show up as very resistive zones in a cross-section. During initial analyses of the data we believed that features such as the large resistive zone in M-61 were anomalies caused by such an air-filled void. To investigate this interpretation Roth and her team drilled wells corresponding to these high resistivity zones (Figure 3). In many cases no void was discovered by the drilling, in other cases voids from 20 cm (0.65 ft) to over 1 m (3.3 ft) were found within the high resistivity zones. It became clear that the depth to bedrock in Metzgar Field varied greatly, between as little as half a meter (1.6 ft) to as much as 12 m (39 ft) within very short distances (Figure 3). It is our interpretation that the high resistivity zones in the cross-sections are not air-filled voids, they are bedrock pinnacles, some of which contain voids. To further test this hypothesis we created and ran some forward models with highly variable bedrock topography, with and without voids in the bedrock pinnacles. In both cases the cross-sections had similar patterns. We believe that highly resistive bedrock shows up in the cross-section but an air-filled void, when present, is masked by the bedrock.

XRD analysis of cores extracted from Metzgar Field revealed that the bedrock at Metzgar is mainly limestone. This agrees with Figure 1, as the Metzgar Field area, when traced along bedrock-strike, correlates with an area beyond the southern end of the Tatamy road-cut where the lithology begins to change from dolomite to limestone.

As resistivity data were being collected at Metzgar Field it was also being collected at Easton Airport. The resistivity lines at the airport were chosen directly up strike of the Tatamy road-cut. Part of the sketch in Figure 1 shows the position of three of these lines (AP-3, 4, and 5). Figure 4A is a calculated resistivity geologic cross-section generated by the Sting/Swift inversion program from resistivity data collected along line AP-3 using a dipole-dipole array. The southern end of AP-3 was about 300 m (981 ft) directly along strike (60°) from sinkhole 1 in the road-cut. The line was run parallel to the outcrop (perpendicular to strike) with an electrode spacing of 5 m (16 ft). A distinct difference between M-26 and AP-3 is evident (Figure 3B, and Figure 4A). In AP-3 the subsurface geology can be interpreted as having bedrock that is less pinnacled and more uniform than the bedrock in M-26. It is our interpretation that the cross-section produced by AP-3 is uniform because it represents subsurface geology consisting mainly of thick uniform dolomite. Most of the data obtained from line AP-3 correlates to the dolomitic section of the Tatamy road-cut.

The southern ends of AP-4 and AP-5 (Figure 4B and C) correlate to 5 and 10 m (16 ft and 32 ft) respectively south of sinkhole 1 in the Tatamy road-cut. The cross-sections produced from these lines resemble the resistivity patterns found in the Metzgar Field cross-sections (M-22, 26, and 61) in that the subsurface geology appears to be pinnacled. We are not sure if this pattern is caused by limestone pinnacles or by the presence of a thin highly resistive unit between some less resistive units. Electrical resistivity values for quartz can be as high as 3×10^6 ohm-m. Consequently a chert bed sandwiched between carbonate rock can have a much higher resistivity value than the carbonate rock. We believe the high resistivity zones picked up by lines AP-4 and AP-5 are the result of nodular-chert. This set of lines correlates with part of the Tatamy road-cut, which contains chert. Drilling would be needed to answer the question.

CONCLUSIONS

Based on stratigraphic evidence, resistivity data, and XRD data we conclude that sinkholes at the site occur primarily in areas underlain by limestone and less frequently, if at all, in those underlain by dolomite. Subsurface bedrock comprised mostly of limestone often takes the form of pinnacles, which seems to be the case here. As the limestone continues to dissolve voids may develop in the pinnacles leading to sinkholes. Characterizing areas prone to sinkhole development in the study region would be routine if outcrops were abundant; one would simply map areas that were primarily limestone as prone to sinkhole development. Since outcrops are never so conveniently or abundantly located (especially in the eastern United States) resistivity sounding provides a useful tool for obtaining a bedrock profile. Of course there are a number of geologic scenarios that can produce similar resistivity profiles. High resistivity zones can be produced by air-filled voids, bedrock pinnacles, or a nested lithology with a high resistivity value. Interpretation requires both resistivity data and an understanding of the local geology. If the resistivity data show an area comprised of dissolution-prone bedrock, then an investigating team would be in a position to pick drill locations based on their interpretation of the resistivity data. This certainly provides a method of interpreting the subsurface geology that is superior to random drilling.

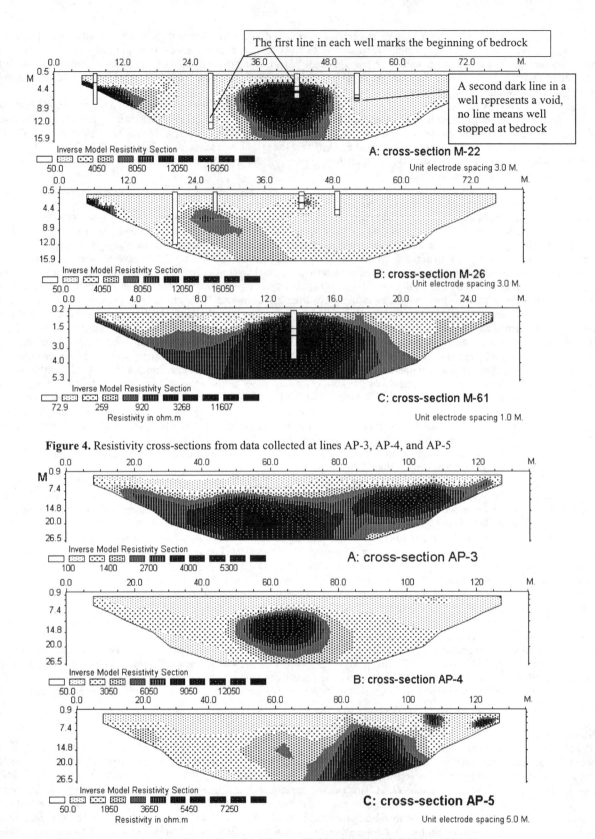

Figure 4. Resistivity cross-sections from data collected at lines AP-3, AP-4, and AP-5

Figure 3. Resistivity cross-sections from data collected in Metzgar Field at lines M-22, M-26, and M-61, well positions have been added.

FUTURE WORK

The main limitation of 2-D electrical resistivity sounding is the influence a 3-D structure can have on a 2-D model. Electrical current will always take the path of least resistance; if it encounters a resistive body it can go under and over it, but it can also go around the sides. When the current finds a path laterally around a resistive body the program still has to account for this decrease in resistivity. The program interprets the lateral path as a deep conductive path. Therefore when viewing resistivity cross-sections constructed using 2-D programs the interpreter must take this phenomenon into account.

Structures can be directly imaged using a 3-D resistivity sounding. A 3-D sounding is accomplished in much the same manner as a 2-D sounding with the electrodes laid out in a grid instead of a straight line. A 3-D inversion program is then able to construct both horizontal and vertical resistivity slices from the grid data.

The main limitation of 3-D sounding is many more electrodes are required. For example to construct a grid with just 10 electrodes on the outer edge would require 100 electrodes. For this reason 3-D work is too expensive for reconnaissance. Instead, 3-D profiling can be used when focusing on appropriate spots previously identified by 2-D sounding. We plan to build on our study by conducting 3-D resistivity soundings in the future.

ACKNOWLEDGMENTS

We would like to thank the Temple University faculty. We would also like to thank Dr. Mary Roth, and Jen Mackey from Lafayette College, for sharing their time and data.

REFERENCES

Chen, L., Roth, M.J.S. 1997, Sinkhole case study: Athletic fields, Lafayette College, Easton, Pennsylvania: The Engineering Geology and Hydrogeology of Karst Terranes, 37-40.

deGroot-Hedlin, C. and Constable, S.C. (1990). Occam's inversion to generate smooth, two-dimensional models from magnetotelluric data, Geophysics, 55, 1613-1624.

Drake, A. A. Jr., 1967, Geologic Map of the Easton quadrangle, New Jersey and Pennsylvania: United States Geological Survey Geological Quadrangle Map GQ-594, scale 1:24,000.

Griffiths and Barker (1993). Two-dimensional resistivity imaging and modeling in areas of complex geology, Journal of Applied Geophysics, 29, 211-226.

Loke, M.H., 1996, RES2DINV ver. 2.1: Rapid 2D resistivity inversion using the least-squares method, Penang, Malaysia.

Roth, M.J.S., Mackey, J.R., Mackey, C. and Nyquist, J.E., 1999, Case study: Characterization Methods in Karst: Seventh Multidisciplinary Conference on Sinkholes and the Engineering and Environmental Impacts of Karst (this volume).

Ward, S.H., 1990, Resistivity and Induced Polarization Methods: Geotechnical and Environmental Geophysics, vol.1, p.147-191.

Deep karst conduits, flooding and sinkholes: Lessons for the aggregates industry

JAMES L.LOLCAMA, HARVEY A.COHEN & MATTHEW J.TONKIN S.S.Papadopoulos and Associates Inc., Bethesda, Md., USA

ABSTRACT

Limestone aggregate quarries in deeply penetrating karst terrain often are at considerable risk of artesian inflow from groundwater or surface water channeled through the karstic aquifer. The inflow occurs through what are likely to be complex conduits which penetrate hundreds of feet into bedrock. Rates of inflow can exceed the operation's pumping capabilities proving uneconomic to manage over the long term. Over time, inflow rates can increase dramatically as turbulent flow through the conduit erodes its soft residual clay-rich fill. One recent investigation observed an inflow rate of more than 40,000 gpm from a surface water source. Flood water persistently laden with sediment is an indicator of conduit washout and implies increasing inflow rates over time.

Conduits carrying flood water can exist in a variety of forms: along deeply penetrating geologic faults, joints, or following the path of preferentially eroded bedding. Preferential structural deformation along faults or bedding can enhance dissolution during subsequent interaction with groundwater. The resulting conduit may be a complex combination of many geological features, making the exploration and remediation of the pathway difficult.

Sinkholes at the site can occur within several contexts. Pre-existing subsidence structures can re-activate and subside further, forming new collapse sinkholes within soil directly overlying the conduit. Cover-collapse sinkhole development can be a direct result of increasing downward groundwater velocities and subsurface erosion associated with enlargement of a conduit. Normal operation events such as a quarry blast can also provide a significant new linkage between the groundwater and the quarry, allowing rapid drainage of the groundwater reservoir. With such drainage and erosion of karst-fill, sinkholes will develop over localized water table depressions, most significantly over enhanced permeability zones associated with fractures. Paradoxically, although the rise in quarry water level will lead to regional reduction in the hydraulic gradients, on local scales, drainage of the groundwater reservoir increases gradients and leads to the development of cover collapse sinkholes.

Recommended methods for preliminary site investigation can include a detailed review of geological literature and drilling logs to compile a conceptual model of the site. A fracture trace analysis with EM geophysics can confirm the locations of major faults and fractures. Fingerprinting of the various water sources to the quarry and the water in the quarry is an inexpensive and effective means of identifying the source and likely direction of the groundwater and surface water flow. Automated geophysical equipment on the market for performing rapid resistivity and microgravity surveys speeds up the site screening process during reconnaissance exploration for deep structure. It is recommended that mine planning fully incorporate this information, so that quarry operators can take proactive measures to avoid catastrophic and costly flooding events.

INTRODUCTION

In the USA, quarrying of limestone aggregate constitutes a sizable chunk of the $7 billion per year aggregates industry. Due to the nature of its target resource, the mining of limestone and dolomite is guaranteed to encounter solution features in rock on some scale at all locations. These features may range from micro-scale stylolitic seams of little economic relevance to groundwater-filled cavern systems with the potential to shut down operations and threaten the extraction of tens of millions of dollars of valuable resource. While blasting and groundwater management associated with mining may appear to cause sudden and catastrophic development of sinkholes and other karstic features, in all cases, it is the karst and its underlying geologic / hydrogeologic foundation which came first. Mining, groundwater extraction, and other activities associated with quarrying merely serve to enhance these features and/or encourage their expression at the surface or in the pit.

Experience has shown that karst is unavoidable in limestone terrains, and that remediation of karstic zones is extremely expensive. As a result, the most cost-effective way to approach this issue is through integration of mine planning with an evolving understanding of local geologic, hydrologic, and hydrogeologic conditions. Traditional practices such as abandonment of flooded pits for new, adjacent pits may only provide short term gains if such factors are not considered. In this paper, we will present some lessons we believe are relevant to the aggregates industry. We base these upon our own experiences, investigating, monitoring, and observing remediation of karst-related groundwater problems in Paleozoic carbonate in the mid-Atlantic Appalachian region. We believe these lessons are applicable to most other areas of the world where aggregate mines penetrate karst bedrock.

NATURE OF THE PROBLEM

Water/karst management problems associated with aggregate mining fall into two categories: 1) those that impact the surrounding areas, e.g. development of induced sinkholes on neighboring properties, and 2) those that directly influence management of the pit, e.g. increased inflow or flooding. While both types may occur catastrophically, they are often managed as independent concerns, rather than related phenomena. Our experience has shown that catastrophic quarry flooding frequently poses a risk of induced sinkhole development. And, while sinkholes themselves might not signal the onset of quarry flooding, their formation may indicate the presence of conditions appropriate for flooding to occur.

What unifies these on-site and off-site features is their origin by erosion of pre-existing karst-fill sediment. Karstic voids in the mid-Atlantic region are commonly filled with reddish-brown clayey residual sediments similar to characteristic karst residual soils (Nutter, 1973; Braker, 1981; White, 1988). Sinkholes that develop rapidly in these areas are due to cover collapse in response to direct erosion or piping of residual soils and void-filling sediments (e.g. Boyer, 1997). Somewhat slower-growing cover-subsidence types are also possible (White, 1988). Similarly, catastrophic flooding of quarries from surface water bodies has been observed to result primarily from sediment removal in response to 1) a lowered water table, and 2) erosion by subsurface turbulent flow. While blasting-induced fracturing, or aperture widening may play a role in initiating flooding events, we have repeatedly observed a correlation between increased inflow volume and suspended sediment, clearly demonstrating a relationship between erosion of sedimentary fill and development of significant karstic "conduits."

Cambro-Ordovician carbonates of the mid-Atlantic region that are widely quarried include limestones and dolomites of the Tomstown Dolomite, Stonehenge Limestone, Conococheague Limestone, Grove Formation and correlative units. All of these units are characterized by an intercalation of limestone and dolomite. Significant siliceous layers are present both as intercalated beds and as adjacent map units. Boyer (1997) and Nutter (1973) have both presented evidence that the presence of siliceous sediments adjacent to a carbonate sequence may enhance karstification by producing pore waters unbuffered by dissolved carbonate. Our investigations have also shown that physical rather than chemical properties of some carbonates can control karstification. Within open- to tightly-folded units, boring logs show a tendency for karst zones (fractured rock, voids, and areas of zones of lost circulation) to develop not within specific beds, but at the contacts between them (Fig. 1). We interpret this to result from the competency contrast between limestones and dolomites and the resulting structural disruption during folding. Subsequent dissolution has preferentially enlarged these sheared and disturbed horizons.

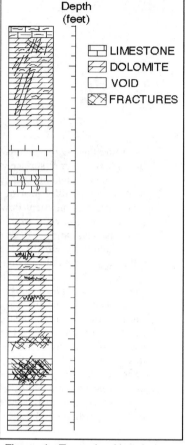

Figure 1 Example of borehole core with voids concentrated along lithologic contacts

SITE EXPERIENCES

During our experience with the aggregates industry, we have investigated karst and collapse sinkholes at quarries in Paleozoic dolomitic limestone which have experienced flooding following a routine blasting event. In one example, flood waters entered through the floor of the quarry from a source that was initially unknown, but later determined to be a conduit connecting the quarry with a karst cavern network outside the pit and extending to a nearby river. Immediately following the blasting event, inflow originated from the dewatering of the karst aquifer, at a rate of about 15,000 gpm. The inflow carried with it eroded karst-fill from the cavern network and the sediment was deposited onto the quarry floor. Over a period of several weeks the inflow was observed to decrease corresponding to the rapid decline of the water table within the karst aquifer. Large areas of the limestone aquifer contained little or no karst and in these areas the water was unaffected by the inflow from karst. During pumping of this initial inflow, a water storage basin located near the river drained rapidly into a new sinkhole. This drainage may have led to enlargement of subsurface voids, creating a continuous connection between the river and the pit which we call the "conduit." Subsequent river inflow to the pit further eroded fill material from the conduit and the rate of inflow was observed to increase over the next several months to over 40,000 gpm.

CONCEPTUAL MODEL

Figure 2 is a schematic depiction of a generic quarry site which is experiencing groundwater seepage, cover- collapse sinkholes, and is at risk of flooding due to interaction of hydrologic conditions and its karstic limestone. The site geology is typical of the Mid-Atlantic Appalachians. The unit being mined is dolomitic limestone, which is folded into a syncline - anticline pair that plunges gently to the north. The quarry is incised into the folded limestone to well below the water table. The water table is located about 50 feet from the ground surface. A major river (3,000 cfs average flow) flows past the quarry, several thousand feet from the pit. A perennial creek flows to the river and passes close to the quarry.

The site geologic map shows the axis of the syncline passing through the quarry and extending beyond the river. A minor thrust fault passes from beneath the river and through the south wall of the quarry. A pervasive fracture system is also present. The creek on the property approximately follows the path of the fracture system. Borehole records reveal karst voids in the limestone concentrated in a stratigraphic unit containing interbedded limestone and dolomite. This unit, found close to ground surface near the river, plunges toward the base of the quarry. A water table depression around the quarry is elongated in the direction of enhanced permeability, or along geologic strike of the bedrock. The water table is depressed locally, at the single fracture, and at the reverse fault. Collapse sinkholes have formed since the start of groundwater seepage, at locations over the syncline axis, over the fault, and over the single fracture.

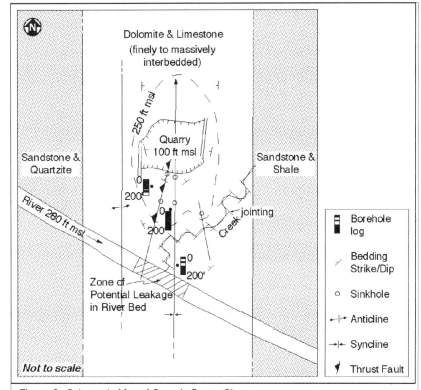

Figure 2 Schematic Map of Generic Quarry Site

The sinkholes have formed where a specific set of conditions exist. Given that most sinkholes develop by collapse of superficial soil, or "cover", into pre-existing voids, activities that reduce the support of such cover materials are likely to promote collapse. For example, seepage of groundwater to a quarry lowers the water table to below the upper elevation of the caverns, causing washout of soft cavern-fill material. Surface water then drains through fissures in soil into the caverns and triggers erosion of a soil cavity over the cavern. The soil cavity grows upward, weakening the support of the shallow ground, which in turn suddenly collapses into the cavity. Figures 3a and 3b show schematically the development of soil cavities and sinkholes. Usually the surface water and entrained soil continue to drain into the collapse sinkhole, causing it to grow in both diameter and depth. Similarly, if the water table is suddenly lowered, for example by excavating into a karst area, groundwater cascades rapidly through the caverns, eroding the soft fill and opening up the caverns to surface water. The collapse sinkholes form within several days of the erosion of the karst fill. Installation and operation of dewatering wells to drawdown the groundwater table around a quarry could have a similar impact on the development of sinkholes.

The property in Figure 2 is at risk of flooding and development of additional induced sinkholes. Potential pathways that could develop into "conduits" include the reverse fault, the interconnected joints and the interbedded stratigraphic interval. Catastrophic discharge from the karst aquifer to the quarry will lower the water table, and extend the depression in the water table south towards the river. Seepage losses from the creek will continue to erode the soft-fill from the karst in this direction, leading to the development of additional sinkholes. Initially, clayey fill within the karst caverns forms a relatively impermeable barrier between the river and the quarry. However as more of the fill is eroded and as the zone of open karst enlarges in the direction of the river, the river may begin seeping directly to the quarry. The pathway for this seepage may lie along the north-south oriented jointing

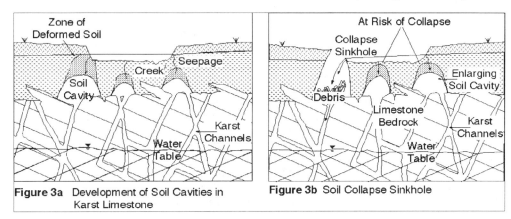

Figure 3a Development of Soil Cavities in Karst Limestone

Figure 3b Soil Collapse Sinkhole

system within the limestone. Our experience suggests that the bed of the river directly overlying the eroded karst will eventually collapse over the conduit (shown in Figure 2 as a wide zone across the intersection of river and geologic structure.) The rate of inflow to the quarry will increase steadily over time as the permeability of the karst increases by erosion of fill material. This scenario assumes a constant gradient across conduit while the quarry tries to maintain their dewatering effort.

REMEDIATION – A VIABLE ALTERNATIVE?

Figure 4 is a generalized example of the costs of grouting remediation versus the resultant flow reduction that might be expected for a difficult inflow remediation problem. The creation of permanent barriers by attempting to construct grout curtains across the conduit, has proven to be a technically challenging problem. Our observations of grouting remediation projects found them to be trying for both the site owner, and the engineering firm implementing the remedy. The primary challenges lie in the erodeability of the soft sediment which remains in the cavern during the placement of a grout plug. During remediation, the plug material (e.g., roofing tar or cement grout) conforms to the surfaces of the fill material in the cavern, temporarily stopping the flow. Fissures immediately form in the soft sediment against the grout curtain. Water flowing through the fissure erodes the soft sediment leaving the plug suspended in sediment and useless. Attempts at forcing grouts (cement-type suspension grouts) into these fissures prior to placing the plug material into the flow conduit are only occasionally successful.

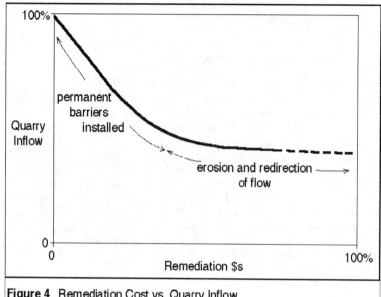

Figure 4 Remediation Cost vs. Quarry Inflow

Psychologically, the lure of grouting success is very powerful. Temporary successes, when achieved, promise the possibility of finally sealing the conduit. Presuming that the sediment erosion problem can be eliminated, the grouting time, materials, and costs mount. The length of time required for a full-fledged grouting program to stop flooding can last months (small voids) to years (massive caverns), and the costs can mount to $10 million or more during that period of time. If we calculate the cost of remediation of flooding inflows, based on several recent experiences, estimates would fall in the range of $500 / gallon per minute of inflow. When we think of flooding inflows on the order of 40,000 gpm, the costs prove to be staggering to an aggregate operation which markets its product at $7 / ton of material.

A PROACTIVE APPROACH TO MANAGING QUARRIES IN KARST LIMESTONE

The risk of ground failure and flooding can be managed using a mine expansion plan that incorporates knowledge of karst geology and hydrogeology. Features must be located which signal risk of ground failure and the potential for flooding. Such features may include existing sinkholes, surface water impoundments, creeks or streams flowing across the property, surface exposures of karst bedrock, and points of entry of large quantities of water into the soil or "swallow holes". Published and unpublished geologic maps will yield information on geologic boundaries, lithology, and structure. Geologic records from existing borings, such as those used for mineral reserve delineation, can be examined to locate karst zones and the water table. Usually the mining of a property has obscured the original ground features. Therefore, historical aerial photographs of the property should be examined to identify fractures, geological faults, interconnected jointing, and traces of sinkholes. The existence of a fault or fracture network can be determined beneath thick soil cover using electromagnetic geophysical methods. Detailed maps of these features should be completed for use in mine expansion planning.

Geophysical logging of groundwater temperature and flow direction in boreholes has, in our experience, been successful at identifying a flow conduit between a creek or river and the quarry when the source of the floodwaters is surface water drainage to the conduit. Similarly, geochemical fingerprinting through the use of naturally occurring tracers and/or isotopes can be invaluable in identifying the source of inflow. Recent advances in resistivity geophysical equipment have also made possible the delineation of caverns to depths of several hundred feet in bedrock. Our experience with automated, high-resolution dipole-dipole resistivity surveys in karst showed the system to be rapidly deployed, and able to survey 1,700 linear feet of terrain per day using 84 electrodes. During exploration surveys, electrodes are spaced at 20 feet. The line can be re-surveyed with electrodes at closer spacings to accurately locate karst features. A geophysical survey completed across the suspected pathways of a conduit at several locations can demonstrate the existence of the pathway. A survey should initially include areas of known subsurface conditions such as areas of massive bedrock, and a cavern associated with a collapse sinkhole.

A drilling program can also be used to explore the resistivity anomalies, geologic faults, and fracture networks, for karst. Open, water-filled karst indicates rapidly flowing groundwater that has eroded the sediment away. Seepage to a quarry can easily follow these pathways. Each borehole should be completed as an observation well. Routine monitoring of the water table in the karst will provide a background against which dewatering of the aquifer into the quarry will be clear. Monitoring well records and groundwater maps can also serve as a useful planning tool to identify areas of the quarry which pose a greater risk for initiating flooding.

CONCLUSIONS: LESSONS FOR KARST MANAGEMENT ON MINE PROPERTIES

- Risk from flooding and collapse sinkholes at aggregate quarries can be managed with a mine planning process that includes detailed information on geology, hydrology, and hydrogeology.

- Blasting and groundwater extraction, in all cases, serve to reactivate and enhance karst, leading to sinkhole development and possible flooding.

- Sinkholes can form by any combination of processes that remove sediment and water from a pre-existing cavern and focus surface water into the soil. Managing quarries in karst terrain is optimized by locating the conditions leading to collapse and mitigating them, as feasible, prior to catastrophic failure.

- The cost of remediation of sinkholes and flooding inflows is staggering as compared to revenues.

- Proper planning requires compilation of information on the geology and hydrogeology, locations of mine water storage ponds, sinkholes, swallow holes, ground lineaments, shallow and deep karst, and the water table. The water table will require regular monitoring as an indicator of aquifer dewatering and indicator of areas of potential concern for collapse or failure.

REFERENCES

Braker, William L., 1981, Soil Survey of Centre County, Pennsylvania, U.S. Soil Conservation Service, Washington D.C., 162 p.

Boyer, Bruce, W., 1997, Sinkholes, soils, fractures, and drainage: Interstate 70 near Frederick, Maryland, Environmental and Engineering Geoscience, 3, 469-485.

Nutter, Larry J., 1973, Hydrogeology of the Carbonate Rocks, Frederick and Hagerstown Valleys Maryland, Maryland Geologic Survey Report of Investigations No. 19, 68 p.

White, William B., 1988, Geomorphology and Hydrology and Karst Terrains, Oxford University Press, New York, 464 p.

A geologic framework in karst: US Geological Survey contributions to the hydrogeology of the Ozarks of Missouri

RANDALL C.ORNDORFF, DAVID J.WEARY, ROBERT C.MCDOWELL, RICHARD W.HARRISON & ROBERT E.WEEMS   US Geological Survey, Reston, Va., USA

STANKA ŠEBELA   Karst Research Institute, Postojna, Slovenia

ABSTRACT

Competing land-use issues related to potential mining of lead and zinc in southeastern Missouri versus protection of large springs and the environment in this area of karst call for a geologic and hydrogeologic framework to understand the potential impact of mining on these natural resources. Information about lithologic units, faults, joints, and karst features (sinkholes, caves, and springs) aids in the understanding of the hydrology of the area. Through detailed geologic mapping, faults have been identified that may have an impact on the hydrogeology either as conduits for or barriers to ground-water movement. Conduits and caves along bedding planes and joints provide avenues for ground-water recharge, movement, and discharge. Joint trend data have been compared with cave passage trends to determine if cave and conduit systems in this part of the Ozarks are controlled by joints. Preliminary results show that cave passages are curvilinear and do not correlate well with measured joint trends; instead, bedding characteristics, bedding-plane dip, and local base level are more important in cave development.

INTRODUCTION

Geologic studies including geologic mapping, evaluation of karst features, and fracture analysis, are being conducted to aid in the decision-making process concerning Federal land-use issues in southeastern Missouri. The study area is focused on the Current River and Eleven Point River drainage basins and includes parts of the Ozark and Eleven Point National Scenic Riverways, the Mark Twain National Forest, several state forests, and some private lands (fig. 1). This area is characterized by many large springs, losing and disappearing streams, caves, and sinkholes. The terrain consists of steep-sided rolling hills and valleys, and entrenched meandering streams; altitudes range from 135 to 400 m (443-1312 ft) and the average relief is 120 to150 m (394-492 ft). The world's largest lead-zinc mining district, the Viburnum Trend, lies on the northern fringe of the study area and exploration for similar deposits has been carried out within the study area. Applications for permits to conduct mineral exploration in public lands have been requested by private industry in the past few years. Federal and State agencies are concerned about environmental impact of exploration and potential mining activities on natural and recreational resources. These competing interests have generated a need for detailed geologic and hydrogeologic studies in order to provide data for informed land-management decisions. Geologic studies that identify karst, stratigraphic, and structural features contribute to the understanding of how aquifers interact and how ground water is transported.

GEOLOGIC AND HYDROGEOLOGIC SETTING

The study area lies within the Salem Plateau of the Ozark Plateaus province of southeastern Missouri. About 750 to 900 m (2460-2952 ft) of gently dipping Upper Cambrian and Lower Ordovician dolomite, sandstone, limestone, shale, and chert overlie Middle Proterozoic rhyolite and granite (fig. 2). Dolomite is the dominant lithology. Of the Upper Cambrian and Lower Ordovician rocks, only the Potosi Dolomite and younger units are exposed in the study area. Middle Proterozoic basement rocks are exposed as knobs that protrude into the Paleozoic section as high as the Gasconade Dolomite.

The Ozark Plateaus Province has the form of a large structural dome roughly bounded by the Missouri, Arkansas, and Mississippi Rivers. In southeastern Missouri, strata overall dip gently to the southeast toward the Mississippi embayment, although structure contours show some broad low-amplitude folds (fig. 3). Locally, strata dip steeply around Middle Proterozoic knobs and near fault zones. Faults are generally steep and trend to the northwest and northeast (fig. 3).

Upper Cambrian and Lower Ordovician strata form three geohydrologic units; two aquifers separated by a confining unit (Imes, 1990) (fig. 2). The lower aquifer, the St. Francois, is 30 to 180 m thick and consists of the Lamotte Sandstone and Bonneterre Dolomite. Overlying the St. Francois aquifer is the St. Francois confining unit (90-110 m thick) formed by shale, dolomite, and

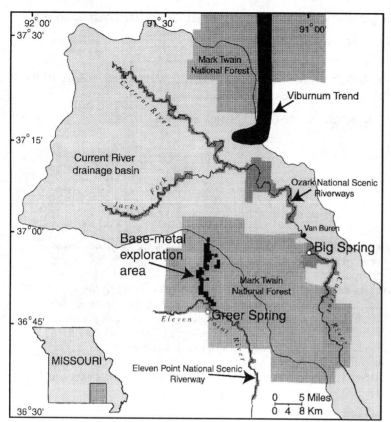

Figure 1. Index map of study area showing Federal lands and minerals exploration area.

Figure 2. Stratigraphic and geohydrologic units of southeastern Missouri.

limestone of the Davis Formation and Derby-Doe Run Dolomite. The upper aquifer, the Ozark (as much as 300 m thick), consists of the Potosi, Eminence, and Gasconade Dolomites, the Roubidoux Formation, and the Jefferson City Dolomite. The Ozark aquifer is the primary source for springs and streams and is used for domestic water supplies.

Distinctive karst features are abundant and underground drainage is significant. Some of the largest springs in the United States are found in the area, including the two largest springs in Missouri, Big Spring (average flow 12 m^3/sec or 282 million gallons per day) and Greer Spring (8 m^3/sec or 183 million gallons per day) (fig. 1) (Vineyard and Feder, 1982). All of the Ozark springs have large fluctuations in discharge related to precipitation events. The prevalence of underground drainage in the area is indicated by many losing and disappearing streams as well as extensive cave and conduit systems. In upland areas, sinkholes in the residual cover are common and collapse sinkholes have been documented related to changes in the ground-water regime (Alley and others, 1972).

LAND-USE ISSUES

An area of potentially economic base metal mineralization has been delineated by exploratory drilling within the Mark Twain National Forest north of Greer Spring (fig. 1). This area is within the drainage basins for the Current and Eleven Point Rivers and dye traces show that this area is also within the recharge basins for Big Spring and Greer Spring (fig. 4). Issues for Federal and State agencies concerned with land management center largely around surface- and ground-water quality and the effects of mine and mill operations on the environment. Of particular concern are the potential effects on water quality of mine tailings, mine dewatering, and dissemination of airborne heavy metal sediments. Lead and zinc mineralization occurs in the Bonneterre Dolomite, part of the St. Francois aquifer. Therefore, it needs to be determined if mining in the St. Francois aquifer will have any adverse effects on the Ozark aquifer, the primary source for springs and domestic water in the area. Faults that cut the St. Francois confining unit may be conduits for ground-water movement and may allow for interaction between the two aquifers. Lithofacies changes within the St. Francois confining unit also need to be evaluated to determine if its impermeability is regionally consistent.

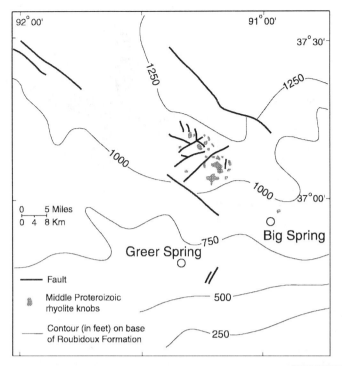

Figure 3. generalized structural features map. structure contours modified from McCracken (1971).

FRACTURE ANALYSIS
Joints

Through detailed geologic mapping at 1:24,000 scale (McDowell, 1998; Orndorff and others, in press), joints were evaluated for spacing, persistence, orientation, and aperture. Most joints in the carbonate rocks are vertical or subvertical, are open, and have narrow apertures (less than 1.3 cm (0.5 in)). However, some joints have been widened by solution and are as much 25 cm (9.75 in) wide. Most outcrops contain at least two joint sets.

Spacing refers to the mean perpendicular distance between parallel joints in a joint set and is a representation of frequency. Because the spacing of joints is related to the number of joints in an outcrop, weighting factors were applied in the statistical analyses of joints, including the generation of frequency azimuth plots (compass-rose diagrams; fig. 5). Closely spaced joints (less than 0.6 m (2 ft) spacing) were weighted by a factor of 3, medium-spaced joints (0.6-1.8 m (2-6 ft) spacing) by a factor of 2, and widely spaced joints (greater than 1.8 m (6 ft) spacing) were used at unit value.

Joint persistence refers to the degree to which joints have propagated through strata. Two levels of persistence were defined for the map area, throughgoing and non-throughgoing. Throughgoing joints are those that cut the entire bedrock outcrop and persist through all beds. Non-throughgoing joints are those that are confined to individual beds and do not extend into adjacent strata. Throughgoing joints provide a more direct avenue for recharge into the ground-water system and tend to be solution widened. Only about 5 percent of joints are throughgoing. In outcrops containing non-throughgoing joints, solution widening has occurred dominantly along bedding planes.

Joints in the Upper Cambrian and Lower Ordovician rocks occur in two dominant sets, N. 20 W. to N-S and N. 70-85 E. (fig. 5). These well-developed preferred orientations are interpreted to have developed under regional stress fields probably related to the Ouachita and Appalachian orogenies.

Faults

Through geologic mapping within the study area, faults have been recognized by stratigraphic offset and the occurrence of fault breccia (McDowell, 1998; Orndorff and Harrison, 1997; Orndorff and others, in press). Kinematic indicators (mullion structures and en echelon fractures) as well as observations of faults in the nearby mines of the Viburnum Trend to the

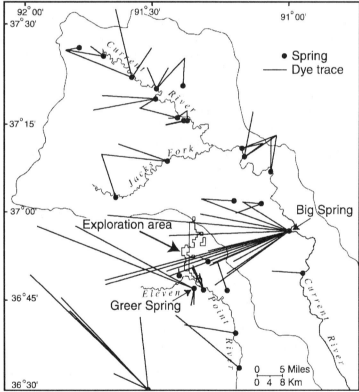

Figure 4. Dye traces from Imes and Kleeschulte (1995) and unpublished data from National Park Service, Ozark National Scenic Riverways, Van Buren, MO.

north, indicate that there may have been significant strike-slip motion on many of the faults. The general trend of faults is northwest and northeast but not parallel to the regional joint trends (fig. 5).

SINKHOLES

All exposed Cambrian and Ordovician carbonate units in the study area contain sinkholes. However, most sinkholes occur on plateau areas underlain by the Gasconade Dolomite, Roubidoux Formation, or Jefferson City Dolomite. Collapse has occurred near the mineral exploration area in the recent past. Alley and others (1972) reported sinkhole collapses in 1968 and 1969 that were related to a surface-water impoundment constructed in 1967 or 1968 near the mineral exploration area. These sinkholes drained the impoundment, and since then it has not held water. One sinkhole, 12.2 m (40 ft) in diameter, collects all of the surface water from the drainage basin. Building an impoundment can effect the surface- and ground-water regime sufficiently to cause collapse by means of (1) localization of surface water over the valley floor, (2) localized recharge of water to the ground-water system, which tends to move fine materials into solution channels and voids, (3) the increased plasticity of residuum, decreasing its structural strength, and (4) increased loading on the valley floor. Events such as these show a high potential for ground-water contamination from impoundments with low water quality.

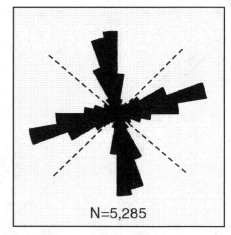

Figure 5. Frequency azimuth plot (rose diagram) of joints measured in study area. Dashed lines are general trends of faults in the region.

CAVE SYSTEM

There are approximately 700 known caves within the study area. Most of these caves have curvilinear cave passages, and much of the development of these branchwork caves appears to be controlled by favorable beds and(or) bedding planes (Palmer, 1991). Frequency azimuth plots (rose diagrams) of cave passage orientations and orientations of joints exposed in caves show some fracture control, although fractures are not the major guiding factor for passage directions (fig. 6). Many caves in the region occur just below sandstone horizons of the Gunter Sandstone Member of the Gasconade Dolomite and the basal sandstone of the Roubidoux Formation, suggesting that these sandstone horizons may be important in cave development; however, the role of the sandstone horizons in caves is not understood at this time. Many caves in this area have been modified by lowering of local base level as can be seen in the downward cutting of cave streams.

CONCLUSIONS

A geologic framework being developed for karst of the Ozarks of southeastern Missouri suggests that bedding characteristics and attitude are the most important factor in cave and conduit development, while joints are of secondary importance. Although there is a low percentage of throughgoing joints in the area, they are important for vertical recharge to the ground-water system. Lateral movement of ground water on a regional scale is controlled mainly by the gentle dip of bedding to the south and southeast toward the Mississippi embayment, occurs along solution conduits related to bedding surfaces, and is possibly influenced by bedding characteristics such as calcite content and grain size. An understanding of the geologic and hydrogeologic processes influencing the karst of the Ozarks will aid in the decision-making process of land managers in this area of potential exploration and mining of lead and zinc. This framework is also important for assessing the hydrologic connection between the exploration area and nearby springs and to identify potential pathways of contaminants.

REFERENCES CITED

Alley, T.J., Williams, J.H., and Massello, J.W., 1972, Groundwater contamination and sinkhole collapse induced by leaky impoundments in soluble rock terrain: Missouri Geological Survey and Water Resources, Engineering Geology Series no. 5, 32 p.

Imes, J.L., 1990, Major geohydrologic units in and adjacent to the Ozark Plateaus Province, Missouri, Arkansas, Kansas, and Oklahoma-Ozark aquifer: U.S. Geological Survey Hydrologic Atlas HA-711-E, 3 sheets, scale 1:750,000.

Imes, J.L., and Kleeschulte, M.J., 1995, Seasonal ground-water level changes (1990-93) and flow patterns in the Fristoe unit of the Mark Twain National Forest, Southern Missouri: U.S. Geological Survey Water-Resources Investigations Report 95-4096, 1 sheet.

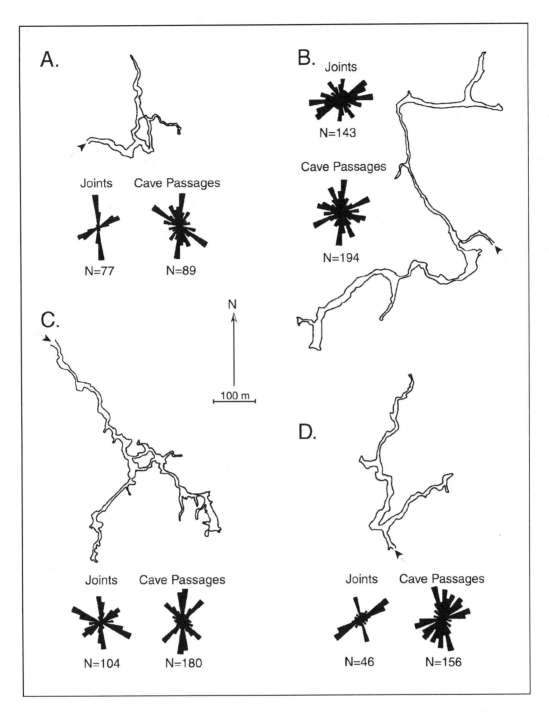

Figure 6. Cave plans and frequency azimuth plots of joints in caves and cave passages for (A) Branson Cave, (B) Round Spring Cave, (C) Wind Cave, and (D) New Liberty Cave. Cave plans reduced from maps produced by the Cave Research Foundation. Arrows indicate cave entrances. Cave passages were weighted with respect to length.

McCracken, M.H., 1971, Structural features of Missouri: Missouri Geological Survey and Water Resources, Report of Investigations 49, 99 p.

McDowell, R.C., 1998, Geologic map of the Greer quadrangle, Oregon County, Missouri: U.S. Geological Survey Geologic Investigations Series Map I-2618, scale 1:24,000.

Orndorff, R.C., and Harrison, R.W., 1997, Preliminary geologic map of the Spring Valley 30' X 60' quadrangle, Missouri: U.S. Geological Survey Open-File Report 97-434, scale 1:100,000.

Orndorff, R.C., Harrison, R.W., and Weary, D.J., in press, Geologic map of the Eminence quadrangle, Shannon County, Missouri: U.S. Geological Survey Geologic Investigations Map I-2653, scale 1:24,000.

Palmer, A.N., 1991, Origin and morphology of limestone caves: Geological Society of America Bulletin, v. 103, p. 1-21.

Vineyard, J.D., and Feder, G.L., 1982, Springs of Missouri: Missouri Geological Survey and Water Resources, Water Resources Report 29, 212 p.

Topographic and hydrogeologic controls on sinkhole formation associated with quarry dewatering

STUART STRUM   North Carolina Division of Water Resources, Raleigh, N.C., USA

ABSTRACT

In August and September 1994, a series of sinkholes developed at a residential property located adjacent to a limestone quarry. Field observations and water levels from the quarry were evaluated to assess the problem. A creek flows between the residential property and the quarry, and this stream occupies a small, but well defined valley, with a steep escarpment of 1.5 to 2.2 m relief between the valley floor and surrounding uplands. The valley bottom is a permanent wetland approximately 25 m across adjacent to the residence. At the time of the field evaluation, the wetland appeared dewatered. The quarry extracts aggregate from the Eocene Castle Hayne Formation, a moldic bioclastic limestone.

During the field visit it was noted that sinkhole development near the quarry occurs along the upland area adjacent to the escarpment that bounds the stream valley. The sinkholes are up to 5 m in diameter and 3 m deep. Concentric cracks tangential to some sinkhole boundaries indicate additional sinkhole growth will probably occur.

A series of 8 transient electromagnetic soundings were performed to develop an apparent resistivity profile of the subsurface. The results show high apparent resistivities, ranging up to 700 ohm-meters, with a steep vertical gradient to the resistivity maxima. The relatively high values for apparent resistivities may reflect voids and enhanced porosity zones that are associated with sinkhole development.

The association of sinkholes and subsidence areas with the upland edge of the stream valley escarpment suggests a topographic control of secondary porosity development in the limestone. Infiltration at the edge of the upland has a relatively short flow path to discharge to the stream valley. The short flow path would have a higher flow velocity because of the hydraulic gradient at the edge of the escarpment, and the ground water would not reach chemical equilibrium with the carbonate matrix of the limestone. Both of these factors would increase secondary porosity development along the valley scarp.

Large changes in head have been observed in the monitoring wells at the quarry as the active pit has been moved across the quarry site. The collapse of the sinkholes concurrent with large changes in water levels at the quarry suggests that head changes in the limestone aquifer may have been a triggering mechanism for sinkhole collapse.

REGIONAL SETTING
Geology and Geomorphology

The study area is located in the outcrop belt of two Tertiary formations that comprise the Castle Hayne Aquifer system in the North Carolina coastal plain, the Eocene Castle Hayne Limestone and the Oligocene River Bend Formation (Figure 1). These units are exposed in the eastern portion of the coastal plain, from New Bern to Wilmington, North Carolina. The coastal plain has been eroded into a series of terraces by eustatic sea level changes during the Cenozoic. The limestone outcrop belt occurs mostly between the Surry and Suffolk Scarps, at elevations between 25 and 3 meters above sea level. A thin veneer of terrestrial, nearshore marine and shelf sediments was deposited across most of the terrace surfaces, and has been partially removed by subsequent erosion. To the north of the outcrop belt, the carbonates are overlain by the Pliocene Yorktown Formation. The southern end of the outcrop belt occurs where the updip limit of the limestones intersects the modern coastline, near the Cape Fear River. The terrace surfaces have been dissected by erosion associated with modern drainage systems. Areas between major streams are often nearly flat, with poorly drained pocosins or upland swamps, while major drainages may be incised up to 12 m below the regional grade. Minor drainages have less pronounced relief, with valleys often less than 5 m below surrounding uplands.

The Castle Hayne Limestone is a sandy fossiliferous limestone that was deposited in shallow marine shelf and bank environments (Jones, 1983). The stratigraphic and paleontologic relationships of the Castle Hayne have been disputed, but the upper part of the Castle Hayne, the Spring Garden Member, is widely recognized as a middle Eocene sandy molluskan limestone with moldic porosity. The overlying River Bend Formation, of Oligocene age, is also a sandy molluskan limestone. This unit may comprise the upper 2 to

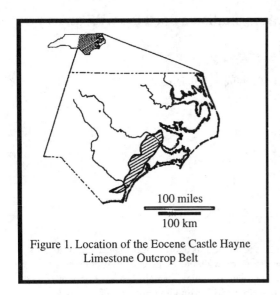

Figure 1. Location of the Eocene Castle Hayne Limestone Outcrop Belt

4 m of the Castle Hayne Aquifer system in the study area. Abundant mollusk tests in the original carbonate sediment have been preferentially dissolved during meteoric diagenesis, resulting in secondary porosity that ranges up to 42 per cent (Thayer and Textortis, 1977). These sediments have been subaerally exposed for significant amounts of time since their deposition in the middle Eocene and Oligocene; meteoric diagenesis has infilled primary interparticle porosity and developed moldic porosity from dissolution of fossil shells . The macrofaunal assemblage of these rocks is dominated by pelecypods, gastropods and bryozoans. The original mineralogy of these fossils was mostly aragonite with some magnesian calcite. Concurrent with the dissolution of the aragonite and high magnesium calcite, low magnesium calcite was precipitated as a fine to medium grained spar cement in the primary pore spaces (Thayer and Textortis, 1977). In the uppermost portion of the limestone section, dissolution features have been observed that are not fabric selective, such as vugs and channels, a result of the dissolution of low magnesium calcite.

Hydrogeology

In the shallow subsurface, ground water generally flows from upland recharge areas to the nearest streams. Interstream areas are usually relatively flat uplands, and hummocky surfaces and mottled soils associated with limestone dissolution are sometimes present. Major and minor stream valleys often have associated wetlands, reflecting discharge along the streams and the interconnection of the surface waters with unconfined aquifers in the recent alluvium. In the middle and lower members of the Castle Hayne Limestone and in the underlying confined aquifers, regional flow is primarily to the east and east-southeast, parallel to regional dip. Some aquifers discharge to the Atlantic Ocean where these units are truncated on the continental shelf. The Castle Hayne and underlying aquifers are water supplies of regional importance where chloride concentrations are below drinking water standards. Because of high aquifer transmissivities, the radii of influence for pumping centers for the Castle Hayne are localized in this area, while regional drawdown is occurring in the underlying confined aquifers beneath and west of the limestone outcrop belt. An extensive cone of depression is present in the Castle Hayne Aquifer System to the northeast of the outcrop belt, where the limestone aquifer is pumped for depressurization at a phosphate mine.

SINKHOLE OCCURRENCE AND FIELD INVESTIGATION RESULTS

In August and September, 1996, seven sinkholes up to 1.75 m in diameter catastrophically collapsed at a residential property located 1 km north of an active limestone quarry in Craven County, North Carolina (Figure 2). In March 1997, the Division of Land Resources (DLR) of the North Carolina Department of Environment and Natural Resources asked the Division of Water Resources (DWR) to evaluate the area impacted by sinkholes and provide recommendations to DLR on possible causes of sinkhole formation and potential mitigation strategies. DWR performed field reconnaissance, transient electromagnetic soundings and reviewed existing hydrogeologic information, including quarry monitoring information, to evaluate the sinkholes.

During the site visits, DWR personnel observed seven sinkholes at the residential property. The largest sinkholes were 5 m in diameter and approximately 3 m deep. The smallest sinkholes were approximately 0.3 m in diameter and up to 0.2 m deep. Two of the larger sinkholes were partially surrounded by concentric cracks in the ground surface, the margin of the sinkhole was underlain by void space, and further lateral growth of these sinkholes appeared to be a strong possibility. Caswell Branch is a permanent stream (mean flow approximately 1 m/sec) that flows between the residential property and the quarry. The stream valley is 1.5 to 2.2 m below the adjacent areas of the residential property and the quarry, with a broad, level valley floor approximately 25 m across. Upland areas near the stream are wooded or open land, with well-drained soils. The small stream valley is densely wooded with mature Black Gum (*Nyssa Sylvatica*) and Bald Cypress (*Taxodium distichum*). Cypress knees were abundant on the valley floor, indicating that the trees had grown in soil that was constantly or frequently saturated (Radford, et al, 1968). No sinkholes were observed in the valley floor. The stream valley is a wetland, as evidenced by phreatophyte dominated flora and a mat of organic debris that had accumulated under subaqueous conditions. At the time of the field evaluation, the wetland appeared dewatered, which was of particular interest, as the field observations were made in early spring, and climatic conditions were not significantly drier than normal. The stream has an anastomosing to braided architecture. The creek flows in relatively well defined channels, and the valley floor was dry, as noted above.

A similar distribution of subsidence features were observed on the quarry side of the stream valley, with a different style of development. Subsidence in this area occurs in a less dramatic fashion, with hummocky depressions up to 5 m across and 0.7 m deep present near the valley boundary. Sinkholes with steep sides and collapse structure were not observed south of the creek. The different style of subsidence probably reflects the control of tree roots on land subsidence. The area is wooded, with mixed white pines and hardwoods up to 40 m tall. This relatively mature forest community has extensive roots in the soil which would prevent development of the pronounced collapse structures present in the open areas on the other side of the creek. Also, it was noted during the field visit that in some of these gentle depressions, a wooden stick 1 cm in diameter could be pushed by hand into the soil approximately 0.8 m, indicating the presence of significant void space in the subsurface.

A series of 8 transient em soundings were performed at the site to develop an apparent resistivity profile of the subsurface along an east-west transect, roughly parallel to the edge of the escarpment of the stream valley. The profile results show high apparent resistivities, ranging up to 700 ohm-meters, with a steep vertical gradient to the resistivity maxima at a depth of 15 to 20 m below land surface. There were two pronounced anomalies detected along the profile, one approximately 21 m east of the west end of the profile and one at 130 m east in the profile. The eastern anomaly extends from just below land surface to an apparent depth of 67 m, with the highest apparent resistivity at 700 ohm-meters (Figure 3). The western anomaly extends from just below land surface to a depth of approximately 30 m, with the highest resistivity at 400 ohm-meters. Both resistivity maxima appear associated with sinkhole development, as they are parallel to the position of two groups of sinkholes north of the transect line, located along the escarpment. The relatively high values for apparent resistivities may reflect voids and enhanced porosity zones that are associated with sinkhole development.

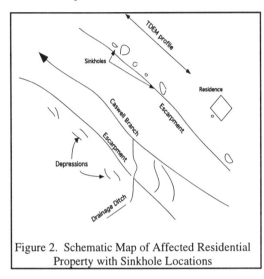

Figure 2. Schematic Map of Affected Residential Property with Sinkhole Locations

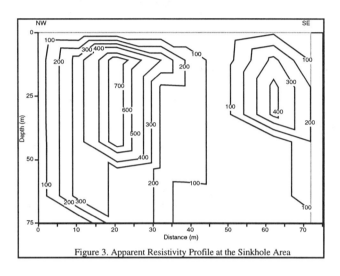

Figure 3. Apparent Resistivity Profile at the Sinkhole Area

QUARRY OPERATION

A construction materials company operates a quarry 600 m southeast of the location where the sinkholes developed (Figure 4). The quarry is an open pit operation. The pit is dewatered by a drainage trench system which runs from the pit face to a pumping station. The drainage system is constructed with a sump approximately 15 m below land surface. Extracted water is routed to settling ponds and then discharged to Caswell Branch. In order to dewater the pit, the drainage system is pumped at a rate of 38 million liters per day. Water levels in monitoring wells on the perimeter of the quarry site have declined by as much as 5 m below pre-pumping conditions. Data from the state of North Carolina's regional monitoring well network indicates that transmissivity of the upper Castle Hayne Limestone in the region is approximately 1.1×10^{-2} m^2/sec.

Water level monitoring performed by the quarry operator shows rapid changes in water levels at the quarry site (Figure 5). Monitoring well 5 at the quarry is approximately 75 m east of the easternmost sinkhole on the residential property. Water level measurements show that the head at well 5 dropped by nearly 3 m in July 1992. After that time the water level at this well varied by no more than 0.1 m until August, 1994, when the water level increased by nearly 3 m. The sinkholes formed on the property north of the quarry in August and September, 1994. The quarry operator's consultant has speculated that there were errors in the water level measurements at the quarry site. However, the consistency of the measurements within the 2 year period between the large changes would not be expected if there were random error introduced by inaccurate measurements. The variance in water levels can be explained by moving the pumping center (the pit face) to the northeastern portion of the quarry property during the time period in question. Concurrent with the recovery of water levels at well 5, water levels decrease at wells 3, 7 and 1, which would be consistent with a southward advance of the quarry pit. Monitoring well 6, located west of well 5, was dry during the 1992-1994 period, reflecting the decrease in water levels at the north side of the site.

INTERPRETATION: CONTROLS ON SINKHOLE FORMATION

The association of sinkholes and subsidence areas with the upland edge of the stream valley suggests a topographic control of secondary porosity development in the Castle Hayne Limestone. This is supported by two aspects of the areal distribution of sinkholes: 1) the absence of subsidence features in the stream valley, and 2) occurrence of only one sinkhole 10 m from the escarpment, with all other subsidence features within 3 m of the escarpment.

Precipitation infiltrating into the ground at the edge of the upland would have a relatively short flow path to discharge to the stream. The steeper hydraulic gradient at the edge of the stream valley causes higher flow velocity for groundwater at the edge of the escarpment. Water infiltrating from soil with significant organic matter often has elevated carbon dioxide content, with pCO_2 for soil water up to 10^{-2} atm (Drever, 1982). The short flow path from the soil zone to the discharge point, combined with the CO_2

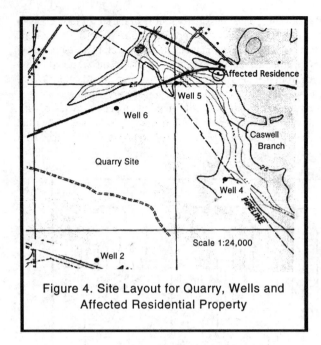

Figure 4. Site Layout for Quarry, Wells and Affected Residential Property

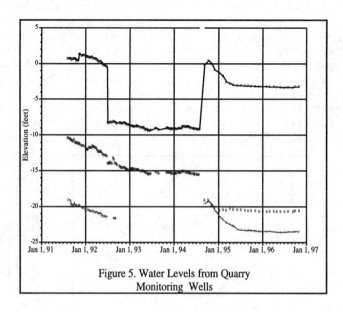

Figure 5. Water Levels from Quarry Monitoring Wells

contribution for soil organic matter, enhances the dissolution of carbonate minerals, as the ground water would not have sufficient time to attain equilibrium with calcite before discharging to the surface drainage. After the less stable carbonates (aragonite and magnesium calcite) were removed by dissolution of the shell material, dissolution of low magnesium calcite in the rock matrix has produced channels and vugs. Interconnection of the non-fabric selective secondary porosity has created larger scale void spaces, which have collapsed to form the sinkholes.

The topography and stream valley flora both suggest that under hydraulic equilibrium, the valley escarpments would be discharge points for groundwater, contributing to base flow of the stream. Wetlands are commonly present at other drainages of similar size and elevation in nearby areas. The wetland was dewatered when the water levels at well 5 were 1 m below sea level. Larger water level declines must have occurred at the stream and sinkhole area when water levels had dropped 3 m at well 5. The relative relationships of the water table, the wetland and land surface during the different pumping conditions are shown in Figure 6. Recovery of the water table in the stream area would have resulted in upward stresses as the water table rebounded to near pre-pumping conditions, and may have triggered the collapse of overlying materials into the pre-existing void spaces that had formed along the edge of the stream valley. DLR is working with the quarry operator to assure that the quarry dewatering does not cause rapid fluctuations in water levels in the aquifer surrounding the site, in order to minimize the potential for further sinkhole development.

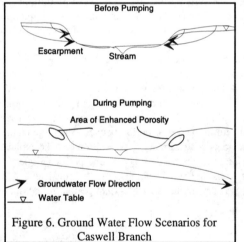

Figure 6. Ground Water Flow Scenarios for Caswell Branch

REFERENCES

Drever, J.I., 1982, *The Geochemistry of Natural Waters* (Prentice-Hall: Englewood Cliffs, N.J.) 388 p.

Radford, A.E., H.E. Ahles, and C.R. Bell, 1968, *Manual of Vascular Flora of the Carolinas* (Chapel Hill, N.C.: The University of North Carolina Press). 1183 p.

Thayer, P.A. And D.A. Textortis, 1977, *Faunal and Diagenetic Controls of Porosity and Permeability in Tertiary Aquifer Carbonates, North Carolina*, Special Publication 7, North Carolina Department of Natural and Economic Resources, Geology and Mineral Resources Section. 35 p.

Applications of GIS technology to the triggering phenomena of sinkholes in Central Florida

DEAN WHITMAN Florida International University, Miami, Fla., USA
TIMOTHY GUBBELS Raytheon Systems, Landover, Md., USA

ABSTRACT

Sinkholes constitute a principal geologic hazard in Central Florida. We use geographic information system (GIS) technology to investigate the spatial relationships between hydrogeologic factors and sinkhole formation near Orlando, FL. Landsat TM imagery, digital topography, and well data are used to construct a model of the head difference between a discontinuous set of surficial aquifers and the Floridan aquifer, a regionally extensive confined aquifer. This model is quantitatively compared to a buffer model of distance to nearest sinkhole constructed from a database of collapse events. Sinkhole occurrence is positively associated with regions where the head difference is between 5 and 15 m. In these regions, sinkholes are more common and more closely spaced than expected. In contrast, sinkholes are less frequent and farther apart than expected in regions of low head difference. This association of sinkhole proximity to high head difference demonstrates the importance of hydrostatic loads in sinkhole hazard.

INTRODUCTION

A geographic information system (GIS) is a collection of computer hardware and software used to manage and analyze spatial data. A GIS provides the ability to organize, visualize, and merge spatial datasets from different sources and allows quantitative spatial analysis and predictive modeling of these data. Recent advances in computer hardware and software have made this technology accessible to a wider community than ever before. Today, even very complex tasks in digital image processing, interpretation, and GIS modeling can be performed with public domain and commercial software running on relatively inexpensive desktop workstations and personal computers. The use of GIS techniques to investigate and mitigate natural hazards is a growing applications area of considerable importance for state and local governments, as well as for the insurance industry.

Central Florida is well known for its numerous karst lakes and landforms. Sinkholes are ubiquitous features of karst terranes and constitute a major geologic hazard in Central Florida. The term "sinkhole" refers to an area of localized land surface subsidence, or collapse, due to karst processes, which results in closed circular depressions of moderate dimensions (Monroe, 1970). Sinkholes are formed by the subsidence or collapse of surficial material into subsurface cavities in regions underlain by limestone and other rocks susceptible to dissolution by ground water. They typically form funnel-shaped depressions ranging in size from meters to hundreds of meters. Although many sinkholes form by processes of progressive subsidence, formation by catastrophic collapse is common and, because of its inherent suddenness, is a major geologic hazard.

Existing studies of sinkhole hazards have traditionally been the province of civil engineers, geologists, and hydrologists accustomed to working at local scales (e.g. Benson and La Fountain, 1984; Wilson and Beck, 1988). New remotely sensed data sets and computer-based techniques developed during the 80s and 90s have opened the opportunity for more synoptic study of the sinkhole hazard. The resolutions of commonly available satellite imagery, 30 m for Landsat TM and 10 m for SPOT, impose firm limits to the detection of small natural objects such as sinkholes and this scale limitation prevents their use for direct local or regional mapping of sinkholes. Instead, these data sets offer potential for understanding sinkhole collapse hazard by permitting spatial analysis within a GIS of such related factors as topography, hydrology and land use.

This study demonstrates the utility of computer visualization and GIS software for investigating the triggering phenomena of geologic hazards. In this study, these tools are used to examine some of the regional geologic phenomena that influence the sinkhole hazard in central Florida. In particular, we apply techniques of spatial analysis to examine the spatial interrelationships between hydrostatic heads of a surficial and a confined aquifer system and the locations of reported sinkholes in Central Florida.

REGIONAL SETTING
Regional Geology and Hydrogeology

The Florida peninsula is the emergent part of a larger feature, known as the Florida platform. In central and northern Florida, the peninsula is composed of generally undeformed Cenozoic carbonate and clastic strata. At the surface, strata from Eocene to Holocene age are exposed. The older part of this section is exposed on the western side of the peninsula and the younger part of this section is

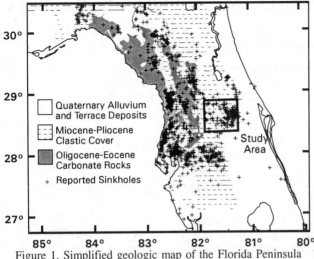

Figure 1. Simplified geologic map of the Florida Peninsula modified (from Florida Geological Survey). Holocene through Miocene units are exposed along the eastern side of the peninsula and Miocene through Eocene units are exposed along the west side. The crossed boxes indicate the locations of reported sinkholes that have occurred between 1960 and 1993 which were compiled by the Florida Sinkhole Research Institute (Spenser and Lane, 1995).

exposed on the eastern side of the peninsula (Fig. 1). Shallow water marine and shoreline processes have been the dominant shaping force of the peninsula, as the platform has been alternately flooded and exposed over the last 10 million years.

The Florida peninsula is underlain by an extensive system of aquifers contained in Cenozoic limestones and clastic sediments. In central and northern Florida region, three distinct hydrostratigraphic units exist: Floridan aquifer, the intermediate confining unit, and the surficial aquifer system (Florida Geological Survey, 1986; Miller, 1997). The Floridan is a regionally extensive aquifer system situated within a thick sequence of Paleocene to early Miocene age limestones which have been subjected to extensive dissolution and cavity formation. The Floridan aquifer system is unconfined in the northwestern parts of the peninsula where these limestones outcrop at the surface or are thinly covered (Fig. 1). Elsewhere, it is confined or semi-confined. The Floridan aquifer provides the principle municipal and agricultural water supply for much of central and northern Florida. Spatial and temporal variations in its potentiometric surface are often related to groundwater withdrawal.

Where confined, the Floridan aquifer is capped by Miocene age, variable thickness, low permeability clay rich clastic sediments of the Hawthorn group. This unit acts as an aquitard and forms the intermediate confining unit. The Hawthorn is covered in places by variable thickness, permeable, undifferentiated sand, silt and shell beds which range in age from Pliocene to Holocene. These units form a set of unconfined, discontinuous aquifers known as the surficial aquifer system. Because of its general low productivity and poor water quality, this aquifer system is rarely used for potable water supplies and in contrast to the Floridan, the water tables of these surficial aquifers are more directly influenced by changes in rainfall, runoff, and landuse.

Sinkholes in Florida

Because large areas of the peninsula are underlain at relatively shallow depths by carbonate bedrock, sinkholes are widespread in Florida. During the 1980s, the Florida Sinkhole Research Institute (FSRI) at the University of Central Florida compiled an excellent computerized inventory of new sinkhole occurrence in the state of Florida (Wilson and Beck, 1992; Spencer and Lane, 1995; Wilson and Shock, 1996). This database contains over 1900 reported sinkholes occurring between 1960 and 1991 and includes information on location, date of occurrence, dimensions and hydrogeologic setting for each sinkhole. This database formed the basis of a number of studies including: Bahtijarevic (1989), Currin and Barfus (1989), Wilson and Beck (1992) and Wilson and Shock (1996). FSRI is no longer operative and the database is currently maintained by the Florida Geological Survey. The database is most complete for the 1980s when the institute was operative.

The locations of sinkholes from this database are shown on Figure 1. Three major types of sinkholes are common to central Florida: limestone-solution, cover subsidence, and cover-collapse (Sinclair and Stewart, 1985). Limestone-solution sinkholes predominate in the northwestern peninsula where limestone is exposed or is thinly covered. They form shallow broad depressions due to the gradual solution of the limestone surface and along joints. Cover subsidence sinkholes tend to occur where the limestone is covered by noncohesive and permeable sand. Subsidence occurs gradually as the cover material spalls downward into solution cavities and depressions in the limestones. Cover collapse sinkholes predominate where the limestone is covered by clay rich (Hawthorn group) sediments with sufficient cohesian to bridge cavities in the limestone. These sinkholes occur abruptly when the cover collapses into the cavities and are the predominant type of sinkhole on the peninsula (Sinclair and Stewart, 1985).

Of the three types of sinkholes, the greatest natural hazard is posed by the cover collapse sinkholes. A simplified developmental model is shown in Figure 2. The sinkhole formation process begins with the formation of a solution cavity within the porous limestone of the Floridan acquifer. As this cavity grows upward, it is bridged by the cohesive Hawthorn group sediments. Collapse occurs when the cohesive strength of the bridge becomes insufficient to support the weight of the overlying material. The clastic materials of the cover unit fail and flow downward into the cavity. As a result, a steep-sided surface depression forms in the cover.

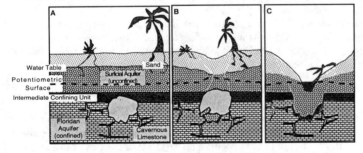

Figure 2 Model of cover collapse sinkhole formation. Modified after Lane (1986).

The local hydrogeology plays an important role in triggering or retarding cover collapse sinkholes. Water in the surficial aquifer system acts as a load causing a downward hydrostatic force on the roof of the cavity. In contrast, water pressure within Floridan aquifer provides an upward force supporting the cavity roof. Thus, a large positive head difference between the surficial and the Floridan aquifer systems can major driving factor in sinkhole collapse. Specific triggering mechanisms include both drops in the level Floridan potentiometric surface caused by excessive well withdrawals, or rises in the water table of the surficial aquifer system caused by increased precipitation or surface water impoundments (Sinclair, 1982; Metcalfe and Hall, 1984; Newton, 1984; Sinclair et al, 1985, Sinclair and Stewart, 1985; Wilson and Beck, 1992).

Study Area

The study area lies adjacent to the rapidly growing Orlando metropolitan area and includes parts of Lake, Orange, and Seminole counties, covering approximately 3500 km^2 (Fig. 3). The most distinct physiographic features are the sand covered Lake Wales, Mt. Dora, and Orlando ridges. These ridges are relict coastal features and correspond to the areas of greatest unconsolidated sediment thickness (Lichtler et. al., 1968; White, 1970). Elevations on the ridges range from 30 to 94 m. The topographic ridges serve to localize the first-order peninsular drainage divides and form the boundaries for the Saint Johns, Kissimmee, and Withlacoochee river drainage basins. The ridges are surrounded by the lower relief Lake Upland and Osceola Plain. The Central Valley lies between the sandy ridges and is occupied by several large shallow lakes including Lake Apopka. The lowest elevations in the study area correspond to the Wekiva basin and St. Johns River valley.

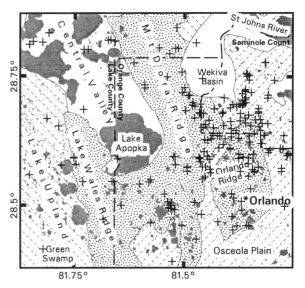

Figure 3. Generalized physiographic map of the study area near Orlando, FL. Areas of high topography indicated in the stippled pattern, intermediate topography in dashed pattern, and lowest topography in white. Sinkholes indicated by crosses are from the Florida Sinkhole Research Institute database (Spencer and Lane, 1995). Physiographic provinces are modified after White (1970).

DATA AND METHOD OF ANALYSIS

Approach

As mentioned above, numerous studies suggest that the hydrostatic head difference between the surficial and Floridan aquifers plays an important role in sinkhole occurrence. For an unconfined aquifer, such as the surficial aquifer, the "hydrostatic head" is defined at the altitude of the water table. For a confined aquifer, such as the Floridan, the hydrostatic head is defined as the level to which water rises in tightly cased wells penetrating the aquifer, and is represented physically by the "potentiometric surface" (Fig. 2) In order to explore the relationships between hydrostatic heads and sinkhole occurrence, a model of "head difference" between these aquifer systems needs to be constructed. This model is then compared quantitatively to the locations of reported sinkholes.

The processing flow is shown in Figure 4. The analysis starts with three basic data sources: topography, Landsat imagery, and well data. First, satellite imagery and a digital elevation model are used to derive a map of lake surface elevations. This map is used as a model of the water table of the surficial aquifer system. Then, well data are used to construct a gridded representation of the Floridan aquifer potentiometric surface. The potentiometric surface grid is then subtracted from the lake elevation map to produce a map of lake elevations relative to the Floridan aquifer potentiometric surface. Finally, a head difference model is constructed and compared with the FSRI sinkhole database.

Surficial Aquifer Model

In the study area, the water table generally lies above the potentiometric surface of the Floridan aquifer (Lichtler et. al, 1968; Boniol et al. 1993; this study). Because the surficial aquifer system is not used for municipal and agricultural water

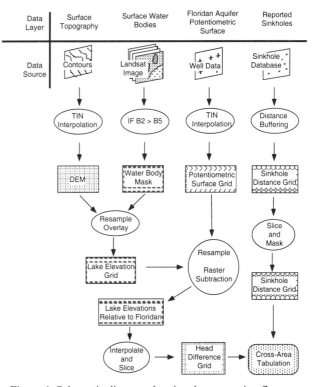

Figure 4. Schematic diagram showing the processing flow used in this study.

supplies, relatively few monitoring wells sample it and precise maps of the water table generally do not exist. Since lake surfaces intersect the water table locally, they provide a convenient surrogate means of mapping lateral variations in water table elevation, especially in areas where they are numerous. In this study, water surface elevations of the numerous lakes in the study area are used as a model of the head of the surficial aquifer system.

Satellite imagery and a digital elevation model are used to determine the locations and elevations of surface water bodies. First, a 30 m horizontal resolution digital elevation model (DEM) is constructed for the study area. This DEM is produced from 5 foot interval topographic contours and water body shorelines digitized from twenty separate USGS 7.5' quadrangle maps. The contour and shoreline vectors are used to construct a triangulated irregular network (TIN) surface model. Elevation values on a regularly spaced grid, with grid spacing interval of 30m, are calculated from the TIN by linear interpolation. This DEM is as good as, or of higher quality than, the comparable 7.5' USGS DEMs which are available in only limited areas in the study area.

Next, a binary raster mask of surface water bodies is extracted from a geocoded Landsat Thematic Mapper image acquired in January 1986. To do this, a band ratio image is computed by dividing band 2 by band 5. Because of the low reflectance of surface water in the middle infrared wavelengths, this 2/5 ratio is an extremely robust way of extracting water from Landsat imagery. The ratio image is constructed using a threshold value of 1.0. In the resultant surface water body image mask, water is assigned a value of 1 and land a value of null.

The water body mask is resampled to the DEM resolution and digitally overlaid on the DEM. The elevation of each water body pixel is assigned the corresponding DEM pixel elevation. The resulting raster map of lake elevations is shown in Figure 5. In general, lake elevations decrease in a northeasterly direction from around 35 m above sea level in the Green Swamp area to near sea level in the St. Johns River.

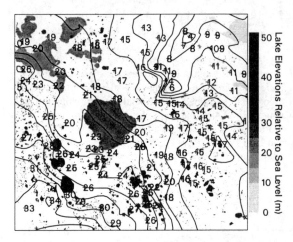

Figure 5. Raster map of lake surface elevations relative to sea level (NGVD 1929). Lake elevations are derived from a digital elevation model and satellite imagery. Spot elevations and contours of the Floridan potentiometric are derived from well data (Schiner and Hays, 1985; USGS 1986b).

The elevations of the lakes and the water table are not static. They change over time in response to seasonal and long term fluctuations in climate and water use. While the locations of the lakes are based on a contemporary data set, their elevations are based on quadrangle maps, most of which date back to the 1950s. More recent lake stage data is available for many of the larger lakes in Central Florida (USGS, 1986a). Within the study area however, stage records exist for less than 15 lakes and this is inadequate for characterizing the complex spatial variations of the surficial aquifer system.

In order to quantify the temporal variation in lake surface elevations, we compared lake stage records within the study area for the Autumn of 1986 (USGS, 1986a) with lake elevations derived from the DEM. In all cases, the 1986 lake stage records are within 2 m of those on the quadrangle maps with the vast majority less than 1m. This relatively small long term change in lake surface elevations suggests that the water table has been relatively unaffected by ground water development in the region since the 1950s (Murray and Halford, 1996). For the purposes of this study, the lake elevation map is assumed to be representative of the long term average water table of the surficial aquifer system.

Head Difference Model

The potentiometric surface for the Floridan aquifer is routinely measured in the field by the USGS and Florida state agencies using a widespread network of monitoring wells. Extensive ground-water development of the Floridan aquifer has led to a drawdown of the potentiometric surface of nearly 5 m in parts of the study area since 1950 (Miller, 1997; Murray and Halford, 1996). For this study, well data from September 1985 (Schiner and Hays, 1985; USGS 1986b) and TIN interpolation are used to construct a potentiometric surface grid. This time period was chosen to correspond to the satellite image used to map the lakes. In addition, this period roughly encloses the median occurrence date of the sinkholes from the FSRI database within the study area. Contours of this surface are shown in Figure 5. The surface generally dips to the northeast from around 35 m above sea level in the southwest part of the study area to less than 10 m in the northeast.

Raster subtraction of the potentiometric surface from each lake elevation pixel produces a map of the lake elevations relative to the Floridan aquifer potentiometric surface. This map is a first order expression of the head difference between the Floridan and surficial aquifers and may be viewed as a discretely sampled estimate of a continuous head difference surface. This surface is reconstructed by interpolation of the head difference values for each lake pixel.

The resulting surface is shown in Figure 6. Because the water table will generally be higher than a line connecting adjacent water bodies, this model should be viewed as a lower bound of the head difference potential within the study area. In most of the study area head difference is low. The areas of highest head difference lie in a northwest trending swath to the east of Lake Apopka and in a diffuse cluster in the southeast corner of the study area.

Spatial Associations

The study area contains 226 reported sinkholes from the FSRI database (Fig. 3). These sinkholes are not distributed randomly throughout the study area, but are instead, clustered in space (Whitman et al., 1999). In general, sinkholes are associated with regions where the head difference is greater than 10 m. Conversely, regions with few sinkholes tend to have a low head difference. Quantitative testing of this interpretation requires the comparison of the spatial co-occurrence of sinkholes, represented as points, with lakes, represented as irregular polygons. Several methods of comparing point distributions do exist (Upton and Fingleton, 1985). It is usually simpler, however, to transform the point distributions to continuous raster layers and then compare the layers on a pixel by pixel basis (Fig. 4). Cross-area tabulation is then used to quantify spatial asociations between maps with summary statistics (Bonham-Carter, 1994)

The first step involves producing a continuous surface from the reported sinkhole locations. Two potential approaches exist. The first approach is to construct a sinkhole density map by applying a moving average kernel to the map shown in Figure 3. The problem with this approach is that the moving average acts as a low-pass filter, reducing the spatial detail of the surface. In addition, the result is extremely dependent on kernel size. An alternative to a density map is to construct a map of sinkhole proximity (Fig. 7). First, a regular 30 m grid is superimposed over the study area. Then for each grid cell, the Euclidean distance from the grid cell center to each sinkhole location within the database is calculated. The minimum distance value is then assigned to that grid cell and the process repeated for the next grid cell. In essence, this is a buffering operation between each cell in the raster grid and the nearest sinkhole point. The key to this operation is that it avoids the bias inherent in a sinkhole density calculation; the resultant map is independent of grid cell size and regridding-induced errors are avoided.

Next, the head difference and distance to nearest sinkhole grids are compared by raster overlay. First, the head difference grid is quantized into 2.5 m head-difference horizontal slabs or "slices" (Fig. 6). Similarly, the sinkhole proximity grid is quantized into 1 km inter-sinkhole distance slices (Fig. 7). Then, a contingency table is constructed where total areas of each unique combination of head difference and distance to nearest sinkhole are tabulated. This contingency table allows direct comparison of the spatial associations between each discrete head difference and distance slice.

The contingency table is shown graphically in Figure 8a. The bulk of the study area is represented by a broad maximum located at

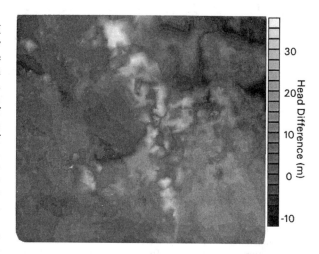

Figure 6. Head difference model interpolated from map of lake elevations relative to the Floridan aquifer potentiometric surface. The grid values are quantized into 2.5 high slices.

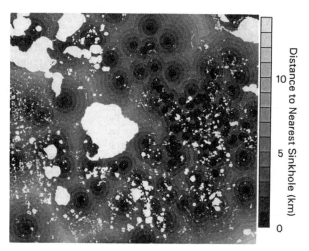

Figure 7 Map of distance to the nearest sinkhole. Map is generated by application of buffer zones to sinkholes shown in Figures 3. Map is quantized into 1 km wide slices. Areas covered by water are shown in white.

head differences of between 0 and 5 m, spanning an interval of distance-to-nearest-sinkhole values from 0 to 8 km (Fig. 8a). A smaller, yet distinct maximum is centered at a head difference of approximately 12 m which is located generally at distance-to-nearest-sinkhole values of less than 4m. This smaller population corresponds to the areas of high head difference situated in the eastern portions of the study area.

Another relevant way to evaluate the association between the maps is to compare the observed contingency table with one expected from statistically independent phenomena (Bonham-Carter, 1994). A comparison of the observed and expected cross distributions is shown in Figure 8. For regions of inter-sinkhole distances greater than 4 km, the observed cross distribution (Fig. 8a) shows greater areas at low head differences ($\Delta h < 5m$) and lesser areas at high head differences ($\Delta h > 5m$) than the expected cross distribution (Fig. 8b). Subtraction of the expected from the observed distribution yields the area residual shown in Figure 8c. The area residual shows the pattern and degree of spatial association between head difference and sinkhole proximity. Regions 2 km or less from the nearest sinkhole show a strong positive association to head differences between 5 and 15 m and a negative association with head differences between 0 and 2.5m. In contrast, areas with inter-sinkhole distances between 4 and 8 km show a strong positive association with head differences between 0 and 2.5 m and a negative association with head differences between 5 and 15 m. In other words, in areas of low head difference, sinkholes are less frequent and farther apart than expected. Conversely, in areas of high head difference, sinkholes are more common and more closely spaced than one would expect.

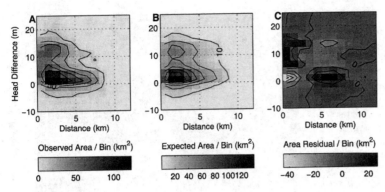

Figure 8. Cross-area residual analysis. A: Observed cross distribution of head difference and distance-to-nearest sinkhole grids. B: Expected cross distribution assuming independence of head difference and distance to nearest sinkhole. C: Residual (observed - expected) cross distribution. Dark areas show positive associations between head difference and distance-to-nearest sinkhole. Light areas show negative associations.

DISCUSSION

The association of sinkhole proximity to head differences between the surficial and Floridan aquifers demonstrates the importance of hydrostatic loads in sinkhole hazard. In particular, sinkhole formation appears to be retarded where head difference is low. This result is consistent with previously reported temporal associations between aquifer drawdown and sinkhole formation (Sinclair, 1982; Metcalfe and Hall, 1984; Newton, 1984; Sinclair et al, 1985, Sinclair and Stewart, 1985; Wilson and Beck, 1992).

Head difference is related to a complex set of factors and reflects the dynamic equilibrium between the surficial and Floridan aquifer systems. A limiting factor is the height of the land surface relative to the potentiometric surface. High head differences cannot exist in lowlands where the ground lies near or below the potentiometric surface. The Lakes Wales and Mt. Dora ridges form the most prominent physiographic features of the study area. Both ridges are underlain by similar thicknesses of clastic cover (Lichtler et al., 1968). Their average heights both lie approximately 15 m above the potentiometric surface. Only the Mt. Dora ridge is associated with head differences high enough to drive sinkhole formation (compare Figures 3 and 6). Apparently, elevated regions of thick surficial cover are not always associated with sinkhole formation. Instead, head difference appears to be the key factor.

Wilson and Beck (1992) observed that 85% of the new sinkholes near Orlando, FL occurred within the high groundwater recharge areas defined by Lichtler et al. (1968). Within their study area, these high recharge areas correspond to the Mt. Dora and Orlando ridges. Their study area was confined to Orange and Seminole Counties and thus did not include the Lake Wales ridge to the west. Their reported association of sinkhole formation with recharge areas overlooked the variability of head difference between recharge areas because the eastern boundary of their study area happened to exclude the Lake Wales ridge. Our decision to include the Lake Wales ridge was driven by the results of synoptic regional terrain analysis using remote sensing and GIS techniques; the results demonstrate the value of these techniques to further sinkhole studies.

Sinkhole occurrence is influenced by many other factors such as urbanization, land use, fracture patterns in the bedrock, and the thicknesses and composition of cover material. Future work will apply spatial analysis techniques to these factors. With this analysis, weights can be assigned to the various factors in order of importance. This information can then be used to construct predictive models of sinkhole hazard.

CONCLUSIONS

GIS and GIS-derived datasets are useful tools in evaluating regional factors associated with sinkhole hazard in central Florida. These methods show that sinkholes within the study area are not distributed randomly but instead appear to be clustered in space. Regions 2 km or less from the nearest sinkhole show a strong positive association to head differences between 5 and 15 m and a negative association with head differences between 0 and 2.5m. In contrast, areas with inter-sinkhole distances between 4 and 8 km show a strong positive association with head differences between 0 and 2.5 m and a negative association with head differences between 5 and 15 m. In other words, in areas of low head difference, sinkholes are less frequent and farther apart than expected. Conversely, in areas of high head difference, sinkholes are more common and more closely spaced than one would expect. This association of sinkhole proximity to head differences demonstrates the importance of hydrostatic loads in sinkhole hazard.

ACKNOWLEGMENT

The authors wish to acknowledge helpful suggestions given by G. Draper, and F. Maurrasse. S. Spenser and W. Wilson provided versions of the sinkhole database. L. Powell constructed the DEM used in this study. This study was supported by NASA Grant # NAG 5-307

REFERENCES

Benson R. C. and L. J. La Fountain, 1984, Evaluation of subsidence or collapse potential due to subsurface cavities, in: B.F. Beck editor, Proceedings of the First Multidisciplinary Conference on Sinkholes, A.A. Balkema, Rotterdam, pp. 201-216.

Bonham-Carter, G., F., 1994, Geographic Information Systems for Geoscientists: Modeling with GIS, Pergamon Press, 398 pp.

Boniol, D., Williams, M., and Munch, D., 1993, Mapping Recharge to the Floridian Acquifer Using a Geographic Information System, Saint Johns River Water Management District, Technical Publication SJ93-5, 41 pp.

Florida Geological Survey, 1986, Hydrogeologic Units of Florida, Florida Geological Survey Special Publication no. 28., 9 pp.

Lane, E. ,1986, Karst in Florida, Florida Geological Survey Special Report no. 29., 100 pp.

Lichtler, W., Anderson, W., and Joyner, B., 1968, Water Resources of Orange County, Florida, Florida Geological Survey Report of Investigations no. 50., 150 pp.

Lichtler, W.F., Hughes, G.H., and Pfischner, FL, 1976, Hydrologic Relations Between Lakes and Aquifers in a Recharge Area Near Orlando, Florida, USGS Water-Resources Investigation 76-65, 54 pp.

Metcalfe, S. J., and Hall, L. E., 1984, Sinkhole Collapse Induced by Groundwater Pumpage from Freeze Protection Irrigation Near Dover, Florida, January, 1977, in Beck, B. F. (ed.), Sinkholes: Their Geology, Engineering and Environmental Impact, A. A. Balkema Pub., Rotterdam, 29-34 pp.

Miller, J., 1997, Hydrogeology of Florida, in The Geology of Florida, A. Randazzo and D. Jones eds., University Press of Florida, 69-88 pp.

Monroe, W., 1970, A Glossary of Karst Terminology, U.S.G.S. Water-Supply paper 1899-K, 26 pp.

Murray Jr., L. C., and Halford, K. J., 1996 Hydrogeologic Conditions and Simulation of Groundwater Flow in the Greater Orlando Metropolitan Area, East-Central Florida, U. S., Geological Survey Water-Resources Investigations Report 96-4181, 100 pp.

Newton, J. G., 1984, Review of Induced Sinkhole Development, in Beck, B. F. (ed.), Sinkholes: Their Geology, Engineering and Environmental Impact, A. A. Balkema Pub., Rotterdam, 3-10 pp.

Sinclair, W.C. and Stewart, J.W., 1985, Sinkhole Type, Development, and Distribution in Florida. Florida Geological Survey Map Series no. 110., 1 pp.

Sinclair, W., Stewart, J., Knutilla, R., Gilboy, A. and Miller, R., 1985, Types, Features, and Occurrence of Sinkholes in the Karst of West-Central Florida, U.S. Geological Survey Water-Resources Investigations Report no. 85-4126, 81 pp.

Sinclair, W., 1982, Sinkhole Development Resulting from Groundwater Withdrawal in the Tampa Area, Florida, U.S.G.S. Water Resources Investigations Report 81-50, 19 pp.

Spencer, S. M. and Lane, E., 1995, Florida Sinkhole Index, Florida Geological Survey, Open File Report 58, 18 pp.

USGS, 1986a, Water Resources Data Florida, Water Year 1986, Volume 1A: Northeast Florida Surface Water, U.S. Geological Survey Water-Data Report FL86-1A.

USGS, 1986b, Water Resources Data Florida, Water Year 1986, Volume 1B: Northeast Florida Ground Water, U.S. Geological Survey Water-Data Report FL86-1B.

White, W. A., 1970, The Geomorphology of the Florida Peninsula, Florida Bureau of Geology, Bulletin no. 51, 164 pp.

Whitman, D.,Gubbels, T. L., and Powell, L., 1999, Spatial interrelationships between lake elevations, water tables, and sinkhole occurrence in Central Florida: A GIS Approach, Photogrammetric Engineering and Remote Sensing , (in press).

Wilson, W., and Beck, B., 1988, Evaluating Sinkhole Hazards in Mantled Karst Terrain, Geotechnical Aspects of Karst Terrains Special Volume, 1-24 pp.

Wilson, W. And Beck, B., 1992, Hydrogeologic Factors Affecting New Sinkhole Development in the Orlando Area, Florida, Ground Water, v. 30, 918-930 pp.

Wilson, W. and Shock, E., 1996, New Sinkhole Data Spreadsheet Manual: A Description of Quantitative Methods for Modeling New Sinkhole Frequency, Size Distribution, Probability and Risk, Based on Actuarial Statistical Analysis of the New Sinkhole Data Spreadsheet, unpublished report, Subsurface Evaluations, Inc., 31 pp.

3 Prediction of sinkhole development and planning studies in karst

Casual hydrofracturing theory and its application for sinkhole development prediction in the area of Novovoronezh Nuclear Power House-2 (NV NPH-2), Russia

ALEXANDER V. ANIKEEV Moscow State University, Geological Faculty, Russia

ABSTRACT

The process of spalling and fissuring of a confining bed under the drop of aquifer head due to ground water withdrawal named casual hydrofracturing is described. The analysis of this phenomenon, or the second form of aquitard destruction, let us derive the failure criterion. Based on some assumptions the formulae available for application in engineering practice are obtained. These formulae are used in the area of NV NPH-2 to estimate the stability of the Upper Devonian and Lower Cretaceous clays joined in a common layer. The latter one serves as a weakly permeable screen dividing the upper water saturateded sands from the underlying carbonate rocks, fissured and karstified. The massif-foundation is a complex natural system having felt the strong anthropogenic impacts due mostly to the long-term pumpage of the karst-fissure water. Taking into account the great environmental and practical significance of this construction the geological conditions have been schematized and reproduced in models from water saturated equivalent materials. Both theory and experiments testify to the complete or partial rupture of the screen under the water head decline. This is very hazardous. However, considering the geological conditions, and the fissure water intake being stopped as well, one can suppose that the fracturing will not express itself at the land surface. As an example of the application of the theory, a map of the failure values of head decline is presented for the site. Its comparison with the real water head descent lets us distinguish the disturbed and stable parts of the confining bed. Such a map is useful to control ground water withdrawal and predict the stability of aquitards and confining beds.

INTRODUCTION

The weakly permeable, mostly, clayey strata of the karst mantle serve as screens isolating groundwater aquifers from contaminated surface water and preventing the transport of water saturated sands into karst collectors and fissure voids (Abstracts, 1992; Beck, 1993). Long time, strong anthropogenic impacts upon the natural hydraulic regime have led to the destruction of aquitards over weakened zones in soluble rocks and unsoluble overburden (Daoxian, 1987; Newton, 1987; Kutepov & Kozhevnikova, 1989; Tolmachev & Reuter, 1990). This is one of the main causes of sinkhole development and groundwater pollution in areas with anthropogenic changes in the hydrogeological environment. Such a complex natural and technogenic system is found in the area of NV NPH-2 situated near Novovoronezh city. Some engineering, geological and environmental problems have arisen there in conjunction with determining the karst surface collapse hazard and construction of new buildings. One of these problems is to estimate and predict the rock mass stability and the possible karst surface collapse hazard.

CASUAL HYDROFRACTURING THEORY

The theory of casual hydrofracturing has been presented earlier in detail (Anikeev, 1991, 1993). Therefore, we shall examine the principal propositions in short.

It was found in experiments that a sudden decline of water pressure beneath a weakly permeable bed caused the failure of water saturated clays above even a small cavity. The development of breakdown fractures, spalling and crumbling are observed inside an arch-shaped zone (Fig.1, a). Nearly in a trice, the rupture-front propagates upward from the bed floor to the surface of equilibrium state. This phenomenon has been named casual hydrofracturing in order to emphasize the difference between this one and that due to premeditated fluid injection into boreholes and mines.

The driving force, which causes the fissuring, results from the difference in hydraulic conductivities of confining layer and aquifer. Generally, the equation for the excess pore pressure inside the layer is written as follows:

$$g\rho_w \Delta H = K_\sigma \sigma_w + K_\tau \tau_w, \qquad (1)$$

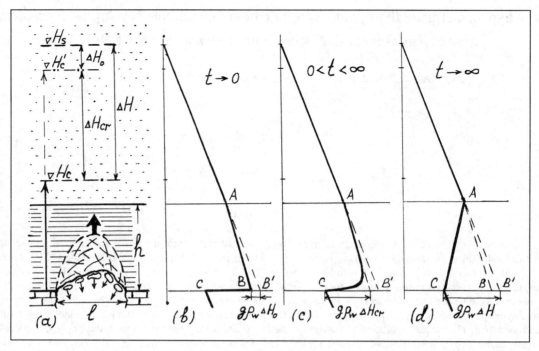

Fig.1. Mechanism of casual hydrofracturing (a), and pore-pressure diagrams at the initial moment (b), intermediate stage of downward seepage (c) and final one (d). ∇H_s: water-table; $\nabla H'_c$ and ∇H_c: potentiometric level before and after water head decline; l: the span of an opening at the base of cover deposits; h: the thickness of confining bed; t: time. Big arrow and small ones show respectively the propagation of spalling front and the "shooting" of clay spalls and pieces from the bed floor.

Where g is the acceleration of gravity; ρ_w is the density of water; ΔH is the magnitude of water head decline; K_σ and K_τ are coefficients, $K_\sigma + K_\tau = 1$; σ_w are the stresses in pore water normal to clay particles, or excess hydrostatic pressure; τ_w are the seepage stresses tangential to the surface of soil graines, hydrodynamic pressure or effective stresses in the theory of aquifer-system compaction (Terzaghi & Peck, 1967; Poland, 1981). In water-saturated rocks with extremely low permeability the shape of the fluid pressure distribution curve is close to that of broken line ABC (Fig.1, b, c) independently on non-steady state percolation or steady state one, i.e. $K_\sigma / K_\tau \gg 1$, and equation (1) is rewritten as follows:

$$g\rho_w \Delta H = \sigma_w. \qquad (2)$$

Over a weakened zone (open-joint fissure, cavern, cave, etc.) not balanced by grain-to-grain pressure, the excess hydrostatic stresses cause the spalling of a confining bed long before the dynamic seepage stresses will be exerted on the grains by viscous drag of vertically moving interstitial water (Fig.1, d). Tensile fractures will occur, if the tensile strength T is

$$T \leq \sigma_w - \sigma_{1,2,3}, \qquad (3)$$

where $\sigma_{1,2,3}$ are the main normal stresses in clay solid above an opening. For simplicity the anisotropy of strength properties is neglected now. Though, the tensile strength of any rock is much smaller than its cohesion (C), we shall assume that $T \approx C$. Other-wise, the utilization of falure criterion (3) presents some difficulties in practice, especially for plastic soils. The next obstacle concerns the identifying of the stress state of an aquitard or confining bed in vicinity of an opening with usually unknown dimen-sions. But it is well known (Glushko & Shirokov, 1967; Avershin et al, 1971) that a relieved arch forms in rocks over a weakened zone. At the base of the arch stresses can be even tensile ($\sigma_{2,3} < 0$, $\sigma_1 = 0$). Near its top they are compressive ($\sigma_{1,2,3} > 0$), but they are small enough in terms of absolute value. In the first assumption let them be equal to zero inside the zone of low stresses. Taking into account the assumptions ($T = C$, $\sigma_{1,2,3} = 0$) and substituting the value of σ_w from (2) into (3) we obtain the simplest relation between the failure or critical value of water head decline ΔH_{cr} and the standard engineering-geological characteristic of disperse rock C:

$$\Delta H_{cr} = C/g\rho_w. \qquad (4)$$

The consideration of the initial distribution of pore water pressure (Fig.1, b) requiers to restrict the height of the spalling front raise. This leads to the new expression for the failure head difference (Anikeev, 1993):

$$\Delta H_{cr} = C/g\rho_w + h, \qquad (5)$$

where h is the thickness of a confining bed. Thus, according to the theory the hydrofracturing will commence at the layer floor un-der condition (4), and a confining layer will be disturbed from the floor to the roof under condition (5).

SITE GEOLOGY AND HYDROGEOLOGY

Geologically in the area of NV NPH-2 the loose covering strata are of Quaternary and Neogene systems, and the basement is composed of Cretaceous and Devonian strata (Fig. 2). The Quaternary sediments are represented by the Holocene alluvial silt, sand, loam and clay of the river flat (aIY in Fig. 2), the Upper Pleistocene alluvial sand with thin loam interbeds of the first and the second terraces (a_2III, a_1III in Fig. 2), and the Middle Pleistocene fluvioglacial deposits (fII in Fig. 2). The last ones widely broadcast, 20 -40 m thick, are composed chiefly of medium sands. In some areas fine sands and thin loam interbeds are found.

The Pliocene alluvial sands with variable size composition (from fine sands to gravel) underlie the Quaternary ones, but the boundary between them is very relative. They fill the dip and wide pre-glacial valley (Fig. 2), their thickness is 20 m at the most.

The Lower Cretaceous clays and sandy clays with the lenses of fine sand have been eroded almost as a whole. They have been preserved only in the form of spots, not more than 7 m thick (Fig. 2).

Lithologically, the Upper Devonian deposits, lying below, are divided into three strata. The upper one is represented by clays with the thin interbeds of argillite, aleurite and limestone. The thickness of the stratum being commonly 4 - 8 m reduces to 2 - 0 m in the slopes and floor of the buried valley (Fig. 2). The middle stratum is composed of limestones with interbeds of clay, 1 - 20 cm (rarely 1 m) thick. The common thickness varies from 4 m to 10 m. The limestones are intensively and non-uniform fissured and karstified, which shows in the output of borehole samples having decreased from 80 - 100 % to 30 - 40 % at different depths. The traces of recent karst and large caves have not been encountered. The lower stratum consists primarily of clays with the interbeds of limestone, argilite and marl.

Suffosion sinks and subsidences, 3 - 300 m in length and less than 1 - 2 m in depth, coinciding with tectonic interruptions are widely developed in the Novovoronezh area. Nearby the building site they are localized in the south slope zone of the buried erosion valley (Fig. 2).

Two aquifers are of practical interest in the area of NV NPH-2. The upper unconfined one is present within the Quaternary and Neogene sands. The Cretaceous and upper Devonian clay strata serve as an aquitard for the subterrane water, and they are the confining bed for the underlying confined aquifer which is formed by the Devonian carbonate rocks. These aquifers unite and form the single aquifer within the buried valley (Fig. 2). Pumping tests showed the fissure permability of the confined aquifer to have been approximately one hundred times the porous conductivity of the surficial one ($k_c = 0.12 – 0.93$ cm/s; $k_s = 10^{-3} - 2 \cdot 10^{-2}$ cm/s). At the building site the depths of the water table and the potentiometric level average 29 - 33 m and 30 -34 m, respectively.

The cone of depression resulted from the long-continued fissure water withdrawal since 1961. Until the last eighty years the productivity of six water-intake wells tended to increase from 3600 m^3/d to 6000 m^3/d. Therefore, the difference in elevations between the water table and the potentiometric level having been commonly 2 - 3 m (Fig. 2) was also growing from 1 m to 5 m. In 1992 the productivity was minimized to 3200 m^3/d, and the water level difference decreased to 1 m. At present, after the water exploitation has been finished, the water levels are re-establishing.

ENGINEERING-GEOLOGICAL SCHEMATIZATION

Considering the previous description one can imagine the central part of the cross section shown in Fig. 2 as a four-stratum rock mass in accordance with the function of every layer. The first stratum (the lower Devonian clay) plays the role of the impermeable basement. The second highly permeable layer (the middle Devonian limestone), 6 - 8 m thick, can accumulate unbound and disturbed soils in the fissure voids. As the elastic and strength characteristics of the clay are much smaller then that of the carbonate rock the latter is convenient to vew as a rigid stratum. Despite the time difference the upper Devonian and Cretaceous clays have surprisingly similar though variable physical and mechanical properties: density of mineral portion $\rho = 2.67 – 2.87$ g/cm^3; bulk density $\rho_s = 1.66 – 1.87$ g/cm^3; porosity $n = 0.29 – 0.48$; saturation ratio $S = 0.74 – 1.00$; cohesion $C = (0.20 – 1.33) \cdot 10^5$ Pa; angle of the internal friction $\varphi = 7° - 21°$. Because of this they can be joined in the single third bed with the thickness of 0 - 12 m. It serves as a relatively impermeable elastic screen protecting the overlying water-saturated sands against the destrucrtion and downward movement. The fourth permeable stratum, 36 - 44 m thick, is formed by the unbound soils of Pliocene and Middle Pleistocene ages. It is charecterized by variable granular size and the follow properties: $\rho = 2.64 – 2.69$ g/cm^3; $\rho_s = 1.46 – 1.74$ g/cm^3; $n = 0.35 – 0.43$; $S = 0.22 – 1.00$; $C = (0.01 – 0.08) \cdot 10^5$ Pa; $\varphi = 27° - 40°$. This layer saturated below the depths of 28 - 33 m is the safe foundation until the fissure bed is screened.

Here the scheme is described as a two-layer overburden with the fractured and karstified rocks at the base. Sinkhole development in the overburden depends upon the the stability of the clayey screen which could be disturbed under the fluctuations of the ground water levels. This should be evaluated.

CALCULATION RESULTS

Formulae (4) and (5) have been utilized to estimate the possibility of aquitard failure. According to the values of clay cohesion ($C = 20 - 133$ kPa) and thickness ($h = 0 - 12$ m) the critical value of fissure-water head decline ranges 0 - 25 m. As an example, the

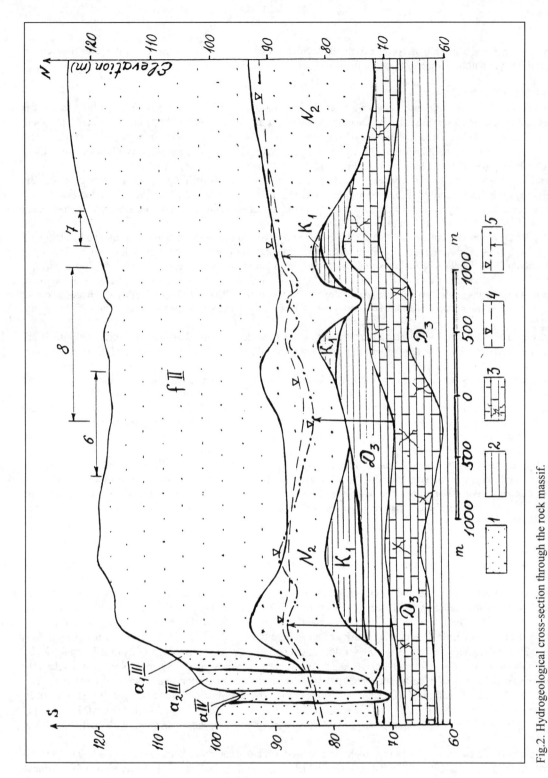

Fig.2. Hydrogeological cross-section through the rock massif.
aIY: the Holocene alluvial sediments of the flood land; a₂III, a₁III: the Upper Pleistocene alluvial sediments of the first and the second stream terraces; fII: the Middle Pleistocene fluvioglacial deposits; N₂: the Pliocene sediments; K₁: the Lower Cretaceous deposits; D₃: the Upper Devonian strata; 1: sands; 2: clays; 3: fissured limestones; 4: water table; 5: piesometric level; 6: water intake (6 in Fig.3); 7: sinkhole occurrence (4 in Fig.3); 8: building site (2 in Fig.3).

results of calculations are shown in Fig.3. It was assumed that the cohesion of the confining bed was minimal and constant within the site. This assumption being not sufficiently correct is explained by the lack of field data. In spite of the great number of boreholes the shear strength of the (D_3+ K_1) clays has been identified in only two well samples obtained within the site (Fig. 3). Therefore, the isoline configuration of the critical water head declines coincide with that of the bed thicknesses in accordance with equation (5).

As has been mentioned, the difference in elevation between the water table and the piesometric level did not exceed 5 m. Com-paring it with the calculated values of the head decrease shows the disturbed parts of the confining bed ($H_{cr} \leq 5$ m in Fig. 3). One can note that this difference does not reflect the real values of the aquifer head drop which have been absent in ground water monitoring data. At the begining of pumpage after its provisional stoppping these values should be expected to have been much more than 5 m, especially in the vicinity of the intake wells. That is why, the minimal magnitude of the clay cohesion have been used in the calcula-tions. Despite of some simplifications and assumptions, the map presented in Fig. 3 shows the parts of the common (D_3 + K_1) layer where it cannot serve as a screen preventing the entrainment of the Pliocene and Quaternary sands into the fissure voids and karst openings. Certainly with consideration of the strength field variations and the aquifer head fluctuations the picture of head dif-ference distribution and aquitard failure would be more comlex, but the main conclusion about the partial or entire disturbance of the clay screen will not change. This could be very hazardous under the other conditions. However, the relatively small thickness of the limestones, the great one of the overburden and the absence of large karst voids, as well as stopping the withdrawal of water let us hope that the downward movement of the disintegrated clays and unbound soils will not show effects on the land surface. But this is the subject of independent investigations related to the evaluation of the accumulation capability of karstified deposits (Anikeev & Fomenko, 1995).

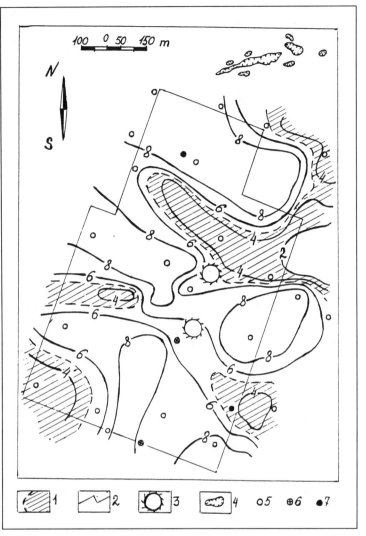

Fig.3. Map showing the failure values of water head descent (isolines of ΔH_{cr} in m) at the building site. 1: entirely fractured parts of confining bed ($\Delta H_{cr} \leq 5$ m); 2: boundary of the site (8 in Fig.2); 3: main buildings; 4: piping sinks and sinkholes (7 in Fig.2); 5: wells; 6: water-intake wells (6 in Fig.2); 7: wells, where the information on the strength properties of (D_3 + K_1) clays have been obtained.

EXPERIMENTAL DATA

In order to verify or to refute the obtained results the geological conditions were reproduced, and the hydrofracturing was studied in models from water-saturated equivalent materials in accordance with the engineering-geological scheme. The technology of the tests and the theoretical foundation for modelling as well (Anikeev, 1988) are not within the scope of this consideration. It should be noted that the experiments were performed in the laboratory flume designed by V.P. Khomenko (1981). They consisted in the measured alteration of water pressure beneath the two-layer models and recording of the induced processes.

It has been found that the hydrofracturing starts but quickly finishes under the values of water head decrease (ΔH_{ex}) being 2.5 - 3 times smaller than those (ΔH_{cr}) obtained from equations (4), (5). Where $\Delta H_{cr} / \Delta H_{ex}$= 1.5 - 2, the spalling leads to the collapse of the relatively thin confining bed (h/l = 2.56). In the thick one (h/l = 6.56) it results in the arch-like cavity with the height approximately equal to the slit span (l) in the rigid base (Fig. 4). If the magnitudes of ΔH_{ex} are between those of ΔH_{cr}, the crumbling of the lower part of the thick bed is accompanied by the roof bending and the cave-in of the upper portion (Fig. 4). Further descent of the water head pressure ($\Delta H_{ex} > \Delta H_{cr}$) causes the downward movement of the disintegrated clays and the overlying water saturated sands into the lower tank of the flume.

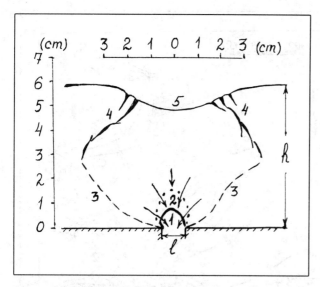

Fig.4. Modelling failure of thick ($h/l = 6,56$) confining bed. 1 and 2: cavity in clay and the region of hydrofracturing at the first stage ($\Delta H_{ex} < \Delta H_{cr}$); 3: the boundary between crumbled clay and undisturbed one; 4: breakdown fractures; 5: the bending of clay roof; h: the thickness of confining bed; l: the span of the slit in the rigid base. Arrows show the flow of spalled clay solid.

Thus, the correspondence between the results of the physical model tests and those of calculations is good. The discrepancy between ΔH_{ex} and ΔH_{cr} at the begining of the fracturing is explained mainly by the first assumtion ($T = C$). The experience testify to the correctness of the second assumption ($\sigma_{1,2,3} = 0$) only for the zone of low stresses. Over the relieved arch the compressive stresses prevent the upward propagation of the spalling front. Also, it is evident that after fissuring the excess hydrostatic stresses (σ_w) transform to seepage ones (τ_w), i.e. the transmition ($K_\sigma / K_\tau \gg 1 \rightarrow K_\sigma / K_\tau \ll 1$) takes place in equation (1). That is why, the increase of the water head difference ($\Delta H_{ex} > \Delta H_{cr}$) is necessary to form the hole in the thick aquitard.

CONCLUSIONS

1. The process of spalling and crumbling of a confining layer or an aquitard over a weakened zone (cave, karst cavity, open-joint fissure, ancient collapse funnel, etc) under water head decrease is named casual hydrofracturing. The theory of this phenomenon explains the formation of an opening in thick weakly permeable beds due to ground water withdrawal. It permits us to obtain the simplest relations between the magnitude of water head decline, the cohesion of relatively impermeable rock, and the thickness of aquiclude, which can be easily used in engineering practice.

2. The rock mass-foundation of NV NPH-2 is a complex natural system which may be imagined as a two-layer unsoluble over- burden which is underlain by fractured and cavernous carbonate rocks. The latter serve as a confined aquifer with the initial elevations of piesometric level approximately equal to those of water table occuring within the upper sand layer of the overburden. Long-term pumpage of the fissure water from 1961 to the mid ninetys' resulted in the difference in elevations between the water table and the level. The maximum difference value of 5 m could be much smaller than the "instantaneous" water head drop. In spite of the close engineering-geological study of the rock mass it has been impossible to describe the strength variability of the lower clay bed of the overburden.

3. Both theory and experiments testify to the partial or complete disturbance of the clay screen at the site. Some assumptions and simplifications taken as a basis of the tests and calculations do not qualitatively change this inference. Taking into account the geological and hydrogeological conditions one can suppose that the failure of the lower horizons of the overburden will produce no sinkholes on the land surface. Compiled before changing the hydrological environment, the map of critical water head decline, shown as an example of the application of the theory, may turn out to be useful in grownd water monitoring and forecasts of sinkhole development.

ACKNOWLEDGEMENTS

Writing the site geology and hydrogeology the author used the investigation data of the Voronezh State University, and the Industrial Research Institute for Building Survey. The author is indebted to V.V.Tolmachev and V.P.Khomenko for the permission to conduct the experiments at the Dzerzhinsk Karst Laboratory, as well as to V.M.Kutepov and I.V.Dudler for the discussion of the obtained results. This paper is written within the theme "Deformation and Destruction of Water Saturated Rock Masses With Structure Defects Under Conditions of Non-steady Hydrodynamic Regime" supported by grant from Russian Ministry of Education.

REFERENCES

Abstracts, 1992, The First International Symp. "Engineering Geology of Karst". USSR, Perm, 175 p.

Anikeev, A.V., 1988, Conditions of Similarity of water saturated model from equivalent materials and rock massif (English translation from Russian, by Allerton Press, Inc.). Inzhenernaya Geologiya, No 2, pp.78-84.

Anikeev, A.V., 1991, Clay collapse over caves and caverns. In: Geological Hazards. Proc. of Beijing International Symp. (Beijing, China, October 1991), pp. 336-342.

Anikeev, A.V., 1993, Two forms of destruction of bound soils over cavity (in Russian). Geoekologiya, No 2, pp. 115-123.

Anikeev, A.V. & Fomenko, I.K., 1995, Subsidence-sinkhole development in sand deposits above karst masses. Proc. of the Fifth International Symp. on Land Subsidence (the Hague, Netherlands, October 1995). Rotterdam, A.A.Balkema, pp. 27-34.

Avershin, S.G., Gruzdev, V.N. and Balalaeva, S.A., 1971, Distribution of Stresses Around Mines (in Russian). Frunze, Ilim Publ., 130 p.

Beck, B.F. (ed.), 1993, Applied Karst Geology. Proc. of the Fourth Multidisciplinary Conf. on Sinkholes (Panama City, Florida, January 1993). Rotterdam, A.A.Balkema, 295 p.

Daoxian, Y., 1987, Environmental and engineering problems of karst geology in China. Proc. of the Second Multidisciplinary Conf. on Sinkholes (Orlando, Florida, February 1987). Rotterdam, A.A.Balkema, pp. 1-11.

Glushko, V.T. & Shirokov, A.Z., 1967, Rock Mechanics and Protection of Mines (in Russian). Kiev, Naukova Dumka, 153 p.

Khomenko, V.P., 1981, Set for investigation of suffosion stability of soils (in Russian). The USSR Author's Certificate, No 851201, cl. G01N15/08.

Kutepov, V.M. & Kozhevnikova, V.N., 1989, Stability of Karst Areas (in Russian). Moscow, Nauka, 151 p.

Newton, J.G., 1987, Development of Sinkholes Resulting From Man's Activities in the Eastern United States. Denver, U.S. Geological Survey Circular 968, 54 p.

Poland, J.F., 1981, Subsidence in United States due to Ground-Water Withdrawal. J. of the Irrigation and Drainage Division, ASCE, Vol 107, No IR2, June, pp. 115-135.

Terzaghi, Karl & Peck, R.B., 1967, Soil Mechanics in Engineering Practice. New York, John Willey and Sons, Inc., 2d ed., 729 p.

Tolmachev, V.V. & Reuter, F., 1990, Engineering Karstology (in Russian). Moscow, Nedra, 151 p.

Karst atlas for Kentucky

JAMES C. CURRENS Kentucky Geological Survey, University of Kentucky, Lexington, Ky., USA
JOSEPH A. RAY Kentucky Natural Resources and Environmental Protection Cabinet, Division of Water, Frankfort, Ky., USA

ABSTRACT

About 55 percent of the area of Kentucky is underlain by carbonate rocks that can form karst, and half of that area, or 25 percent of the State, is underlain by mature karst. The topography of mature karst is characterized by land forms such as sinkholes, springs, karst windows, and caves, but, more important, by a general absence of above-ground drainage. The hazards associated with karst are vulnerability of ground water to contamination, flooding of sinkholes, and potential for cover-collapse sinkholes. The karst atlas project is a long-term effort to develop maps depicting karst geologic hazards and related phenomena. The proposed series of maps making up the karst atlas will show where these features or hazards may occur and provide planners and builders with the information necessary to avoid or mitigate future problems.

The first element of the karst atlas to be undertaken is a series of maps depicting karst ground-water basins. Five 30 x 60 minute quadrangle maps at a scale of 1:100,000 have been completed. The maps were compiled using data from dye traces performed by numerous researchers. The maps depict the estimated path of ground-water flow and the ground-water basin boundary. They are useful in establishing the relationship between the surface catchment area and the spring to which the ground water flows. To protect a water-supply spring, the disposal of waste within its drainage basin must be managed. Unless the ground-water basin boundary is known, protecting the water supply of a karst spring becomes an uncertain and exceedingly expensive task. The maps can be used to estimate the base-flow discharge of a spring being considered as a water supply and evaluate its vulnerability to pollution from existing development.

The second planned element of the karst atlas is a series of maps showing karst flood-prone areas. In mature karst, bedrock has dissolved to form conduits and caves that route runoff underground during most flow conditions. The abandoned courses of most formerly surface-flowing streams have been disrupted by the formation of sinks and are no longer efficient hydraulic channels. During major storms the drainage capacity of the conduit underlying a sinkhole, or the sinkhole's outlet, can be exceeded, causing water to pond in the sinkhole. During prolonged storms, ground water can rise above the lower elevations of a sinkhole and be temporarily stored in the sink. Consequently, structures built below surface overflow levels in karst valleys and sinkholes are frequently damaged by flooding. Buildings are often constructed in sinkholes because people do not recognize the features as closed depressions and hence flood plains. This may be because the sinkhole is shallow or too large to view in its entirety from a single vantage point and because it may only flood infrequently. Atlas maps will depict individual karst features that flood and give explicit warning of that possibility.

Structural damage from cover-collapse sinkholes is occurring at an ever-increasing frequency in Kentucky because of expanding urbanization. Data must be gathered on the geology and frequency of cover-collapse events. With these data, and the geologic quadrangle maps already available, maps showing the distribution of historical collapses and the probabilities of future collapses could be prepared. Such maps are needed to give planners, developers, potential homeowners, and insurance companies a basis upon which to make risk assessments.

The monetary loss to Kentucky residents from karst-related hazards is not currently quantifiable and may be substantial. Unlike major river floods, or most other natural disasters, karst-related hazards affect scattered groups of one or two families, and they do not receive significant media attention. To the families affected, however, who may face replacing a water supply or a home, the economic impact can be devastating.

INTRODUCTION

Approximately 55 percent of the area of Kentucky is underlain by carbonate rocks, and half of that area is mature karst. The karst atlas program at the Kentucky Geological Survey (KGS) is a multifaceted, multiyear effort to map the karst ground-water basins, geologic hazards, and related karst phenomena of Kentucky. The priority goal of the karst atlas program is the creation of karst ground-water basin maps. In 1997 KGS received U.S. Environmental Protection Agency funding, indirectly through the Kentucky Division of Water (DOW) via Section 319(h) of the Water Pollution Control Act of 1993, for development of four new karst ground-water basin maps. The four quadrangles mapped (Campbellsville, Harrodsburg, Beaver Dam, and Somerset), along with the Lexington quadrangle prototype, cover a large percentage of the karst regions of Kentucky (Fig. 1). Several quadrangles for which significant data are available remain to be completed, however.

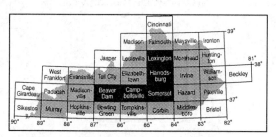

Figure 1: Locations of the 30 x 60 minute quadrangle maps covering Kentucky. The shaded quadrangles show completed karst ground-water basin maps.

PURPOSE

The goal of this project is to compile existing ground-water tracer data into regional maps of karst ground-water basins. In mature karst, bedrock has dissolved to form conduits and caves that route runoff underground during most flow conditions. The abandoned courses of formerly surface-flowing streams have been disrupted by the formation of sinks. In karst areas, the boundaries of watersheds contributing flow to springs are difficult to map because of the disrupted surface drainage. This makes it difficult to determine the source of a pollutant identified at a spring without tracing data. The karst ground-water basin maps can be used by various agency staff, private citizens, environmentalists, industrial managers, and agricultural operators for managing and mitigating sources of pollution of ground water in the karst areas of Kentucky.

METHODOLOGY

Hundreds of ground-water tracer tests have been conducted by karst researchers in Kentucky over the past 50 years to delineate karst ground-water basins. These data are frequently located in theses, consultants' reports, and unpublished files. We have been systematically collecting these tracer-test data for several years and have combined these data with unpublished ground-water tracing results from our research files. New ground-water traces were not conducted specifically for the maps.

The locations of springs, swallow holes, and other karst features associated with ground-water dye traces were plotted on 1:100,000-scale, 30 x 60 minute topographic quadrangle base maps. When adequate, the locations shown on maps in theses or reports were transferred directly to the base map to avoid errors, and as a labor-saving device. The precision of the point locations provided in some reports was poor, however, because of the small scale of printed maps or imprecise coordinate locations. Data from these latter reports were plotted on 7.5-minute, 1:24,000-scale topographic quadrangle maps to clarify locations and expedite interpretation.

The databases of the Kentucky Geological Survey and the Kentucky Division of Water were searched for existing records of springs and other karst features. The format of the karst ground-water basin maps specified the use of Assembled Kentucky Ground-Water (AKGWA) identification numbers, assigned by the Division of Water, to identify springs and their basins. In some instances, records for springs associated with documented dye traces were not found in the databases. For those springs, database forms were completed using information from the relevant publication and our personal knowledge, AKGWA numbers were assigned, and the forms were sent to the DOW database manager.

Inferred ground-water flow routes were interpreted from dye-trace vectors, and ground-water basin boundaries were estimated and drafted on the 30 x 60 minute quadrangles maps. In addition to the ground-water basin boundaries and ground-water flow routes, significant data were added to the maps to characterize the hydraulics of some flow routes. Known surface and underground overflow routes were shown, and overflow and under-flow springs (Worthington, 1991) were identified. Intermittent lakes were also shown where they are known to occur.

An early attempt was made to assemble spring-location data from all available databases so that all known spring locations could be plotted on each 30 x 60 minute quadrangle map. The plan was abandoned because when several databases were queried, the same spring would be reported several times, with slightly different coordinates; also, springs would plot in seemingly illogical places such as in the middle of a river or in the center of nonkarst areas. Furthermore, most of these springs did not have AKGWA numbers, which would have required the wholesale assignment of numbers to springs about which essentially nothing was known except the location. It quickly became apparent that resolving the problems with springs in the databases will require a systematic, labor-intensive process of plotting locations on large-scale maps that would then be reviewed by hydrogeologists. Any residual ambiguities would then have to be resolved by field work. This effort would require a substantial investment of human resources, which was clearly beyond both the available resources and work plan of the ground-water basin map project. Therefore, the spring locations shown on the maps are restricted to those mapped in conjunction with ground-water traces.

The spring and dye-injection locations, the ground-water basin boundaries, and the inferred flow routes were digitized using the 1:100,000-scale hand-drawn map and compiled with a digital representation of the 1:100,000-scale topographic base map. The collection of files was then migrated from AutoCAD into Macromedia FreeHand graphics software for the addition of text and cartographic enhancement (highway numbers, etc.). The agency that funded the project required that the maps also be made available as a geographic information system (GIS) product. The funding provided was insufficient for data compilation and digitization at a scale of 1:24,000, however. We recognize that any future GIS products for the karst atlas should use location data that are as precise as current field methods allow. This means that the locations (swallow holes and springs) associated with dye traces reported in theses and other documents, which do not meet current standards for precision, accuracy, or general completeness of the records, should be verified by field studies. This process should be part of an ongoing database compilation and inventory of karst features. Also, for future maps, the ground-water flow routes and ground-water basin boundaries should be drawn and digitized at a scale of 1:24,000. A section of the Harrodsburg quadrangle map is presented in Figure 2 as an example of a finished map.

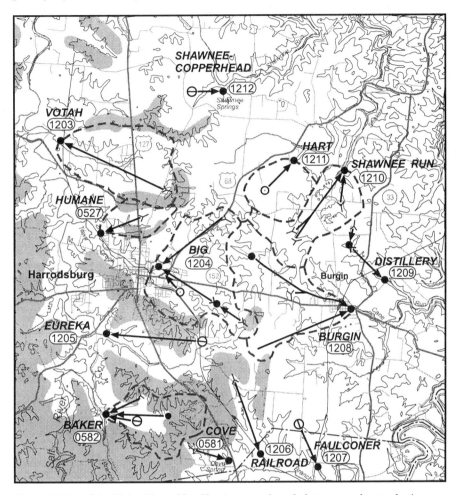

Figure 2: Part of the Harrodsburg 30 x 60 minute quadrangle karst ground-water basin map.

FUTURE WORK

The karst atlas is conceived of as consisting of several maps for each 30 x 60 minute quadrangle. The ground-water basin maps are perhaps the most critical to the ultimate success of the karst atlas because they are the most urgently needed for the protection of water resources. Among the other maps, the most important are those showing sinkholes prone to flooding and with potential for cover collapse. Other maps, such as those showing areas underlain by mapped cave passages, lineaments, and aquifer characteristics (for example, recession coefficients of springs), are also envisioned, but are lower priority and will require many years of data acquisition. Because locations of underground streams can add significant detail to karst ground-water basin maps, a map showing cave passages with flowing streams is the most important among the lower priority maps. Because of the ethical issues relevant to the conservation of cave resources, such maps may require significant generalization. Quinlan and Ewers (1985) make convincing arguments that cave maps showing underground streams are not only useful, but essential to the correct interpretation of tracer data.

The second most important type of map in the karst atlas is one showing karst flood-prone areas. During major storms, the flow capacity of the conduit draining a sinkhole can be exceeded, causing water to pond in the sinkhole. During prolonged storms, ground water can rise above the lower elevations of a sinkhole and be temporarily stored in the sink. Consequently, structures built below surface overflow levels in karst valleys and sinkholes are frequently damaged by flooding. Buildings are often constructed in sinkholes because people do not recognize the features as closed depressions containing local flood plains. This may be because the sinkhole is shallow, or too large to view in its entirety from a single vantage point, or because it may only flood infrequently. Atlas maps will identify individual karst features that flood, and give explicit warning of that possibility.

The methodology under consideration for the sinkhole-flooding maps is to digitize water bodies imaged on aerial or satellite photographs taken during both dry periods and floods. The pre-flood image would be digitally subtracted from the flood image, leaving only the outline of water bodies that were a result of the storm. Several remote-sensing products are potential sources of this imagery. According to the Federal Emergency Management Agency (FEMA), not only is it desirable to identify sinkholes that flood, but a 1 percent annual flood probability elevation line should also be established. Because of the huge number of sinkholes to be mapped, and the paucity of hydrologic data with which to directly determine flood return frequencies, establishing the elevation of a 1 percent annual flood probability for every sinkhole is a daunting task.

Another goal of the karst atlas program is to predict the frequency of the development of cover-collapse sinkholes. Structural damage from cover-collapse sinkholes is becoming more common in Kentucky because of expanding urbanization. The maps will show areas where a cover-collapse event may occur at a specified time interval per unit area. No attempt will be made to pinpoint the exact location of the next cover collapse. Each major carbonate unit will probably have a different cover-collapse frequency. Fortunately, a program to digitize the 1:24,000-scale geologic quadrangle maps of Kentucky, and compile them into 30 x 60 minute, 1:100,000-scale quadrangle maps, is currently under way at KGS. The digitized geology from the 1:100,000-scale geologic maps would be used to define the areas of each probability zone. A critical aspect of cover-collapse mapping is the collection of case histories. All reports of cover-collapse sinkholes in KGS files and the files of other agencies that can be accurately located on a map will need to be assembled. If the case history data set is too small for statistical analysis, new field work or remote-sensing analysis will be required. Therefore, aerial photography or satellite imagery may also be used to create these maps.

FUNDING FOR THE PROGRAM

Our hope is that as the ground-water basin maps become more widely used, and demonstration projects are completed for the other atlas maps, funding for the statewide program will become available. We think that the reason funding has been difficult to obtain for karst hydrogeology and geologic hazards projects is that the cost of karst geologic hazards is largely obscured within the overall economy. Karst geologic hazards tend to affect small groups of people during any one episode and therefore do not receive the publicity from the media that a major river flood receives, for example. Also, karst-related economic losses have never been quantified on an annual basis, so they receive relatively little attention by State agencies or the Kentucky legislature. Finally, many people minimize the importance of geologic hazards in general because they are an abstract concept of a disaster that might happen at some nebulous time in the future. The economic loss to Kentucky residents from karst-related geologic hazards is not currently quantifiable, but is probably substantial and much greater than the projected cost of the entire karst atlas.

RESULTS

Compilation of the karst ground-water basin maps was straightforward. Digitization and cartographic work proceeded rapidly once the hand-drawn maps were completed. Cartographic conventions were defined for all future karst ground-water basin maps for the State. Each ground-water basin quadrangle map is accompanied by an explanatory text, a list of traced springs, and bibliographic references directing the user to the source of the ground-water trace data. In addition to showing the boundaries of karst ground-water basins and inferred flow routes, each map shows the area underlain by carbonate rock susceptible to dissolution. The completed quadrangles (Harrodsburg, Somerset, Campbellsville, and Beaver Dam) cover about half of the major karst areas of the State. The maps also show topography, urban areas, highways, streams and lakes, county boundaries, and other physiographic data. The maps were published in September 1998.

The completion of the first four karst ground-water basin maps (and the Lexington quadrangle prototype) has drawn considerable attention from karst specialists and the staff of environmental agencies. We think that as the value of these maps becomes known, it will become easier to obtain additional support for both the karst ground-water basin maps and the other maps of the atlas. The three types of karst atlas maps planned will address the principal economic impacts from karst-related geologic hazards in Kentucky. The ground-water basin maps will be used for assessing ground-water pollution and resources, the sinkhole-flooding maps will be used to identify locations prone to flooding, and the cover-collapse sinkhole maps will be used to estimate the probability of damage to structures and highways. Completion of these maps has the potential to significantly reduce monetary losses that may result in the future from karst geologic hazards.

REFERENCES

Ahlers, T.W., Cobb, William, Coons, Don, Knutson, S.M., O'Dell, P.W., Schwartz, R.A., and Taylor, R.L., 1976, Unpublished cave mapping data: Mammoth Cave National Park, Ky.

Anderson, R.B., 1925, Investigation of a proposed dam site in the vicinity of Mammoth Cave, Kentucky, *cited in* Brown, R.F., 1966, Hydrology of the cavernous limestones of the Mammoth Cave area, Kentucky: U.S. Geological Survey Water-Supply Paper 1837, 64 p.

Crawford, N.C., 1994, Unpublished ground-water tracing data: Crawford and Associates, Inc., Bowling Green, Ky.

Currens, J.C., 1989, Unpublished ground-water tracing data: Kentucky Geological Survey.

Currens, J.C., and Graham, C.D.R, 1993, Flooding of the Sinking Creek karst area in Jessamine and Woodford Counties, Kentucky: Kentucky Geological Survey, ser. 11, Report of Investigations 7, 33 p.

Ewers, R.O., 1992, Unpublished ground-water tracing data: Ewers Water Consultants, Inc., Richmond, Ky.

Ewers, R.O., Onda, A.J., Estes, E.K., Idstein, P.J., and Johnson, K.M., 1991, The transmission of light hydrocarbon contaminants in limestone (karst) aquifers, in Proceedings of the Third Conference on Hydrogeology, Ecology, Monitoring and Management of Ground Water in Karst Terranes: U.S. Environmental Protection Agency and Association of Ground Water Scientists and Engineers, December 4–6, Nashville, Tenn., p. 287–306.

Harmon, D.L., 1992, Hydrology and geology of the Buffalo Creek area on the north side of Mammoth Cave National Park: Richmond, Eastern Kentucky University, master's thesis, 70 p.

Hopper, W.M., Jr., 1985, Karst hydrogeology of southeastern Mercer County and northeastern Boyle County, Kentucky: Lexington, University of Kentucky, master's thesis, 122 p.

Hopper, W.M., Jr., 1995, Unpublished ground-water tracing data: Lexington, Ky.

Leo, D.P., 1990, Hydrogeology of a limestone spring and its recharge area in southeastern Rockcastle County, Kentucky: Richmond, Eastern Kentucky University, master's thesis, 81 p.

Leo, D.P., 1995–97, Unpublished ground-water tracing data: Kentucky Division of Water, Frankfort, Ky.

Maegerlein, S.D., and Dillon, Clarence, 1976, Unpublished spring diving data (12-28-76): Graham Springs Complex, Ky.

Meiman, Joe, 1997, Unpublished ground-water tracing data: Mammoth Cave National Park, Ky.

Meiman, Joe, and Capps, A.S., 1995, Unpublished ground-water tracing data: Mammoth Cave National Park, Ky.

Meiman, Joe, and Ryan, M.T., 1990–94, Unpublished ground-water tracing data: Mammoth Cave National Park, Ky.

Miotke, F.D., and Papenberg, Hans, 1972, Geomorphology and hydrology of the Sinkhole Plain and Glasgow Upland, central Kentucky karst: Caves and Karst, v. 14, no. 4, p. 25–32.

Morris, F.R., IV, 1983, Karst hydrogeology of Cedar Creek and adjacent basins in east-central Pulaski County, Kentucky: Richmond, Eastern Kentucky University, master's thesis, 126 p.

Murphy, J.D., 1992, Determining the hydrology of the Turnhole Spring groundwater basin of Mammoth Cave National Park, Kentucky, using quantitative dye tracing: Durham, England, University of Durham, undergraduate dissertation, 51 p.

Powell, R.L., 1979, Unpublished ground-water tracing data: Warren County, Ky.

Quinlan, J.F., 1974, Unpublished ground-water tracing data: Mammoth Cave National Park, Ky.

Quinlan, J.F., 1986, Application of dye tracing to dam-site evaluation in a Kentucky karst area, U.S.A.: Proceedings, International Symposium on Karst Water Resources, Ankara, Turkey, unpaginated.

Quinlan, J.F., and Ewers, R.O., 1985, Ground-water flow in limestone terranes: Strategy, rationale and procedure for reliable, efficient monitoring of ground-water quality in karst areas, in Proceedings of the 5th National Symposium and Exposition on Aquifer Restoration and Ground-Water Monitoring: National Water Well Association, p. 197–234.

Quinlan, J.F., and Ray, J.A., 1981, Groundwater basins in the Mammoth Cave region, Kentucky, showing springs, major caves, flow routes, and potentiometric surface: Friends of the Karst, Occasional Publication 1, scale 1:138,000.

Quinlan, J.F., and Ray, J.A., 1981–89, Unpublished ground-water tracing data: Mammoth Cave National Park, Ky.

Quinlan, J.F., and Ray, J.A., 1989, Groundwater basins in the Mammoth Cave region, Kentucky, showing springs, major caves, flow routes, and potentiometric surface [rev. ed.]: Friends of the Karst, Occasional Publication 2, scale 1:138,000.

Quinlan, J.F., and Ray, J.A., 1995, Normalized base-flow discharge of groundwater basins: A useful parameter for estimating recharge areas of springs and for recognizing drainage anomalies in karst terranes, in Beck, B.F., ed., Karst geohazards: Proceedings, 5th Multidisciplinary Conference on Sinkholes and the Engineering and Environmental Impacts of Karst, Gatlinburg, Tenn., p. 149–164.

Quinlan, J.F., and Rowe, D.R., 1977, Hydrology and water quality in the central Kentucky karst: Phase I: University of Kentucky, Kentucky Water Resources Research Institute, Research Report 101, 93 p.

Quinlan, J.F., and Rowe, D.R., 1978, Hydrology and water quality in the central Kentucky karst: Phase II, part A: Preliminary summary of the hydrogeology of the Mill Hole sub-basin of the Turnhole Spring groundwater basin: University of Kentucky, Kentucky Water Resources Research Institute, Research Report 109, 42 p.

Quinlan, J.F., and Schafstall, Tim, 1988, Unpublished ground-water tracing data: Mammoth Cave National Park, Ky.

Ray, J.A., 1994, Surface and subsurface trunk flow, Mammoth Cave region: Proceedings, 3d Mammoth Cave National Park Science Conference, Mammoth Cave, Ky., p. 175–187.

Ray, J.A., 1994, Unpublished ground-water tracing data: Kentucky Natural Resources and Environmental Protection Cabinet, Division of Water, Frankfort, Ky.

Ray, J.A., Ewers, R.O., and Idstein, P.J., 1997, Mapping water-supply protection areas for leaky, perched karst groundwater systems, in Beck, B.F., ed., Proceedings, 6th Multidisciplinary Conference on Sinkholes and Engineering and Environmental Impacts of Karst, Springfield, Mo., p. 153–156.

Robertson, S. E., 1996, Unpublished ground-water tracing data: Kenvirons, Frankfort, Ky.

Romanik, P.B., 1986, Delineation of a karst groundwater basin in Sinking Valley, Pulaski County, Kentucky: Richmond, Eastern Kentucky University, master's thesis, 101 p.

Ryan, M.T., 1987, Using newly developed quantitative dye-tracing techniques to determine the karst hydrology of the Buffalo Spring groundwater basin of Mammoth Cave National Park, Kentucky: Richmond, Eastern Kentucky University, master's thesis, 121 p.

Ryan, M.T., and Meiman, Joe, 1996, An examination of short-term variations in water quality at a karst spring in Kentucky: Ground Water, v. 34, no. 1, p. 23–30.

Saunders, J.W., 1980, Grady's Cave, in Baz-Dresh, John, and Mixon, Bill, eds., Speleo Digest 1973: National Speleological Society, p. 26–31.

Saunders, J.W., 1981, Buckner Spring Cave, description and comparison to Grady's Cave, in Mixon, Bill, ed., Speleo Digest 1974: National Speleological Society, p. 76–77.

Saunders, J.W., 1984, Pruett Saltpeter Cave, Warren County, Kentucky: Central Kentucky Cave Survey, Bulletin 1, p. 208–209.

Schindel, G.M., 1984, Enteric contamination of an urban karstified carbonate aquifer: The Double Springs drainage basin, Bowling Green, Kentucky: Bowling Green, Western Kentucky University, master's thesis, 141 p.

Schindel, G.M., Quinlan, J.F., and Ray, J.A., 1994, Determination of the recharge area for the Rio Springs groundwater basin, near Munfordville, Kentucky: An application of dye tracing and potentiometric mapping for determination of springhead and wellhead protection areas in carbonate aquifers and karst terranes: Project completion report, U.S. Environmental Protection Agency, Groundwater Branch, Atlanta, Ga., 25 p.

Sendlein, L.V.A., Dinger, J.S., Minns, S.A., and Sahba, Arsin, 1990, Hydrogeology and groundwater monitoring of the John Sherman Cooper Power Station, Burnside, Kentucky: University of Kentucky, Department of Geological Sciences, 68 p.

Smith, J.H., and Crawford, N.C., 1989, Groundwater flow in the vicinity of the Curtis Peay landfill, Warren County, Kentucky: Unpublished consultant's report, Bowling Green, Ky.

Thrailkill, John, Spangler, L.E., Hopper, W.M., Jr., McCann, M.R., Troester, J.W., and Gouzie, D.R., 1982, Groundwater in the Inner Bluegrass karst region, Kentucky: University of Kentucky Water Resources Research Institute, Research Report 136, 108 p.

Uhlenbruch, C.R., 1993, Hydrogeology and groundwater quality of the Maple Springs area, Mammoth Cave National Park, Kentucky: Richmond, Eastern Kentucky University, master's thesis, 138 p.

Wells, S.G., 1973, Geomorphology of the Sinkhole Plain in the Pennyroyal Plateau of the central Kentucky karst: Cincinnati, Ohio, University of Cincinnati, master's thesis, 108 p.

Worthington, S.R.H., 1991, Karst hydrogeology of the Canadian Rocky Mountains: Hamilton, Ontario, McMaster University, Ph.D. dissertation, 380 p.

Geohazard map of cover-collapse sinkholes in the 'Tournaisis' area, southern Belgium

OLIVIER KAUFMANN & YVES QUINIF Faculté Polytechnique de Mons, Belgium

ABSTRACT

This paper reports the methodology developed to draw up a geohazard map of cover-collapse sinkhole occurrences in the 'Tournaisis' area. In this area, carboniferous limestones are overlain by a mesocenozoïc cover mainly consisting of marls, sand and clay. The thickness of this cover ranges from a few meters to more than one hundred meters. The surficial morphology of the area does not show any karstic evidence except for the occurrence of these collapses.

From a paleogeographical point of view, a developed quaternary karst is not conceivable in the area. Recent works suggested that the collapses are set off from reactivated paleokarsts. The paleokarsts studied in the area proved to be the result of a particular weathering of the limestone. The organization of these paleokarsts seems very low and mainly guided by the limestone fracturing. As for most induced sinkholes, the reactivation of these paleokarsts is linked to the lowering of piezometric heads.

In most of the area a thick cover and intensive land use mask potential surface hints of the buried paleokarsts and of the fracturing of the bedrock. Aerial photographs and remote sensing techniques have therefore shown little results in delineating collapse hazard zones up to now. The study of the surficial morphology is also of little help.

In order to draw up the geohazard map in such a difficult context, hydrogeological data and geological mapping information only could be used. These information are based on a limited number of boreholes and piezometers thus only valid on a regional scale. Records of former collapses were also available. These records were of great interest since sinkholes distribution is obviously clustered in the area.

Bedrock roof and cover formations floor altitudes were digitized and adapted to produce digital thematic maps. Piezometric heads were imported from a calibrated groundwater model of the aquifer. These data and a digital elevation model of the area were integrated into a GIS to produce a coherent 3D description of the area on a regional scale. Parameters such as the dewatering of the limestone, thickness of cover formations where sinkholes occurred were then estimated. Density of former collapses were also computed. This showed that zones of high sinkholes occurrences coincide with zones of heavy lowering of piezometric heads. Combining the density of former collapses with the dewatering of the limestone enabled us to delineate zones of low, moderate and high collapse hazard.

INTRODUCTION

The 'Tournaisis' area is situated in southern Belgium (Europe) near the French border (cf. figure 1). The studied area stretches over ~700 square kilometers (~270 square miles) with altitudes ranging from 15 m to 130 m (49 ft to 427 ft) above sea level.

From time to time steep-walled circular collapses open up in the area. Records of former collapses suggest that they have been much more numerous since the fifties. Up to now about two hundred collapses have been reported. Their size ranges from one meter up to 30 meters (~3.3 ft up to ~100 ft) in diameter and some exceed ten meters (~33 ft) deep.

The 'Tournaisis' area lies over fairly flat lying carboniferous limestone marked by East to West shear faults. The bedrock is overlain by an unconformable mesocenozoïc cover that mainly consists of Turonian marls, Tertiary sands and clays and Pleistocene deposits in the Escaut valley. The thickness of the cover varies form a few meters to more than one hundred meters (~330 ft). The Escaut river flows northwards and incise an East to West horst structure in the neighbourhood of Tournai.

The carboniferous limestone of the' Tournaisis' area is an aquifer that is a major water resource for Belgium and northern France. The bedrock structure and the nature of the cover delimit two hydrogeologic entities in the limestone: in the East and South, a free water table and in the North a confined aquifer. In the latter, piezometric heads have fallen by at least one meter (~3.3 ft) a year since the early fifties. The limestone roof was progressively dewatered over large areas.

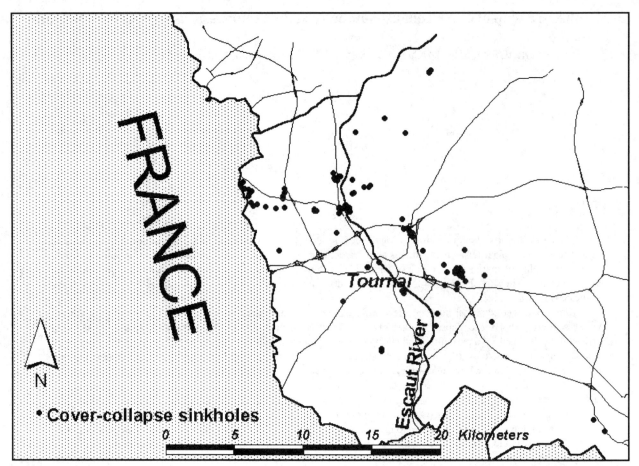

Figure 1: Map of the 'Tournaisis' area showing the city of Tournai, the Escaut river, the French border and the situation of cover-collapse sinkholes.

PALEOKARSTS

Observations at the bottom of sinkholes, in excavation work over former sinkholes and in quarries indicated that the collapses were set off from reactivated paleokarsts. These paleokarsts are the result of a particular weathering of the limestone described in Vergari A., 1997 et Vergari A., 1998. The weathering residuum is a soft material called 'ghost-rock' as carboniferous fossils and cherts beds are often preserved. The organization of these paleokarsts seems very low and mainly guided by the fractures in the limestone. Different paleokarstic morphologies have been described in the area. In the 'couloir' morphology, an elongated weathered zone stretches at the roof of the limestone. These 'couloirs' are usually several meters wide, up to hundreds meters long and tens of meters deep. They often form a network aligned on the bedrock fractures. In the 'pseudo-endokarst' morphology, the soft zone is totally inside the limestone massif and a limestone roof is present (Vergari A., 1998). Very large areas where nearly all the rock has been deeply weathered and only 'ghost-rock' remains are also encountered. All these morphologies are linked up together but cover-collapse sinkholes are mainly expected to occur over 'couloir' morphologies.

DETECTION OF POTENTIAL COLLAPSES
Surface hints

Surface hints of potential collapses over the whole area would obviously be invaluable. This is why aerial photographs and remote sensing data seem so attractive at first. However comparison of SPOT images taken at different times over zones where collapses occurred did not show any significant hint. Another test using ERS images and soils samples has not given convincing results up to now. It appeared that spatial and temporal resolutions were not suitable in both cases. Aerial photographs have a far better spatial resolution. Good temporal coverage could be organized from now on. However attempts to detect hints on pictures taken before reported collapses opened up were disappointing. Even finding former collapses on aerial photographs is tricky.

Intensive land use mask potential surface hints of the buried paleokarsts or the effects of a growing void in the cover. It also quickly erases every single mark of sinkholes in the landscape. Moreover, the bedrock is overlain by a thick cover in most of the area. For theses reasons, the study of surface is of little help as a straight link does not exist between paleokarsts and present day relief. It is therefore not surprising that, in this area, aerial photographs and remote sensing techniques have shown little results in delineating collapse hazard zones up to now.

In practice, the records of former cover-collapse sinkholes are the most useful surface hints of potential collapses on a regional scale. This is backed up by the obvious clustering of the collapses. Moreover filled sinkholes often reopen some years later.

Geologic and hydrogeologic parameters

Geologic description of the area should give clues to delineate zones were collapses may occur.

Given the geographical and geological context of the area, the nature, expanse and thickness of underground terranes are inferred through a limited number of boreholes and quarries. Analysis based upon these data therefore should be carried out on a regional scale.

The fracturing of the bedrock seems to have strongly conditioned the orientations of the paleokarstic features. But it is difficult to assess as limestone outcrops are not present in most of the area.

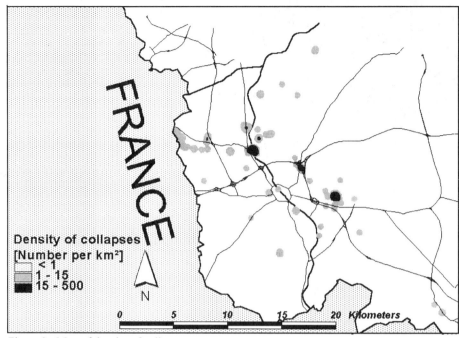

Figure 2 : Map of density of collapses.

Hydrogeologic parameters are also of importance. Like in many other cases of induced sinkholes, dewatering is to blame for the occurrence of cover-collapse sinkholes as the lowering of the carboniferous limestone water table triggered the reactivation of the paleokarsts.

GEOHAZARD MAPPING
Integrating regional data

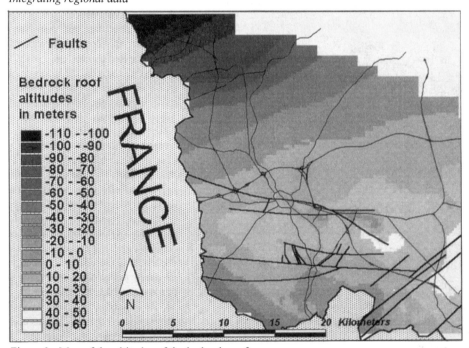

Figure 3 : Map of the altitudes of the bedrock roof.

To draw up the geohazard map, regional parameters that might have an impact on collapse occurrences had to be mapped. Geographical information systems (GIS) offer an integrated environment to process these data and then test hypotheses and scenarios.

First of all, records of former collapses were fed into a database. Then a collapse density map based upon the number of collapses reported by area was computed (see figure 2). Other density measures such as density computed on estimated volume of the collapses could have been chosen. However, estimated volume of older collapses is often unknown. Moreover, small sinkholes have frequently preceded bigger ones. Thus, it seems good practice to consider all collapses as evidences of buried paleokarst regardless of their size and to estimate the geohazard level on the

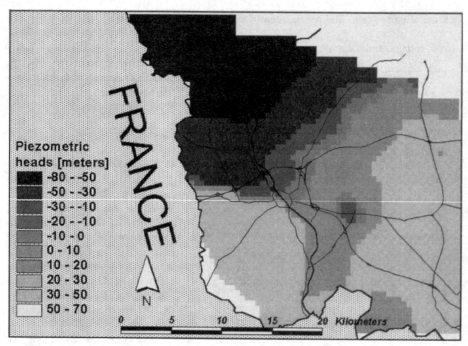

Figure 4 : Map of the piezometric heads in the Carboniferous limestone.

density of the number of events.

The second step was to produce a three dimensional representation of the geological context. To achieve this, simple interpolation methods between known points are usually not suitable. Hand-drawn maps from cartographers were preferred as input. The elevation of the bedrock roof and of the floor of cover formations were digitized. The resulting data was then interpolated taking boundaries such as faults and expanse limits into account. Finally, the resulting surfaces were adapted to ensure full coherence between the different layers. This lead to digital thematic maps which can be used in the GIS (see figure 3).

Thirdly, the dewatering was dealt with. To compute the dewatering, piezometric heads had to be known over the whole area both at the initial state (before pumping) and at the time of the analysis (named 'final state'). The present piezometric heads were imported from a calibrated groundwater model of the carboniferous limestone aquifer of the area (Rorive A. & Squerens P., 1995) (see figure 4). To estimate piezometric heads at the initial state, hypotheses had to be made because data recorded before industrialization are lacking. However, the bedrock roof was most likely below the water table in most of the studied area at that time. Another parameter dependent on the lowering of piezometric head but that does not require to known the initial state can be used in a first approach. It is the height of bedrock above the water table at final state (see figure 5). Comparisons were performed between this parameter and dewatering computed with likely simple hypotheses on initial water table levels. This showed that both approaches were coherent and did not led to large differences. However, a better estimation of the initial piezometric heads will probably be required to refine the model further.

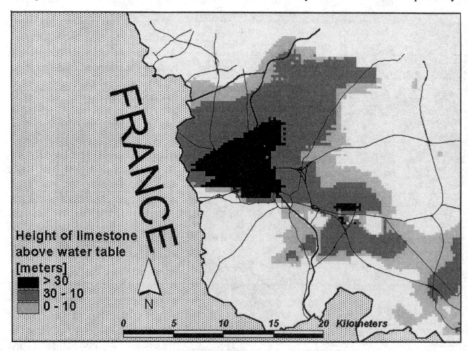

Figure 5 : Map of height of Carboniferous limestone above the water table.

Selecting parameters

The input data and maps were combined and crossed in order to select relevant parameters.

In this analysis, the information available about the cover revealed itself of little help. On the one hand, bigger collapses tend to occur where the cover is thicker. (However, the size of collapses have not been taken into account in this approach.) On the other hand, a mitigation effect in areas where thick marls deposits are laying right over the limestone might be expected. (But could not be demonstrated.) The most significant influence of the cover is related to its permeability since the recharge of the water table depend on it.

A good temporal and geographical match have been found between the density of known collapses and the height of limestone above water table (for

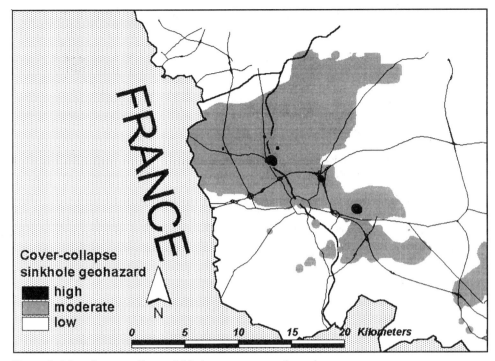

Figure 6 : Map of cover-collapse sinkhole occurrence geohazard.

example see figure 2 and figure 5). Moreover collapses tend to occur in clusters and reopen several times at the same place. Thus, density of reported collapses and height of limestone above water table have been selected to estimate the collapse geohazard.

To draw up the map, thresholds were chosen for both parameters keeping the accuracy of data in mind. The retained classification is the following : Zones were the density of collapses is less than one per square kilometer and no more than ten meters of limestone are above the water table were classified as low collapse geohazard. Zones were the density of collapse is between one and fifteen collapses per square kilometer and / or where the water table is more than ten meters below the limestone roof were classified as moderate collapse geohazard. Zones were the density of collapse exceeds 15 collapses per square kilometer were classified as high collapse geohazard.

In the South of the area, this classification might led to over-estimate the geohazard. When considering estimated dewatering instead of height of limestone above the water table, the moderate sinkhole geohazard areas tended to shrink in this part of the map. A more specific analysis should be carried out to back up these results.

CONCLUSIONS

The methodology exposed in this paper allowed us to delineate zones corresponding to three collapse hazard levels on a regional scale. This map is based on piezometric heads and records of former collapses. Both will evolve. Periodic update of this geohazard map is therefore required. The use of the GIS model will ease these updates. This model is also an effective environment to test scenarios and to refine the geohazard delineation.

ACKNOWLEDGEMENTS

This study is part of a research program with the support and funding of the 'MINISTERE DE LA REGION WALLONNE, DIVISION GENERALE DES RESSOURCES NATURELLES ET DE L'ENVIRONNEMENT'.

REFERENCES

Delattre N., 1983 , Les puits naturels du Tournaisis, contribution à l'étude de leur génèse, Mémoire de licence en sciences géographiques, Université de Liège (Belgium), 127 p.

de Roubaix E., Derycke F., Gulinck M., Legrand R., Loy W., 1979, Tournaisis '77'-'78' Effondrements à Kain et Evolution récente de la nappe aquifère profonde : Service Géologique de Belgique, Professional paper 1979|1, Nr157, 47 p.

Hennebert M., 1998, L'anticlinal faillé du Mélantois – Tournaisis fait partie d'une "structure en fleur positive" tardi-varisque, Ann. Soc. Géol. du Nord, T. 6 (2ème série), p. 65-78.

Hennebert M., Doremus P., Carte géologique de Wallonie à 1:25 000ème, pl. Sartaigne-Rongy 44/1-2, Ministère de la Région Wallonne, Namur, Belgium, *in press*.

Kaufmann O., Quinif Y., 1997, Cover-collapse sinkholes in the 'Tournaisis' area, southern Belgium, The Engineering Geology and Hydrogeology of Karst Terranes, B.F. Beck & J; Brad Stephenson, Editor, p. 41-47.

Legrand R., Nybergh H., 1981, Présentation des cartes situant la base des différentes formations géologiques recouvrant le socle dans le Tournaisis, Ann. Mines Belg., 6/1981, p.493-502.

Rorive A., Squerens P., 1994, Les grandes nappes aquifères du Hainaut et L'exhaure des Carrières : Craies et calcaires en Hainaut, de la géologie à l'exploitation, Faculté Polytechnique de Mons, p. 54-58.

Rorive A., Squerens P., 1995, Simulation mathématique de l'aquifère du calcaire carbonifère dans sa partie ouest (Tournaisis), rapport d'étude effectuée pour le compte du Ministère de la Région Wallonne, Direction Générale des Ressources naturelles et de l'Environnement, Division de l'Eau, Service des eaux souterraines.

Vergari A., Quinif Y. and Charlet J-M., 1995, Paleokarstic features in the Belgian carboniferous limestones - Implications to engineering, Karst Geohazards, B.F. Beck Editor, p. 481-486

Vergari A., Quinif Y., 1997, Les paléokarsts du Hainaut (Belgique), Geodinamica acta, 10, 4, Paris, p. 175-187.

Vergari A., 1998, Nouveau regard sur la spéléogenèse : le pseudo-endokarst du Tournaisis (Hainaut, Belgique), Karstologia Nr 31-1/1998, p. 12-18.

Predicting the distribution of dissolution pipes in the chalk of southern England using high resolution stratigraphy and geomorphological domain characterisation

JOHN LAMONT-BLACK   Department of Civil Engineering, University of Newcastle, UK
RORY MORTIMORE   School of the Environment, University of Brighton, UK

ABSTRACT

Much of the Chalk of southern England is, or has been in the recent geological past, mantled with poorly consolidated mixtures of sands, clays and gravels of Palaeogene to Pleistocene age. Engineering operations on the Chalk frequently encounter potentially costly problems caused by the karst system represented by the interaction of these materials and their chalk host. Detailed geological mapping of sites in southern England has demonstrated a relationship between syn- and post-Cretaceous tectonic events with the preservation of overlying strata and the development of different styles of dissolution pipes. Construction of a major highway in southern England is used to show how different dissolution pipe characters such as shape, type of fill and weathering effects on the chalk host are associated with different tectonic domains. These associations are then used to widen an existing system of chalk landscape classification.

INTRODUCTION

The most abundant karst features in the Upper Cretaceous Chalk are dissolution pipes (formerly solution pipes, e.g. Prestwich, 1854). These occur as cavities in the upper surface of the chalk, filled with material which has gradually slumped into the chalk as dissolution has progressed (Higginbottom, 1966). These features cannot be classified as dolines or sinkholes because they frequently lack a surface expression.

Chalk dissolution pipes can be hazardous to Highway construction. Earthworks on chalk which cross the grain of the hilly terrain are especially vulnerable because large cut/fill operations oscillate around the level of maximum dissolution pipe density. In an attempt to predict the presence of dissolution pipes a number of databases have been developed. These maintain records of chalk dissolution features encountered but, lacking a process-related basis, their resolution is poor and they can dramatically underestimate concentrations of dissolution pipes.

Mortimore (1990) introduced the concept of scientific frameworks to the understanding of chalk engineering geology. He advocated the collection and interpretation of site investigation data with respect to the frameworks of stratigraphy, sedimentology, structural geology and geomorphology. Stratigraphy is crucial to the understanding of the other frameworks, and Mortimore (1993) suggested a routine correlation of stratigraphic marker horizons to a resolution of one metre to be both necessary and achievable for engineering design. This paper shows how high resolution stratigraphic mapping of chalk in southern England enables terrain characterisation and improves the understanding of the distribution and variety of dissolution features in chalk terrain.

ENGINEERING EFFECTS OF DISSOLUTION PIPES

Before describing the above approach it is useful to review the impacts of dissolution pipes on engineering. These include:

- increased mass compressibility leading to excessive settlement or subsidence
- susceptibility to being washed away by water (e.g. broken water supply pipes) leading to subsidence
- reduction in volume of chalk available for earthworks (may require selective excavation and disposal of unsuitable material)
- reduction in intact dry density and hence earthworks suitability of chalk in the vicinity of dissolution pipes because of karstic 'softening' of the chalk
- associated presence of unstructured uncemented chalk silt ('jackets' of putty chalk, see below) reducing earthworks stability and foundation performance
- contamination of chalk with clay in the vicinity of dissolution pipes
- slope instability

- reduction in rock mass quality of proximal chalk i.e. increased fracture apertures and fracture frequency, implications for the stability of tunnel portals and slopes
- loci of potentially increased surface water recharge to the Chalk with implications for contaminant migration

For more detailed accounts see Higginbottom (1966) Higginbottom and Fookes (1970), Edmonds *et al.* (1987), McDowell, (1989), Lord *et al.* (1995) and Lamont-Black (1995, 1998).

PHYSICAL CHARACTERISATION OF DISSOLUTION PIPES ON A MAJOR HIGHWAY IN SUSSEX, ENGLAND
The Chalk

The A27 Brighton Bypass (Figure 1) is constructed predominantly in the Seaford and Newhaven Members of the Upper Chalk in Sussex (Mortimore, 1986; Bristow *et al.*, 1997). Dissolution pipes are characterised according to their size, shape, spatial and stratigraphic distribution and the nature of their fill materials. These characterisations are described below.

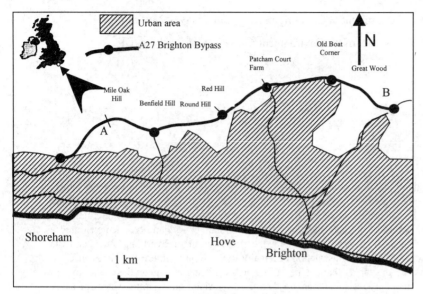

Figure 1: Location of the study area; A27 Brighton Bypass, Sussex, UK.

Spatial distribution of dissolution pipes

It is seen on the long section from A to B along the line of the highway (Figure 1) that dissolution pipes are present only on the tops of hills, principally at Round Hill, Red Hill, Old Boat Corner and Great Wood (Figure 2). The altitude of these localities varies between about 125m AOD and 150m AOD and scrutiny of the presence or absence of these features on all the hill tops reveals no simple relationship with altitude.

Spatial density of dissolution pipes varies between locations. Red Hill shows the highest density with approximately 265 features per hectare; Round Hill at about 200/ha, 150/ha at Old Boat Corner; 30/ha at Great Wood, 8/ha in Foredown Hill and 3/ha in Marquee Brow. The paucity of features in the latter two localities disallow their inclusion in the general characterisations that follow.

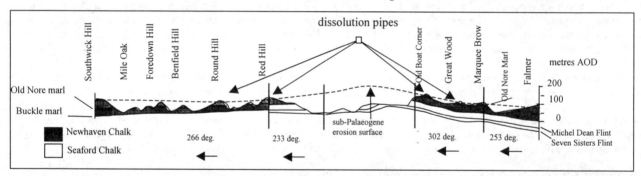

Figure 2: Long section of the A27 Brighton Bypass (construction level not shown) showing the stratigraphy and the position of the sub-Palaeogene erosion surface and dissolution pipes. Note how the erosion surface cuts down the stratigraphy towards the Old Nore marl central portion of the section (Hollingbury Dome, see Figure 4) and ascends the stratigraphy eastwards and westwards towards the Caburn syncline and the Sussex trough respectively (Figure 4). Dissolution pipes were not recorded in Southwick Hill because tunnels carried the highway several tens of metres below the zone of potential dissolution pipes.

Fill material in dissolution pipes

During the mapping of dissolution features, four types of fill material were recognised, these included:

- Clay-with-flints *sensu stricto* (Loveday, 1962)
- sandy clay, which comprised all cohesive sediments other than Clay-with-flints including clayey silty loams and green and purple variegated clays
- fine to coarse sand
- basal Palaeogene alteration mineral suite including aluminite, basaluminite and nordstrandite

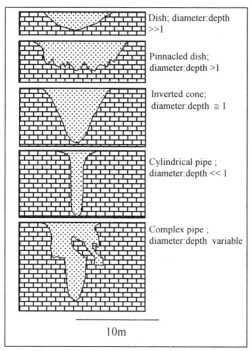

Figure 3: Shape characterisation of dissolution pipes on the A27

Fill type varied between location with Clay-with-flints and cohesive material dominating dissolution pipes in Round Hill and Red Hill. Dissolution pipes in Old Boat Corner and Great Wood, however, contained sandy fills.

In addition to the material within dissolution pipes, the quality of the chalk surrounding dissolution pipes was described. Softening of chalk close to dissolution pipes is well known and results in a material, described as putty chalk (Higginbottom, 1966; Mortimore, 1990). Putty chalk forms 'jackets' around dissolution pipes wherein the chalk material is entirely degraded and has very low strength (Lamont-Black, 1998). Putty jackets were well developed around dissolution pipes in Red Hill with a mean thickness of approximately 0.5m and maximum thickness of just over 2m. In all other settings putty jackets were generally 0.1-0.2m thick.

Shape of dissolution pipes

Owing to the nature of the construction project, good sections through dissolution pipes were frequently exposed thus allowing the discrimination of different morphotypes (Figure 3). The numbers of dissolution pipes in Foredown Hill and Marquee Brow were too few to draw any general characteristics. Red Hill and Round Hill settings were dominated by inverted cone and pinnacled dish shapes. However, Old Boat Corner and Great Wood were characterised by cylindrical pipes. Thus two broad groupings emerged:

- Red Hill Character: densely packed inverted cone pipes filled with cohesive material and exerting a markedly softened chalk.
- Old Boat Corner character: less densely packed cylindrical pipes filled with sandy material and limited softening on the surrounding chalk.

STRUCTURAL ASSOCIATIONS OF DISSOLUTION PIPES

The sub-Palaeogene erosion surface (Figure 2; Wooldridge and Linton, 1955), represents the approximate level at which erosion ceased prior to the deposition of Palaeogene strata. It is seen that this surface varies in altitude and stratigraphic level. Passing from Southwick Hill to Red Hill the sub-Palaeogene erosion surface cuts down through the Newhaven Chalk. It reaches its lowest exposed level at Old Boat Corner where it is almost at the level of the Old Nore marl. Eastwards from Old Boat Corner it ascends the stratigraphy towards Falmer.

Comparing this information with Figure 4 reveals definite associations of dissolution pipe development with structural tectonic features (folds). The dissolution pipes at Old at Corner and Great Wood are at the western end of the Caburn syncline, which preserves a sequence of *in situ* Palaeogene strata at Falmer. Those at Red Hill are associated with the, much smaller, Coney Hill syncline which preserves Clay-with-flints, reaching a maximum thickness of 4m at Red Hill. Separating these two synclinal areas is the Hollingbury Dome. Changes in chalk lithostratigraphy and sequence thickness in the area (Mortimore and Pomerol, 1991; Lamont-Black, 1995) indicate growth of these structures during the Upper Cretaceous. Variations of the sub-Palaeogene erosion surface in terms of stratigraphy and altitude indicate their continued activity, prior to, and after, Palaeogene sedimentation.

DISCUSSION

The association of chalk dissolution pipes (and other features e.g. swallow holes) with the contact of overlying Palaeogene material, informally referred to as the 'feather edge', is well known (e.g. Prestwich, 1854; Dowker, 1866; Edmonds *et. al.*, 1987). Structural downwarps in the underlying chalk create zones of reduced erosion where Palaeogene strata have been protected from erosion during subsequent landscape evolution. Chalk from which overlying material such as Palaeogene strata has long since been removed generally develops a thin residual soil or rendzina.

The identification of structurally associated dissolution pipe characters has not been made before. On parts of the A27 construction contract, some of the largest disruptions to the critical path were due to earthworks difficulties caused by dissolution pipes and their effects on the surrounding chalk. The worst cases occurred in the Red Hill and Round Hill cuttings. Dissolution pipes in the Old Boat Corner area had a less severe effect because the pipes were fewer in number, there was much less soft chalk and putty associated with the pipes, and the chalk was not contaminated with plastic clay, as at Red Hill.

These observations have concentrated on dissolution pipe associations with Clay-with-flints and Palaeogene sediments. Both these materials can be seen as the remnants of Palaeogene strata that previously covered the entire chalk outcrop (Hodgson et. al., 1972). Catt (1986) showed that Clay-with-flints is developed at the periphery of Palaeogene outliers on the Chalk. In this way, as the 'feather edge' is traversed the materials immediately beneath the surface change from Palaeogene sediments to a mixture of Palaeogene and Clay-with-flints to Clay-with-flints to a thin chalk rendzina. For landscape development these contacts can be viewed as ranging from an immature contact beneath a thick sequence of overlying Palaeogene material (largely protected from weathering effects) through to a mature contact with chalk rendzina.

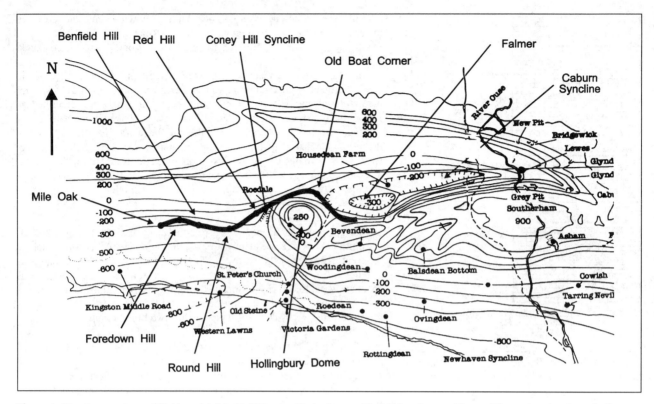

Figure 4: Structure contour of Bridgewick Marl1 (Kingston Beds, Lewes Chalk) showing positions of the major structures in the Brighton area (Mortimore and Pomerol, 1991). These include the Hollingbury dome (uplift) and the Caburn Syncline and Coney Hill Syncline. To the south west of the Hollingbury Dome, structure contours indicate a broad synclinal feature, the South Downs trough (Mortimore, 1983). The A27 Brighton Bypass is indicated by the bold line.

A small tectonic structure, the Coney Hill syncline, preserving only a thin deposit above the Chalk, can be considered to have a relatively mature contact with the underlying Chalk. Construction records showed this situation to be the worst case scenario in terms of chalk quality. The thin Clay-with-flints at Red Hill, prevented the erosion of Chalk which has suffered near surface weathering effects such as periglacial freeze-thaw and dissolution. Such chalk is removed from more mature contacts containing rendzinas, and is absent from immature contacts which are either beyond the reach of near-surface weathering effects or have experienced them for a relatively short period of time. In addition, contacts may be complicated by materials creeping downslope, as is noticed with many Clay-with-flint deposits on hill slopes around Brighton. A mature contact, therefore, may at a later stage become, overlain by materials that characterise a less mature contact. Such contacts are not characterised by large dissolution pipes (e.g. Lamont-Black, 1995, p.48).

The evidence suggests that the Coney Hill syncline, being a much smaller structural feature than either the Caburn syncline or the South Downs trough, has lost most of the Palaeogene strata that it would have contained in the past. The deposits are a mature weathering product comprising Clay-with-flints and occasional small pockets of sand, clay and alteration minerals, which are recognisable as Palaeogene in origin. It is suggested, here, that this has a bearing on the nature of dissolution pipes and the condition of the surrounding chalk.

Expanded geomorphological domain characterisation

Mortimore (1979, 1990) classified chalk outcrops into geomorphological domains including; (I) unweathered hilltops, (II) steep hill slope domains, (III) dry valleys and (IV) deeply weathered hilltops. The hilltops containing dissolution pipes on the Brighton Bypass all fit into the deeply weathered hill top domain. It is suggested here that this particular classification should be widened to include the structural associations, material types and weathering maturity conditions noted on the Brighton Bypass. Such a classification is shown in Table 1. Types 1 and 2 are exemplified by the Red Hill and Old Boat Corner characters. Type 3 is exemplified by Clay-with-flints which has crept downslope (probably caused by periglaciation).

CONCLUSION

Two basic dissolution pipe characters in the Chalk are identified and these are associated with different engineering impacts. Tectonic structures in the sub-Palaeogene erosion surface were identified using high-resolution stratigraphy. The two dissolution pipe characters are spatially separated and associated with particular synclines in the Chalk, identified using high-resolution stratigraphy at the sub-Palaeogene erosion surface.

Table 1: Classification of dissolution pipe geomorphological domains

Geomorphlological domain	Materials present	Engineering implications
Deeply weathered type 1	Dominantly Clay-with-flints, Palaeogene materials very highly disturbed, much clay in chalk fractures, much softening of chalk and putty development	Present severe problems for earthworks due to high density of dissolution pipes and very poor quality chalk of low earthworks suitability. Bearing capacity of chalk reduced.
Deeply weathered type 2	Dominantly Palaeogene materials – sands and clays, Clay-with-flints present as generally thin, stiff, slikensided, manganiferous lining to dissolution pipes.	Potential problems for earthworks volumes calculations; dissolution pipes will require spatial evaluation for the design of foundations.
Deeply weathered type 3	Thin highly disturbed layer of Clay-with-flints mixed with unstructured chalk and a highly undulating but small amplitude contact with intact chalk.	Upper surface of chalk very soft but increases in quality rapidly with depth. May experience instability and cause problems if used as a haulage route.

This approach can be applied to the contact of Chalk with other materials such as loess, brickearth, glacial till etc. to build a comprehensive picture of the interaction of Chalk with its overlying materials and their structural associations. In this way the distribution of different types of chalk dissolution feature can be evaluated and the engineering risks, created by the weathering domains, reduced.

ACKNOWLEDGEMENTS

We are particularly indebted to the support given by L. G. Mouchel and Partners Limited and the Highway Agency, during the tenure of and EPSRC funded research programme at the University of Brighton from which the data in this paper partly derives.

REFERENCES

Bristow, R., Mortimore, R. N. and Wood, C. J., 1997, Lithostratigraphy for mapping the Chalk of southern England. Proceedings of the Geologists Association, 109, 293-315.

Dowker, G., 1866, On the junction of the Chalk with the Tertiary Beds in East Kent. Geological Magazine, 3, 210-213

Edmonds, C. N., 1983, Towards the prediction of subsidence risk upon the Chalk outcrop. Quarterly Journal of Engineering Geology. 16, 261-266.

_____, 1987, Induced subsurface movements associated with the presence of natural and artificial openings in areas underlain by Cretaceous Chalk. In, Bell, F. G., Culshaw, M. G. and J. C. Cripps, (eds.) Engineering geology of Underground Movements, Proceedings of the 23rd. Conference of the Engineering Group of the Geological Society of London, Nottingham, 1987. 239-257.

Higginbottom, I. E., 1966, The engineering geology of chalk. In Proceedings of the Symposium on Chalk in Earthworks and Foundations, April 1965, Institution of Civil Engineers, London. 1-14.
_____, and. Fookes, P. G 1970, Engineering aspects of periglacial features in Britain. Quarterly Journal of Engineering Geology, 3(2), 85-117.

Hodgson, J. M., Catt, J. A. and Weir, A. H 1967, The origin and development of Clay-with-flints and associated soil horizons on the South Downs. Journal of Soil Science, 18, 85-102.

Lamont-Black, J. 1995, The Engineering Classification of Chalk with special reference to the origins of Fracturing and Dissolution. Unpublished PhD thesis, University of Brighton, 438p.
Lamont-Black, J., 1998, A ;'whole rock' approach to the engineering geology of the Chalk of Sussex, in Bennett, M. R. and Doyle, P. (eds.) Issues in Environmental Geology: a British Perspective, 122-172.

Lord, J. A, Twine, D. and Yeow, H., 1995, Foundations in chalk. Funder's Report 13, CIRIA Project Report, 11. 189 pp.

Loveday, J., 1962, Plateau deposits of the southern Chiltern Hills. Proceedings of the Geologists' Association, 73, 83-102.

Mortimore, R. N., 1990. Chalk or chalk. In Chalk, Proceedings of the International Chalk Symposium, Brighton 1989. Thomas Telford, London. 15-45.

_____, 1993. Chalk water and engineering geology. In (Downing, R., A., Price, M. and G. P. Jones Eds.) The Hydrogeology of the Chalk of North-West Europe. Clarendon, Oxford. 67-92.

_____ and Pomerol, B., 1991, Upper Cretaceous tectonic disruptions in a placid Chalk sequence in the Anglo-Paris Basin. Journal of the Geological Society, 148 391-404.

McDowell, P. W., 1989. Ground subsidence associated with doline formation in chalk areas of Southern England. In (Beck, B. F. Ed.) Engineering and Environmental Impact of Sinkholes and Karst, Proceedings of the third multidisciplinary conference on sinkholes and environmental impacts of karst, Florida. Balkema, Rotterdam. 129-134.

Prestwich J., 1854. On some swallow holes on the chalk hills near Canterbury. Quarterly Journal of the Geological Society of London, 10, 222-224.

Sweeting, M. M., 1972. Karst Landforms. Macmillan, London, 362 pp.

Wooldridge, S. W. and D. L. Linton, 1955. Structure, Surface and Drainage in south-east England. George Phillip & Son Ltd, London. 176pp.

Is human activity accelerating karst evolution?

BERNARDAS PAUKŠTYS Hydrogeological Company 'Grota', Vilnius, Lithuania

ABSTRACT

Karst in Lithuania is developing in the Upper Devonian gypsum dolomites overlain by thin (0-10 m) Quaternary deposits. On the land surface a dense network of sinkholes, land subsidence and karst lakes expresses the underground karst processes. During the last decades acceleration of karst is being observed. Human activity: groundwater abstraction and agriculture are considered to be the main reasons for the increased rate of karst development. Is the present karst development in Lithuania accelerated? Is man really triggering the occurrence of new sinkholes? What is the role of regional and global environmental changes for karst evolution? These are the main issues discussed in the paper.

INTRODUCTION

Karst is the most important present day geological process in Lithuania. It occurs on an area of more than 1000 square kilometers and possesses a dense network of sinkholes (dolines), land subsidence and karst lakes. This landscape is noted as the karstic region of the country. Surface karstic features, both old and new, have developed due to the collapse of cavities, which have formed in the Upper Devonian gypsum dolomites. These features impact human activity in the region, causing structural failures in construction and engineering works, as well as affecting the landscape and limiting the use of the land for agriculture (Paukstys et all 1997). On the other hand, human activities disturb the natural equilibrium of the vulnerable karst eco-system and increase the rate of karst development. According to the review of aerial photos on the site of active karst with an area of 3 sq. km 61 new sinkholes appeared during the 23 years from 1967 to 1990 (Paukstys 1996). This has led to a greater need for geological and hydrogeological investigations of the karstified area.

Detailed geological-hydrogeological investigations of karst region of Lithuania have been carried out since 1979. Various methods were employed for evaluation of environmental changes of the region: groundwater monitoring (observations of level fluctuations and changes of chemical composition), geophysical investigations, experimental works, geodetic survey, large scale (1:50,000) mapping, airborne photography, hydrodynamic and hydrochemical modeling. A lot of data has been collected and included into a large database. The conclusions were made that present karst evolution is being accelerated by human activity. Groundwater abstraction and agricultural pollution were considered to be the main factors for the increased karstification rate.

It was believed that enhanced groundwater exploitation and unlimited use of fertilizers during the time of the Soviet regime developed regional groundwater drawdown and increased infiltration of fresh, unsaturated and polluted surface waters into karstic aquifers, accelerating process evolution. When in 1990 Lithuania has separated from Soviet Union and became independent, industrial production declined causing a dramatic decrease in centralized groundwater consumption and reduced amount of fertilizers. However, the appearance of new sinkholes on the farm fields and in the settlements of karst region is continuing.

Is karst acceleration a natural or human impacted process? Is present karst development accelerated or it is occurring on a natural rate? Most geologists in Lithuania support the idea that anthropogenic impact triggers karst process. They are probably right, but sometimes it seems that we, environmental scientists, are too quick to accuse human beings for natural disasters paying less attention to the role of regional and global environmental changes on the Earth.

An example from the Lithuanian karst region will be presented and discussed below.

GEOLOGICAL-HYDROGEOLOGICAL DESCRIPTION OF THE KARST REGION

The karst region is situated in the northern Lithuania (Fig. 1). The karst is developed in the Upper Devonian gypsum formation overlain by a thin Quaternary cover (0-10m) of glacial sands, sandy loam and till. Both the Devonian and Quaternary formations are water bearing and form a single interconnected hydraulic system consisting of one unconfined and three confined aquifers. All the aquifers are being exploited to varying degrees for domestic and industrial use. The quality of the water in the various aquifers is quite different. The Quaternary glacial sands contain mainly fresh calcium bicarbonate water (TDS 0.5 - 0.8 g/l). The water is often heavily polluted by nitrogen and organic compounds. The karst aquifer typically has slightly mineralized but very hard water (TDS 1.5 - 2.4g/l with hardness up to 35 meq/l). Calcium and sulfate are the main chemical constituents in this water. The aquifer is very vulnerable to pollution and as a result the level of contaminants (nitrogen and organic) often exceeds the maximum allowable concentration for drinking water, (Zakutin and Paukstys, 1993).

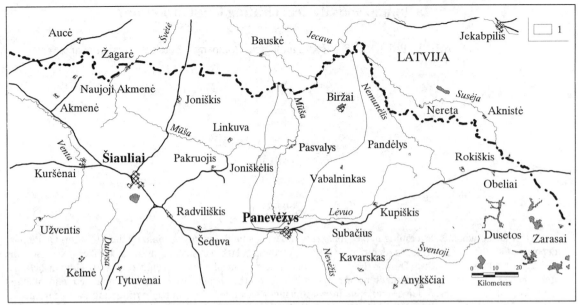

Figure 1: Location map of the karst region of Lithuania. 1.Boundaries of karst region.

Most of the region is a covered karst. The main surface karst features are sinkholes, with sometimes several small ones joining to form one large collapse feature. Near Kirkilai village there are some 300 karst lakes of different shapes and sizes (Kilkus, 1977).

The density of sinkhole distribution varies in the region. The highest number of them, 200 units per square kilometer, has been recorded in the Karajimiskis Geological Reserve. Karst sinkholes occupy more than 30% of the territory of the Reserve and serve as direct paths for surface and groundwater connection. The density of the sinkholes is inversely proportional to the thickness of the overlying deposits. Most of the karstic region has a Quaternary cover of up to 5m. The depth of the sinkholes varies from 2 to 12m while their diameter range from several meters to 60m (Buceviciute and Marcinkevicius, 1992).

HUMAN FACTORS AFFECTING KARST DEVELOPMENT

As has already been mentioned, the main human activities influencing karst development are groundwater abstraction and agriculture. Minor impacts include land reclamation, civil and hydrotechnical construction, mining, wastewater discharges etc.

The interconnected system of aquifers is being exploited for drinking water supplies by dug and drilled wells. Some 600 bored wells, tapping different aquifers, draw water from beneath the karstic terrain. The average water consumption from an individual well range between 10 and 50 cubic meters a day (Paukstys et al, 1997). Besides the individual wells, two waterworks in the towns of Birzai and Pasvalys provide, respectively, 2000 and 3600 cubic meters a day of groundwater. Previous estimates for the two towns to the year 2010 envisaged increases to 9000 and 14000 cubic meters a day (Paukstys, 1991). However, recent political-economic changes in Lithuania, including a decrease in industrial production, cast doubt on the accuracy of these projections. As a result of the abstractions, the water level at the Pasvalys waterworks since 1970 has fallen by 7.5m. At Birzai the drawdown has increased by 8m over the period since 1961. The additional drawdown in the aquifers has further complicated their hydraulic relationships. The vertical flow velocity has increased and the groundwater head in the aquifers has fallen and been redistributed. This in turn has led to increased recharge from precipitation and polluted surface water to the karst aquifer.

Agriculture is considered to be the other important human activity influencing karst development on a regional scale. Farming is highly developed in the karst region as in other parts of Lithuania. Indeed, except for the geological reserves, almost all the karstic region is given over to agriculture. In order to improve yields inorganic and organic fertilizers, herbicides and pesticides were widely used. The usual modern range of nitrogen, phosphorus and potassium (N, P, K) fertilizers were applied together with a variety of trace elements; both liquid and solid manure were also widely used as is the addition of lime. The results have been increased fertility of the soils but also regional groundwater pollution. The shallow unconfined water in some dug wells is heavily polluted by nitrates (NO_3 concentration reaches 400 mg/l, while the maximum allowable concentration (MAC) for drinking water is 50 mg/l), nitrites -7.5 mg/l (MAC – 0.1 mg/l), ammonium - 100 mg/l (MAC - 2.0 mg/l), permanganate oxidation - 19 mg/l (MAC – 6.5 mg/l). The water in the karst aquifer is also heavily polluted in some areas (Klimas & Paukstys, 1993). Groundwater pollution creates an obvious risk to the quality of potable groundwater resources, but is contamination accelerating gypsum dissolution?

DISCUSSION

Three main issues will be discussed below:

- What are the main reasons for the regional karst water level decline?
- Does contamination increase the aggressiveness of karst waters?
- Is the present karst developing on a natural or accelerated rate?

DECLINE OF KARST WATER LEVEL

Prognostic calculations based on the groundwater consumption corresponding to the needs of 2010 show that the deepest drawdown will develop in a major production aquifer around the well fields of the centralized water supplies in Pasvalys (12m) and Birzai (18.6m). Groundwater abstraction boosts gypsum solubility and accelerates the removal of dolomite debris from subsurface cavities into the rivers and/or adjacent openings. Outside the areas of the well fields, on rather remote areas, karst water drawdown will vary from 0.5 to 2.0 m. Critical review of groundwater production data creates questions as to whether forecasted water extraction can influence regional groundwater decline, particularly keeping in mind that major groundwater production is from the aquifer below the karstic layers. Drawdown development, without a doubt, creates a disturbance of the subsurface environmental equilibrium. The question remains, however, is groundwater abstraction the only reason for water level decline. Analysis of long term hydrologic data from the region indicates natural periodic water level fluctuations related to various cycles of the planet Earth. Since 1960 mean annual precipitation and surface runoff in the area is significantly reduced (Paukstys and Taminskas 1992), (Fig. 2). Aren't natural regional and global environmental changes the main explanation for karst water level decline?

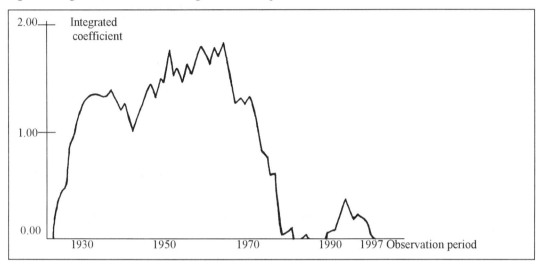

Figure 2: Deviation of amount of precipitation from mean values

POLLUTION AND GYPSUM SOLUBILITY

How is groundwater pollution influencing solubility of karst sediments? For a better understanding of the interactions in the system of groundwater-gypsum, experiments of gypsum solubility in a laboratory and under ambient conditions were arranged. Gypsum samples of similar shape, form and weight were cut from a piece of gypsum rock and immersed into solutions of fertilizers widely applied in the karstic region: potassium chloride (KCl), carbamide ($CO(NH_2)_2$), liquid manure and various natural water sources: observation wells containing water of different chemical composition, the lake of a karst cave, surface stream, etc. Gypsum samples were weighed once a month. Results of experiments showed diverse solubility of gypsum in waters of different chemical compositions.

The fastest dissolution of a gypsum sample was observed in the natural medium of a fresh water stagnant stream. A gypsum sample of 20 g dissolved completely in 65 days there. In the same water in laboratory conditions (5-liter bottle) the sample lost 10 g of weight in 110 days. However, in the saturated groundwater of the Karves Ola cave, with TDS of 2.2 g/l, the weight of the gypsum sample reduced by only 1.5 g after 100 days of the experiment. Tests also showed different solubility of the gypsum samples in fertilizer solutions. 12 grams of gypsum were dissolved in the carbamide solution in 100 days, while only 4.8 g of the sample vanished in concentrated manure from the farm of a karst territory. Although the results of experiments can not be transferred directly to the natural situations, they are helpful for understanding the general regularities of water-gypsum interaction in different solutions. The experiments showed the importance of water saturation by $CaSO_4$ on further gypsum solubility and karst process development. There was no direct influence of pollution on gypsum solubility observed.

IS KARST DEVELOPMENT ACCELERATED?

Is present karst evolution really accelerated? How can we know what was the karstification rate in the Devonian period and before or even after the last glaciation? These are questions easily asked but difficult to answer.

Most Lithuanian geologists (Klimas and Paukstys, 1993, Paukstys 1996, Paukstys and Narbutas 1996, Paukstys et al 1997) support the idea that anthropogenic impact has increased the karstification rate. Their conclusions are based on historical analysis of karst evolution. Although it is difficult to say precisely when dissolution of gypsum commenced it obviously happened in the Devonian period and has actively continued after the last deglaciation. Paleoenvironmental studies provide evidence of sinkholes existing before the last glaciation, since the age of peat filling the holes is 9.8 thousand years (Paukstys 1996). A shrinking glacier cut the covering layer from the underground cavities exposing gypsum deposits to the surface. Climatic conditions in the region were changing from humid to dry and this had a different influence on karst development. Active peat sedimentation was again observed 8-7 thousand years ago. 1.5 meters of peat were deposited in the observed sinkholes at that time. The periods of humid and hot climate were also favorable for karst development. Another humid climate period started in the Holocene and lasts up to present. If karst development would have occurred at the present rate, at least from the deglaciation time, then in the course of the last 10 thousand years it would hardly be possible to find gypsum layers in the geological cross-section of the region. This leads to the conclusion that before the start of active economical activity natural karstification was much slower.

CONCLUSION

Both natural and human impacted factors should be evaluated before accusing only human beings for all disasters on the Earth. Natural environmental reasons usually are the main driving force for karst development.

REFERENCES:

Buceviciute, S. and Marcinkevicius, V. 1992. Karst morphology. Proceedings of the Lithuanian High Schools. Geologija. No. 13. 49-58p.p. [in Lithuanian].

Kilkus, K. I. 1977: Karst of North Lithuania. Karst lakes. (Proceed. of Academy of Science of Lithuania) B (103) 125-133 p.p. [in Russian].

Klimas, A. and Paukstys, B. 1993: Nitrate contamination of groundwater in the Republic of Lithuania. Bulletin of Geological Survey of Norway 424, 75-85 p.p.

Paukstys, B. 1992: The risk of karst activization and limitation of economic activity. Proceedings of the Lithuanian High Schools. Geologija. 13, 131-140 p.p. [in Lithuanian].

Paukstys, B. 1996: Hydrogeology and groundwater protection problems in karst region of Lithuania. Scientific papers of the Geological Society of Lithuania. No. 6. 1-72. Vilnius.

Paukstys, B. and Narbutas, V. Gypsum karst of the Baltic Republics. Gypsum karst of the world. International Journal of Speleology. A. Klimchouk, D. Lowe, A. Cooper and U. Sauro (eds.). Vol. 25(3-4), 1996. 279-285 p.p.

Paukstys, B., Cooper, A.H. and Arustiene J. Planning for gypsum geohazards in Lithuania and England. The engineering geology and hydrogeology of karst terranes. Barry F. Beck and J. Brad Stephenson (eds.).1997. 127-137 p.p.

Paukstys, B. and Taminskas, J. 1992. Surface and groundwater regime. Karst in Lithuania: Hydrogeology and Groundwater Protection. Proceedings of Lithuanian High Schools. Geologija. No. 13. 49-58 p.p. [in Lithuanian].

Zakutin, V. P. and Paukstys, B. P. 1993: Effets geochimiques de la pollution agricole des eaux souterraines dans la region karstique de la Lituanie. Hydrogeologie. No.1. 65-70 p.p.

Site selection and design considerations for construction in karst terrain/sinkhole-prone areas

BASHIR A.MEMON, ABNER F.PATTON, LOIS D.GEORGE & THOMAS S.GREEN
P.E.LaMoreaux & Associates, Inc., Tuscaloosa, Ala., USA

ABSTRACT

The occurrence of subsidence and sinkholes are reported and documented each year in parts of the United States underlain by carbonate rocks. Sinkhole related failures not only can cause significant property and environmental damage but also pose serious threats to the lives and economic and emotional well being of those involved.

Great potential for damage exists when any structure including buildings, pipelines, roads, railroads, airports, lakes, waste lagoons, stockpiles, and landfills are built in areas of karst terrain. A karst terrain is subject to subsidence and/or failure, unless measures are taken to prevent, mitigate or minimize the mechanisms of sinkhole formation and the natural and anthropogenic forces that drive the mechanisms. To select a suitable location and design for a project, a thorough understanding of geologic structure; patterns and interconnection of fractures; cavities and solution features; and subsidence or collapse triggering factors is necessary. It is also necessary to investigate the degree of dissolution, slot and pinnacle geometry, the extent of karst related features and the potential for their further development. Careful evaluation of subsurface data along with comprehensive planning and engineering design can overcome sinkhole related problems and avoid or minimize damage due to future sinkhole collapses.

The potential for subsidence and sinkhole formation must be considered in land-use planning and design of associated improvements including roads and pipelines. Consideration must also be given to avoiding or minimizing ground-water level changes that can trigger sinkhole collapses. Structures whose failure could cause significant environmental damage or impact human health and safety should be designed to be insensitive to sinkhole development. A realistic model can be developed which represents typical conditions as well as the extremes that are likely to occur at each site. The model should focus on the peculiar features of carbonate terrain that will influence design and performance of the proposed project.

INTRODUCTION

Subsidence and sinkholes of various sizes and shapes develop in many parts of the United States and around the world where soluble rocks, such as evaporites and carbonates, are partially dissolved by ground water. The resulting landform, karst terrain, is characterized by irregular surfaces with pinnacles and depressions, voids, losing streams, sinkholes, cavities and caves. Karst terrain/sinkhole-prone areas present special problems and challenges for the design and construction of a development because of the variable and changeable nature of the soil and rock that may support the structure. Concern related to subsidence and/or sinkhole failure must be addressed and preventive measures taken to mitigate or minimize the natural and anthropogenic forces that drive the mechanisms of sinkhole formation.

Environmental issues that may be exacerbated by the presence of sinkhole-prone carbonate formations must be addressed at the design stage. Building, utility, and highway designs must accommodate the karst conditions, because sinkhole related failures can cause significant damage to the environment and property and pose serious threat to the lives and economic well being of those involved.

Environmental and engineering problems in limestone terrain are largely the result of realized rock dissolution. Periodically, however, most of the problems develop or are realized in the overlying residual (regolith) or deposited soils that cover rock solution features (voids and cavities). To address these problems, including inherent defects and weaknesses of the soil and rock, it is important to understand and characterize both the rock and any overlying soil as well as their interaction to physical changes. The major factors are natural and anthropogenic activities such as surface water infiltration, ground water fluctuations with respect to the rock-soil interface, site excavation, or filling and impoundments, landfills, stockpiles, buildings and other construction activities.

INVESTIGATION/DESIGN CONSIDERATIONS

Development of a site in a karst terrain or sinkhole-prone area requires multi-disciplined team efforts. A professional team, at a minimum, consists of competent and experienced geologists/hydrogeologists, geotechnical engineers, civil engineers, and assorted required specialists. Team members with demonstrable experience in karst, can bring invaluable perspectives to the design process. Facts uncovered during each phase of an investigation provide design guidelines that enhance the quality of design efforts.

To select a suitable location and design for a facility a thorough understanding is necessary of the geology and geologic structure, patterns and interconnection of fractures, cavities and solution features, and their aerial extent, slot and pinnacle geometry, and the potential for further development and triggering factors of subsidence and/or sinkhole formation.

The predevelopment site investigation phase identifies the degree of dissolution and the pattern and extent of hazards such as sinkholes, soil raveling and erosion domes and the potential for their further development. The thickness and strength profile of soil overburden, particularly the soft zone over rock, location of rock collapses, or the potential for soil dome collapse, all of these factors influence design and use of shallow foundations. Adding fill on top of rock or the soil overburden, as well as excavating soil and rock, alter the present and future integrity of the karst system because these activities change the stress in the underlying formations. Existing ground water conditions and potential changes that occur naturally or by human activities are evaluated so that steps can be taken to minimize adverse effects. Careful evaluation of subsurface geologic and hydrologic data and the physical changes that effect solutioning rock or soil dome collapses along with comprehensive planning and engineering design can overcome sinkhole related problems and avoid or minimize damage. Understanding and evaluating these critical factors aids in developing realistic models representing typical conditions as well as the extremes that are likely to occur at a site and also aid in designing structures to be insensitive to sinkhole development; and assist in estimating the risks involved with long term use of a site.

DESK TOP INVESTIGATIONS

Investigations start with the collection and review of existing fundamental data including published and unpublished information and maps (topographic, geologic, structural and property, etc.) by federal, county, state and local agencies, consultants reports and USGS topographic maps. Topographic maps, black and white and color and false-color infrared aerial photography. Satellite imagery are used to identify structural features, stressed vegetation, potential locations of sinkholes, springs, surface water bodies, depressions and drainage features and lineaments (shallow erosion gullies, wet weather streams, etc.). Geomorphic features are studied to identify karst terrain as well as pinpoint locations that are most likely to become engineering problems related to carbonate rock solution features (sinkholes and solution depressions).

It is also important that a field reconnaissance survey be performed to ground truth the features identified on the topographic maps and aerial photographs during desk top investigations. Particularly suspicious terrain details that are difficult to identify from review of aerial photography and maps, because of tree and vegetative cover as well as overhangs and other obstructions or because of their small size or low relief, are to be investigated during field activities. This phase is often referred to as a karst inventory and should include not only the site itself but also any adjacent areas which may be relevant (for example, springs which may be impacted by the site).

FIELD INVESTIGATIONS

Field investigations can include vertical and horizontal test borings; rock coring, test pits or test trenches; collection of representative samples of soil, overburden rock, filling material in rock voids and cavities. Borehole geophysics, video imaging of borehole walls, and appropriate geophysical exploration methods (resistivity, radar, gravity surveys) are performed to characterize the configuration of the bedrock surface, randomly oriented fissures, large size voids and other solution features.

Vertical borings in carbonate rocks are drilled to assist in defining the orientation and size of fractures and openings, voids, and cavities. Horizontal and directional borings may be used to identify and explore the solution enlargement of vertical fractures, fissures and joints and their width and fill material.

A down-hole color camera is a direct means to evaluate a borehole with closely fractured rock that produces no core recovery, to determine if fissures are open or clay filled and to determine if the voids encountered are isolated enlarged joints or large continuous caverns.

Sometimes, even with large diameter boreholes, it is not possible to define the extent of void and pinnacle formation. Test trenches and pits are dug to the rock surface to delineate the orientation and extent of such features and gain information that cannot be obtained by boreholes. The size and depth of trenches and pits may be limited by the availability and capability of equipment, strength of soil overburden, extent of soft zones and depth to water. After collection of required data, pits should be filled with compacted soil and cement grout.

GEOPHYSICAL EXPLORATION

Geophysical surveys are used to explore shallow (±30 meters) subsurface (within tens to hundred feet) conditions. Geophysical methods are used to determine indirectly the extent and nature of geologic material beneath the surface. The thickness of residuum, probable void zones or cavities, depth to the basement rock and water table can also be determined.

Geophysical exploration involves measuring a force system (geophysical properties) in the earth and inferring boundaries between zones of similar responses to those forces, and the patterns of force change, known as anomalies. The force system can be natural (gravity and the earth's magnetic field) or induced (electrical current, electromagnetic or shock wave).

Geophysical exploration involves making measurements at many different locations, either in traverses or grid patterns. Results are interpreted from the variation of the geophysical properties relative to location and, in some cases, with the spacing between the sensing points. Various geophysical methods are employed to determine the depth to rock and probable void zones, cavities, etc. Successful interpretation of the subsurface using geophysical procedures occurs only under ideal conditions. In general, the interpretation of geophysical methods can be misleading unless used in conjunction with other data from exploratory drilling and/or test trenches for confirmation or correlation.

Gravity Survey

Measurement of the gravimetric fields of the earth is a standard geophysical method used to study the structure and composition of the earth. Gravity exploration is useful in the early stages of site investigations to locate areas that are likely to exhibit significant near surface dissolution features or where there are large shallow raveling erosion domes in the soil. An anomaly in a gravity survey data may indicate the presence of dissolution feature, thus warranting additional investigation.

Ground Penetrating Radar

Ground-penetrating radar (GPR) is based on the transmission of repetitive pulses of electromagnetic waves into subsurface material. The pulses are reflected back to the surface when the radiated waves encounter an interface between two materials of differing dielectric properties.

The depth of penetration of electromagnetic energy varies in different lithologies. In sandy materials, above the water table, penetration of more than 30 meters is possible whereas in clay, below a water table, penetration of only 3 to 6 meters may be possible. The data are analyzed and interpreted to determine the depth and the lateral extent of change in subsurface material.

Electrical Resistivity

This method measures the electrical conductivity (inverse of resistivity) of soil and rock. The extent and resolution of material tested depends on the spacing and patterns of electrodes or induction arrays compared to the geometry of the high and low conductivity strata or lenses. The depth and geometry of zones of different conductivity are deduced by empirical or mathematical analysis of data. Cavities filled with air have no conductivity, cavities filled with clay or mud are relatively more conductive than cavities with water. Electrical resistivity methods are useful in estimating the average depth to rock, and usually, the depth to ground water.

Seismic Refraction

Seismic refraction involves the measurement of velocity of a compressive shock wave induced at the ground surface or in a borehole by a hammer blow or a small explosion. The wave velocity depends on the density, modulus of elasticity, and Poisson's ratio of the soil or rock. The type of soil or rock sometimes can be inferred from the shock wave velocity, particularly if the geophysical results are correlated empirically with soil and rock test data from a nearby boring. Interpretation of data is subjective in solutioned limestone with irregular boundaries. This method seldom can detect large cavities in the overburden because the shock waves through the subsurface soil travel faster than through the cavity. It can not define cavities in rock because the higher wave velocity in solid rock above cavities obscures the slower wave return of the cavity. The ability of the method to sense irregularities decreases with increasing depth below the surface.

DESIGN AND CONSTRUCTION CONSIDERATIONS

Planning, design, and construction considerations focus on overcoming difficulties inherent in karst terrain. These difficulties include features and weaknesses of the soil and rock such as sinkholes, solution depressions, troughs and pinnacles in the soil-rock interface, dome cavities in the soil overburden, and collapse-prone caverns in the underlying rock. Existing karst-related problems can be aggravated by changes and activities such as frequent significant seasonal changes in ground water level, excavation, stockpiling, impoundment, dewatering, poor surface water drainage, improper project management during construction, poor maintenance and waste management, and leaking pipes during and after completion of the construction. Offsite activities such as ground water level fluctuations or excessive ground water withdrawal, or excessive soil erosion could act as triggering mechanisms for subsidence or sinkhole development. Therefore, planning, design, and construction in a karst terrain must involve all activities that influence the conditions that aggravate a karst terrain, including the carbonate rock and related overburden voids on site and adjacent property controlled by others.

Site preparation in karst terrain also includes remedial activities to improve and fix any existing sinkhole or solution-related problems, such as voids or cavities or erosion domes that might impact construction, as well as measures to minimize the future development of karst features during the life time of the project.

Additional investigations are warranted before construction operations commence to identify erosion domes, depressions, sinkhole throats and openings in the rock surface and collect information so that remedial steps can be taken prior to construction. Also the potential for collapse of an existing cavity or cavities under the weight of equipment that could delay construction, damage equipment, and, more important injure workers, should be thoroughly investigated.

Altering the surface topography and/or surface drainage by construction activities such as excavation can redirect water flow thereby concentrating it and aggravating downward seepage. Consequently raveling erosion and dome development or reactivation of sinkholes can cause project delay and added cost, or development of karst features on adjacent properties.

It is often necessary to make changes in design and construction in karst terrain because of significant fluctuations of water levels, or the occurrence of irregular but abrupt soil-rock interfaces, or inverted residual soil strength profiles. These changes may result in additional project costs and consequently becomes a "point of contention" among owner-sponsor, designer and contractor.

These conflicts essentially end up in delaying the project and adding expenses, and may jeopardize the entire project. Thus it is prudent to promptly identify, evaluate and correct potential problem areas.

RISK ASSESSMENT

It should be acknowledged that the risks of development in karst terrain sinkhole-prone areas involves unforeseeable site conditions that may need specialized geotechnical investigations to minimize additional construction costs and future problems. A thorough understanding and differentiation between observed facts and interpretation of data is necessary to facilitate decision-making processes regarding technical remedial measures.

The factors that should be included in risk assessment are both the natural and anthropogenic conditions that drive failure mechanisms. Natural conditions to be considered are climate, geology and ground water. Human activities include regional or local environmental disturbances such as excessive ground water pumping, quarry dewatering, diversion of natural drainage, and other construction activities.

Land surface damages may range from subtle movement and subsidence to fissure development and sudden collapse. The results of these occurrences range from minor inconvenience to catastrophic loss and environmental disaster.

CONCLUSIONS

The risk-potential for subsidence and/or sinkhole development in karst terrain can be minimized and/or mitigated by (1) thorough understanding of a geologic/hydrogeologic setting including geologic structure, pattern and interconnection of fractures, solution features and cavities, degree of solutioning, geometry of voids and pinnacles and triggering factors; (2) mitigative measures to minimize natural and anthropogenic factors that drive failure mechanisms, (3) comprehensive planning and design, site preparation, construction and optimal remedial engineering.

BIBLIOGRAPHY

Fisher, J.A., Greene, R.W., 1993, "Roadway design in karst": Proceedings of the Conference on Applied Karst Geology, edited by Barry F. Beck, Balkema, Rotterdam.

Fisher, J.A., Greene, R.W., Ottoson, R.S., Graham, T.C., 1987, "Planning and Design Consideration in Karst Terrain": Proceedings 2nd Multidisciplinary Conference on Sinkholes and Environmental Impacts of Karst, Orlando, Florida.

Jennings, J.E., Brink, A.B.A., Louw, A., and Gowan, G.D., 1965, "Sinkhole and subsidences in the Transvaal dolomite of South Africa': Proceedings of the 6th International Conference on Soil Mechanics and Foundation Engineering, University of Toronto Press, Toronto, Canada, 51-54.

Newton, J.G., 1976, "Early Detection and Correction of Sinkhole Problems in Alabama, with a Preliminary Evaluation of Remote Sensing Applications": HPR Report No. 76, U.S.G.S.

Siegel, T.C., and Belgerie, J.J., 1995, "The importance of a model in foundation design over deeply weathered pinnacled carbonate rock": Karst Geohazards, Beck, B.F., ed., A.A. Balkema, Rotterdam, The Netherlands, 375-382.

Sowers, G.F., 1996, "Building on Sinkholes": American Society of Civil Engineers, New York, N.Y. 10017-2398.

Troitzky, G.M., Tolmachyov, V.V., Khomenko, V.P., 1993, "Sinkhole danger – engineering problem of covered karst": Proceedings of Conference, Applied Karst Geology, edited by B.F. Beck, Balkema, Rotterdam.

White, W.B., and White, E.L., 1995, "Thresholds for soil transport and the long term instability of sinkholes": Karst Geohazards, Beck, B.F., ed., A.A. Balkema, Rotterdam, The Netherlands, 73-79.

Not your typical karst: Characteristics of the Knox Group, southeastern US

JAMES C. REDWINE   Southern Company Services, Incorporated, Birmingham, Ala., USA

ABSTRACT

The Knox Group, a thick package of Cambro-Ordovician rocks, occurs over a wide geographic area in the southeastern U.S. The characteristics of the Knox which are not typical of other karst terrane include strong structural control on porosity and permeability; deep near-vertical solution features; great depth of water circulation; dolomite, rather than limestone, hosting the karstic features; extreme anisotropy and heterogeneity; the regional post-Knox unconformity; and anomalous geochemistry due to mineralization. Many of these characteristics relate to the complex geologic history of the Knox, which includes ancient, as well as modern karstification.

INTRODUCTION

The Knox Group of the southeastern U.S. consists of a thick sequence (up to 1280 meters [4200 feet] or more) of limestones, dolomites, and associated cherts and sandstones (Figure 1). The Knox is an important water-supply aquifer, as well as a host for oil, natural gas, and Mississippi Valley-type ore deposits. As a substrate for diverse land uses, the Knox has received environmental pollutants, and has created numerous foundation problems for roads, buildings, dams, and other types of engineered structures.

The Knox has been extensively studied at its southeasternmost outcrop extent, in central Alabama U.S.A. The raw data for this study came primarily from the Logan Martin Dam site, about 56 kilometers (35 miles) east of Birmingham, which has experienced more than 30 years of foundation leakage and subsidence problems. The Logan Martin data base includes thousands of meters of core drilling, thousands of piezometer readings, grouting records, results of dye tests, photogeologic studies, geologic mapping, water chemistry studies, and other types of information. This data was extensively analyzed, including statistical analysis, and a hydrogeological conceptual model developed for the lower Knox Group. Controls on porosity and permeability were determined, and ground-water flow velocities were calculated. Some of the concepts developed in central Alabama were tested at a site underlain by the upper Knox Group in northwest Georgia.

SOLUTION CAVITY DEVELOPMENT

During the Logan Martin study, 34 cores (2793 meters [9,157 feet] total) from critical areas were logged to provide detailed structural and stratigraphic information. Cores were selected based on location and depth, for example, to investigate lineaments and suspected geologic structures, as well as to expose the maximum stratigraphic interval. Dolomite ($CaMg(CO_3)_2$) is the predominant rock type in the study area, making up about 70 percent of the core logged. Other rock types, in order of abundance, include intermingled dolomite and chert (15 percent); breccias, all types (7 percent); chert (6 percent); sandstone and sandy dolomite (2 percent); and limestone (0.2 percent).

Solution cavity development is widespread, though not random, at the Logan Martin site. Dolomite is the host rock for the solution cavity development. Patterns of cavity development mimic structural features, such as faults, or follow stratigraphic horizons along the limbs of folds. As Figure 2 shows, solution cavities occur to depths of greater than 194 meters (635 feet) below the surface (Elevation -66 meters [-215 feet]), even though the area has only about 122 meters (400 feet) of relief. (Note that on Figures 2, 4, 6, and 7, stations and elevations are expressed in feet.) Groundwater circulation (flow) was detected in one boring to a depth of 148 meters (486 feet) (Elevation -6 meters [-19.5 feet]) with a borehole flowmeter.

Near-vertical solution features, extending a few meters to 46 meters (150 feet) or more, occur at both the central Alabama and northwest Georgia sites. At the northwest Georgia site, one boring drilled on a lineament intersection extended for 46 meters (150 feet) without encountering the top of rock. The boring appeared to be located on an ancient solution feature or doline. The boring penetrated at least 32 meters (105 feet) of stratified silts and clays with plant fossils. The deposition of this material may have occurred at a time when base level was significantly lower than the present, because the deposits occur at depths at least 30 meters (100 feet) lower than a nearby river.

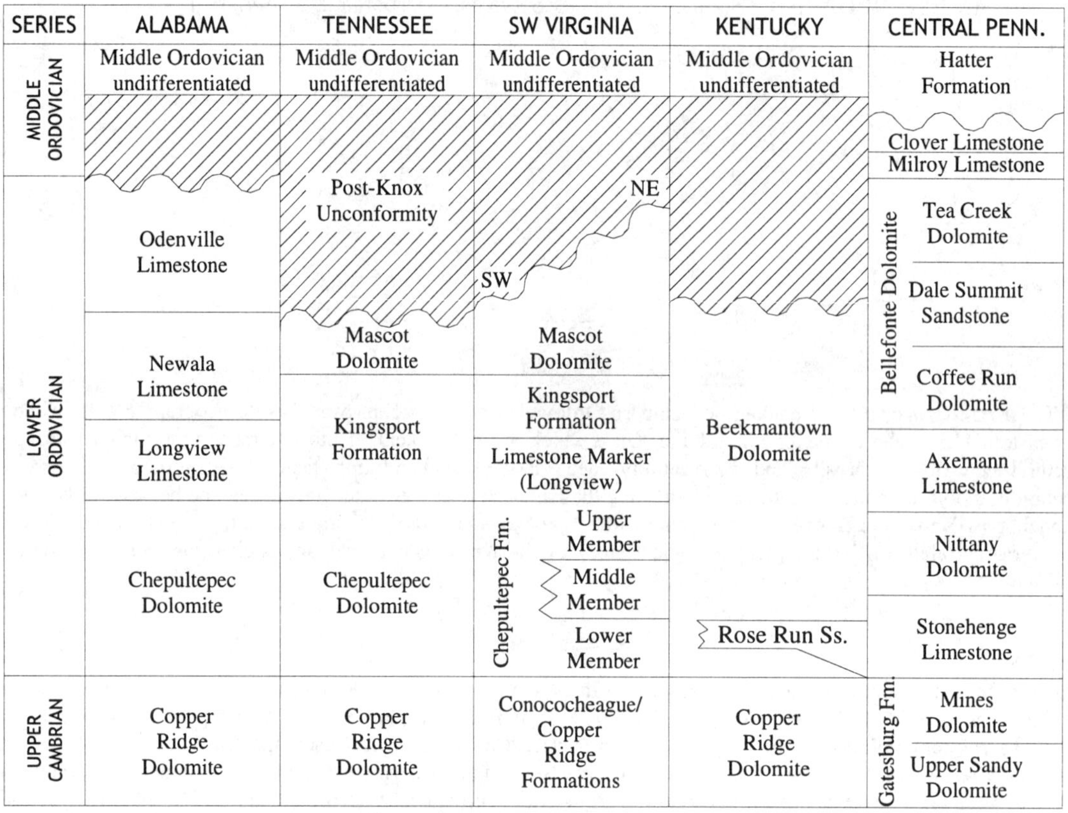

SERIES	ALABAMA	TENNESSEE	SW VIRGINIA	KENTUCKY	CENTRAL PENN.
MIDDLE ORDOVICIAN	Middle Ordovician undifferentiated	Middle Ordovician undifferentiated	Middle Ordovician undifferentiated	Middle Ordovician undifferentiated	Hatter Formation
					Clover Limestone
					Milroy Limestone
LOWER ORDOVICIAN	Odenville Limestone	Post-Knox Unconformity	NE / SW		Tea Creek Dolomite (Bellefonte Dolomite)
					Dale Summit Sandstone
	Newala Limestone	Mascot Dolomite	Mascot Dolomite		Coffee Run Dolomite
		Kingsport Formation	Kingsport Formation	Beekmantown Dolomite	
	Longview Limestone		Limestone Marker (Longview)		Axemann Limestone
	Chepultepec Dolomite	Chepultepec Dolomite	Chepultepec Fm. Upper Member		Nittany Dolomite
			Middle Member		
			Lower Member	Rose Run Ss.	Stonehenge Limestone
UPPER CAMBRIAN	Copper Ridge Dolomite	Copper Ridge Dolomite	Conococheague/ Copper Ridge Formations	Copper Ridge Dolomite	Mines Dolomite (Gatesburg Fm.)
					Upper Sandy Dolomite

Figure 1: Generalized Stratigraphy of the Knox Group, Alabama to Virginia, and Age-equivalent Rocks of Central Pennsylvania (modified from Childs and others, 1984, *in* Raymond, 1993; Bova, 1982; and Parizek *in* Parizek and others, 1971).

CONTROLS ON POROSITY AND PERMEABILITY

Normal Faulting

Several lines of evidence indicate that a normal fault zone oriented N70°E enhances porosity and permeability (Table 1 and Figure 3). As Figure 4 shows, the hanging wall of this normal fault contains about 33 percent cavities, whereas the footwall contains only about 17.5 percent cavities. A statistical analysis of cavity data as well as rock mechanics theory support this observation, that is, greater porosity on the hanging wall (downthrown block) of the N70°E normal fault.

A test grouting program was initiated across the N70°E fault zone. The grouting, piezometric, and other data from this program indicate that the N70°E zone is a preferential flow path for ground water to depths of at least 150 meters (500 feet) below the surface. Weirs on discharge points (boils or springs) provide another method for direct flow measurement. To date, the flow in Weir 27 has been reduced more than 73 percent (from about 2840 to 757 liters per minute [750 to 200 gpm]) due to the present grouting program across the N70°E normal fault zone (Figure 5).

Thrust Faulting

Thrust faults in the Knox Group are identified by zones of well-cemented breccias, microbreccia (cataclasite), intensely fractured rock, and repetition of marker beds above and below the thrust fault zone. Based on packer testing, grout takes, and cavity distribution, thrust faults at depth do not appear to increase porosity and permeability by themselves.

Table 1
Evidence for Enhanced Porosity and Permeability along the N70°E Normal Fault

Major regional lineaments that appear on several sets of photographs and remotely-sensed imagery

Cavernous rock as determined from the original grout curtain and subsequent drilling

Disappearance of rock outcrops aligned N70°E across the Coosa River

Alignment of springs and boils downstream of Logan Martin Dam

Development of chimney-shaped sinkhole in the Logan Martin earth embankment

Original sinkholes in reservoir

Dye and tracer tests results

Historical piezometric high in the vicinity of piezometer 173

Geophysical anomalies along the N70°E trend in Logan Martin Lake

Large irrigation well along N70°E trend southwest of Logan Martin Lake

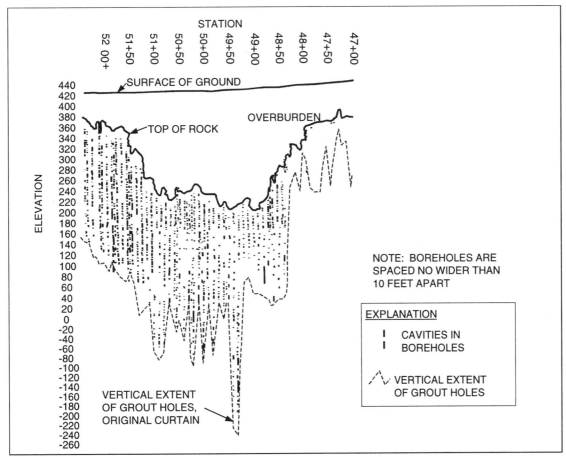

Figure 2: Cavity Section From Station 52+35 to 47+00, Logan Martin Dam.

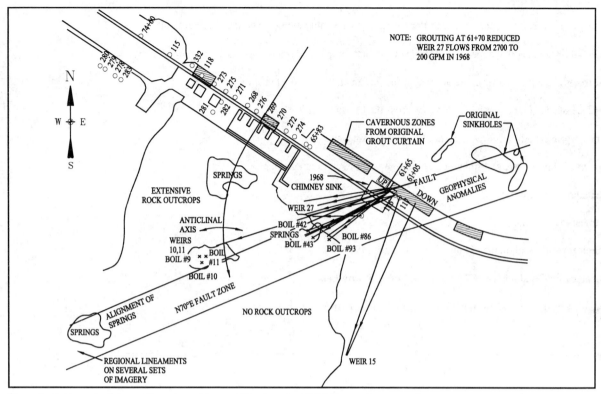

Figure 3: Evidence for Enhanced Porosity and Permeability along the N70°E Normal Fault Zone.

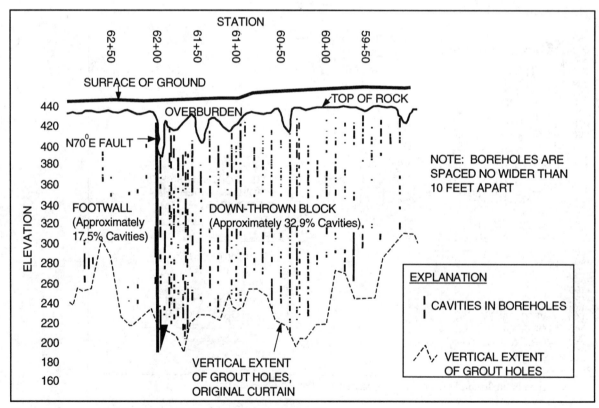

Figure 4: Location of N70°E Fault with Respect to Cavities, Section View.

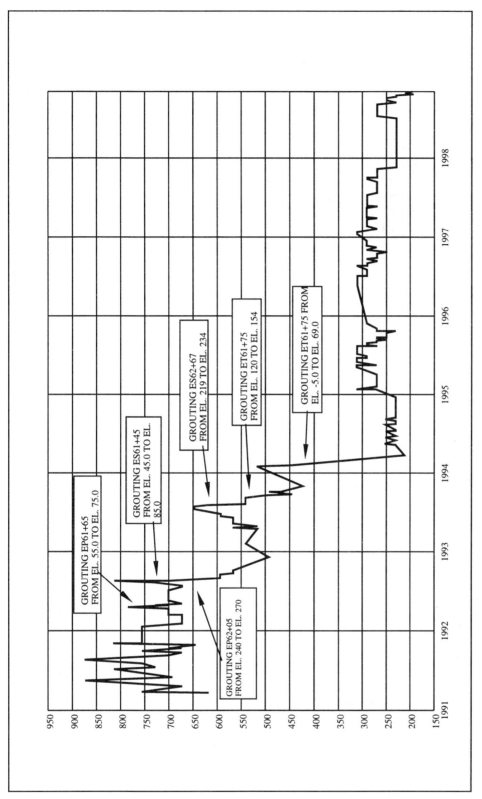

Figure 5: Effect of Grouting on Weir 27 Flows, Logan Martin Dam (After B.E. Williams and R.L. Robinson, 1998)

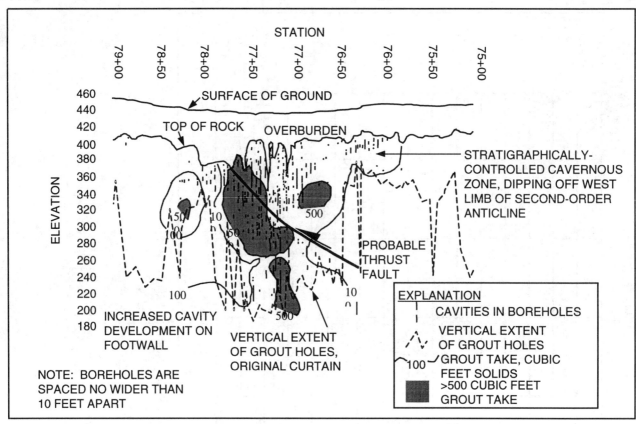

Figure 6: Effects of Near-Surface Thrust Fault on Secondary Porosity and Permeability.

At depths less than about 75 meters (250 feet), thrust faults appear to create secondary porosity and permeability, because water can enter the thrust-related fractures more readily at shallow depths and dissolve the dolomites. Cavity distributions suggest that porosity is preferentially increased on the footwall (Figure 6). Porosity and permeability is also increased along thrusts and associated structures where the younger, tensional normal faults cut them and provide ready access for water.

Stratigraphic Controls

During core logging for this study, a quartz sandstone bed was observed 35 places in core. When the sandstone bed occurred at a depth of 34 meters (110 feet) or less, a cavity occurred immediately below the bed in 78 percent of the observations (seven out of nine times). In contrast, a cavity was associated with sandstone beds deeper than 34 meters (110 feet) in only seven out of 26 occurrences (27 percent). The shallow cavity ranges in thickness from 3 cm to 3.6 meters (0.1 to 11.8 feet), with a median thickness of 30.5 cm (1.0 foot). Figure 6 suggests other stratigraphic controls are operative as well.

A subhorizontal, highly porous (cavernous) and permeable zone occurs from approximately elevation 30.5 to 0 meters (100 to 0 feet) east of the N70°E normal fault, extending from the fault at station 61+78 to at least station 56+45. (Note that dam stationing is expressed in feet). The zone is defined by very high grout takes in the present grouting program; for example, the interval from elevation 21 to 9 meters (70 to 30 feet) has received 6,077 bags of cement and 152 metric tons (168 short tons) of aggregate through boring ES 56+25, and 56,598 bags of cement, 1348 metric tons (1486 short tons) of aggregate and 433 cubic meters (566 cubic yards) of pea gravel through boring EP 56+45.

The subhorizontal orientation of this zone suggests a stratigraphic control on porosity and permeability. Based on core logging, the predominant rock in this zone is cherty dolomite, with algal structures (thrombolites and stromatolites). Another possible control includes thrust faulting, though insufficient deep boreholes exist through this zone to determine the presence of a thrust fault. Figure 7 integrates all the observed contols on porosity and permeability into a hydrogeological conceptual model for the lower Knox Group in central Alabama.

GEOCHEMISTRY

In conjunction with an environmental study at a site in northwest Georgia, a comprehensive literature review on mineralization in the upper Knox was performed. The upper Knox Group hosts economic deposits of zinc, barite, and lead, and can contain other heavy metals such as molybdenum, nickel, cobalt, mercury, silver, and arsenic (Bridge, 1956; Hoagland and others, 1965; Maher, 1970; Saunders, 1992). Occurrences of these metals should not be surprising, particularly in association with brecciation and the post-Knox unconformity.

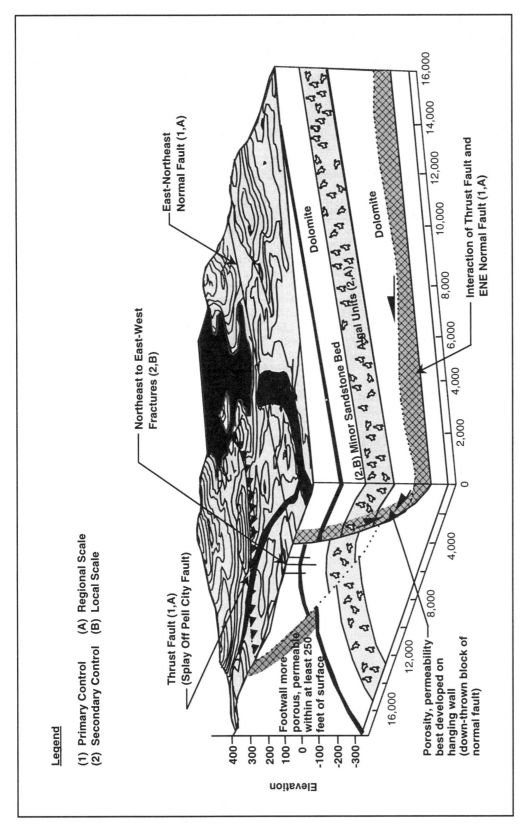

Figure 7: Hydrogeological Conceptual Model, Knox Group Rocks, Pell City Thrust Sheet, Alabama U.S.A.

The post-Knox unconformity separates the Knox Group rocks from overlying, middle Ordovician limestones (Figure 1). This regional unconformity occurs throughout the craton and Appalachian strata, and often shows paleokarstic features (Raymond, 1993). The post-Knox erosional surface marks emergence of the carbonate platform rocks, and is probably due to the combined effects of eustatic sea level fall and shelf uplift that preceded the Taconic orogeny (Roberson, 1988). Volcanism associated with the Taconic orogeny likely served as the source of the metals in the upper Knox Group near the post-Knox unconformity.

CONCLUSIONS

Based on the work performed in central Alabama and northwest Georgia, the following conclusions can be drawn with respect to the Knox Group:

1. The Knox Group contains major solution features, including deep vertical fissures as well as dolines. The solution features can occur in dolomite, as well as limestone.
2. Solution cavity development in the Knox Group dolomites is closely associated with, and mimics the geometry of, structural features such as faults. When scale is taken into account, post-Alleghanian normal faults and Alleghanian thrust faults emerge as dominant controls on porosity and permeability development in the Logan Martin area. Both these features may be represented by lineaments on aerial photography and remotely-sensed imagery in the southern Appalachians.
3. In the case of normal faulting, porosity appears to be best developed on the hanging wall (down-thrown block). In contrast to normal faults, porosity and permeability associated with thrust faults appears to be better developed underneath the fault, that is, on the footwall. Porosity and permeability associated with thrust faulting appears to be best-developed within about 75 meters (250 feet) of the surface, or when thrusts are connected to other permeable features such as normal faults.
4. Deep-seated secondary porosity and permeability, associated with post-Alleghanian normal faulting, can occur in the Knox Group. Even though the Logan Martin area has relatively low relief (about 120 meters [400 feet]), solution cavities occur to depths of greater than 194 meters (635 feet) below the surface (Elevation -66 meters, -215 feet), and groundwater circulation (flow) has been detected to depths as great as 148 meters (486 feet) (Elevation -6 meters, -19.5 feet). It is proposed that extension, associated with rifting and the opening of the Atlantic in the Mesozoic, created the initial deep-seated secondary porosity and permeability which was subsequently enhanced by dissolution of the carbonate rocks.
5. Based on 71 individual dye tests performed in the Logan Martin area, ground-water flow velocities range from 228 to 10,607 meters per day (748 to 34,778 feet per day), with a mean and standard deviation of 3465 and 1738 meters per day (11,361 and 5,699 feet per day), respectively.
6. Heavy metals can occur in the upper Knox Group and its overlying residuum. Documented metal anomalies include barium, lead, zinc, molybdenum, nickel, cobalt, mercury, silver, and arsenic.

ACKNOWLEDGEMENTS

I would like to thank Professors Richard R. Parizek and William B. White, The Pennsylvania State University, for teaching me how to perform the investigations described in this paper. I would also like to thank Alabama Power Company for funding the work and sharing their data. I would also like to thank Joe Abts and Jackie Keyman for drafting the figures, and Dortha Bailey for preparing the manuscript for publication.

REFERENCES

Bridge, J., 1956, Stratigraphy of the Mascot-Jefferson City zinc district: U. S. Geological Survey Professional paper 277, 76 p.

Bova, J. A., 1982, Peritidal cyclic and incipiently drowned platform sequences: Lower Ordovician Chepultepec Formation, Virginia [M.S. thesis]: Virginia Polytechnic Institute and State University, Blacksburg, Virginia, 145 p.

Childs, O. E., Steel, Grant, and Salvador, Amos, project directors, 1984, Southern Appalachian region, Correlation of Stratigraphic Units of North America (COSUNA) project: American Association of Petroleum Geologists Correlation Chart Series, 1984.

Hoagland, A. D., Hill, W. T., and Fulweiler, R. W., 1965, Genesis of the Ordovician zinc deposits in East Tennessee: Economic Geology, 60, p. 693-714.

Maher, S. W., 1970, Barite resources of Tennessee: Tennessee Division of Geology Report of Investigations 28, 40 p.

Parizek, R. R., 1971, Hydrogeologic framework of folded and faulted carbonates *in* Parizek, R. R., White, W. B., and Langmuir, D., 1971, Hydrogeology and geochemistry of folded and faulted rocks of the central Appalachian type and related land use problems: Earth and Mineral Sciences Experiment Station, Mineral Conservation Series Circular 82, College of Earth and Mineral Sciences, The Pennsylvania State University, University Park PA 16802.

Raymond, D. E., 1993, The Knox Group of Alabama: An overview: Alabama Geological Survey Bulletin 152, 160 p.

Roberson, K. E., 1988, The post-Knox unconformity and its relationship to bounding stratigraphy, Alabama Appalachians: Tuscaloosa, Alabama, University of Alabama unpublished M.S. thesis, 148 p.

Saunders, J. A., 1992, Geology of the Shiloh church Mo-Zn-Pb deposit, Polk county, Georgia: Georgia Geological Society Guidebooks, 12(1), p. 81-87.

Williams, B. E., and Robinson, R. L., 1998, Personal communication, Southern Company Services, Inc., P. O. Box 2625, Birmingham AL 35202.

4 Engineering and remedial treatment of sinkholes and other karst features

Detection and treatment of karst cavities in Kuwait

WALEED A. ABDULLAH Kuwait Institute for Scientific Research, Kuwait
MOHAMMAD A. MOLLAH Soil Probe Limited, Toronto, Ont., Canada

ABSTRACT

A comprehensive study consisting of geotechnical, geological and geophysical investigations was carried out in a recently developed desert terrain turned hosing suburb in Kuwait to identify the causes and extent of sinkhole development in the area and to recommend remedial and preventive measures against their reoccurrence. The investigations indicated that the geological profile consists of 31-39m of overburden soil underlain by limestone bedrock. A network of thin interconnected karst cavities were found to exist at depths from 39 to 63 m, within the top 28 m of the limestone bedrock. Analysis of the vertical distribution of cavities indicated that there were three general depths or levels at which cavities were encountered in the area: from 39.7 to 42.6m deep, from 49.6 to 53m deep, and from 56 to 63 m deep, with thickness from 0.2 to 5m. To reduce the risk of sinkhole development in the area, injection of cavities with sand-cement grout was found to be the best economical and most practical treatment measures.

INTRODUCTION

Four major ground subsidences occurred in the form of sinkholes in Al-Dhahar suburb, Kuwait, during 1988-1989. A comprehensive investigative study was undertaken by the Kuwait Institute for Scientific Research (KISR) to investigate the causes of these ground subsidences and to recommend remedial and preventive measures against the reoccurrence of such subsidence. During the course of the investigation, various detection methods of underground cavities were used which included aerial and topographical, geophysical, and geotechnical and geological methods. This paper presents, in brief, a description of sinkhole incidents, a review of the regional geological and hydrological conditions, and a description of the investigation program. The karst cavities encountered and their treatment measures designed are also discussed. A detailed description of the comprehensive investigation study and preventive and treatment plans recommended are found in Abdullah and Mollah (1996), Al-Mutairi, et al, (1998) and Abdullah et al, (1998).

SEQUENCE OF THE SINKHOLE INCIDENTS AT AL-DHAHER AREA

All four of the sinkhole subsidence occurred in the Southwest corner of Sector A-1 of the Al-Dhahar area. Figure 1 shows the locations of the sinkholes (SH-1 through SH-4), labeled in order of occurrence. Table 1 lists data on the sinkhole development, dimensions and present conditions. Details of the sinkhole development history in Al-Dhahar area are available in Al-Mutairi et al (1998).

LOCAL GEOLOGICAL AND HYDROLOGICAL CONDITIONS

The geological profile of interest in this study consists of a sequence of unconsolidated terrigenous sediments of sands and gravel with occasional evaporite phases, which are known as the Kuwait group. The Kuwait group overlies the Dammam limestone formation which is composed of dense dolomitic and numulitic limestone at the base, grading upwards into chalky limestone with irregular dark chert nodules. Burdon and Al-Sharhan (1968) indicated that the Dammam limestone was subjected to extensive erosion and karstification from late Eocene to the end of the Oligocene period to form a sharp unconformity with the overlying Kuwait Group sediments. The Dammam formation crops out in Saudi Arabia and dips gently towards the North and Northeast. The thickness of Kuwait group increases with the general dip of Dammam formation from about 35-40m at the Al-Ahmadi ridge to over 350m in the northwestern part of the country. A general structural high exists below the Al-Dhahar area. This is reflected in the surface topography and thus the term "Al-Dhahar", the Arabic vocabulary for high relief, was used to describe this relatively high area (Al-Refaiy, 1990).

The karstified upper zone of the Dammam limestone, underlying impermeable cemented sand layer, is considered as a good aquifer Mukahopadhyay et al. (1996). (Sayed et al., 1992) indicated that an aquifer system exists within the upper sub-layers of the Kuwait Group on top of the cemented sand layer which acts as an aquitard separating the aquifer system within the Kuwait Group and the aquifer system within the Upper Dammam Formation.

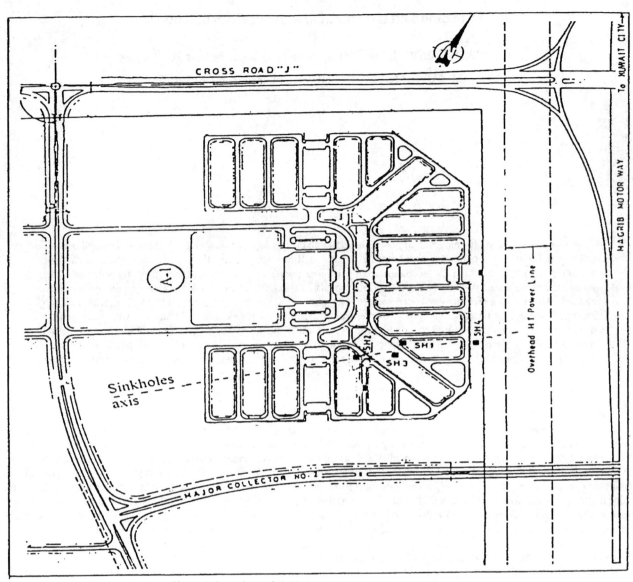

Figure 1. Locations of sinkholes in Sector A1 in Al-Dhahar area.

Table 1. Sinkhole history data.

Sinkhole #	Date day/month/year	Size Width/diameter (m)	length (m)	depth (m)
SH1	17/4/1988	15	----	31
SH2	21/4/1988	4	----	7
SH3	31/10/1988	7	----	9
SH4	5/8/1989	8	2	1.5

INVESTIGATION PROGRAMS

The investigation programs used in the affected area included:

1. Aerial and topographical: The aerial and topographical investigation consisted of reviewing aerial photographs of the area for periods before and after construction, topographic elevation contour maps before and after construction, surface and subsurface drainage patterns, and groundwater flow paths. A reconnaissance survey of the recent geological history of the area was also

conducted in which unofficially reported cases of land subsidence (obtained from the dwellers) that occurred before the development of the area.

2. Geophysical: The geophysical investigation consisted of gravimetric testing using a La Coste Romberg Model-D microgravity meter.

3. Geological and geotechnical: The geological/geotechnical investigation consisted of both field and laboratory testing. Field investigation consisted of boreholes drilling well within the underlying limestone bedrock to recover soil and rock samples for laboratory testing, and to perform in-situ testing, which included Standard Penetration Test (SPT), packer tests and pressuremeter tests. The soil samples and rock cores collected from the field were tested in the laboratory to determine the physical, index and strength properties of the sub-surface soils and underlying bedrock at the site.

A total of 28 boreholes were drilled in the area down to average depths ranging from 40 to 83 m from the ground surface. The borehole locations were carefully selected to cover zones containing different sinkhole development potential risk factors. These factors were assigned after reviewing the microgravity contour maps which indicated the areas of possible existence of underground cavities within the limestone and the overlying soil. Incidents of loss of water circulation and rapid drop downs of drill rods were considered as indicators for the existence of fractures and cavities within the formation. The existence of cavities within a borehole was also confirmed by lowering a video camera lowered inside selected boreholes to show clearly the nature and locations of the cavities at the depths where drops of drilling tools had been recorded. Details of the investigation program including boreholes locations and depths, sampling and coring schedule, and in-situ testing procedure and locations are included in Abdullah et al. (1998).

RESULTS
Lithological profile
Figures 2 present a generalized three-dimensional lithological profile through the investigated area. The profile was drawn by extrapolating the information collected at the specific boreholes to the areas in between, using sound geological and engineering judgment. The distances between the boreholes were drawn as close as possible to represent the actual distances in the field.

The three-dimensional lithological profile reveals that at Al-Dhahar area, the overall geological profile of the Al-Dhahar area consists of 31 to 39 m of overburden, known as the Kuwait Group, underlain by the bedrock, the Dammam limestone Formation. The Kuwait Group is comprised mainly of dense to very dense, predominantly quartz sand, which is further characterized as fine-grained to medium-grained with minimal coarse-grained and/or gravel. These coarse-grained particles are virtually cemented aggregates and are often characterized by calcite precipitation and gypsum recrystallization. The Unified Soil Classification System (USCS) for these soils ranged between an extreme SP-SM (non-plastic poorly graded silty sand) and SC (clayey sand with low to moderate plasticity).

Although the overburden lithology comprises mainly sandy materials with composition and properties varying randomly, a two-zone division, is apparent as follows: an upper non-plastic sandy zone (~20 to 25 m thickness) containing discontinuous bands/lenses/pockets of calcareous and weakly cemented sands which is classified as an upper aquifer in the hydrogeological terms; and a lower zone (~ below 20 to 25 m) that is characterized by being more consistently plastic, and weakly to moderately cemented which is classified as an acqiutard in hydrological terms.

The underlying Dammam Limestone Formation varied from an upper (depth varies depending on location with a maximum of 67 m) karstified, weathered, shaley siliceous dolomitic limestone with irregular chert concentrations, containing often fosiliferous infillings, to a lower massive compact, sometimes chalky, limestone porous, and often fossiliferrous, shaley limestone to massive limestone as depth increases. In many zones, rocks are compact and massive, and re-crystallized, sometimes dolomitized. Vugs are intensive with apertures ranging between 2 and 4 cm although sometimes these range between 5 and 10 cm (moderately wide). Consequently, infillings are found to accumulate within the apertures which comprises a mixture of coarse and fine soils in less weathered zones, but becomes increasingly finer and plastic with increased weathering. Available records indicate that the area is extensively faulted and shear zones are common (Al-Rifayi. 1990). It is, therefore, likely that the weathering is extensive along the faulted zones.

The soil-rock interface contains dark chert nodules and sandstone. The latter is composed primarily of quartz although other minerals, e.g., feldspar, calcite, opaques and heavy minerals are also present. Some samples also show mud matrix, purple red in color (possibly stained with iron oxide), and calcite cement. Considerable irregularities in the rock heads are indicated by the field observations of varying depths to rock surface in borings drilled in close proximity.

The groundwater table was encountered during drilling at depths of 32.5 to 39.5 m from the ground's surface, varying around the soil-rock interface. No water table was encountered within the upper Kuwait Group in the Al-Dhahar area.

Karst cavity existence
During coring incidents of loss of circulation accompanied by frequent drop down of drilling tools were experienced at various depths within the upper layers of the limestone bedrock in several boreholes. These incidents were recorded and documented in the lithological profiles and were interrupted as an indication of the existence of cavities within the limestone bedrock.

Based on the extent of the rapid drops of drilling tools recorded, the karst cavities encountered ranged in thickness from 0.2 to 5 m, and were generally located at depths from 38.9 to 63 m, within the top 28 m of the limestone bedrock. Analysis of the vertical distribution of cavities indicated that there were three general depths or levels at which cavities were encountered in the area: from

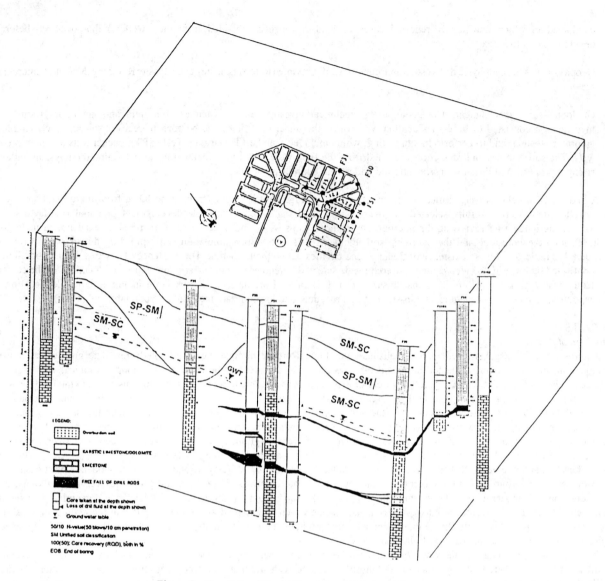

Figure 2. Generalized three-dimensional lithological profile.

39.7 to 42.6 m deep, from 49.6 to 53 m deep, and from 56 to 63 m deep. Taking this findings into consideration, it might be reasonable to conclude, as shown in Figure 2, that the cavities encountered in a borehole are connected to other cavities encountered in other boreholes within the same depth or level, given that the boreholes are within the same anomaly. A further analysis of the cavities' distribution lead to the conclusion that thin, interconnecting tunnel networks can be assumed to prevail for the cavity system in the Al-Dhahar area which makes it likely that one or more boreholes would fail to hit the network if they were drilled just away from the cavity's tunnel.

Confirmation of the existence of cavities was also established by lowering a video camera inside two boreholes. The video camera identified the depths within the borehole profile where cavities exist which correlated well with the depths where drop downs of drilling rods have been recorded. A careful review of the video tape records also confirmed the thin inter-connecting tunnels network nature of the Karst cavities system.

In-situ testing

Single packer tests were performed between depths of 30 and 40 m from the ground's surface to investigate the permeability of the strongly cemented layer covering the karst cavities at the top of the limestone bedrock. The permeability coefficient, k, along the depth of one borehole. The tests showed that k for the strongly cemented sand layer, varied between 0.2×10^{-6} m/sec and 1.2×10^{-6} m/sec which classifies the layer as having poor drainage to being impervious (Lee et al., 1983). This finding confirms the geological profile described above for the Kuwait Group which describes this layer as an acquitard dividing the deeper Dammam limestone formation acquifer and the upper Kuwait Group acquifer. The very low values of k imply that the flow of water in this layer is dependent largely on the distribution and volume of macro-pores (fractures) in the rock formation instead of primary porosity (micro-fabrics). The former act as preferential pathways for the instant movement of water, causing loss of circulation.

DISCUSSION
Sinkhole Development

Sinkhole development in the Al-Dhahar area is caused by movement of the soil cover downward into the karst cavities at the top of the limestone bedrock. A major triggering mechanism for the soil movement was the continuous downward flow of water from irrigation and leaking water supply pipes through the uncemented sand layer into the cemented sand layer at the bottom of the Kuwait Group. The lower cemented calcareous sand layer, being of low permeability coefficient, acted as an aquitard upon which the downward seeping water accumulated creating a perched water table. This resulted in a gradual increase in the moisture content, and, therefore, increased the weight of the cemented sand layer bridging over the karst cavities. The increased moisture in the cemented sand also caused softening and/or dissolution of the cementation bonds, therefore, reducing the cohesive strength of the cemented layer. Lowering of the groundwater table, which was believed to have been in operation in close vicinity of the area, to levels below the bottom of the cemented sand layer may have caused a reduction in the buoyant upward pressure from the groundwater table and, therefore, increased the effect of the added weight on the cemented sand layer. Also the lowering of the groundwater table may have induced water erosion of the cemented sand layer, due to the hydraulic head between the upper and lower water tables. Ultimately, due to the combined effect of downward seeping water and the lowering of groundwater table, the cemented sand layer was no longer able to bridge over the karst cavities at the top of the limestone leading to collapse of this layer into the cavities and consequently raveling of the overlaying uncemented sand layer leading to the development of a sinkhole at the ground surface.

Karst cavities treatment

To reduce the risk of future sinkhole development at Al-Dhahar area, soil movement downward into the karst cavities has to be stopped. This can be implemented by adapting either one of the following three measures: i) sealing the soil-rock interface, ii) compaction grouting, or iii) cavity filling. Due to the poor drainage characteristics of the cemented soil overlaying the karst cavity layer, classified as being impervious with a permeability coefficient (k) between 0.2×10^{-6} m/sec and 1.2×10^{-6} m/sec, sealing the soil-rock interface would only be feasible using expensive chemical grouts which was forecast as uneconomical for the project. Since the soil covering the karst cavities was described as dense to very dense with SPT values not less than 50, compaction grouting was concluded not to be feasible. Analysis the nature of the karst cavities encountered in the area: being at depths from 38.9 to 63 m, distributed at three general depths or levels, and having the form of thin, interconnecting tunnels network, indicated that the most economical and practical treatment measure that can be applied successfully is filling of the cavities in upper level of the three cavities levels, from 39.7 to 42.6 m deep with a cement/sand/mortar mix. A thorough and detailed analysis of the treatment measure selected is available in (Al-Mutairi, et al. 1998).

ACKNOWLEDGMENTS

The work described in this paper forms part of the project 'Evaluation of sub-soil conditions in Al-Dhaher area, which was carried out by KISR, as requested by the Council of Ministers, Kuwait. The authors gratefully acknowledge both KISR and Council of Ministers who have granted permission to use the data in this paper. The authors are indebted to Dr. Al-Mutairi Al-Mutairi, Acting Director, Engineering Division and, Manager, Civil and Building Department, KISR, for critically reading the initial manuscript and giving a number of valuable suggestions.

REFERENCES

Abdullah, W., Al-Mutairi, N., Mollah, M. Al-Fahad, F., and Mussallam, H., 1998, Subsurface exploration study in the Al-Dhahar area: Final Report, Vol. II, EB001K, Evaluation and Treatment of Underground Cavities at Al-Dhahar Area.

Al-Mutairi, N., Eid, W., Abdullah, W., Misak, R., Mollah, M., Awny, Randa, and Al-Fahad, F., 1998, Evaluation and treatment of subsurface conditions at Al-Dhahar area: Final Report, Vol. I, EB001K, Evaluation and Treatment of Underground Cavities at Al-Dhahar Area.

Al-Rifaiy, A., 1990, Land subsidence in the Al-Dhahar residential area in Kuwait: a case history study: Quarterly Journal of Engineering Geology, London, 23:337-346.

Burdon, D. G., and Al Sharhan, A., 1968, The problem of paleokarstic Dammam limestone aquifer in Kuwait: Journal of Hydrology, 6: 385-404.

Mukhopadhyay, A., Al-Sulaimi, J., Al-Awadi, F, and Al-Ruwaih, F., 1996, An overview of the tertiary, geology, and hydrology of the northern part of the Arabian Gulf region with special reference to Kuwait: Earth and Science Reviews. 40: 259-295.

Sayed, S. A. S., Saeedy, H. S. and Szekely, F., 1992, Hydraulic parameters of a multilayered aquifer system in Kuwait City: Journal of Hydrology, Amsterdam, 130: 49-70.

Engineering approaches to conditions created by a combination of karst and faulting at a hospital in Birmingham, Alabama

TONY COOLEY   Department of Geological Sciences, University of Kentucky, Lexington, Ky., USA

ABSTRACT

Foundations for a major expansion and modification of a multi-story hospital in Birmingham, Alabama were founded on faulted and karst dissolutioned dolomite. The foundation approach had to accommodate a high degree of uncertainty concerning local conditions due to limited access for exploration and extremely variable rock conditions. The scope of the construction included excavation of a sub-basement into rock with associated tiebacks to support adjacent foundations, installation of rock-bearing shear walls and rock anchors under the existing hospital, and installation of rock bearing caissons and wall foundations outside the existing hospital. Local complications included areas of highly shattered rock, a generally pinnacled rock surface with 3 to 6 meters (10 to 20 feet) average relief, locally very deep cutters and pits, areas where dolomite was weathered to sand or weak rock up to 3 meters (10 feet) thick, and pockets of flowing sand and mud near the rock surface.

Because of the complexity of site conditions and limited initial access to the site, on-site geotechnical services required innovative approaches to gather additional information on the highly variable and ambiguous rock conditions and adapt detailed foundation design and foundation approaches to the actual conditions encountered. These approaches included triple tube coring of shattered rock at selected caisson locations; development of a technique for installation of rock anchors into shattered rock, determination of required undercut depths and remediation at individual foundations where rock was shattered, disaggregated, or steeply pinnacled; characterization of individual cutters by airtrack probing for remediation information in wall foundations; low-angle coring for cutter characterization in the tieback area; change in foundations from walls to caissons or caissons to mat foundations in select areas; and above all, careful judgment-based design. Limitations of characterization methods are also discussed. A fundamental understanding of karst processes and three-dimensional conceptualization was an essential part of the engineering required for this project.

INTRODUCTION

In 1984 and 1985 the VA Medical Center in Birmingham, Alabama was both expanded and was stiffened to resist potential earthquake damage. The Medical Center was originally built in 1951 of concrete construction with rock-bearing caisson foundations. Later expansions were also founded on rock. The new construction consisted of an expansion area founded on caissons and rock bearing walls, a sub-basement excavated immediately adjacent to and below the bearing elevation of a three story section of the hospital that required lateral support of existing foundations, and the construction of shear walls with rock anchors in both old and new portions of the hospital for earthquake resistance.

GEOLOGY

The hospital is located in the northeast-trending Birmingham Valley of the Valley and Ridge geomorphic province in the thrust-faulted southern Appalachian Mountains. It is underlain by the Ketona Dolomite of Cambro-Ordovician age that is generally a medium to thick bedded, relatively pure dolomite that dips about 20° to the southeast at the site. This dolomite is relatively strong as an intact rock with unconfined compression strengths commonly 70 to 100 MPa (10,000 to 14,000 psi). Bedding is indistinct and the bedding planes are relatively strong. However, a combination of faulting and dissolutional weathering has produced highly variable foundation conditions that required special engineering approaches.

Dissolutional weathering has produced a pinnacled rock surface on the relatively unfaulted dolomite areas. Figure 1a. shows typical conditons encountered. The general relief of the bedrock surface is commonly about 3 to 4.5 meters (10-15 feet) with steep-sided pinnacles sloping 60° or more. Many nearly vertical cutters extend below this general pinnacled surface to depths of 25 meters (80 feet) or more. At the site, the dominant cutter direction was skewed about 15° east of the strike of the bedrock. Most dissolution is concentrated in these cutters (called "slots" by local engineers), but some solution channel development along bedding planes also occurs. The rock in the pinnacles and between cutters is generally intact and strong except near the upper portion of the rock. At many locations, this upper rock has disaggregated to a dense dolomitic sand that can often be excavated by a hand shovel or rock pick. This

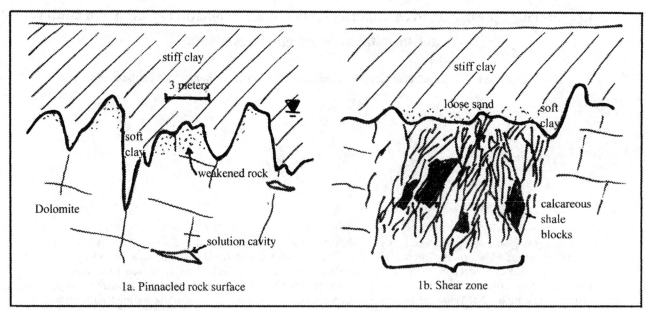

Figure 1 Conceptual illustrations of pinnacled and shear zone areas

disaggregated material varies in thickness from a rind of a few millimeters to locally more than 3 meters (10 feet) with an abrupt transition to intact rock. The pinnacled rock surface is overlain by a silty clay residual soil commonly of about 4.5 to 6 meters (15 to 20 feet) thickness which tends to be stiff to very stiff in consistency except in the cutters, where it may be soft to very soft. Ground water is commonly encountered at a depth of 3 to 5 meters (10 to 15 feet) and affects construction of caissons and rock walls, but usually has little effect on large excavations beyond an inital pumping phase due to limited storage in the rock.

Cutting through the site at about 15° east of the bedding strike is a sub-vertical shear zone about 9 to 30 meters wide (30 to 100 feet). Conditions in this area are illustrated in Figure 1b. This area has many areas of shattered rock with an abrupt transistion to the relatively unbroken areas to either side. The rock in the shear zone is relatively uniform in depth without pinnacling and cutters, but does include local areas of disaggregated dolomite sand and soft mud. Steeply dipping blocks of dark blue calcareous shale are sometimes found in the shear zone that may have weathered to a brick-red clay, while the bedding of the dolomite itself is also steeply dipping. The crushed zones are commonly very permeable to ground water.

At the interface between the shattered zone and relatively intact rock, two features formed by relatively concentrated solution were encountered, one a circular pit filled with clay over 9 meters (30 feet) deep (excavation did not reach bottom). These are interpreted to form at this location as a result of the different water regimes in shear vs. intact rock areas. This was discussed in Cooley (1993).

ENGINEERING APPROACHES
Introduction

Engineering the foundations for this sort of site required dealing with highly variable conditions. This involved three components: 1) the initial exploration, including borings, combined with the accumlated local experience in this formation, 2) use of conservatism in design recommendations to allow for common adverse conditions to reduce the amount of redesign required by encountered condtions, and 3) careful on-site field engineering to customize foundations to the observed conditions. At the outset, the engineer's mindset must incorporate the understanding that a large degree of ambiguity is inevitable in karst due to its complexity and the limited information that can be gained from borings. As an example, two borings 20 centimeters apart (8 inches) encountered rock at elevations differing by about 1 meter (3.5 feet) and could easily have differed by 6 meters. The faulting further complicated conditions in the shear zone because of the erratic clay blocks within the zone. An HQ triple-tube core boring in the center of a 3-foot diameter caisson was insufficient in itself to characterize the foundation conditions for that caisson. It was only by combining specific on-site observations during construction, experience elsewhere in this formation in Birmingham, the available boring information, and an engineering approach that used design that would accomodate the worst likely conditions that at reliable foundation structure was attained. Over-reliance on data gathered prior to construction for design would have resulted in a rude awakening should each boring have been considered to represent its vicinity. In such an area, all information must be considered simultaneously as a sampling of overall conditions, recognizing that any single boring may be very misleading as to its immediate surroundings.

The data gathering methods available at this site were the inital core borings prior to construction and supplemental core drilling, some using an HQ triple-tube core barrel, low angle core drilling, air track and hand drill percussion probing, and direct observation. Initial core borings were very limited by poor accessibility. Existing facilities and hospital activities required drilling to be confined to open parking areas drilled at night. Even with this, the shear zone was encountered in one boring and the recommended bearing capacity for that area reduced to allow more options for conditions encountered during construction. Once construction began and improved access to certain areas was obtained, the addtional exploration was done as needed for both pre-planned exploration, such as the low angle core holes in the sub-basement, and exploration added in response to observed conditions. That additional exploration during construction that was required in response to observed conditions was anticipated and represents the most rational response to the highly variable conditions. Apart from the limited access problem, this exploration during construction targeted very localized features that could not have been known to exist prior to construction and would not reliably have been located by any practical pre-construction exploration program.

Problem areas

Three different geologic domains existed at this site with different engineering characteristics: 1) pinnacled areas of generally intact rock, 2) the shattered and erratic rock of the shear zone, and 3) the concentrated dissolution features at the interface between the intact rock and shear zone areas. The foundation challenges differ between these areas as do the luck of the draw concerning the foundations that had to placed in each area. In the pinnacled areas, the problems related to installation of caissons, shear walls, and tiebacks as these interacted with the cutters and other dissolution features. In the shear zone, the problems related to installation of caisson and shear wall foundations and the required tension anchors in the shattered rock. At the transition zone, problems with very deep rock and lack of lateral support were encountered.

PINNACLED AREAS
Caissons

In pinnacled areas, a major problem is uncertainty of the required depth to bearing materials and difficulty obtaining a secure bottom. Figure 2 illustrates the problems. To address the uncertainty of caisson depth, the caisson contractor commonly uses telescoping steel casing, beginning the hole oversize through the soil, casing that about 30 cm. (a foot) oversize, continuing 15 cm. oversize (6 inches) and then finally stepping down to design size if needed. The caisson drill commonly encounters rock on one side of the hole 1 to 25 meters (3 to 80 feet) before it would be encountered on the other side of the hole. Fortunately, the strength of the rock in the pinnacles allows completion of the caisson bearing surface as a partial bottom as long as at least 66% of the diameter of the bottom is strong rock and probing of the bottom does not reveal weak rock or voids. Prior to probing, all soil is cleaned from the bottom of the caisson excavation. This probing is typically done using a hooked rod "feeler bar" probing the sides of holes 2 to 3 meters deep (6 to 10 feet) drilled with a hand air drill by the contractor. The field engineer directs the location of the probe holes, probes them with the feeler rod, and directs the drilling of additional holes if judged appropriate. When the rock is relatively intact, the sides of the hole will be fairly smooth and an experienced field engineer can reasonably judge conditions. Additional drilling is required if rock is not judged sufficient in quality or quantity, is underlain by voids, or the water inflow cannot be sufficiently controlled.

Shear walls

Shear walls were present both beneath and outside the existing building. The excavations for the shear walls were shored excavations about 1.2 meters (4 feet) wide and 3 to 6 meters (10 to 20 feet) long, representing the bay width between caissons. Beneath the building, these were hand excavated in the utility crawlspace and were evaluated by probing holes drilled with a hand air

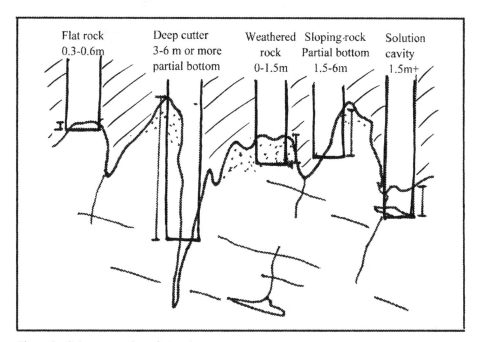

Figure 2. Caissons on pinnacled rock

drill for exploration and for rock anchors. Additional holes were drilled if needed by the engineer for evaluation, including at angles to the vertical. Often holes were drilled at an angle below the bounding existing caissons to check on their bearing depths. The most difficult problem of interpretation was evaluating the significance of voids found in the probe holes. Because rock excavation was being done by hand without explosives under the building, it was important to assess carefully what rock needed to be removed due to voids. Many voids on excavation proved to be narrow, vertically oriented features of little bearing capacity signficance. Small dissolutionally enlarged features (minicutters) were "healed" for foundation purposes by removing the soil fill for a depth of twice their width and a concrete plug used to bridge them by arching stresses to the sidewalls of the feature. Features strongly skewed to the excavation may require widening the excavation locally for appropriate abutments for the plug. The bearing requirements of the shear walls had been reduced during the original design phase to allow accomodation of imperfect bearing conditions.

Shear walls outside the existing building were evaluated similarly, except the additional option of checking select areas for depth to rock using a airtrack prior to excavation was used. This would have allowed substitution of caissons if required in some areas. This probing displayed a disadvantage of airtrack drilling in steeply pinnacled rock, however. The thread type used on airtrack drill rods is very flexible and tolerant of deviations. When a cutter was actually excavated that had been probed with the airtrack, it was significantly narrower than indicated by assuming vertical holes. The airtrack rods had followed the steep walls of the cutter some distance before biting into the rock, giving a misleading result. Another larger cutter was excavated and cleaned out using high pressure fire hoses and Wilden pump for placement of a bridging concrete plug and an airtrack drilling angle holes used to evaluate the continuity of the remaining rock. The drilling behavior of the airtrack, as well as probing the holes with a feeler rod, was used to assess the foundation conditions.

Tiebacks

Tiebacks were needed both to retain the soil above rock along the outside perimeter of the subbasement excavation and as a permanent support for existing foundations along one side of the excavation. The impact of pinnacling on the soil tiebacks was limited to bending the steel H soldier piles as they encountered the steep pinnacles at the bottom of the wall and additional hand work at the bottom of the walls to retain soil at the irregular surface. The tiebacks themselves were shallower than the pinnacles and temporary and so were not affected by the pinnacles. The permanent tiebacks for support of the existing foundations were another matter. These were placed into rock and the pinncales had to be assessed in their placement.

As seen by Figure 3, these tiebacks supported a pair of supplemental piers placed opposite each existing caisson to replace lateral support removed by the excavation. The tiebacks were about 7.5 to 9 meters (25 to 30 feet) long and were individually stressed to 550 to 690 MPa (40 or 50 tons). Rock bedding dipped about 20° into the cut face. An analysis indicated that a cutter present within the tensioned zone would result in an inward failure of the existing caisson toward the cutter. Alternately, a cutter near to and nearly parallel to the anchor zone of the tieback would potentially compromise its pullout plug by allowing a lateral pop-out into the clay-filled cutter. To evaluate the conditions for tiebacks, a planned coring program was done; 8 vertical borings located one per caisson immediately outside the existing building and 13 additional low angle core holes when excavation reached the level at which the first row of tiebacks would be installed. Slope inclinometer casing was placed in the vertical borings to check for movement during construction and following loading of the tiebacks. The low angle borings were drilled by a skid rig with a rotating head to core a series of holes at angles of 20 to 32° to the horizontal. The holes were sited and evaluated in the field by the field engineer, the results of each boring evaluated and incorported into the location and trajectory of the next boring. Unfortunately, when more than one soil-filled cavity was encountered in each boring, many ways to connect the dots and interpret the results were present. The clay-filled voids were not unique markers; one cutter did not differ in its filling from another. Careful three-dimensional thinking was required to interpret the results, as well as reference to observations on cutters elsewhere on the site. The trajectory of the tiebacks at one caisson was modified to avoid one identified cutter, the remaining tieback trajectories were OK as originally planned.

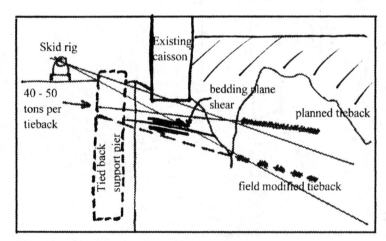

Figure 3. Tieback exploration coring

SHEAR ZONE
Caissons

Caissons in the shear zone were a major problem. Fortunately, the depth to rock was shallow enough that water inflows were not a problem as shattered rock commonly is very permeable and shutting off water for bottom preparation is commonly a limiting factor on depth in the Birmingham area. However, evaluation of the bearing capacity could not be done by the normal methods used in the pinnacled area. Drilling in the shattered rock with an air percussion drill was difficult, with significant chance of hanging up the drill steel, and the resulting hole was too rough to evaluate with a feeler rod. Should a drill steel or feeler rod be trapped in the hole and not be recovered, this would greatly complicate further removal of rock. The highly fractured condition of the rock made it very vulnerable to bearing failure due to punching into a deeper cavity, say one produced by mixing corrosion at a past water table, or by lateral failure of fractured rock into a weathered shale block by extruding its clay upward. The design bearing capacity of 275 Mpa (20 tsf) had been selected for this area in anticipation of problems, but the difficulty of assessment required an additional response.

It was decided to core drill each of the affected caisson locations using an HQ triple-tube wireline corebarrel. Because the shattered rock was too unstable to allow raising the core bit for adding drill rods and we lacked a sufficient quantity of 5-foot-long wireline H rods, a Mobile drill rig with a 2.4 meter (8 foot) long clearance between power head and ground was brought to the site. This was then jacked up on timber packs an additonal 0.6 meters (2 feet) at each hole to allow addition of 3 meter (10 foot) drill rods while lifting the bit less than 2 to 5 cm off the bottom. Very careful drilling technique was used to maximize recovery and the drilling behavior was also watched for evidence of voids. Generally good recovery was attained even in shattered rock once below the top few feet of disaggregated dolomite sand. No deeper voids were encountered, but pockets of steeply dipping calcareous shale or red clays derived from weathering of those shales were encountered. Because caisson installation was going on during the day and the area was an excavation with limited room, the caisson rig had to be removed each night, coring was done through the night, and the caisson rig returned the next day for the next caisson. Interpretation of the core was done to set the planned termination depth for the caisson and this was cleaned for observation. Several times, after cleaning, evidence of a clay block was seen in the excavation that was not encountered by core boring in that caisson and the hole had to be deepened or otherwise modified to adjust for that. Conditions were very erratic in that shear zone, changing within 30 cm (one foot) or less due to the incorporated clay blocks. The unweathered calcareous shale was suitable for the bearing conditions, the problems came when the weathered red clay was encountered.

Shear walls

In part of the area, shear walls had been planned for an elevator tower. In this area, a mass excavation to expose the top of the rock was done to allow use of an unreinforced mat to bridge flaws and spread the load out below the walls. This revealed a modest-sized pocket at the boundary between the shear zone and adjoining intact rock that was dug out for placement of a concrete plug to arch across it. The greater excavation actually saved time and expense by expediting the work and limiting rock excavation.

Remedial measures for residential structures damaged by sinkhole activity

RAMANUJA C. KANNAN Williams Earth Sciences, Incorporated, Largo, Fla., USA
N. SANDY NETTLES N.S. Nettles and Associates, Incorporated, New Port Richey, Fla., USA

ABSTRACT

Residential structures damaged by sinkhole activity have been traditionally repaired by either compaction grouting or underpinning. The method used to implement such remedies has always resulted in additional settlement and further damage. Based on a review of damage to residential structures and repairs effected in the Tampa Bay area, the authors recommend alternative methods of structural repairs. These methods aim at retarding the solution activity in the limestone, stabilizing the overburden soil and then effecting the structural remedies. Investigation of other successful methods of stabilization is also recommended.

CASE HISTORIES ON SINKHOLE DAMAGE

The Tampa Bay region in Florida has attracted some attention for the past two decades as an area of accelerated sinkhole activity. This region encompasses a five-county area, including Hernando, Hillsborough, Pasco, Pinellas and Polk counties. More sinkholes are reported in these five counties than any other metropolitan region. The primary reason for this phenomenon is two fold. Firstly, this region is one of the fastest growing areas of the state and there has been a steady draw-down from the Floridan Aquifer. Secondly, there are more published reports of sinkholes concentrated in this region, resulting from studies conducted by the former Florida Sinkhole Research Institute. The awareness of sinkholes has resulted in more people reporting sinkholes and filing claims against their insurance policies. The resulting investigations conducted by the insurance companies has documented numerous cases of sinkholes, especially in Pinellas and Pasco Counties. However, as these investigations have a financial consequence to the insurance companies, the results of the investigations have not always been consistently interpreted as sinkholes. Thus there is a perceived increase in the number of cases of reported sinkholes, and perhaps two thirds of the investigated cases are actual sinkholes. In any case, the fact remains that there is an increased level of sinkhole activity.

Since the early reports of sinkhole damage to structures in Dunedin in Pinellas County emerged about two decades ago, an average of ten new sinkholes are reported per square mile, per year in the five-county Tampa Bay region. These range from localized damage to single homes to large sinkholes that affected entire subdivisions. There have been some dramatic ones in the last two to three years, one involving a gypsum stack in Hillsborough County and the other, more recent one involving a residential subdivision in Hernando County. In the Hillsborough County case, the sinkhole was stabilized by extensive grouting. The impact of gypsum entering the aquifer was also a topic of study. This case study has been extensively documented (Fuleihan, Cameron & Henry, 1997). However, in the case of the residential subdivisions, recommendations for remediation varied from doing nothing, strengthening existing foundations and taking pro-active measures in foundation design prior to construction (Mishu, Godfrey & Mishu, 1977).

SETTLEMENT STUDIES ON RESIDENTIAL STRUCTURES

In the Tampa Bay region, the authors have been involved in over sixty cases of residential structural damage due to reported sinkhole activity in the past two years and more are scheduled for the near future. When the homeowners made claims against their insurance carriers for repairs, the insurance carriers conducted extensive studies to the extent required by law, to establish the validity of the claim. In most of these cases extensive studies have been conducted using available geophysical and engineering methods. The outcome of such studies, though not open for discussion with specifics, fall into three categories:

(1) Settlement is not caused by sinkhole activity.
(2) Settlement is caused by sinkhole activity.
(3) There is an irreconcilable difference of opinion.

Settlement is not caused by sinkhole activity

When the homeowner files a claim, the insurance carrier engages the services of a geotechnical engineering firm to conduct subsurface exploration to determine the cause of the settlement. Typically, the subsurface exploration consists of conducting two to four standard penetration test (SPT) borings, a few hand-augered borings, one or more test pits to determine the size of the footings and some laboratory tests for soil classification. The cause of settlement is determined to be unrelated to solution activity in the limestone formation, even if solution activity is found to exist; and is related to problems near the surface. Settlements are attributed to shrink-swell clay, organic matter, presence of trees and poor construction.

Settlement is caused by sinkhole activity

A direct relationship between solution activity and settlement is established on basis of SPT tests and geophysical surveys, which may include ground-penetrating radar (GPR) or electrical resistivity (ER) surveys. If the geotechnical engineer interprets the soil profile to indicate a raveling zone, whether or not confirmed by the geophysical tests, the settlements are related to sinkhole activity. Settlement of the claim may involve remediation or condemnation of the structure, depending on the economics. If remediation is agreed upon, the geotechnical engineering firm typically recommends the use of compaction grouting or the use of some type of under-pinning system. In the experience of the authors, both remedies fail, because the settlement cracks reappear. This is discussed in a latter section.

There is an irreconcilable difference of opinion

The opinions expressed by the insurance carrier's geotechnical engineer are disputed by the owner. This happens when the owner is aware of sinkhole activity in the vicinity and has a neighbor who either obtained a financial settlement or the neighbor's structure stabilized by grouting, which affected the claimant's structure. In such cases the same tests are conducted by two to four geotechnical engineering firms, each reaching a different conclusion on the cause of the settlement. Eventually, the relationship between the settlement and sinkhole activity is decided by a jury, and the more persuasive panel of experts prevail. About ten percent of the cases the authors have handled in this category have ended in a jury trial, and of those, one case was decided in favor of the insurance carrier. That decision too, was reversed on appeal. After the trial, the insurance carrier's options are to either remedy the structure or to condemn the structure. If repaired the residence becomes un-insurable, and therefore the owner opts for a condemnation. The insurance carrier has in a few cases auctioned the residence to be repaired and used as rental property. As mentioned earlier, this is a very small number.

REMEDIAL SOLUTIONS

If the cause of the settlement is related to sinkhole activity, irrespective of the effect of such activity, remediation is attempted. Presence of a sinkhole or solution activity that may potentially cause sinkhole formation, is sufficient cause to attempt remediation. The most commonly used methods of remediation attempted in this region are:
(1) Compaction grouting, and
(2) Underpinning.

The authors have recommended strongly against both of these methods as these methods do not provide a long-term solution. Further, the home owner is left with a home that is not repaired satisfactorily, not marketable and not insurable. The geotechnical engineering firm that recommended the remedy, the contractor that executed it and the insurance company that paid for it all hold themselves indemnified by their contractual obligations. The authors therefore have recommended the following remedies, some of which have been successfully carried out:
(1) Stabilize the limestone by grout sealing and supporting the structure on a piled foundation system.
(2) Stabilize the limestone by grout sealing and support the structure on a post-tensioned mat, placed on stabilized ground.
(3) Re-construct the residence after stabilizing the ground.

In the discussion that follows, we have cited our experience with the remediation work and suggest directions for future experimentation and research.

Compaction Grouting

This is the most frequently attempted remedial measure and is cited in a standard reference book on this subject (Greenfield & Shen,1992). This is the only ground modification system developed in the United States and has remained the most common method of sinkhole rectification. When used as method of sealing cavities in limestone formations, it works successfully (Welsh, 1988). Perhaps for this reason, it remains the most economical method. However, when used to stabilize overburden soils, compaction grout has caused more damage. The authors have encountered at least six cases of residential structures stabilized by compaction grouting, where settlements have either reappeared or have accelerated after compaction grouting. The reason for this appears to be two fold. Firstly, the criteria used for grouting do not control what is targeted but only how grout volume is controlled. Secondly, the sinkhole which was the source of the settlement is not remedied.

When used to stabilize the overburden, compaction grouting has been unsuccessful. There have been reports of 160 to 350 cubic yards (120 to 280 m3) of grout pumped under residential structures around 2,500 square feet (230 m^2) in area. The depth of grout penetration is between 50 and 80 feet (15 to 25 m) below the ground surface. This is the equivalent of 4,000 to 9,000 lineal feet (1 200 to 2 800 m) of 14-inch diameter auger-cast piles, or roughly 60 to 130 piles to 70 feet depth under the residence. Despite these figures, the residences still settle further, normally within two months to a year. There are two primary reasons for this. Firstly, the grout travels in all directions, controlled only by the contractor's guidelines which specify that grouting must be

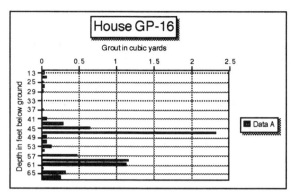 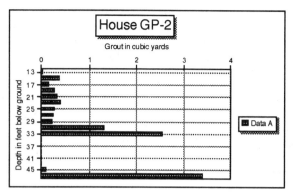

FIGURE 1. The two graphs above show the volume of grout pumped at various depths under a house at two points of injection in the overburden. The house suffered additional settlement problems within months of grouting.

stopped at each grout point when the volume exceeds a pre-set quantity or when the pressure exceeds a certain pre-set value. Secondly, the added weight of the grout spread unevenly throughout various strata causes uneven settlements. When a residence is grouted, it is not uncommon for the grout to surface or to cause damage (primarily uplift) at a neighbor's residence. In an attempt to control such uncontrolled grouting, attempts have been made to develop a rational design method for compaction grouting, especially in overburden soils, to form a compaction grout "mat" (Schmertmann & Henry, 1992). However, at this time it is not clear whether any design methods are used in remedial grouting of sinkhole damaged structures other than the grout-control criteria established by the grouting contractor.

Underpinning

Supporting the structure on a pin-pile or a similar underpinning system has been practiced in the Tampa Bay area for over 30 years, mostly by small contractors. In most cases, this involves installing a steel pipe to a depth of 10 feet to 30 feet (3 to 10 m), pouring grout inside and outside the pipe, and designing a bracket to support the footing on the thus-formed pile. This system has had an average life of five years before settlements re-appear. The reason for the re-appearance of the settlements appears to be the fact that the underpinning system supports the structure mostly in friction, and as the ground settles or as regional ground subsidence occurs due to sinkhole activity, the pile system moves with the ground. A variation of the pin-pile system uses a small diameter, flush jointed steel pipe, jetted to a depth of 50 to 80 feet. Once the pipe encounters adequate resistance, perhaps from a clay or a weathered limestone formation, it is grouted inside from bottom up, which fills the pipe and may jacket the pipe. The foundation loads are again transferred by brackets to the pin-pile system. This system appears to fail because of the slenderness ration of the pile, which may bend, buckle or dislodge at the joints of the pipe sections.

Having seen sufficient failures of these systems (one would be enough) the writers have strongly recommended against the use of these two methods of structural remediation. We therefore recommend that the limestone cavity or the sinkhole be first repaired, the ground stabilized and then the foundation system be improved.

Grout sealing limestone cavities

In this method of stabilization, the first step is to use a fluvial (high slump) grout with sand, cement and flyash for sealing the known void spaces within the limestone formation, not in the overburden. To limit volume of grout used, a method of pre-grouting, similar to pre-splitting of rock is recommended. Some contractors use the terms primary and secondary grouting to describe the same procedure. The primary grout acts as a partial barrier to the secondary or intermediate grout injection. A grout seal thus formed substantially reduces the solution activity and reduces the potential for the formation of ravelling zones and cover-collapse mechanism. It is not clear whether grout sealing has any effect on controlling ground subsidence as no instrumented settlement studies have been conducted. Grout sealing is particularly effective when used prior to construction of a structure and has been successfully used in this manner in this area by the authors (Kannan, 1997).

Recently, grouting with clay has been reported to be successfully applied for a case in the Ukraine and has been used experimentally in Montana in the United States (http://192.188.119.21, 1998). Instead of a cement grout, a site-specific mix consisting of kaolinite/illite and other additives is used. This is reported to be very effective and relatively inexpensive. Using a natural substance to seal the limestone is also considered environmentally safe. However, reports of successful application of the clay grouting method have not been published, though it appears that the method has been used successfully in the United States.

Settlement tolerant foundation systems

Typical spread footings used in residential construction are sensitive to differential settlements and numerous other factors, such as shrinkage in concrete, presence of clay layer near the footing, fluctuations in ground water table, etc. Localized anomalies such as solution pipes and surficial depressions could be bridged using either post-tensioned mats or piled foundations.

Post-tensioned mats could be designed to tolerate 2 to 3 inches (50 to 75 mm) of differential settlement. They also perform well in soft and expansive soils. However, post-tensioned mats are difficult to repair in case of a large collapse or if a corner of the slab fails due to subsurface solution activity, surficial erosion or slope failure.

Piled foundations are feasible both to control settlements in new construction as well as retard settlement of sinkhole damaged structures. The cost of supporting a typical residential structure on piles is comparable to the cost of compaction grouting. However, driving or drilling piles around an existing structure is an expensive and difficult process. Auger-cast piles have been recommended in a few cases.

Ground improvement is a feasible alternative if it could be carried out before a residential structure is built. At a site in Pasco county located near State Road 52 adjacent to a paleo-sink, two to three story town homes were built on stabilized ground. The subsurface soils to a depth of about 30 feet (9 m) were improved by installing a row of auger-cast piles under the spread footings. The additional friction and end bearing afforded by these small-diameter (14-inch) piles was a preferred alternative to compaction grouting, as the volume of grout could be controlled. The cost over-run due to additional volume of grout was less than 15% in this case, while compaction grouting has often resulted in 50 to 100% cost over-run.

Another variation of ground improvement is to stabilize the ground by vibro-compaction or by vibro-replacement. This method uses a probe which is advanced into the ground by jetting water. As it is pulled up, the probe vibrates and compacts the soil around it, forming a densified column. To further improve the soil, additional sand or gravel may be introduced as the probe is vibrated and pulled up. When additional soil is introduced, the method is described as vibro-replacement, and if gravel is introduced, it may be termed a stone column. Ground improvement by vibro-compaction or vibro-replacement is comparable in cost to a piled foundation system. This method has been successfully used on two to ten story structures.

If the solution activity is accompanied by the presence of a high-plasticity clay near the surface, two methods of settlement control could be used. The first and the easier one is to over excavate the foundation trenches by about an additional three feet below the bottom of the footing and replace the clays with granular material such as sand. The sand thus placed, should be provided with proper drainage and a positive outfall for the water that collects in the sand-trench.

The second alternative is to stabilize the clays with lime columns. This method was originally developed in Sweden (Broms & Boman, 1979), and has been used in the United States (Hardianto & Ericson, 1994) and other countries to stabilize phosphatic, expansive, soft and quick clays. A lime column is formed by mixing hydrated lime with the clay using an auger similar to a screw-plate. The clay-lime mixture gradually gains strength, while the plasticity is lost because of partial substitution of calcium ions for the sodium and potassium ions in the clay mineral lattice. The lime column gains strength with age.

CONCLUDING REMARKS

(1) Residential structures damaged by solution activity could be repaired by a wide range of available methods. The two most common methods of stabilization, compaction grouting and underpinning, are known to fail if not carried out properly.

(2) Instead of compaction grouting, grout-sealing of the limestone voids should be specified. The overburden soils should then be stabilized by an appropriate method or the foundations should be supported on a settlement-tolerant foundation system such as a post-tensioned mat or a piled foundation system. As a further improvement of this method, clay grouting should be investigated, though it is in an experimental stage at this time.

(3) In most methods of repair which involve ground improvement, there will be some additional settlement or ground subsidence. This will impact the residence by causing additional settlement cracks. Hence the most feasible solution should involve either moving the residence out of its foundations and re-positioning the house after the ground is stabilized. It may be more economical to demolish the residence and rebuild it on stabilized ground. Some insurance claims for sinkhole damage have been settled on this basis.

(4) New construction should be planned on basis of subsurface exploration conducted prior to the construction of the residence. Any ground improvement that is necessary should be completed before the residence is built.

(5) Research in the use of clay grouting, post-tensioning as a method of repairing the failed foundation, applications of lime columns and other methods of deep soil compaction should be researched.

ACKNOWLEDGEMENTS

The authors wish to thank the many homeowners who provided the data used in discussion of the case histories. The authors also wish to thank Ms. Adriana Loncar of Williams Earth Sciences, Inc. for proof reading the manuscript.

REFERENCES

Broms, B. B., and Boman, P., 1979, Lime Columns - A New Foundation Method, Journal of the Geotechnical Engineering Division, ASCE., Vol.105 GT4, pp.539-556

Fuleihan, N. F., Cameron, J. E., & Henry, J. F. 1997, The Hole Story: How a sinkhole in a phoshogypsum pile was explored and remediate, Proceedings Of the Sixth Multidisciplinary Conference on Sinkholes and the Engineering and Environmental Impacts of Karst, Springfield, MO.p.363-370.

Greenfield, S. J. and Shen, C.K., 1992, Foundation Problems in Soils, Prentice Hall, Englewood Cliffs, NJ., pp.149-163

Hardianto, F. S., and Ericson, W. A., 1994, Stabilization of Phosphatic Clay Using Lime Columns, Florida Institute of Phosphate Research, Pub. No. 02-088-102.

http://192.188.119.21/env/site95/demo/ongoing/morrison.html., February 1998, MK/STG Morrison Knudsen Corporation/ Spetstamponazhgeologia Enterprises :Clay-based Grouting Technology, (Available on the web)

Kannan,R. C., 1997, Proceedings Of the Sixth Multidisciplinary Conference on Sinkholes and the Engineering and Environmental Impacts of Karst, Springfield, MO.pp.313-318

Mishu,L. P., Godfrey, J. D., and Mishu, J. R., 1997, Proceedings Of the Sixth Multidisciplinary Conference on Sinkholes and the Engineering and Environmental Impacts of Karst, Springfield, MO. pp.319-328.

Schmertmann, J. H., and Henry, J. F., 1992 A Design Theory for Compaction Grouting, Grouting, Soil Improvement and Geosynthetics, Proceedings of the Geotechnical Division, ASCE, New Orleans, LA. pp.215-228

Welsh J. P., 1988, Sinkhole rectification by Compaction Grouting, Geotechnical Aspects of Karst Terrains, Proceedings of the ASCE National Conference, Nashville, TN. pp.115-132.

Hydrogeological monitoring strategies for investigating subsidence problems potentially attributable to gypsum karstification

JOHN LAMONT-BLACK, PAUL L.YOUNGER & RICHARD A.FORTH Department of Civil Engineering, University of Newcastle upon Tyne, UK

ANTHONY H.COOPER British Geological Survey, Kingsley Dunham Centre, Keyworth Nottingham, UK

JASON P.BONNIFACE Babtie GeoEngineering, Glasgow, UK

ABSTRACT

Karst regions, especially gypsum ones, are prone to subsidence; this can cause severe problems in urban areas. However, this subsidence may have causes other than active karstification. A decision-logic framework designed to tackle this issue is presented. It comprises subsidence description; identification of causal mechanisms; construction and evaluation of conceptual models; evaluation and parameterization of fundamental processes and development of a management strategy. This framework is applied to an area of active subsidence in the UK underlain by gypsiferous rocks. In this example, particular attention is paid to the evaluation of gyspum dissolution using four criteria: presence of evaporite; presence of undersaturated water; energy to drive water through the system, and an outlet for the water. Gypsum palaeokarst was identified from borehole evidence and contemporary karstification is indicated by groundwaters containing up to 1800 mg/l of dissolved sulphate. Strontium-sulphate ratios enabled the discrimination of gypsum and non-gypsum derived sulphate ions and correlation with the hydrostratigraphy. Continuous measurement of groundwater levels showed differential potentiometric surfaces between stratigraphical horizons and indicated a complex pattern of groundwater movement. Integration of these data in a physically- and chemically-based groundwater model, incorporating a void evolution capability, is suggested.

INTRODUCTION

Ground subsidence is a frequently encountered geological hazard. It can result from one or a combination of processes. These include: removal of fluids; shrinkage of organic or clay-rich materials; hydro compaction of sediments; creep or catastrophic collapse of materials into voids created by mining activities or natural processes like karstification and periglaciation. Subsidence classification schemes (e.g. Prokovich, 1978; Waltham, 1989), can be a useful first step in identifying possible causes. However, given the diverse settings and variety of processes at work, adopting a general classification scheme can lead to assumptions that limit applications or fail to take full account of potential hazards.

As in the more widespread limestone karsts, the most obvious and profound surface manifestations of gypsum karst are collapse sinkholes or dolines. These have been recorded in many parts of the world (Cooper, 1995; Benito et al., 1995; Johnson, 1996; Paukštys et al., 1997; Yaoru & Cooper, 1996; Martinez et al., 1998). In the UK, cover collapse sinkholes related to gypsum karst are common around the city of Ripon (Cooper, 1986, 1998). However, gypsum dissolution need not lead to a sudden collapse, and a sinkhole can form by a more gradual subsidence process. Gradual subsidence can have origins other than bedrock dissolution and these must be considered in any long-term management strategy. The rate of gypsum dissolution in water is approximately 30 - 70 (Klimchouk et al., 1996) to 100 - 150 (Martinez et al., 1998) times that of limestone. Therefore, it must also be recognised that changes in groundwater circulation (which may be caused by groundwater abstraction) may initiate dissolution or increase existing dissolution (Paukštys et al., 1998; Cooper, 1988).

The research results discussed below are drawn from a study of an area in the UK that suffers from subsidence problems. It is underlain by 40 to 50m of Quaternary till and a substantial thickness of Permian gypsiferous strata. The surrounding areas contain limited indications of gypsum dissolution and the area has hitherto been classified as one of low subsidence risk. Owing to an agreement of confidentiality we are not able divulge the precise location of the site.

To develop an effective long-term hazard management strategy, it was considered that a thorough process-based understanding of subsidence mechanisms should be sought. It was realised that it is a mistake to assume automatically that an area in which karstification is known will have subsidence that is caused by karstification. This is especially so in areas with a complex hydrogeological and geomorphological history. However, to assume that if karstic voids have caused few problems in the past, they are unlikely to do so in the future is equally mistaken. This paper presents a working framework for identifying and evaluating a variety of component processes responsible for subsidence in areas underlain by gypsiferous strata. The results are drawn from an EC funded study entitled ROSES (Risk of Subsidence due to Evaporite Solution).

DECISION-LOGIC FRAMEWORK

Subsidence over gypsiferous strata can have a number of different causes including some that may not be related to dissolution. Klimchouk (1996) classified gypsum karst into 8 speleogenetic types. In the UK at Ripon, Cooper (1998) illustrated 16 sinkhole variations that fall within the subjacent, entrenched and, possibly, mantled categories of Klimchouk, (1996). Thus, it is seen that gypsum karst is complex. In addition, numerous subsidence mechanisms (Table 1) can be present within a given setting. A decision-logic framework was developed to enable the characterisation and classification of the observed gypsum karst and the design of a monitoring strategy. The framework follows the six steps listed below. These are followed by an example of its application to a UK site.

1. Describe the observed subsidence accurately
2. Identify the possible causal mechanisms, making reference to local geology and Table 1
3. Construct conceptual ground models incorporating the processes driving the above mechanisms
4. Collect data to evaluate the different fundamental processes
5. Parameterise the important fundamental processes
6. Develop a management strategy based on understanding these parameters

Table 1: Possible causes of subsidence that may be present in an area underlain by gypsiferous rocks

Action or group of processes	Subsidence mechanisms	Examples
Removal of ground fluid.	Loss of buoyant support for soil particles; compaction of sediment; possible failure of material bridging a void, thus causing catastrophic collapse.	Sinkholes overlying gypsum karst in Lithuania (Paukštys et al., 1997).
Shrinkage of organic or clay-rich materials.	Loss of moisture; consolidation of sediment; nett loss of organic material.	Shallow depressions from desiccated swelling clays (Biddle, 1983; Driscoll, 1983). Subsidence of peat in subsidence hollows (Cooper, 1998).
Hydrocompaction of sediments.	Softening and yield of metastable interparticle bonds by the introduction of water.	Building subsidence caused by compaction of gypsiferous silt in alluvial fan, e.g. Calatayud Spain (Gutiérrez, pers. comm).
Fluvial karstification of gypsum bedrock.	Loss of mechanical support – syn- and post-sedimentary subsidence of alluvial deposits into depressions forming at the alluvium/gypsum contact.	Subsidence depressions on alluvial flood plains, Calatayud Graben, Spain (Gutiérrez, 1995), river valley 'subrosion' (Ford, 1997).
Downwashing of unconsolidated sediments.	Downward movement of sediments into existing voids or breccia pipes resulting in gradual growth of shallow depressions.	Sinkholes in Lithuania (Paukštys et al., 1997); sinkholes in Ripon (Cooper, 1986).
Gradual collapse of unconsolidated materials overlying gypsum karst.	Sagging of materials overlying a void leading to concomitant lowering of the cover/gypsum interface, and growth of a surface depression.	Sinkholes in west Ripon (Cooper, 1998).
Catastrophic collapse of unconsolidated sediments into existing voids (may be preceded by a period of gradual subsidence).	Collapse that probably requires a triggering mechanism; ravelling of a void upwards through unconsolidated sediment possibly accompanied by removal of sediment by rapid water flow through the void.	Sinkholes in Ripon (Cooper, 1986, 1998); sinkholes in China (Yarou and Cooper, 1996); sinkholes in glacial drift overlying gypsum karst, Nova Scotia (Martinez and Boehner, 1997).
Catastrophic collapse of competent strata into voids (breccia pipe propagation).	Failure of material bridging void; development of sinkhole generally requires a triggering mechanism.	Sinkholes in Ripon (Cooper, 1986), New Mexico (Martinez et al., 1998), NW territories, Canada, Ford (1997).

DESCRIPTION OF THE OBSERVED SUBSIDENCE (SITE IN THE NORTH OF ENGLAND)

Prior to the 1970s the recorded subsidence around the northern England site was limited to a few shallow closed depressions and a group of four sinkholes. These sinkholes yield sulphate-rich water and were formed by a catastrophic collapse during the Twelfth Century. In recent years, however, subsidence has been experienced on a much wider scale. It takes the form of shallow ground depressions with maximum widths of a few tens to approximately 100 metres and has mostly affected residential properties. Although building subsidence rarely exceeds 300mm, the total cost of remedial work over recent years has been in excess of £1m.

IDENTIFICATION OF POSSIBLE CAUSAL MECHANISMS

To identify the possible mechanisms involved, an initial desk study of the area's geology, hydrogeology and land-use history was made. Geological data were established from British Geological Survey maps and boreholes. The area is gently undulating, and the sequence comprises 40 – 50m of glacial till (boulder clay) and laminated clay overlying Permian dolomites, dolomitic limestones, gypsum and marl (Figure 1). From previous investigations it was known that the glacial sequence was likely to contain a variety of materials including lenses of soft, water saturated, silty and sandy material.

The Permian dolomitic limestones and dolomites form an important regional aquifer, which is in a confined state beneath the area. Increased exploitation of this resource in recent years may be influencing gypsum karst genesis. Consideration of the above ground conditions, the genetic types of sinkholes and the broad types of subsidence mechanisms listed in Table 1, led to the following mechanisms being postulated:

- Clay shrinkage including:
  - clay shrinkage owing to localised desiccation by large trees
  - clay shrinkage owing to climatically driven desiccation
  - localised lowering of the water table in the till and consolidation of lacustrine clays
- Gypsum karst including:
  - gypsum dissolution by groundwater
  - upward migration, through the till, of karstic voids originating in the gypsum
  - localised enhanced dissolution at the till/gypsum interface and sagging of overlying till
  - downwashing of saturated sands or clays in the till towards voids in gypsum, with consequent surface depression growth

Mining activity is completely unknown in the area, so mining related subsidence was discounted. Therefore, the above mechanisms were used to construct conceptual models of the observed subsidence (Figure 1) thus providing a number of hypotheses to be tested.

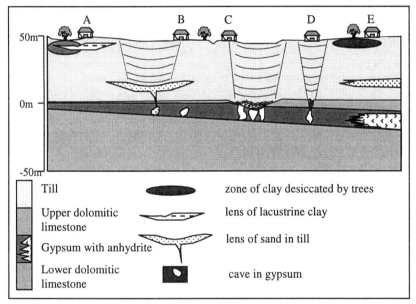

Figure1: First conceptual model of subsidence mechanisms: (A) desiccation of lacustrine clays in dry periods (amplified by trees); (B) Downwashing of sand and silt in the till into karstic voids in the gypsum; (C)gradual lowering of till/gypsum interface; (D) upwards migration of karstic voids in the gypsum; (E) localised desiccation and shrinking of clays due to large trees. Not to scale.

CONCEPTUAL MODEL CONSTRUCTION
Clay shrinkage as a conceptual model

Localised building subsidence damage can be caused by clay shrinkage due to tree roots removing soil moisture (type E Figure 1). Biddle (1983) and Driscoll (1983) examined the effects of different tree species on soil moistures in the UK. They noted that elm, oak, willow and, in particular, poplar, have the greatest effect in reducing soil moisture. They also noted that for a given set of environmental conditions, clay type has little effect on the degree of desiccation that will occur, but it is very important in terms of the amount of shrinkage that can occur. The Building Research Establishment (1996) highlighted the importance of stress history (degree of overconsolidation) in calculating the degree of desiccation from observations.

Clay shrinkage, caused by climatically driven soil desiccation, is a well known British phenomenon that has occurred widely on clay soils following very dry summers such as those in 1976 (Driscoll, 1983) and 1990. The worst damage generally occurs on soils rich in swelling clays, in areas where soil moisture deficits are highest. This effect is linked closely to the effects of trees. A dry period with gradual lowering of the water table could also give rise to consolidation of deposits, such as glacio-lacustrine clays, and cause subsidence (type A Figure 1).

Gypsum karst as a conceptual model

The first stage in evaluation is to decide the genetic type of karst. Klimchouk's (1996) classification includes every type of karst (not necessarily gypsum) within a complex multi-stage development framework. This classification provides an important framework for understanding the processes involved and the physical types of karst that can be expected. The northern England potentially contains subjacent, mantled and buried karst (Klimchouk, 1996). If the karst is buried it could suggest that karstic processes are no longer active. Johnson (1996) and Martinez et al., (1998) list four conditions that must be met for evaporite karstification to be considered active. These include:

1. an evaporite deposit in the subsurface
2. water that is unsaturated with respect to the evaporite mineral;
3. an outlet for the escape of solvent water, and
4. energy to cause water to flow through the system.

It is important, first, to evaluate whether the above conditions were met, and then if any of the conditions had been altered (e.g. increased groundwater heads resulting in higher rates of water flow through the system). Only if these conditions exist is it sensible to consider the mechanisms that might be operating in the overlying deposits. All these mechanisms are shown in Figure 1 (constructed from desk study data) and include, slow upward migration of a void such as a breccia pipe in limestone overlying the gypsum (type D, Figure 1),downward illuviation of loose saturated sediments (type B, Figure 1), and gradual lowering of the till/gypsiferous unit interface (type C, Figure 1).

EVALUATION AND DEVELOPMENT OF CONCEPTUAL MODELS

Clay shrinkage as a conceptual model

Clay soils of different types underlie the whole of the study area. Subsidence was recorded in a variety of locations overlying different soils with different tree types and planting densities. There was no systematic correlation between incidences of subsidence and either extended dry periods or proximity to trees. Moreover, many of the buildings affected were built in the 1950s but did not begin to experience subsidence until the late 1970s. Similar housing stock, on similar Quaternary deposits, with similar tree densities, but with no underlying gypsum, have not suffered subsidence. Although the indications are that clay shrinkage is not the main cause of the subsidence, this can only be confirmed by a systematic analysis of the trees with respect to their species, proximity to buildings and maturity. The results must then be related to the underlying geology, particularly the presence or absence of gypsum.

Gypsum dissolution as a conceptual model

The desk study showed that additional information was required to allow a full evaluation of the possible effects of gypsum dissolution. This information comprised detail of the solid and drift geology and the hydrogeology, including water levels and water quality. In one of the areas most severely affected by subsidence, four boreholes were drilled to a depth of approximately 100m. Data from these boreholes and data from the British Geological Survey borehole archive were used to construct a series of digital elevation models (DEMs) describing the geological surfaces, and data from the Ordnance Survey was used to construct a DEM of topography.

Groundwater levels were monitored in all four boreholes by installing standpipe piezometers at specific points related to the stratigraphy. This was done in order to measure seasonal variations and differential pressure heads in the strata. Falling head permeability tests were performed to estimate the hydraulic conductivities of the strata. Water quality analyses for major ions and strontium and were performed on all the samples to determine mineral saturation indices. In addition, records of borehole logs, water quality and piezometric data for the surrounding area were obtained from the British Geological Survey and the Environment Agency. These data allowed the four essential conditions for gypsum karstification to be considered.

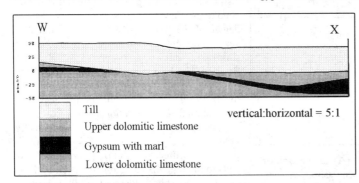

Figure 2: Geological cross-section along line W-X (see Figure 4 for location)

Is Gypsum present?

Figures 2 and 3 are geological cross-sections constructed from the archive and ground investigation data (for locations see Figure 4). The gypsiferous horizon, present under much of the area, attains a maximum penetrated thickness of 19.6m. It comprises alabastrine gypsum with marl horizons near the top and base. Farther to the east, where the unit is thicker and deeper, gypsum is replaced by anhydrite. The unit may be absent from part of the area (Figure 4) which is coincident with a depression in the rock head surface where the gypsum may been removed by dissolution or other erosion mechanisms. However, this idea cannot be validated without additional borehole data.

The gypsiferous unit is significantly thinner in the central south-eastern parts of the area (Figures 2 and 3) and this is coincident with a depression in its top surface (Figure 4). However, this depression is not reflected in the rockhead surface. If the thinning of the gypsum here is due to dissolution, it appears likely to have dissolved prior to the erosion of the present rockhead surface during the Pleistocene. The borehole records consistently recorded deposits of cave-fill type (up to 3.7m thick) in the basal part of gypsiferous unit where it is in contact with the underlying dolomitic limestone aquifer.

Is there water which is undersaturated with respect to gypsum?

The typical water chemistry of the lower dolomitic limestone aquifer is of a Ca-Mg-HCO$_3$ type with typical values for HCO$_3^-$, Ca^{2+}, Mg^{2+} and SO$_4^{2-}$ of 34, 85, 28 and 75mg/l respectively. Analysis of such data shows that the groundwater is saturated with respect to calcite, aragonite and dolomite, but undersaturated with respect to gypsum (SIg $\cong$ −1.7). However, water sampled from piezometer tips sealed off close to the gypsiferous unit provided samples with sulphate concentrations of 1700mg/l and a corresponding saturation index of -0.05 i.e. almost fully saturated with respect to gypsum, thus indicating that gypsum dissolution has occurred.

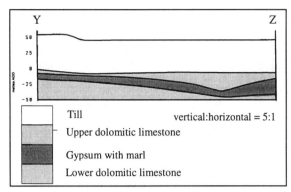

Figure 3: Geological cross section along line Y-Z (see Figure 4 for location)

Is there energy to drive water through the system?

The energy required is generated by regional groundwater gradients and differential piezometric heads between different lithological units. In the simplest terms, the groundwater hydrostratigraphy comprises the Quaternary till overlying the confined lower dolomitic limestone aquifer. Falling head permeability tests, performed in the piezometer tubes, indicated hydraulic conductivites of 1 to 10md^{-1} for the lower dolomitic limestone and 0.01 to 0.04md^{-1} for the till. In reality, the till is likely to have a more variable hydraulic conductivity than the dolomitic limestone. This is because of its inherent heterogeneity, including lenses of water-lain clay, silt, sand and gravel. In all four boreholes the lower dolomitic limestone aquifer exhibited a lower piezometric level than the till.

Klimchouk (1996) showed that the most common and rapid development of gypsum caves occurs in the interstratal setting, when water under artesian pressure is driven through a layer of gypsum in a direction roughly perpendicular to the bedding. Such conditions were obtained in Borehole A (Figure 5), where a head difference of 0.2m was present across the gypsum layer. This head difference has reduced with the onset of water quality sampling in July 1998. This reduction could be a response to well development as large quantities of water have been removed during purging operations. There appears, however, to be some divergence in the hydrograph levels of the blue and brown piezometers in Borehole A, indicating that the initially observed head difference was real and this was disturbed by the sampling; a longer period of observation will be required to assess this.

In contrast, Borehole B yielded piezometric head data that showed almost zero head difference across the gypsiferous unit, but higher water pressures in the till above and in the dolomitic limestone aquifer beneath (Figure 6). This can be interpreted as the gypsiferous unit acting as a drain to the system. If this is the case then it is reasonable to assume that gypsum dissolution is ongoing and that the water containing high levels of dissolved gypsum is not sampled because it is confined within the gypsiferous unit. It is important to note that Borehole B is located in the area where the gypsiferous unit thinned markedly and there is a depression in its upper surface (Figures 2-4). The draining activity may be related, therefore, to reactivation of the palaeokarst interpreted from the lithostratigraphy.

The hydrograph of Borehole A (Figure 5) shows a difference in potentiometric level between the brown piezometer (situated at the base of the gypsiferous unit) and the yellow piezometer (lower dolomitic limestone). The absence of any significant clay strata in the lower dolomitic limestone aquifer, and the presence of only a 3m-thick bentonite seal between the two packed zones, suggests that the gypsiferous unit has a low hydraulic conductivity. However, falling head tests were unsuccessful, because the water levels fell too quickly to enable a measurement to be taken. This suggests open voids, perhaps related to the cave deposits proved by boreholes at the base of the gypsum. A possible explanation for these observations, however, might be that the clay in the cave deposits has formed an 'armour' or 'filter cake' at the base of the gypsiferous unit thus offering an effective hydraulic barrier at the top of the lower dolomitic limestone.

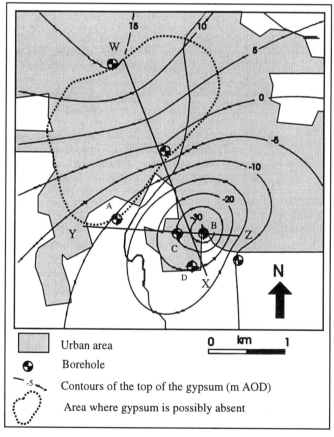

Figure 4: Contoured interpolated upper surface of the gypsum unit. Note the depression overlying Borehole B and the zone where the gypsum is absent.

Is there an outlet for the water?

To maintain a consistently lower pressure in the gypsiferous unit in Borehole B, water must be flowing laterally within the unit. As Figure 2 shows, the gypsiferous unit continues with a shallow southerly dip suggesting that an outlet might exist to the south. Possible confirmation is provided by the presence of sulphate rich waters in the sinkhole flashes 2km to the south and sulphate rich springs 3.5km to the south. Borehole A is adjacent to a river (Figure 4) that gains from the groundwater in the late summer and autumn. A larger river, 4km to the south, gains groundwater from the dolomitic limestone throughout the year.

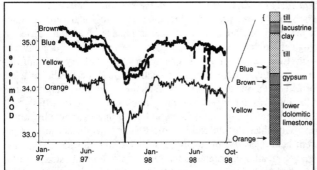

Figure 5: Groundwater levels from standpipe piezometer tubes in Borehole A

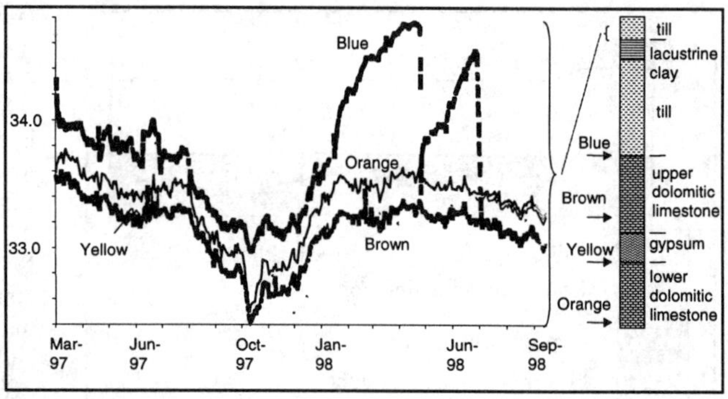

Figure 6: Groundwater levels from standpipe piezometer tubes in Borehole B

Non-gypsum sources of dissolved sulphate

It is important not to rely simply on sulphate levels as an indication of gypsum dissolution. Groundwater from the till in Boreholes C and D contains sulphate concentrations of between 660 – 890 mg/l; ordinarily this might indicate gypsum dissolution. However, the existence of a higher potentiometric surface in the till makes it difficult to imagine a mechanism that would enable groundwater, containing dissolved gypsum, to be found 30m above the rockhead surface. Data presented in Figure 7 shows that strontium concentrations correlate generally well with those of sulphate. This would be expected if the sulphate is from gypsum, as strontium is concentrated progressively during the evaporation of seawater that leads to gypsum precipitation. However, Figure 7 also shows that the sulphate and strontium levels from the till in Boreholes C and D plot on a different trend, thus indicating a different source. To the north of the area, the Coal Measures outcrop, which was traversed by the glacier that deposited the till, contains several pyrite-rich horizons. Incorporation of material from this stratum into the till and its subsequent oxidation would provide the high sulphate levels independent of gypsum dissolution. This possibility is supported by the frequent inclusion of coal fragments in the till.

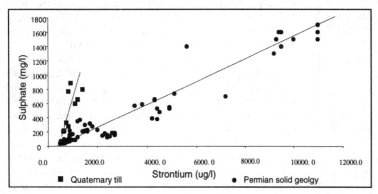

Figure 7: Plot of strontium versus sulphate for all analyses showing two distinct trends relating to gypsum and non-gypsum sources for sulphate

GENERAL CONCEPTUAL MODEL

A general model of gypsum dissolution can now be derived for the study area. All the components for gypsum dissolution exist: gypsum is present, undersaturated water is present, there is energy to drive water through the system, and there appears to an outlet for the water. The situation, shown schematically in Figure 8, allows for two mechanisms of subsidence at the surface, i.e. upward migration of a void developed by gypsum karstification or, downwashing of loose sediments. It has not yet been established which of these is the more appropriate.

PARAMETERIZATION OF FUNDAMENTAL PROCESSES

Initially, two groups of possible subsidence mechanisms were identified: clay shrinkage and mass wasting relating to gypsum karstification. The approach described above has identified several important parameters, especially those relating to the hydrogeology, which describe karstification processes. The parameters include:

- magnitude of piezometric heads in geological units
- direction of water flow in relation to the gypsiferous strata
- concentrations of dissolved sulphate and mineral saturation indices
- relationship between groundwater and surface water
- lateral variations in thickness of gypsiferous strata

Values of all parameters will vary over different time scales. It is intended to perform continuous monitoring of the first four parameters to establish the long-term trends. However, this study has been hindered somewhat by the unusually wet summer of 1998. Therefore neither piezometric levels nor groundwater/surface interactions have been representative, and their effect on groundwater quality is not clear.

This study shows that dissolution is occurring in the gypsiferous strata beneath the study area. Without more detailed hydrogeological work, the dissolution cannot be quantified confidently. Therefore, the process has not been related directly to the subsidence that is being observed at the ground surface. Continued monitoring of ground subsidence and a full evaluation of the subsidence data must be undertaken. The latter is necessary to test the initial conclusion of a non-systematic relationship between instances of subsidence and soil types or trees.

DEVELOP A MANAGEMENT STRATEGY BASED ON UNDERSTANDING THE CONTROLLING PARAMETERS

Although it can be costly to repair damage caused by near-surface clay shrinkage, the process has few long term implications for buildings after their foundations have been deepened to reach below zones of high soil moisture deficit. The long-term effects of gypsum karst, however, would be much more difficult to control. The fullest possible evaluation is needed because the consequences of long-term damage can be severe, as demonstrated in Ripon, Yorkshire (Cooper, 1986, 1996, 1998; Thompson et al., 1998).

Initial work suggests that karstification may be driving the subsidence in the present study area. If this is so, then a more detailed investigation should be planned as the first step in the management strategy. This would be necessary to improve several areas of knowledge i.e. geological aspects such as variations in lithology, structural features and, in particular, the solid-drift interface and stratigraphy of the Quaternary deposits; groundwater flow and hydraulic conductivites. With this information, it should prove possible to construct a representative groundwater flow model using the EVE (Evaporite Void Evolution) code (newly developed under the ROSES project), which has added turbulent conduit flow and dissolution in gypsum caves to the basic capabilities of the well-known MODFLOW® code (McDonald & Harbaugh, 1988). EVE is designed to model the effects that groundwater abstraction may have on flow vectors. The present investigations indicate that some aggressive groundwater flows from the glacial deposits downwards. If the potentiometric level in the lower dolomitic limestone aquifer fell because of further groundwater abstraction, then the pressure difference across the gypsiferous unit would be increased thus possibly accelerating gypsum dissolution rates, provoking additional subsidence.

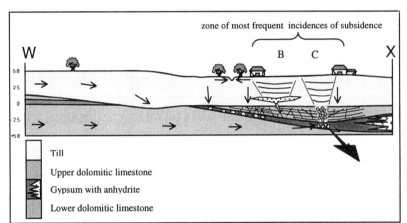

Figure 8: Generalised conceptual model of karstification and subsidence in the study area developed along section W-X. Large arrow indicates postulated outflow of water along the gypsum unit, small arrows indicate suggested groundwater flow direction in the till and lower dolomitic limestone.

ACKNOWLEDGEMENTS

The authors would like to thank Dr. Alex Klimchouk, Professor Mateo Gutiérrez-Elorza, Dr. Francisco Gutiérrez-Santolalla, Dr. Martin Sauter and Dr. Rudolph Liedl for their valuable discussions on gypsum karst. Dr. Dave Lowe and Tim Charsley are thanked for reviewing the manuscript. AHC publishes with the permission of the Director British Geological Survey (NERC). The research presented above is part of the ROSES project funded from under the European Commission Framework IV Programme (Contract number ENV4-CT97-0603).

REFERENCES

Benito, G., Pérez del Campo, P., Gutiérrez, F., Elorza, M. and Sancho, C., 1995, Natural and induced sinkholes in gypsum terrain and associated environmental problems in NE Spain: Environmental Geology, 25, 156-164.

Biddle, P. G., 1983, Patterns of drying and moisture deficit in the vicinity of trees on clay soils: Géotechnique, 33, 107-126.

Building Research Establishment, 1996, Digest 412, "Desiccation in Clay Soils", HMSO, London, 11p.

Cooper, A. H., 1986, Subsidence and foundering of strata caused by the dissolution of Permian Gypsum in the Ripon and Bedale areas, North Yorkshire: in Harwood, G. M. and Smith, D. B. (eds.). The English Zechstein and Related Topics, Geological Society London, Special Publication, 22, 127-138.

------------------, 1988, Subsidence resulting from the dissolution of Permian Gypsum in the Ripon area; its relevance to mining and water abstraction: in Bell, F. G., Culshaw, M. G., Cripps, C. J. and Lovell, M.A. (eds.) Engineering Geology of Underground Movements, Geological Society, London, Engineering Geology Special Publication, 5, 387-390.

------------------, 1995, Subsidence hazards due to the dissolution of Permian gypsum in England: investigation and remediation: in Beck, B. F. (ed.) Karst Geohazards: Engineering and Environmental Problems in Karst Terrane. Proceedings of the 5th

Multidisciplinary Conference on Sinkholes and the Engineering and Environmental Impacts of Karst, Gatlinburg, Tennessee, 2-5 April, 1995. Balkema, Rotterdam, 23-29.

------------------, 1998, Subsidence hazards caused by the dissolution of Permian gypsum in England: geology, investigations and remediation. In, Maund, J. G. and Eddleston, M. (eds.), Geohazards in Engineering Geology. Geological Society, London, Engineering Geology Special Publications, 15, 265-275.

Driscoll, R. , 1983, The influence of vegetation on the swelling and shrinking of clay soils in Britain: Géotechnique, 33, 93-105.

Ford, D. C., 1997, Principal features of evaporite karst in Canada: Carbonates and Evaporites, 12 (1) 15-23.

Gutiérrez, F., 1997, Gypsum karstification induced subsidence: effects on alluvial systems and derived geohazards (Calatayud Graben, Iberian Range, Spain): Geomorphology, 16, 277-293.

Johnson, K. S., 1996, Gypsum karst in the United States of America: International Journal of Speleology, 25 (3-4), 1996, 184-194.

Klimchouk, A., 1996, The typolology of gypsum karst according to its geological and geomorphological evolution, International Journal of Speleology, 25 (3-4), 1996, 49-60.

Klimchouk, K., Cucchi, J., Calaforra, J., M., Askem, S., Finocchiaro, F. and Forti, P., 1996, Dissolution of gypsum from field observations, International Journal of Speleology, 25, (3-4), 37-48.

Martinez, J. D., Johnson, K. S. and Neal, J. T., 1998 Sinkholes in Evaporite Rocks, American Scientist, 86, 39-52.

Martinez, J. D and Boehner, R., 1997, Sinkholes in glacial drift underlain by gypsum in Nova Scotia, Canada: Carbonates and Evaporites, 12 (1) 84—90.

McDonald, M. G. and Harbaugh, A. W., 1988, A modular three-dimensional finite difference ground-waterflow model: U. S. Geol. Surv., Techniques of Water Resources Investigations 06-A1, 576p.

Paukštys, B., Cooper, A. H. and Arustienne, J., 1997, Planning for gypsum geohazards in Lithuania and England: in Beck, B. F. and Stephenson, J. B. (eds.), The Engineering Geology and Hydrogeology of Karst Terranes: Proceedings of the 6th Multidisciplinary Conference on Sinkholes and the Engineering and Environmental Impacts of Karst, Springfield, Missouri, 6-9 April, 1997, 127 – 135.

Prokovich, N. P., 1978, Genetic classification of land subsidence: in Saxena, S. K., (ed.) Evaluation and Prediction of Subsidence, Proceedings of the International Conference on Evaluation and Prediction of Subsidence, Florida, January, 1978, 389-399.

Thompson, A., Hine, P., Peach, D. W., Frost, L. and Brook, D., 1998, Subsidence planning assessment as a basis for planning guidance in Ripon, in, Maund, J. G. and Eddleston, M. (eds.), Geohazards in Engineering Geology. Geological Society, London, Engineering Geology Special Publications, 15, 415-426 .

Yaoru, L., and Cooper, A. H., 1997, Gypsum karst geohazards in China: in Beck, B. F. and Stephenson, J. B. (eds.) The Engineering Geology and Hydrogeology of Karst Terranes: Proceedings of the 6th Multidisciplinary Conference on Sinkholes and the Engineering and Environmental Impacts of Karst, Springfield, Missouri, 6-9 April, 1997, 117-126.

Waltham, A. C., 1989, Ground subsidence, Blackie, Glasgow, 202p.

Engineers challenged by Mother Nature's twist of geology

CHRISTOPHER M.REITH, ALLEN W.CADDEN & CHARLES J.NAPLES III Schnabel Engineering Associates, Inc., West Chester, Pa., USA

ABSTRACT

A $30 million corporate headquarters complex was constructed along the north side of the Schuylkill River in Conshohocken, Pennsylvania. Preliminary exploratory revealed that the site is underlain by phyllitic limestone of the Ordovician period Conestoga Formation that dips near vertically, resulting in extreme variation in weathering and depths to bedrock ranging from about 10 ft to over 100 ft. A Triassic period diabase dike also crossed through the east side of the site. Associated with this dike were thin sills and locations of contact metamorphism of the surrounding rock. The Cream Valley fault is located just east of the site, and the localized crush zone is also believed to have affected the site conditions.

The office towers were supported on shallow spread footings. However, the highly variable conditions underlying the southern half of the site where the garage was proposed presented a much more complicated challenge. Numerous foundation systems were evaluated for this project. Since the project was being developed under a guaranteed maximum price contract, the construction manager desired a foundation system where a contractor could provide a lump sum price. Given the site conditions, this was nearly impossible without contractors including excessive contingency costs. In order to reduce the unknowns in foundation pricing, more exploration was performed. Test borings with rock coring and air percussion holes were drilled at each column location, spaced five feet apart, to confirm bearing grade and variation of the rock, thus reducing the unknowns for the foundation alternative.

The subsurface exploration consisted of 218 borings and air rotary percussion holes that were drilled during design. Foundation alternatives including end bearing caissons, rock socketed caissons, micropiles, driven piles of many varieties, pressure injected footings, mat foundations, and compaction grouting were evaluated. Compaction grouting was selected and proved to be flexible enough to handle the highly variable geologic conditions on time and within budget. This was possible through a comprehensive exploration program, project involvement during design and bidding, and construction quality control.

INTRODUCTION

This paper discusses the difficult site conditions associated with the complex geology and karst related features for a $30 million corporate office complex located in Conshohocken, Pennsylvania. The site is about 4 acres and is bounded by state and local roads on all sides of the property. Two office towers and a parking garage were planned, with a gross footprint totaling about 110,000 square feet, which essentially occupied all of the useable space on the site. The site topography was relatively steep with grades sloping from a high of about EL 102 to a low of about EL 58. The southern portion of the site was located within the 100-year flood plain of the Schuylkill River. The 100-year high water level is EL 64.9 and the 500-year high water level is EL 70.0. A schematic site plan is shown in Figure 1.

As part of the geotechnical engineering study, a preliminary subsurface exploration was conducted consisting of a series of test borings and test pits. The borings revealed that the subsurface conditions on the southern half of the property were highly variable and karstic which were considered unsuitable for shallow foundations for support of the parking garage. The borings and test pits also revealed that the site was underlain by miscellaneous urban fill to depths of 5 to 10 feet from demolition of previous structures. Fill was encountered to depths of up to 30 ft along an abandoned railroad right-of-way which traversed the site. A second exploration program was conducted using additional test borings and air rotary percussion drilling to better define the subsurface conditions within the parking garage footprint. The ultimate goal was to select a foundation system that could economically and safely support the structure while enabling a lump sum or guaranteed maximum price bid to be obtained for construction.

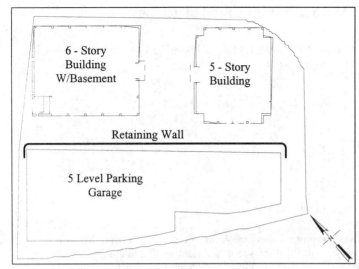

Figure 1: Schematic site plan.

PROPOSED CONSTRUCTION

The office buildings consisted of a six-story structure with a basement and a five-story structure with the first level on grade. The basement level was about EL 88 and the first floor level was about EL 102. This required cuts of up to 12 ft for the basement along the north side of the site and fills of up to 23 ft at the south end of the buildings. Maximum column loads of about 700 kips were estimated for the office buildings.

The parking garage consisted of a five level pre-cast concrete structure with the lowest level at about EL 62. This required cuts of up to 20 ft along the north side of the garage. The difference in floor grades resulted in a 40 ft change in grade between the office buildings and parking garage over a horizontal distance of about 45 ft that also required room for an access drive. This required a retaining wall or a deep basement wall along the north side of the garage to accommodate the grade change. A segmental block retaining wall with a maximum exposed height of 34 ft and 3H:1V slopes at the top and toe was ultimately used to achieve this grade change. Maximum column loads for the garage were estimated to be 1600 kips.

The short construction schedule dictated that the work on all three structures essentially take place at the same time. The construction contract was for a guaranteed maximum price of about $30 million. This required that the foundation system be economical and not subject to large time delays or extra costs.

GEOLOGIC CONDITIONS

Regional geology

The site is located in the Gettysburg-Newark Lowland Section of the Piedmont Physiographic Province in Montgomery County, Conshohocken, Pennsylvania (Berg, 1989). The Norristown, Pennsylvania Geologic Quadrangle indicates that the project area is underlain by albite-chlorite schist of the Wissahickon Formation with a diabase dike crossing through the site. The Wissahickon Formation is believed to be from the Lower Paleozoic age and the diabase is from the Triassic age. A contact with the Conestoga Formation, a carbonate geology with known solution features and sinkhole problems, is shown just north and south of the site. The Conestoga is believed to be from the Cambrian or Ordovician age. Also, the Cream Valley fault is located just to the south (Berg, 1981). An interpretive cross section shows the Conestoga Formation to underlie the Wissahickon Formation and that these formations are folded and steeply dipping (Berg, 1980). The regional geology is shown in Figure 2.

The Conestoga Formation is described as impure limestone that includes micaceous limestone, phyllite, and alternating limestone and dolomite. The rock is described as crudely bedded to poorly bedded, thin, and highly crumpled. The fracturing is described as having an irregular pattern, poorly formed, moderately abundant, widely spaced, having an uneven regularity, and many are open but some are filled with quartz or calcite. The weathering is described as moderately resistant, slightly weathered to a shallow depth, variably weathered, large irregularly shaped fragments result, overburden thickness is highly variable and may be extremely thick, and the interface between bedrock and soil is pinnacled in most places (Geyer, 1982).

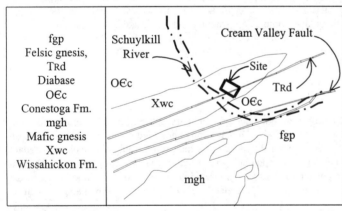

Figure 2: Regional geology.

Site specific geology

A subsurface exploration program was designed to evaluate the site conditions as part of the geotechnical engineering study. The initial program consisted of 54 test borings and 16 test pits within the proposed building areas. Test borings were advanced with hollow stem augers and Standard Penetration Testing was generally conducted on 5 ft intervals. Rock coring was performed using NQ_{II} size wireline core barrels and generally five foot runs. Percent recovery and rock quality designation was recorded for each core run. The borings and test pits revealed that the northern portion of the site was underlain by a residual soil matrix derived from weathering of the Wissahickon Formation schist. However, the southern portion of the site was underlain by recent alluvial deposits and residual soils with a highly variable bedrock surface.

At the location of the office towers, the overburden depths generally ranged from about 20 ft to greater than 80 ft. The diabase dike was present at the ground surface along the southeastern corner of the five-story office building. There appeared to be a ridge of high rock between the buildings that paralleled the orientation of the diabase dike and the fault. Rock coring was performed where the overburden soil was generally less than 30 ft thick and revealed that the underlying bedrock was limestone of relatively good quality.

At the location of the parking garage, the overburden depth ranged from a few feet to over 80 ft, due to a highly variable and pinnacled bedrock surface. The overburden soil in several areas consisted of very soft, unconsolidated materials. Rock coring was attempted at every location, but the steeply sloping and pinnacled rock often deflected the augers too far to allow the core barrel to pass. The rock cores recovered indicated that the rock type was generally limestone and that the quality varied dramatically. Where the diabase dike was cored, it revealed that there were thin sills ranging from a few feet to 15 or more feet thick that were underlain by soft soil for at least 10 to 15 ft before encountering limestone bedrock.

A geotechnical design report was issued based on the results of this subsurface exploration which recommended spread footings for support of the office towers and spread footings on compaction grout improved soils for the parking garage. Subsequent to completion of the report, the parking garage layout was changed and expanded. Also, the construction manager wanted to explore the use of caissons as an alternate foundation scheme for the parking garage. A second subsurface exploration program was conducted in the parking garage area with test borings to bedrock, and ten feet of rock coring at every column location. In addition, an air rotary percussion hole was drilled at each column location that was offset five feet from the boring. This program was designed to provide estimates of bearing grades at each caisson location and to evaluate the extent of the sloping rock conditions so that a lump sum price could be obtained for the caisson alternative.

Ninety-seven test borings and 67 air rotary percussion holes were drilled for the second subsurface exploration. The test borings were conducted in a similar manner as the first program, and the air rotary percussion drilling was performed using a 4 inch diameter down hole hammer with a button bit in an uncased borehole. The drilling encountered problems with obstructions and sloping rock in almost every hole, which resulted in numerous off-sets and re-drills, and breakage or loss of drilling equipment in the hole. Running sand was also encountered in several of the holes, which plugged the drilling equipment and extended several feet up into the augers and/or core barrel. During drilling of the air rotary holes, air/mud/water was often expelled from several of the surrounding holes. The rock surface was observed to be extremely variable with changes in elevation of tens of feet over a few feet horizontally.

The rock cores recovered indicated that the limestone and diabase were highly fractured across most of the garage footprint, with the bedding plane and fracture orientation nearly vertical. This resulted in a highly pinnacled rock surface with troughs, ledges, and several deep soil filled crevices that cross the site generally parallel to the strike of the fault and dike. At least four deep zones of soft soil and highly weathered rock were identified by the test borings. Inferred rock surface bearing profiles from the borings and from the air rotary holes along the north side of the parking garage are shown in Figure 3.

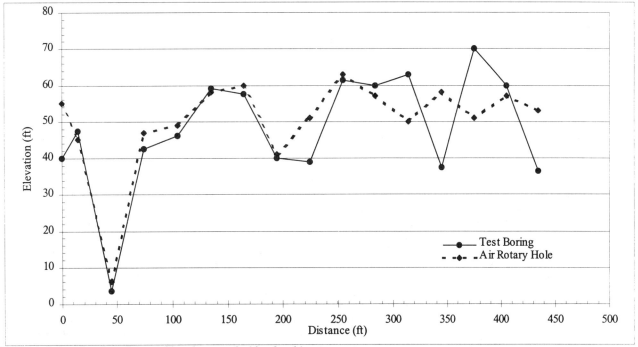

Figure 3: Inferred rock surface profiles along north side of parking garage.

FOUNDATION SYSTEM CONSIDERATIONS

Multiple foundation systems were evaluated during the design development phase of the project based on construction feasibility, base costs, subsidence risk, schedule, and the potential for cost overruns or time delays. These systems included a mat foundation, various driven piles, caissons, auger cast piles, pressure injected footings, micro piles, and various ground improvement methods, including: deep dynamic compaction, jet grouting, stone columns, and compaction grouting. The options were narrowed to ground improvement using compaction grouting based on technical feasibility and overall costs. The construction manager also wanted to have a caisson alternate bid since it appeared to have a favorable initial cost and it was more traditionally used in soft ground conditions in the greater Philadelphia region.

The final foundation systems selected for the garage during the construction drawing development phase of the project consisted of spread footings supported on grout improved soils and rock, and a caisson alternate. Two separate foundation designs were developed to address the specifics of each system with the intent of obtaining a lump sum price for both options and then choosing the one with the least risk in terms of initial costs, schedule, and the potential for extras and/or time delays.

Caisson alternate

Traditional straight shaft end bearing caissons, and small diameter skin friction rock socketed caissons were evaluated for this alternate. An allowable end bearing capacity of 30 tsf was given for straight shaft end bearing caissons founded on the fresh to slightly weathered, massive to moderately fractured limestone or diabase. A higher bearing pressure was not considered feasible based on the highly variable and generally low rock quality designations observed during drilling. Thus, the maximum caisson size would have been about 6 ft in diameter. As an alternate to end bearing caissons, smaller diameter caissons socketed into rock were also considered as an acceptable foundation system. A bond value of 50 psi was recommended for concrete to the competent rock as described above. In this method the caissons are designed to carry the entire load in skin friction so that cleaning and inspection of the bottom can be eliminated (McVay, 1992).

The following items were considered to present significant difficulties in installing caissons: 1.) The upper 10 to 15 ft encountered construction debris and rubble, which will be partially removed when the site is cut to grade. However, some obstructions will be present which will create difficult drilling conditions. 2.) Extensive dewatering or construction using slurry methods will be required for caissons. Loss of ground due to running sands and dewatering will occur and will be difficult to control. Similarly, loss of drilling fluid into karst features or fractures is likely. 3.) Drilling through variable depths of fresh to highly weathered rock, five to 25 ft or more, will be required to reach suitable rock. This was not considered practical using rock augers or core barrels. Cluster drills or large diameter down hole hammers would likely be more practical considering the amount of rock removal that will be required. 4.) Sloping rock will create difficulties maintaining shaft alignment and plumbness. 5.) Permanent casing or concreting and redrilling of shafts will be required where soft material is encountered and loss of temporary casing is also likely to occur.

Preliminary pricing by local caisson contractors ranged from about $500,000 to $800,000, but they were heavily qualified with regard to rock excavation, inspection, and dewatering extras. When the alternate was formally bid, the lowest responsive lump sum price was about $1.6 million, and was still qualified with regards to rock excavation and dewatering. The potential for time delays and additional extra costs was considered relatively high for this option.

Compaction grouting

A shallow foundation system was considered feasible for this project when combined with limited undercutting and replacement with crushed stone and compaction grouting. The compacted dense graded crushed stone fill served to limit stress concentrations and differential settlements where the bearing material transitioned from rock to soil. In addition, the stone fill allowed dissipation and spreading of the foundation loads over the compaction grouted soils. Given this subgrade preparation method, continuous foundation elements designed for a 6000 psf net allowable bearing pressure were recommended. The use of continuous foundations was specified to limit the risk of differential movements between foundation elements should future ground subsidence occur.

The compaction grouting was the key element to the success of this system. To limit the magnitude of the grouting program, the soil improvement was isolated beneath the foundation elements. A minimum of two rows of grout elements was specified for the exterior footings and three rows for the interior footings. The grout holes were required to extend at least two feet into the underlying rock but should be at least as deep as three times the width of the strip footings.

The type of grout selected was critical. By utilizing a low mobility (compaction) grout with a measured slump of less than two inches, the location of the grout is more controllable (Werner, 1992). Given the goals of the ground improvement of filling voids and densifying or displacing soft soils beneath the foundations, it was critical to use a material and injection process tailored to control the grout flow. The use of high mobility (water-cement) grout to fill fractures and seal the top of the rock would not achieve these goals. Furthermore, past experiences have shown that this type of grouting is completely uncontrollable and often unsuccessful in sealing openings in the surface of carbonate rock to a level required for foundation support. A high mobility grout would also not be suitable for improvement of the overburden by permeation since the material is primarily fine grained in nature.

Utilizing compaction grouting provided the flexibility to accommodate varying subsurface conditions by adding injection locations, varying the volume and pressure of the grout injected, or adjusting the location of grout elements beneath the continuous foundations. The added benefit was that this flexibility was provided at a relatively small cost when compared to more traditional deep foundation systems.

Preliminary pricing was solicited from grouting specialty contractors that ranged from about $700,000 to $1,200,000 with various limitations. Given this information, one contractor was selected for further negotiations. A complete grout location as well as specifications detailing mix design, hole depth, cutoff criteria, and conditions where additional holes would be required was established in conjunction with this contractor. Based on the agreed upon, project specific criteria, a lump sum price of $900,000 was established. This system was chosen as the foundation scheme for the project.

FOUNDATION CONSTRUCTION
Equipment and materials

The grout injection points were drilled with a Gill Beetle air-powered track mounted rotary drill with a four-inch diameter downhole hammer. Following drilling, a drive point was afixed to the end of a 2 inch I.D. pipe which was manually lowered to the bottom of the hole. The grout pipes had flush joint couplings to facilitate installation and removal. A pneumatic percussion hammer was used to advance the pipe where it could not be lowered to the bottom of the drilled hole. The grout injection pipe was lifted about one foot and the drive point was then knocked out of the end of the pipe. The annular space remaining between the casing and drilled hole was filled with ¼ in. open graded aggregate. Actual depths of drilling and casing reached 92 ft in some locations before encountering rock. A total of 17,337 lf of grout pipe was installed in 559 locations. By monitoring the drilling, a relative indication of the consistency of the subsurface material was obtained for future correlation with grouting volumes and pressures so that field adjustments could be made if necessary.

The grout specification called for a minimum compressive strength of 1,000 psi after 28 days. The specified grout for this project consisted of a mixture of cement, soil or sand, mineral filler and water reducing agent. The soil was specified to classify as silty sand having a Liquid Limit of less than 15, less than 30 percent by weight passing the No. 200 sieve, and 100 percent passing the No. 4 sieve. Sand was specified as processed mineral aggregate having less than five percent passing the No. 200 sieve and 100 percent passing the No. 4 sieve. Mineral filler such as flyash was acceptable for use with sand as required to produce a suitable grout. The final grout mix was not to have more than 30% fine grained material by weight including the weight of the cement.

Compaction grout was mixed on-site using two 15 cubic yard capacity mobile batch plants. The contractor used a mixture of Type I Portland Cement, screened topsoil, No. 10 limestone screenings and ¼ inch open graded stone to produce the grout. The approximate proportions were 20 percent cement to 20 percent topsoil and 60 percent of a mixture of two parts limestone screenings to one part stone by volume. The actual mix was adjusted throughout the project as delivered materials varied slightly. Constant monitoring of the grout was conducted to allow for necessary adjustments, thus providing a suitable, pumpable grout, and to interpret variations in ground response as indicated by pressure measurements and grout takes.

Soil laboratory testing of the basic grout mix used indicated that the blend of materials classified as silty sand having 27 percent by weight passing the No. 200 sieve, 100 percent passing the No. 4 sieve and was non-plastic. Compressive strength testing of grout cylinders indicated an average 28 day compressive strength of 2,800 psi, which was significantly greater than the 1,000 psi specified.

Compaction grouting was performed in one foot stages using two DGS 2015 duplex piston positive displacement pumps, each with a calibration value of 0.25 ft^3/stroke. The grout was pumped at pressures ranging from 0 to over 600 psi through the grout injection pipes. The grout was generally pumped at a rate of two ft^3/min or slower.

Process

The final grouting program design called for three rows of grout elements on a 6 to 8 ft triangular spacing. A total of 555 grout probes, 12,200 linear feet of drilling, and 37,200 cubic feet of grouting were specified as part of the base contract. A total of 550 of the design grout locations was installed within the building footprint. Five of the design locations were abandoned where drilling or pumping data indicated that these probes were not necessary, and nine additional locations were added where drilling and pumping data indicated a need for additional improvement and/or verification. Grout was injected into each of the 559 pipes utilizing a bottom up procedure and maximum one foot stages. Measured pressures during injection ranged from about 0 psi, indicating no resistance from the surrounding soils, to over 600 psi, which was considered refusal or maximum resistance from the surrounding soils. There was considerable variation in the relative pressures observed which was associated with the highly variable subsurface conditions. It was imperative that the field inspection personnel continuously monitor the process for such changes and modify the injection criteria accordingly. A profile showing grouted depths for the three rows of grout elements along the north side of the garage with the grout volumes injected is shown in Figure 4. Rock was present at the proposed undercut grade where data for only one row of grout elements is shown.

The following criteria were utilized to evaluate the injection process and determine when a grout stage was complete: grout flow ceases at a maximum pressure of 500 psi; ground surface movement was observed; a minimum injected volume of 2.5 ft^3/linear feet at a pressure of at least 100 psi; a maximum of ½ yd^3 was injected in a stage. During production, the injection criteria were modified to account for several site specific effects observed. In order to account for the high water table and frictional losses within the grout pipes, the minimum pressure to allow a cutoff of 2.5 ft^3 for any stage was increased from 100 psi to 180 psi for grouting performed at a depth greater than 40 ft.

Altogether, 42,418 ft^3 of grout were injected over 12,982 linear feet in the 550 design grout locations. A total of 591 ft^3 of grout were injected over 199 linear feet in the 9 additional grout locations. Grouting was not performed above the level of the crushed stone that was specified below the foundations. Given these data, an average grout take of 3.2 ft^3/linear feet was achieved.

CONCLUSIONS

The subsurface conditions at this site were highly variable due to the complex geological features related to the limestone, diabase intrusion, Cream Valley fault, and the Schuylkill River flood plain. These conditions required an extensive subsurface exploration program specifically coordinated with the proposed construction to evaluate foundation systems that would be cost effective while reducing the risk for time delays, cost over runs, and future subsidence. The cost for subsurface exploration in karst areas is typically higher than for other geologic conditions, but for this site the cost savings realized through this added exploration was likely greater than 10 to 1.

The highly variable depth to rock, steeply dipping surface, and diabase sills created a condition that would make estimation of costs and installation of any foundation system very difficult. Ground improvement using compaction grouting provided flexibility to deal with the highly variable conditions while reducing risks for time delays, cost overruns, and future subsidence. This system enabled the use of spread footings and was also the least expensive initial cost of any systems evaluated. The flexibility of the system along with the high quality subsurface data allowed a lump sum price to be negotiated for the foundation contract which ultimately was completed on time and in budget.

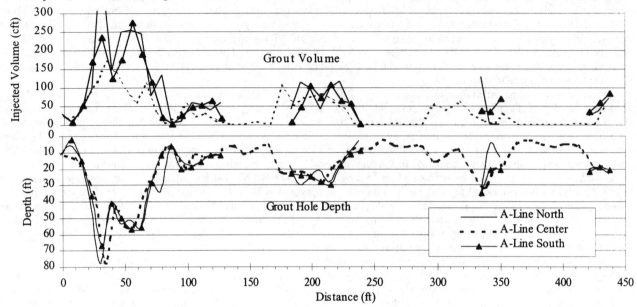

Figure 4: Grout hole depth and grout volumes injected.

This foundation system was successful due to the high level of involvement of the geotechnical engineer during the design development, construction drawing, bidding and negotiation, and construction observation, inspection, and testing phases of this project. This enabled the owner to benefit from a cost effective foundation system while reducing their financial risk. The overall cost of the geotechnical engineering services from conceptual design through construction monitoring of this project was less than 1 percent of the total project cost.

ACKNOWLEDGEMENTS

The authors wish to thank the design and construction management team of BLM Architects, O'Donnell & Naccarato, Timothy Haahs & Associates, Pennoni Associates, Granary Associates, Keating Building Corporation, and Mercy Keystone Heath Systems. Incorporation of the geotechnical engineering consultant as an integral partner in the design and construction process allowed for the timely and open communication necessary so that the owner could benefit from a cost effective foundation solution on a complex site while managing the risks. We also wish to thank Schnabel Engineering for allowing us the time and providing the support to prepare this paper.

REFERENCES

Berg, T.M., ed., 1980, Geologic Map of Pennsylvania, Map 1, Plate 1 East Half, Pennsylvania Bureau of Topography and Geological Survey

Berg, T.M., ed., 1980, Interpretive Geologic Cross Sections, Map 1, Plate 2, Pennsylvania Bureau of Topography and Geological Survey

Berg, T.M., and Dodge C.M., 1981, Atlas of Preliminary Geologic Quadrangle Maps of Pennsylvania, Map 61, Pennsylvania Bureau of Topography and Geological Survey

Berg, T.M., et al., 1989, Physiographic Provinces of Pennsylvania, Map 13, Pennsylvania Bureau of Topography and Geological Survey

Geyer A.R., and Wilshusen J.P, 1982, *Engineering Characteristics of the Rocks of Pennsylvania, Environmental Geology Report 1*, Pennsylvania Bureau of Topography and Geological Survey

McVay, M.C., et al., 1992, "Design of Socketed Drilled Shafts in Limestone", *Journal of Geotechnical Engineering*, Vol. 118, No. 10, October 1992, pp. 1626-1637

Warner, J., Schmidt, N., et al., 1992, "Recent Advances in Compaction Grouting Technology", *Grouting, Soil Improvement and Geosynthetics*, Proceedings, ASCE Conference, New Orleans, Louisiana, pp. 252-264.

Schnabel Engineering Associates, 1997, *Geotechnical Engineering Report, Keystone Mercy Health System*, 975103, West Chester, Pennsylvania

Schnabel Engineering Associates, 1998, *Additional Geotechnical Consultation, Keystone Mercy Health System*, 975103, West Chester, Pennsylvania

Compaction grouting versus cap grouting for sinkhole remediation in east Tennessee

TIMOTHY C.SIEGEL & JAMES J.BELGERI   S&ME, Inc., Blountville, Tenn., USA
MICHAEL W.TERRY   Hayward Baker Inc., Atlanta, Ga., USA

ABSTRACT

A major portion of East Tennessee is underlain by carbonate geology consisting of dolomite and limestone rocks. Due to the composition and weathering pattern of these carbonate rocks, the subsurface profile generally includes an upper clayey crust layer which transitions into the soil/rock interface. The formation and remediation of sinkholes typically occurs within the soil/rock interface.

Compaction grouting and cap grouting are two different approaches for the remediation of sinkholes. The former is typically performed from the soil/rock interface up through the overburden soil at controlled high pressures using a widely-spaced grout hole pattern. In contrast, cap grouting is performed only at the soil/rock interface at controlled low pressures using a closely-spaced grout hole pattern. Selection between these two approaches should be based on the subsurface conditions and the project's objective(s). In general, compaction grouting has the advantage in projects that involve lifting of nearby structures, improvement or stabilization of the soil overburden, and/or a rock surface profile that is favorable to a widely-spaced grout hole pattern. In comparison, cap grouting may have an advantage if the project involves a sufficient upper crust layer, no need for lifting, and an irregular, pinnacled, bedrock surface. Overall, the cost-effectiveness of either grouting approach to remediate sinkholes will depend upon properly applying geotechnical engineering and grouting principles and effectively engaging drilling and grouting equipment. This paper discusses the relative merits of both grouting approaches and the considerations when selecting whether to compaction grout or cap grout. Also, several case histories are presented to illustrate the importance of selecting the appropriate grouting approach based on the specific subsurface conditions.

INTRODUCTION

A major portion of East Tennessee is underlain by carbonate geology consisting of dolomite and limestone rocks. Due to the composition and weathering pattern of these carbonate rocks, the subsurface profile generally includes an upper clayey crust layer which transitions into the soil/rock interface. The dissolution of carbonate bedrock and progressive downward movement of overburden soil within the soil/rock interface leads to the formation of sinkholes. To remediate developing sinkholes and to reduce the potential for further sinkhole development, grouting techniques can be used to fill voids, improve soils, and plug openings in the bedrock. Compaction grouting and cap grouting are two different approaches to the remediation of sinkholes. The former is typically performed from the soil/rock interface through the overburden soil at controlled high pressures using a widely-spaced grout hole pattern. In contrast, cap grouting is performed only at the soil/rock interface at low pressures using a closely-spaced grout hole pattern.

KARST TERRAIN IN EAST TENNESSEE
Geology

East Tennessee is located within the boundaries of three physiographic provinces; the Blue Ridge, the Valley and Ridge, and the Appalachian Plateau (DeBuchananne and Richardson, 1966). The Valley and Ridge province is predominately composed of carbonate Paleozoic rocks, namely dolomite and limestone. This province is characterized by elongated, northeasterly-trending ridges formed by more resistant cherty limestone and sandy shale. Between the ridges, broad valleys and rolling hills are formed in areas underlain by more soluble limestone and dolomite. Within East Tennessee, the average width of the Valley and Ridge is about 65 km (about 40 mi). Prominent geologic groups of the Valley and Ridge are the Chickamauga, Knox, and Conasauga. The limestone and dolomite that make up these groups are characterized as very dense and hard. For example, testing by the authors of carbonate rocks of the Knox group has indicated a unit weight of up to 27.5 kN/m^3 (175 pcf), and an unconfined compressive strength in excess of 210 MPa (31.5 ksi).

General Subsurface Conditions

The general subsurface profile in the karst of East Tennessee consists of four zones (Siegel and Belgeri, 1995). From the ground surface, these are the 1) upper crust, 2) soil/rock interface, 3) weathered rock, and 4) unweathered rock. A graphical presentation of this general subsurface profile is shown in Figure 1. Sinkhole remediation is performed in the upper crust and soil/rock interface zones.

The upper crust is near-surface, clayey residual soil derived by the in-place weathering of the parent bedrock. This residual soil is typically overconsolidated, and is stiff to very stiff in consistency. Soil moisture content at the near-surface can be near the plastic limit, but generally increases with increasing depth. The gradation and moisture content are particularly important in determining whether or not the soil can be densified by compaction grouting. The thickness of the upper crust can vary depending on the extent of weathering and solutioning. In most grouting projects, the thickness of the upper crust is at least 5 m (16.5 ft), which is also the excavation limits of a medium-sized conventional backhoe.

The soil/rock interface is immediately beneath the upper crust and is characterized as a variable mixture of rock boulders, rock lenses, bedrock pinnacles, and soft, wet, residual soil. The rock surface in this zone is typically uneven with openings and/or soil-filled "throats". The soil is typically soft and wet, since it has not been subjected to as much desiccation and has been shielded from overburden pressures by nearby rock. Within the soil/rock interface, moisture content can approach the soil's liquid limit. It is within this zone that initial soil cavities originate as rock weathers to soil and the volume of the solids is reduced. Soil migrates into voids in the bedrock which leaves the overburden soil unsupported. Progressive ravelling of the upper crust into the underlying void can lead to collapse of the ground surface or roof collapse. The process of soil-ravelling/erosion dome formation and collapse is more thoroughly described by Sowers (1996). The thickness of the soil/rock interface can vary greatly, and it is not uncommon for soil-filled or open slots in the bedrock to extend 30 m (100 ft) or more below the ground surface. In many cases, it is near or at the locations of these deep slots that sinkholes develop at the ground surface.

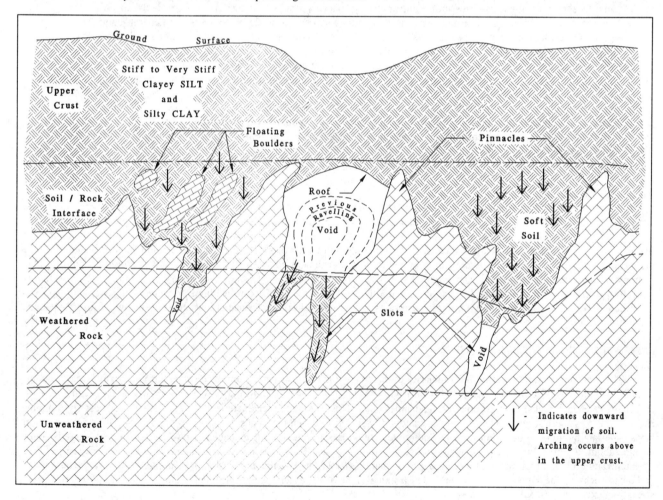

Figure 1: General Subsurface Profile in East Tennessee Karst

SELECTION OF GROUTING TECHNIQUE

Compaction grouting for sinkhole remediation is the injection of low slump (less than 50 mm or 2 in) grout to fill voids, compact/reinforce the soil overburden, and, if necessary, lift structures by upward soil displacement (Welsh, 1988; Shaefer, 1997). Cap grouting is the injection of grout to fill voids and to form a barrier at the top of a porous rock surface (Sowers, 1996). There are apparent similarities between the compaction grouting technique and the cap grouting technique. Most notably, they both involve drilling into the subsurface and pumping grout. Selection between compaction grouting and cap grouting should be based on the subsurface conditions and the project's objective(s). But the cost-effectiveness of the sinkhole remediation program will depend upon properly applying geotechnical engineering and grouting principles and effectively engaging drilling and grouting equipment. As pointed out by Fischer and Fischer (1995), grouting in a "haphazard approach" without a well-defined objective is rarely cost-effective.

Compaction Grouting

The ideal compaction grouting project has a relatively deep rock surface and an objective of overburden reinforcement/improvement. Because of the wide spacing associated with compaction grouting, this technique may be less effective in pinnacled rock, since the sinkhole feature(s) may be located between the grout hole locations and grout may not be delivered to the sinkhole feature(s). In addition to filling voids and fissures at the soil/rock interface, compaction grouting can improve overburden soil that is near or below its optimum moisture content. However, in wet silts and clays, displacement only is likely. It is important to note that although very limited improvement possibly occurs in saturated silts and clays, the overall overburden profile is "reinforced" by placement of a substantial grout column during compaction grouting. Compaction grouting also has an advantage in projects that require lifting of structures.

Cap Grouting

The ideal cap grouting project has a sufficient upper crust layer, no need for lifting, and may have an irregular, pinnacled, bedrock surface. Sinkhole remediation in a stiff, relatively thick, upper crust may only consist of filling the zone of ravelling along the soil-rock interface to support the upper crust. The low pressure associated with cap grouting is sufficient to fill voids and displace soft soil at the soil-rock interface with little risk of uncontrolled lifting nearby structures. The closely-spaced grout hole pattern provides more information concerning the features of the soil-rock interface allowing better definition of areas underlain by slots.

EQUIPMENT

Compaction grouting and cap grouting are similar, and, in fact, some equipment is capable of performing either method. Cap grouting, which is the older method, can be performed with low cost rotary drill rigs through hollow-stem augers. Grout pumps need only to achieve a pressure in the range of 140 kPa (20 psi) and only need to move a moderate slump grout, though a low slump is preferable. In contrast, compaction grouting equipment is more specialized and technically advanced. Typically, compaction grouting equipment is designed to achieve and withstand high grout pressures, drill at various angles, grout with precise volume control at incremental depths, and penetrate into bedrock. The greater flexibility associated with specialized compaction grouting equipment can be very helpful in difficult karst subsurface conditions and limited access situations.

COMPARISON OF COMPACTION/CAP GROUTING TECHNIQUES

As previously stated, it is the overall objective of the particular project that dictates the most effective grouting approach. Table 1 compares the compaction grouting technique to the cap grouting technique. Figures 2 and 3 illustrate the results of these two approaches.

CASE HISTORIES OF GROUTING TO REMEDIATE SINKHOLES

Compaction Grouting to Treat a Sinkhole Area in Morristown, Tennessee

At an industrial facility in Morristown, Tennessee, the original repair of 6 m (20 ft) diameter sinkhole consisted of backfilling the depression with graded stone and then capping the stone with soil to create an inverted filter. Approximately three years after the original repair, the inverted filter failed during a period of heavy rainfall. Also during this heavy rainfall, several new sinkholes developed over a 5,015 m² (54,000 ft²) area. The development of these sinkholes and the risk they presented to nearby structures (i.e., a railroad spur and high voltage electric station) were concerns to the owner.

The subsequent investigation indicated that the depth of the underlying limestone bedrock ranged from 4.5 to 20 m (15 to 65.5 ft) below the ground surface with significant voids at the soil/rock interface. Using this information, compaction grouting to remediate the sinkholes by filling voids and improving the soil overburden was chosen for the following reasons:

- Test borings confirmed the presence of voids and loosely filled slots at the soil/rock interface. It was deemed that these features would be best treated by compaction grouting from the ground surface to bulk fill these voids.
- Due to the relatively large area affected by sinkhole development, it was economically desirable to limit the compaction grouting to the more critical areas. The potential for voids in the soil overburden and the associated risk of future settlement were concerns. To address these concerns, improvement of the overburden soil by reinforcement by compaction grouting was an objective.

Table 1: Comparison of Compaction Grouting and Cap Grouting

Compaction Grouting	Cap Grouting
• Grouting is typically performed at the soil/rock interface and above in the affected soil overburden.	• Grouting is only performed directly above the bedrock surface.
• Improvement of the overburden soil is possible.	• Improvement of overburden soil is not an objective.
• Relatively high grout pressure of 1380 kPa (200 psi) or greater.	• Low grout pressure of 140 kPa (20 psi) or less.
• Primary grout hole spacing is typically 3.5 to 5 m (10 to 15 ft).	• Grout hole spacing is typically 0.9 m (3 ft).
• Grout is placed at high pressure to fill voids, plug slots over a larger less focused area per hole, and displace/improve overburden soil.	• Grout is placed at low pressure at the rock surface to fill voids, plug slots, and displace soft soil to provide support to the upper crust.
• Higher grout pressures, greater grout hole spacing, higher grout quantity refusal criteria, and overburden treatment generally results in larger grout takes.	• Closer grout hole spacing requires greater drilling footage. Cap grouting philosophy generally results in lower grout takes.
• Greater primary grout hole spacing may not intersect sinkhole feature(s). Drilling tools usually allow penetration through boulders and rock lenses. Secondary holes may be required.	• Closer grout hole spacing provides better coverage to intersect sinkhole feature(s). Auger drilling may not extend to bedrock due to shallow refusal on boulders, etc.
• Allows for controlled lifting of structures (care required).	• Very little potential for lifting nearby structures.

For this project, drilling was initiated at the approximate center of each depression area. In some cases, nearby structures necessitated angle drilling to intercept the soil/rock interface based on a presumed depth. Smooth wall casings were installed through the soil overburden to the bedrock surface using rotary percussion drilling methods. The percussion drilling was extended to a minimum of 1.5 m (5 ft) of bedrock to reasonably verify continuous bedrock. Because the objective for the initial locations was to bulk fill voids rather than improve the overburden soil, a relatively high quantity cut-off criteria was used. Most of the initial locations accepted between 23 and 38 m^3 (30 to 50 yd^3) of grout without developing the cut-off pressure of 22 kPa (150 psi). After grouting at the initial locations was completed, a series of 3 to 6 additional grout locations on 3 to 4.5 m (10 to 15 ft) centers were drilled and grouted both around the initial grout holes and around nearby critical structures. This grid pattern, with secondary locations where necessary, was used to complete the treatment of each sinkhole feature and overburden profile in an effort to prevent further, sinkhole-induced settlements at the pre-determined critical structures.

The grout holes intended to improve the overburden soil (rather than fill voids) had the following cut-off criteria per 0.6 m (2 ft) of depth as the grout casings were extracted:

1. A maximum grout volume of 1.5 m^3 (2 yd^3)
2. A maximum grout pressure of 29 kPa (200 psi)
3. Any detectable heave of the ground surface or nearby structure(s)

For the entire project, approximately 498 m^2 (650 yd^3) of grout was injected to remediate 6 sinkhole locations and several other critical areas. The total treated area was approximately 1394 m^2 (15,000 ft^2). Secondary grout locations were added to the project in areas of high grout "take". Typically, the secondary grout locations accepted only small volumes of grout prior to reaching the cut-off pressure indicating that no further grout acceptance at the soil/rock interface was possible and that no further improvement of the soil overburden was feasible. In summary, the bulk void filling combined with compaction grouting provided a technically successful and cost-effective treatment to remediate active sinkhole conditions and to reinforce the soil overburden.

Cap Grouting a Construction-Induced Sinkhole

The project involves an industrial facility overlying carbonate bedrock in Johnson City, Tennessee. During construction, a sinkhole with a diameter of approximately 1.8 m (6 ft) developed inside the building. Design plans indicated that the area of the sinkhole had been excavated to a depth of about 5.5 m (18 ft) below the original ground surface to prepare the building subgrade. The geotechnical exploration report indicated that soft soil was encountered in the test borings at the subgrade level and that auger refusal (presumed bedrock) was encountered approximately 9 m (30 ft) below the subgrade level. The construction records

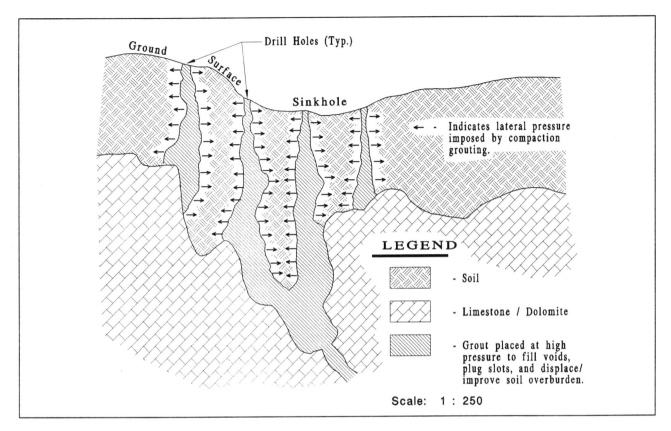

Figure 2: Idealized Profile of Compaction Grouting Results

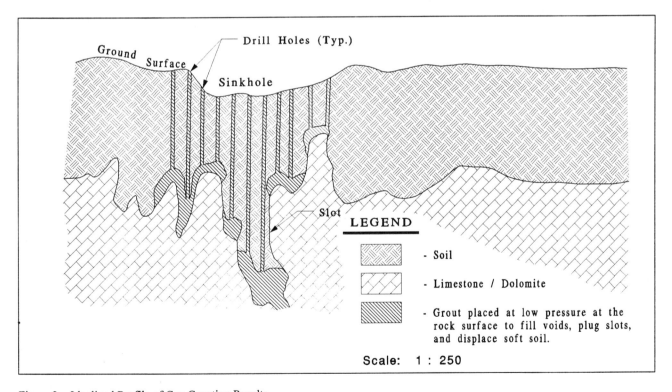

Figure 3: Idealized Profile of Cap Grouting Results

confirmed that soft soil was encountered at the subgrade level and that the near-surface subgrade conditions were improved by a limited undercut of soft soil and replacement with compacted crushed stone.

Cap grouting was considered for remediation of the subject sinkhole for several reasons. First, although excavation to expose the rock surface is possible in some circumstances, for this project it did not appear feasible because of the relatively deep bedrock. Furthermore, any excavation may have endangered nearby foundations. Finally, filling voids to support the overburden appeared to be the most effective approach because the upper subgrade had already been improved by undercut/replacement.

The cap grouting program consisted of pumping low slump grout at a depth of approximately 0.3 m (1 ft) above auger refusal. The sinkhole area was grouted through twenty grout holes using a center-to-center spacing of 0.9 m (3 ft). The actual auger refusal depth during drilling of the grout holes ranged from 1.5 to 17.5 m (5 to 58 ft), but primarily ranged from 7.5 to 10.5 m (25 to 35 feet). Low pumping pressures were maintained, and eight of the twenty grout holes accepted the cut-off grout volume (1.5 m^3 or 3 yd^3) before reaching the cut-off pressure (140 kPa or 20 psi). A total of 21 m^3 (27 yd^3) of grout was placed at the soil-rock interface as defined by auger refusal. On the basis of the drilling and grouting results, it was the engineering interpretation that the subsurface conditions consisted of a relatively thin upper crust underlain by very soft soils with "toothpaste" consistency and a very irregular rock surface. The cap grouting program was considered successful in remediating the developing sinkhole conditions by plugging the rock surface, displacing very soft soil, and supporting the upper crust.

Emergency Cap Grouting of a Sinkhole under a Sewer Line Outfall

A series of surface collapses occurred at a site in East Tennessee with a history of sinkhole development and occupied by a Publicly Owned Treatment Works (POTW). One of the collapses exposed the side of a final settling tank critical to the continued operation of the plant. Another collapse was centered directly under the main outfall for the plant. Failure of either the settling tank or the main outfall unit would have cut-off the sewer service to a population of over 30,000. Continued operation by both units was vital. Urgent repair was therefore critical. However, more conventional remediation approaches (such as excavation and replacement) were not considered feasible. The most expedient approach was to use locally available geotechnical equipment and products. This equipment dictated the use of a moderate to high slump rather than the typical, preferred low slump mix.

The cap grouting program consisted of pumping moderate to high slump grout at a depth of approximately 3 m (10 ft) above auger refusal if no void was found, or at the immediate top of any void encountered by the drilling. The relatively high slump of the grout ensured that the mix flowed into all available openings and overlapped with adjacent grout plugs. The collapsed areas were grouted by approximately 15 primary holes on a radial pattern with an approximate center-to-center spacing of 3 m (10 ft) by 3.5 m (12 ft). Sixteen secondary holes were also used resulting in an ultimate center-to-center spacing of 1.5 m (5 ft) by 1.8 m (6 ft). The actual auger refusal depth during drilling of the grout holes ranged from about 6.4 to 15.2 m (21 to 50 ft). Low pumping pressures (70kPa or 10 psi) were used. Grout takes were large, as might be expected for a higher slump grout, and pumping was generally continued in the secondary holes until some indication of return to the surface from adjacent primary or secondary holes was observed. It is estimated that in excess of 45 m^3 (60 yd^3) was pumped into the subsurface using progressive cavity and diaphragm pumps through hollow-stem augers. The cap grouting program was considered successful in remediating the developing sinkhole conditions by plugging the rock surface, displacing very soft soil, and supporting the upper crust. The site of the POTW has experienced several episodes of surface collapse since and most have been remediated using essentially the same methods. To date, there has not been any recurrence of surface collapse in any of the remediated areas. Nevertheless, the authors believe that the use of low slump mixes in either the cap grouting or compaction grouting methods is superior to the moderate to high slump mix dictated by the equipment available for this project.

CONCLUSIONS

Both compaction and cap grouting techniques involve drilling into the subsurface and placing grout. Each has its advantages and disadvantages. Selection between these two approaches should be based on the subsurface conditions and the project's objective(s). In general, compaction grouting has the advantage in projects that involve lifting of nearby structures, improvement or stabilization of the soil overburden, and/or a rock surface profile that is favorable to a widely-spaced grout hole layout. In comparison, cap grouting may have an advantage if the project involves a sufficient upper crust layer, no need for lifting, and an irregular, pinnacled, bedrock surface. Overall, the cost-effectiveness of either grouting approach to remediate sinkholes will depend upon properly applying geotechnical engineering and grouting principles and effectively engaging drilling and grouting equipment.

ACKNOWLEDGEMENTS

The authors gratefully acknowledge Basil Barkley for the preparation of the illustrations, Michele Cole for preparation of the final manuscript, and Chris Herridge for editorial assistance.

REFERENCES

DeBuchananne, G.D., and Richardson, R.M., 1966. "Ground-water resources of east Tennessee," *Bulletin 58, Part 1*, State of Tennessee Department of Conservation, 393 p.

Fischer, J.A., and Fischer, J.J., 1995, "Karst site remediation grouting," *Karst Geohazards*, Beck, B.F., ed., A. A. Balkema, Rotterdam, The Netherlands, pp. 325-334.

Schmertmann, J.H., and Henry, J.F., 1992, "A design theory for compaction grouting," *Grouting, soil improvement, and geosynthetics*, ASCE Special Publication No. 30.

Shaefer, V.R. Editor, 1997, *Ground improvement, ground reinforcement, ground treatment: developments 1987-1997*, ASCE Geotechnical Special Publication No. 69, 619 p.

Siegel, T.C., and Belgeri, J.J., 1995, "The importance of a model in foundation design over deeply-weathered, pinnacled, carbonate bedrock," *Karst Geohazards*, Beck, B.F., ed., A. A. Balkema, Rotterdam, The Netherlands, pp. 375-382.

Sowers, G.F., 1996, *Building on sinkholes: design and construction of foundations in karst terrain*, ASCE, New York, 202 p.

Welsh, J.P., 1988, "Sinkhole rectification by compaction grouting," *Geotechnical aspects of karst terrains*, ASCE Geotechnical Special Publication no. 14, pp. 115-132.

The role of the geological engineering consultant in residential sinkhole damage investigations

TED J.SMITH & WAYNE A.ERICSON BCI Engineers & Scientists, Inc., Lakeland, Fla., USA

ABSTRACT

Since passage of legislation requiring Florida insurance companies to cover sinkhole damage to residential properties, the consultants' role in performing sinkhole damage investigations has evolved and expanded to include a number of multi-disciplined aspects. The current standard of practice in Florida requires a comprehensive, site-specific investigation to identify or eliminate sinkhole activity as the cause of damage to a structure. In addition to investigating the subsurface geological conditions, evaluating a sinkhole damage claim involves several engineering questions, including the structural condition of the house and the potential for defects in construction and materials. At sites with identified sinkholes, a site-specific remedial plan must be developed. Project costs are estimated to assist the insurance company in adjusting the claim relative to available policy limits. Engineering oversight and inspection services are provided to ensure that remediation is completed in accordance with project specifications. Follow-up evaluations are often completed to address re-occurring damage that may be due to unrecognized site conditions. In cases where rapid development of cover-collapse sinkholes threatens a residence, the focus shifts to emergency response services to stabilize the collapse and reduce damage to the structure. Throughout the process, assistance to the homeowner is essential to address concerns regarding the level of threat to life and property and to prevent misconceptions regarding the investigation process and the effectiveness of any recommended remedial repairs. Several case histories will be examined illustrating these points.

INTRODUCTION

Florida statutes require insurance companies to provide coverage for sinkhole damages as part of most residential homeowners' policies. These statutes also require that properties with purported sinkhole damages be investigated by a "qualified" professional - typically a geologist or a geotechnical engineer. The continued growth of development in Florida has been accompanied by a substantial increase in subsidence damage to residential structures, which may, or may not, be the result of sinkhole activity. Common, "non-karst" causes of subsidence include construction in unsuitable areas (former wetlands and filled areas), inadequate construction practices, and the presence of unsuitable soils (organic soils, shrink/swell clays). Much of Florida consists of a mantled karst terrain where irregular "karstified" limestone bedrock is overlain by variable thicknesses of clastic sediments (sands and clays). Consequently, a variety of sinkhole morphologies can occur, ranging from obvious, abrupt cover-collapse sinkholes to slowly developing cover-subsidence sinkholes that are often difficult to recognize. Such factors have placed heavy demands on insurance adjusters when evaluating a potential sinkhole claim, requiring the geologic engineering consultant to provide a range of multi-disciplinary services. These include investigation of settlement damage claims, preparation of remedial plans for sites with identified sinkhole conditions, engineering oversight and inspection services, reassessment of previous claims and repairs, emergency response to deal with imminent sinkhole hazards, and general assistance to the homeowner.

KARST GEOLOGY AND SINKHOLE MECHANISMS

In west-central Florida a mantled karst terrain is formed by shallow Tertiary limestone bedrock that is overlain by variable deposits of unconsolidated clastic sediments and soils. Figure 1 shows the occurrence of the various shallow limestone units in this study area and associated sinkhole morphologies, which are directly related to the composition and thickness of the cover sediments (Beck and Sinclair, 1986). The geology of Florida with respect to karst and sinkhole development has been described by Schmidt and Scott (1984) and by Sinclair et al. (1985). In west-central Florida shallow limestone bedrock units include the Eocene Ocala Limestone, the Oligocene Suwannee limestone, and the Tampa Limestone member of the Arcadia Formation (Miocene Hawthorn Group). These limestones are unconformably overlain by variable thicknesses of clayey and sandy sediments of the Upper Hawthorn Group (Miocene) and younger undifferentiated Plio-Pleistocene marine sands and clays.

Sinclair et al. (1985) has recognized two primary mechanisms of sinkhole development that dominate in west-central Florida - cover-subsidence and cover-collapse. Cover-subsidence sinkholes form by the gradual downward raveling of non-cohesive sediments (usually sand and silt) into solution enlarged fissures and fractures in the limestone bedrock. Significant cavities in the limestone usually do not develop since they are continually filling with sand. This raveling propagates to the surface producing a broad area of

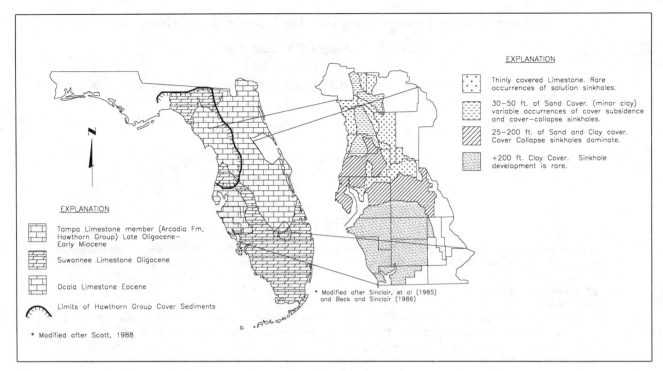

Figure 1: Generalized limestone bedrock map and distribution of sinkhole types in west-central Florida

subsidence. In areas where the cover sediments consist of clay (or damp, semi-cohesive sand) the developing cavities will be bridged until the bearing strength of the clay is exceeded. The clay will fail by upward stoping, followed by sudden downward movement of the overlying soils (Culshaw and Waltham, 1987). The result is a steep-sided collapse feature at the surface. Given that karst processes have been operating over substantial periods of geologic time, buried or paleo-sinkholes are also a consideration due to the potential for re-activation of buried sinkholes by both of these processes.

SINKHOLE DAMAGE INVESTIGATIONS – THE PROCESS
Initial Site Investigations

Sinkhole investigation methods have been described by Wilson and Beck (1988) and Smith (1997) and recommended guidelines have been developed by the Florida Geological Survey (1997). Table 1 is a summary of the investigation methodology for sinkhole damage evaluations utilized by our firm for the past several years. A multi-disciplined approach is taken involving an assessment of the geological conditions at the site along with the general structural integrity of the residence. The initial site visit typically involves an interview with the homeowner to determine the location and history of the damage, as well as other relevant site data. This is followed by a site inspection and damage assessment where an inventory of the damage to the residence is completed along with detailed mapping and photographs of damaged areas. One objective at this point of the investigation is to determine if the pattern of damage is consistent with a localized area of foundation settlement. Hand auger borings, probe rods, and shallow test pits are utilized to investigate shallow soil conditions and to examine the house foundation. A geophysical survey, typically ground penetrating radar (GPR), is also completed during the initial site visit. GPR provides particularly useful information of the continuity of shallow soil units and can identify subsurface anomalies such as down-warped soil layers, soil cavities, or columns of disturbed or raveled soils.

Areas of potential settlement or concern are then tested through drilling of standard penetration test (SPT) borings. During drilling particular care is given to accurate recording of soil types, circulation losses, and any soft or loose soil zones. The holes are subsequently back-filled with a cement-bentonite grout mixture. Laboratory soil tests are utilized to classify soil types and to determine critical soil engineering properties. Typical test methods include natural moisture content, percent passing a 200-mesh sieve, organic content, and Atterberg limits. The final data analysis in determining a cause of damage entails several aspects involving consideration of the pattern and extent of damage to the structure, as well as an assessment of the subsurface conditions with respect to evidence for raveling and downward transport of surficial soils into solution features in the limestone.

Remediation and Engineering Oversight

At sites where a sinkhole is identified as the cause of damage an appropriate site-specific remedial plan needs to be developed. In Florida, subsurface compaction grouting is the most common ground improvement technique used to remedy sinkhole conditions at residential sites. As discussed by Henry (1986), compaction grouting involves the high-pressure injection of a low-slump grout to densify the subsurface soils and to help seal and fill solution features in the limestone bedrock surface. When properly employed, compaction involves four stages: 1. plugging and sealing of limestone cavities, 2. densification of loose and raveled overburden soils, 3. lifting of overburden soils and the house structure, and 4. densification of the near-surface soils.

Table 1.: Site Investigation Methodology

1. INITIAL SITE VISIT	3. SUBSURFACE ASSESSMENT
<u>Interview Homeowner</u>: determine damage history, house construction, site data, etc. <u>Damage Assessment</u>: site map, exterior & interior inspections, photographs <u>Initial Subsurface Assessment</u>: hand auger borings, soil probings, test pit excavations <u>Ground Penetrating Radar (GPR)</u>: exterior & interior transects, define anomalous areas	<u>SPT Borings & Sampling</u>: log borings, collect samples, record soft drilling zones, circulation losses, rod drops, water levels <u>Laboratory Testing</u>: determine soil classifications and engineering properties
2. PRELIMINARY DATA ANALYSIS	4. DATA ANALYSIS & REPORTING
<u>Subsidence Indicators</u>: damage indicative of settlement, surface depressions, drainage conditions <u>GPR Anomalies</u>: location and type, relation to damage areas <u>Soils Maps</u>: organic soils, shrink/swell clays, high water table <u>Historical Information</u>: site conditions prior to house construction, former wetlands, landfills, dredged or filled areas <u>Drill Holes</u>: select borehole locations & depths, drilling equipment	Compile and evaluate SPT data Confirm soil classification with lab data Evaluate geological site conditions Evaluate cause of damage Prepare project report (include remedial plan if needed)

Estimating the costs to repair a sinkhole-impacted house is done to determine if grouting if feasible with respect to available policy limits. Most grouting jobs are bid on a unit rate basis due to the uncertainty in subsurface conditions. Grout quantities are estimated based on the anticipated density increase (from grouting) and the number and depth of the grout injection points. In practice, this formula is reasonably accurate, except at sites where significant limestone solution features or soil cavities are encountered during grouting.

During the grouting program oversight and inspection services are essential to provide information to the homeowner and the insurance company that the remedial plan is being implemented in accordance with project specifications. Standard inspection requirements include monitoring of depths and locations of grout pipes, grout slumps, injected grout quantities and pressures, and collection of grout samples for compressive strength tests, and site clean-up.

Follow-up Evaluations

Florida insurance rules require that properties be reinsured following repairs for sinkhole damages. Some sites require follow-up investigations to evaluate continued settlement damage after subsurface grouting of a sinkhole condition. While similar investigation methods are utilized (see above), the focus shifts to determining the effectiveness of the remedial program and the potential for unrecognized site conditions that are contributing (or causing) the on-going damage. The presence of shallow shrink/swell clay soils are one example and may be addressed through the installation of moisture barriers, site drainage controls, or underpinning in extreme cases. In other situations additional grouting may be appropriate, such as sites where the original grouting did not treat all the affected areas or the grout injection holes were not completed deep enough.

Emergency Response

Although rare, cover-collapse sinkholes do occur and can pose an immediate threat to a house structure. In these situations one of the homeowner's first actions is to contact the insurance company. Prompt action by the consultant can be important in providing needed assistance to the insurance adjuster and in preventing further property damages. Depending on the size and location of the collapse relative to the house, back-filling of the sinkhole may be sufficient to temporarily stabilize the structure. Shoring of the foundation and emergency grouting may also be necessary. Once the house is stabilized, a site investigation can be conducted to determine the extent of sinkhole-related damage and the scope of a final remedial plan.

Homeowner Assistance

Due to some widely publicized cases in Florida where large catastrophic sinkholes have swallowed entire homes (or even an entire city block), most homeowners with sinkhole damage claims have understandable concerns regarding the threat to their property and their personal safety. Common issues include the level of threat to life and property, typical layman's misconceptions regarding karst activity such as "underground rivers or caves" under the house, and the effectiveness of recommended remedial repairs. During the investigation process, the consultant should act as a responsible representative of the insurance company. Prompt and non-technical responses to questions will help reduce misunderstandings and conflicts.

CASE HISTORIES

Pinellas County Site

The subject property is a single-story, masonry block structure located in southeastern Pinellas County, in the city of St. Petersburg, Florida (Figure 2). The homeowner filed an insurance claim following extensive exterior and interior cracking to the house. The insurance company requested an investigation to determine the cause of the damage. During the initial site visit, a site

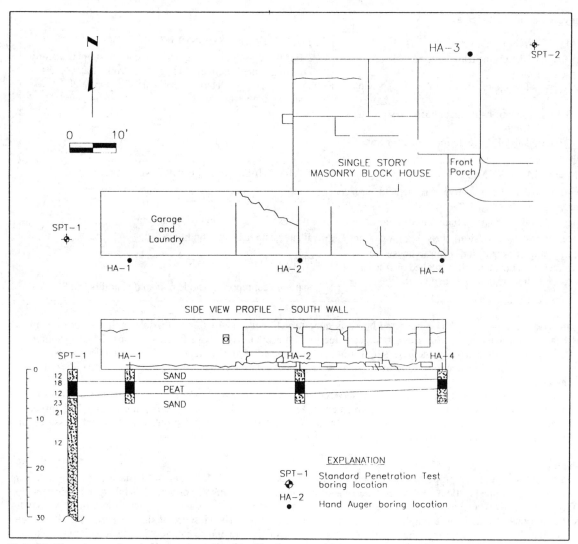

Figure 2: Plan view and geologic cross-section, Pinellas County residence

inspection/damage assessment, hand auger borings, and a GPR survey were completed. The foundation consisted of a stem-wall and crawl space with the interior floor supported on joists. The homeowner reported that similar damage was repaired two years prior to the inspection. Four shallow hand auger borings were completed during the inspection. Three of the borings along the south exterior wall encountered a layer of highly organic, fibrous peat at depths ranging from two to five feet below grade. The GPR survey indicated the presence of two distinct, subsurface reflections over most portions of the site. Based on the hand auger borings, the reflections were interpreted to represent the interface between the surficial sandy soils and the peat unit. The top of the peat was estimated to range in depth from 0.5 to 5.4 feet, with a gentle dip to the west side of the property. No anomalies suggestive of possible subsurface sinkhole or karst features were identified.

Area soil survey maps and a series of historical maps were examined following the site inspection. Mapped soil units at the site include the Made Land designation, which consist of mixed sand, clay, shell and rock generated during land-filling activities. The historical aerial photographs verified this condition. The 1926 aerial photograph shows the site as a low-lying, marshy wetland.

Two SPT borings were completed adjacent to the southwest and northeast corners of the house, as shown in Figure 2. Both borings encountered a thick section of generally medium dense quartz sands (Δ 10 blows per foot). The shallow peat layer was present in the boring at the southwest corner of the house. A unit of mixed clayey sand and limestone was encountered at the bottom of boring SPT-1 at a depth of 85 feet. The presence of abundant limestone fragments was interpreted to represent the transition from the undifferentiated surficial sands to the underlying Arcadia formation of the Hawthorn Group. Laboratory soils tests verified the highly organic nature of the shallow peat unit, with organic contents ranging from 24 to 80 percent. The results of the investigation clearly indicated the settlement damage was the result of the decay and compaction of the shallow peat soils along the south wall of the house. These organic soils are within the zone of seasonal fluctuation of the water table and are subject to the effects of oxidation and decay. As shown by the historical aerial photographs and the soil survey maps, the peat was likely deposited in a former wetland that was present at the site prior to construction of the house.

Hillsborough County Site

This site is a single-story, slab-on-grade, masonry block structure located in northern Hillsborough County, Florida. A geotechnical investigation consisting of hand auger borings, test pit excavation, and SPT borings was completed at the site in 1996. At the time, sinkhole activity was not found to be the cause for the settlement cracking and distress to the east exterior wall of the house. Following the investigation the damage was repaired and the house re-painted.

The damage re-developed shortly thereafter. Cracking in the garage wall and around the front door resulted in separations of up to one inch in width. During the following site inspection and damage assessment the extensive settlement cracking was documented and photographed. The pattern of the damage suggested an outward rotation of the east exterior wall, with an apparent locus of settlement at the southeastern corner of the garage. Shallow hand auger borings advanced through the garage floor slab indicated the surficial soils consist of fine-grained quartz sand. Review of soil survey maps and pre-construction aerial photos indicate the site was a former citrus grove, with no indications of unsuitable soil conditions. In addition, no obvious construction defects were found to explain the settlement. A GPR survey was completed along exterior portions of the site and within the garage. Two broad anomalies were identified at the northeast corner of the garage and in the front yard, south and southeast of the area of the damage at the front entryway. Two SPT borings were then drilled near these two anomalies. Both borings encountered a stratigraphy similar to the 1996 SPT borings, consisting of a thick layer of loose to medium dense surficial sands, underlain by clayey sand and clay. Limestone bedrock was encountered at depths of 36 and 42 feet. Although a narrow zone of very soft clay (2 blows per foot) and a circulation loss was recorded in boring SPT-2 near the limestone bedrock, no obvious raveled soil zones or limestone solution features were found.

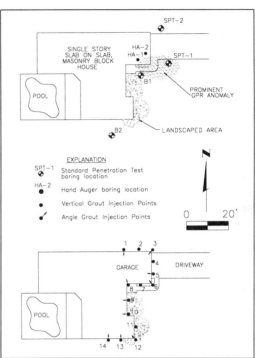

Following the SPT drilling, a second GPR survey was completed by another contractor to further delineate the previous subsurface anomalies. This survey concentrated on the anomaly at the area of the damage near the front entryway. This GPR survey clearly delineated the anomaly, which was interpreted to represent a possible buried depression in the subsurface clayey sand layer as indicated by an area of down-warped reflections. The apparent depression had a diameter of 27 to 35 feet and was centered near the southeast corner of the garage. Even though an SPT boring was drilled approximately 10 feet northeast of this location, it did not reveal a typical karst related stratigraphy. Using the GPR data and the magnitude and rate of the renewed settlement damage, sinkhole activity was determined to be on-going. Drilling results in this case were inconclusive, due to the proximity of the sinkhole to the house.

A program of subsurface compaction grouting was recommended to stabilize the house against further sinkhole-related subsidence. Underpinning of the foundation in the area of damage was also recommended if the grouting was not effective in re-leveling the structure. As shown in Figure 3, 14 vertical and angled grout injection points were specified along the eastern portion of the house to address the sinkhole condition. A grouting and underpinning plan, project specifications, and a cost estimate was included in the final project report.

Figure 3: Plan View, Hillsborough County residence

Hernando County Site

This site is located in western Hernando County, Florida (Figure 4). A single-story, slab-on-grade, masonry block house is located on the property, which is owned by a couple who recently moved to Florida. In late May 1998 a sudden sinkhole collapse formed along the east side of the residence. At the time, the homeowners were on vacation. A neighbor happened to be inside the house at the time and reported a loud rumbling noise. When he looked out a side door, he reported seeing the 'bushes and shrubbery disappearing into a cloud of dust.' Following extensive media coverage and the return of the homeowners, emergency response activities were initiated.

The sinkhole collapse occurred along the east exterior wall near the southeast corner of the house, and had stabilized with a surface diameter of approximately 20 feet and a depth of 15 to 20 feet. The west side of the collapse exposed and undermined the foundation of the house over a length of approximately 15 feet. At the time, minimal settlement damage was present, consisting of minor buckling of the soffit and minor separation cracking at the southeast corner of the house. Over the next three days, the sinkhole collapse was back-filled with approximately 190 cubic yards of natural sand/clay mixture. A temporary shoring was then constructed to support the exposed portion of the foundation, consisting of an 8 by 8 inch wooden beam with hydraulic jacks extended tightly against the footer. The emergency response activities were followed by a GPR survey. In general, the GPR profiles depicted a series of gently dipping subsurface reflections at depths of seven to 57 feet. The reflections were interpreted to represent various soil layering related to cross-bedding in the thick section of sand underlying the site. One significant anomaly was identified at the southeast corner of the screened pool enclosure.

A remedial plan for surface compaction grouting was then prepared. The plan consisted of seven angled grout injection points along the east side of the house, with four additional points within the back-filled sinkhole and the GPR anomaly at the corner of the pool enclosure. The grouting was completed 21 days following the sinkhole collapse. A total of 76 cubic yards of grout was pumped. The highest grout takes were recorded in the three injection points within the area of the sinkhole. Upon completion of the project, the house had sustained minimal settlement damage.

ACKNOWLEDGMENTS

Many thanks to Connie Rossman for preparing and drafting the figures.

REFERENCES

Beck, B. F., and Sinclair, W.C., 1986, Sinkholes in Florida, an Introduction: Florida Sinkhole Research Institute, Rpt. 85-86-4, Orlando, Florida, University of Central Florida, 79p.

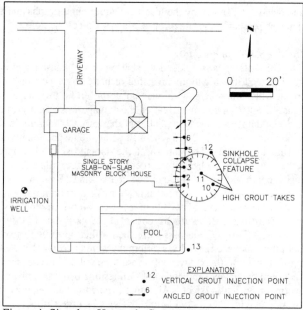

Figure 4: Site plan, Hernando County residence

Culshaw, M.G., and Waltham, A.C., 1987, Natural and artificial cavities as ground engineering hazards: Quarterly Journal Engineering Geology, vol. 20, p.139-150.

Florida Geological Survey, 1997, Geologic and geotechnical assessment for the evaluation of sinkhole claims; Florida Geological Survey, Open File Report No. 72.

Henry, J.F., 1986, Low Slump Compaction Grouting for Correction of Central Florida Sinkholes: in Proceedings, National Well Water Association Conference, Bowling Green, Kentucky, October 1986.

Schmidt, W., and Scott, T.M., 1984, Florida karst – Its relationship to geologic structure and stratigraphy: in Proceedings of the First Multidisciplinary Conference of Sinkholes, A.A.Balkema Publishers, Rotterdam, Netherlands, pp. 11-16.

Scott, T.A., 1988, The Lithostratigraphy of the Hawthorn Group (Miocene) of Florida: Florida Geological Survey Bulletin No. 59.

Sinclair, W.C., Stewart, J. W., Kuntilla, R.L., Gilboy, A.E, and Miller, R.L., 1985, Types, Features, and Occurrence of Sinkholes in the Karst of West-Central Florida: U.S. Geological Survey Water Resources Investigation Report 85-4126.

Smith, T.J., 1997, Sinkhole damage investigations for the insurance industry in west-central Florida; The Engineering Geology and Hydrogeology of Karst Terranes, Beck & Stevenson eds., A.A.Balkema Publishers, Rotterdam, Netherlands, pp. 299-304.

Wilson, W.L., and Beck, B.F., 1988, Evaluating Sinkhole Hazards in Mantled Karst Terrane: in Geotechnical Aspects of Karst Terrains: Exploration, Foundation, Design and Performance, and Remedial Measures, ASCE Geotechnical Special Publication No. 14, p. 1-24.

Karst and engineering practice

VLADIMIR V.TOLMACHEV State Venture Antikarst and Short Protection Institute, Dzherzhinsk, Russia

ABSTRACT

Economic development in karst terrane gives rise to a number of problems, which are solved by engineering karstology. Twenty types of problems can be cited on the basis of the geological and engineering/building disciplines. The nature of karst hazards should only be considered within the framework of "karst construction". The probable types of karst impacts upon structures are classified into four groups, which determine preventative measures against karst impacts (antikarst measures). The efficiency of various predictive models of karst collapse is also a function of concrete engineering and economics.

The problem of how to provide safety to the population and economic enterprises in karst terranes has always been a reality in Russia. This is due to the following circumstances:

- Karst terrane makes up a considerable portion of the land—approximately 1/5th of the country as a whole and approximately 1/4th of the European part (Figure 1). Therefore, 1/3rd of all populated sections in one or another part of the country are prone to karst processes.

- Economic activity, from the past to the present, essentially stimulates karst processes. The reason is that geological environmental protection is rarely conducted; activity proceeds without considering the karst-specific processes. Human's influence upon the ecological environment almost seems to attract karst formations to economic activity. In addition, the influences of humankind have a great period of post-activity—up to decades.

- The number of industrial concerns calling for perfect safety is increasing—nuclear power stations, large enterprises that are extremely dangerous to the ecology, etc. In these situations, maximum antikarst protection must be provided. As a general rule, perfect protection against karst subsidence (especially under conditions of intensive man-made influences upon the geological environment, the occurrence of deep soluble rocks and karst conduit collapse) cannot be provided even for these dangerous enterprises. In these cases, we must not restrict ourselves to the single long-term prediction of karst hazards. The prediction should be of medium term—up to 50 years, short term—up to 10 years, and operational—up to one year.

Operational prediction must be permanent and carried out based on special karst monitoring. Only in this way will a case of "Karst Chernobyl" not occur.

- The speed of train traffic is increasing on railways that are very sensitive to karst deformation. In this regard, karst sinkholes of any size beneath the railway bed almost always cause train wrecks, or at best long interruptions in train traffic. Many railways in Russia were built without properly considering karst processes. Capital antikarst protection of constructions of large extent, such as railways, is extremely ineffective from the geological engineering and the economic point of view.

- The length of oil and gas pipelines is increasing. Pipelines are generally the least sensitive to the effects of sinkholes; in most cases the "critical span" of a sinkhole underneath a pipeline is more than 20 m. However, because of the decreasing quality of construction, building defects, and decreasing capital repairs, gas and oil pipelines are now becoming the most dangerous objects in karst terrane. In some cases, serious damage to gas pipelines occurred when sinkholes were only 6 m in diameter.

- For various reasons, most structures that have been built in karst terrane do not have proper antikarst protection. This is especially true of construction more than 30-40 years old.

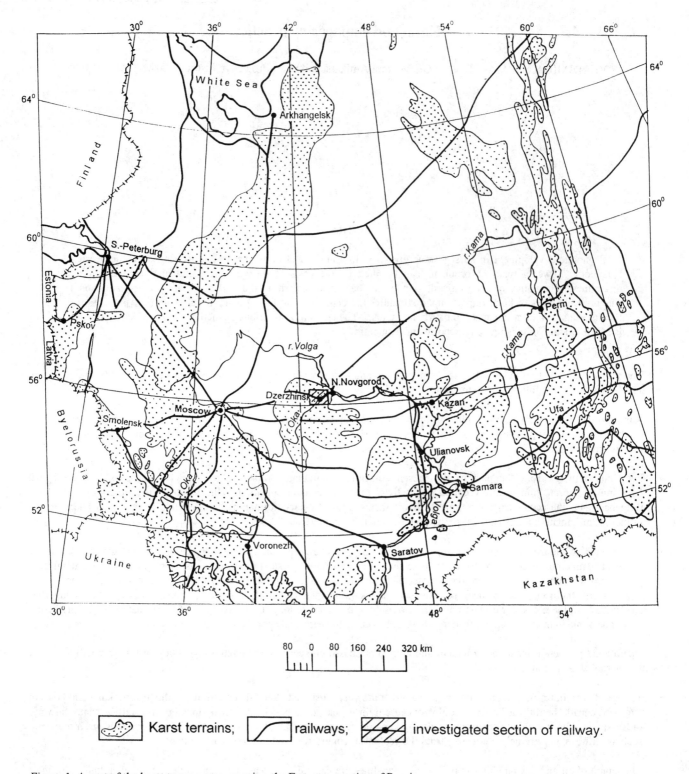

Figure 1. A part of the karst terrane map covering the European portion of Russia.

- Where waste disposal and industrial sites are located in karst terrane, conditions exist for severe deep pollution of rocks and groundwater. Until the present time, this problem was not addressed well in Russia. It is important to devise special methods of monitoring the pollution dynamics in karst terrane, to carefully and rationally choose sites for waste-disposal, and to carefully choose engineering methods for protection against geological environmental pollution at new sites and for sites already in operation.

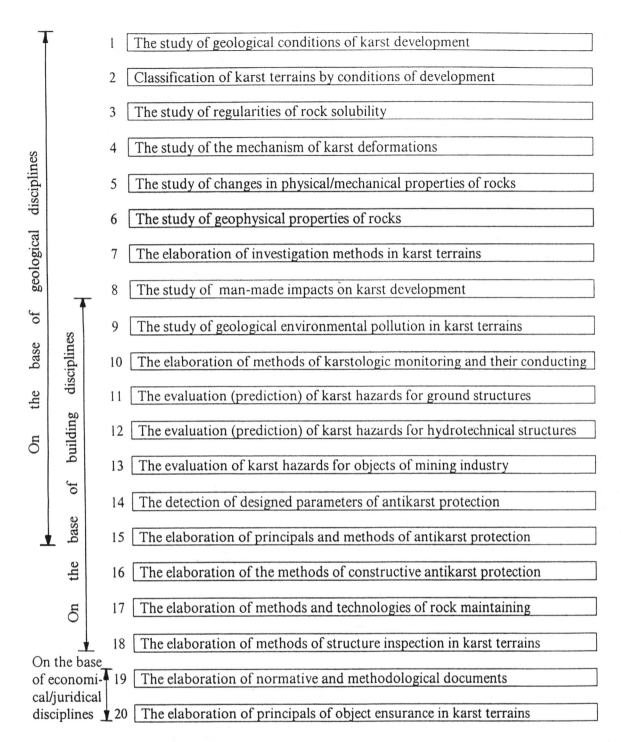

Figure 2. Main directions of engineering karstology.

From the brief review above, it is evident that investigations of karst phenomena must not only be purely informative but should also reflect the problems resulting from economic activity. Economic development in karst terrane gives geological engineers, engineers, ecologists, economists, lawyers, and other specialists theoretical, methodological, and administrative tasks. The most complicated problems appear when karstology proper overlaps with the disciplines of engineering and economics. Solving these problems is the aim of engineering karstology. It should be noted that by virtue of its specific goals, engineering karstology is distinct from traditional karstology. Engineering karstology is forming its own approach to the study of karst. This approach is affected by the necessity to arrive at concrete design and administrative decisions. Those decisions are influenced by the stochastic nature of the karst process, insufficient data, poorly accounted for man-made influences to karst stimulation, the limited time of

Table 1. Four groups of failures of structures, caused by karst deformations.

The type of failure	Common characteristics	Versions of structural failure
I	Disastrous damages which cause total work stoppage or long interruptions of operation	The loss of static stability of the structure. Damage to individual components preventing continuation of operation. Damage to processing equipment, which can in turn destroy the structure, hurt humans, cause unacceptable releases of chemicals or radioactivity.
II	Damages which cause long interruptions of operation	Impermissible leaning of a structure Deformation of structures reaching the stage near damage stage. Deformation of structures reaching sizes hazardous for operating.
III	Damages which cause disruption to normal operations	Danger to or damage to individual components of a structure. Deformation of structures outside of elastic stage.
		Formation of wide cracks in reinforced concrete or stone structures, but those cracks do not cause a lessening of carrying capacity. Deformations which restrict normal operations but are repairable within a short amount of time. The proximity of ground deformations or their symptoms to structures that have not lead to visible damage but cause fears for safety.

investigations and technical probabilities, the lack of money for antikarst protection, the individual structures, etc.

Presently engineering karstology is developing in such countries as Russia, Germany and the United States. This has resulted in publications of monographs on various applied problems of economic development in karst terrane (Lyicoshin, 1968; Kuharev, 1975; Zlobina, 1986; Khomenk, 1986, Tolmachev, etc., 1986; Kutepov, etc., 1989; Reuter, etc., 1990; Tolmachev, etc., 1990; Dublyanskaya, etc., 1992; Lyicoshin, etc., 1992; Sowers, 1996). Another result are regular international and national conferences on various problems of engineering karstology (Russia: Perm, 1947; Gorky, 1965; Dzherzhinsk, 1978; Podolsk, 1984; Kuybishev, 1990; Perm, 1990, 1992; Dzherzinsk, 1996. Germany: Hannover, 1973. USA: Orlando 1984, 1987; St. Petersburg, 1989; Panama City, 1993; Gatlinburg, 1995; Springfield, 1997; Harrisburg, 1999).

Within a certain degree of convention, about 20 scientific aspects of engineering karstology are noted. They are divided primarily on the basis of ecological and building disciplines (Figure 2), although, in recent years we have also had to deal with new disciplines for karstology—economics, jurisprudence, ecology, etc. These various aspects are all related to each other, and as a rule, represent the field of professional activity of individual specialists.

Karst formation and its influence upon engineered structures vary greatly. Karst subsidence beneath the foundation of structures can act in a variety of ways upon the structure's integrity. Using terminology of "reliability theory", we call the consequences of such actions "failures", by which is meant the partial or total loss of the structure's integrity. As the analysis of accidents and structural damage due to karst deformation has shown, it is appropriate to classify all types of failures in karst terrane into four groups (Table 1).

The probability of one or another type of failure depends first of all on the type of structure and the type of karst deformation in the compressible rock mass of the ground (Table 2). In Table 2 are shown predictions of the types of failure ranging from the worst case in the numerator to the best case in the denominator.

From Table 2 it may be concluded that the impact of karst deformation on structures may vary radically according to the type of karst deformation and the type of structure. Any local deformation of the bed of frame buildings on detached foundations always gives rise to disasters, Type I failure. For frameless buildings, as a general rule, such failure can occur only if a karst collapse takes place beneath the bed of a building and has a great span, more than 10-12 m. One can read about the determination of the design span of a collapse under a foundation in published papers (Sorochan, etc., 1985; Tolmachev, etc., 1986).

The data of Table 2 can be used when evaluating necessary antikarst protection. It can be used to determine the minimum amount of protection necessary compared to the available monetary funds (Tomachev, etc., 1986; Tolmachev, 1997).

The amount and nature of antikarst protection are dependent upon combinations of the parameters of karst deformation and design structure. By way of example, let us analyze the degree of impact of sinkhole frequency and some parameters of the designed structure (Figure 3):

λ = the time unit of the number of sinkholes formed
t_n = the schedule and deadlines for utilization of a structure

Table 2. The most probable types of failures of structures upon exposure to karst deformations.

Type of structure		Types of karst deformations at the bed of structures				
		Sinkholes				Subsidences on a large scale
		Collapses		Raveling sinks (shallow depths)	Local subsidences	
		With small diameters	With large diameters			
Frame buildings on detached foundations		I/II		I/III		II/III
Frame buildings with belt or plate foundations		II/III	I/II	II/III	I/II	II/III
Frameless buildings		II/III	I/II	II/III	II/IV	III
Tower buildings and structures		I/II				I/III
Bridges	Statically determinable	I/II		II/III		III
	Statically undeterminable	I		I/II		I/III
Railways		I/II				I/III
Roads		II/III	I/IV	III/IV	II/III	II/III
Pipelines		II/IV	I/III	III/IV	I/III	I/IV

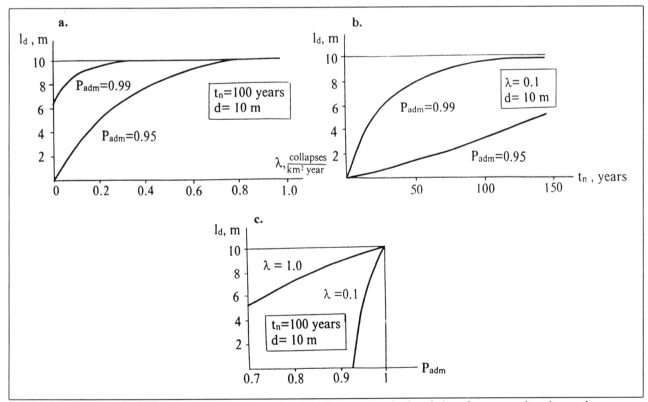

Figure 3. The nature of the dependence of designed span λ_d of a collapse under the foundation of a structure based upon the parameters of sinkholes and structures.

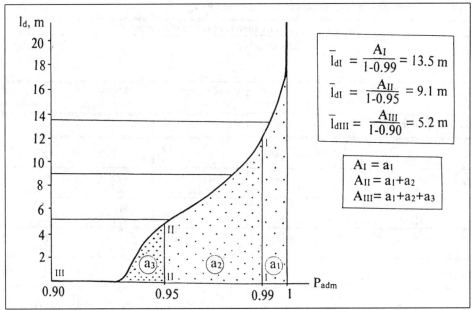

Figure 4. An example of the determination of the mid-weighted meaning of the designed span I_d for structures which fall under the responsibility types I, II and ($P_{adm\ I}$ = 0.99, $P_{adm\ II}$ = 0.95, $P_{adm\ III}$ = 0.90)

d = a constant meaning of the diameter of a sinkhole and its area in the plan of the structure
I_d = designed span of a karst sinkhole at d
P_{adm} = the degree of structure responsibility per I_d at d

P_{adm} is the minimum acceptable structural safety factor, ranging from zero to a maximum of 1. The design span is defined as the median sinkhole diameter corresponding to the interval P_{adm} max P_{adm} (Figure 4).

Though the given analysis has been made with reference to an industrial frameless building with belt foundation in karst terrane in Dzherzhinsk, Russia, these general principles can be extended to other conditions in a qualitative sense. The dependence of I_d on the discussed factors (at d=constant=10m) is shown in Figure 3. The general principles can be followed from the nature of this relationship. From the relationships developed, it is obvious that by combining the detailed data on the parameters of karst sinkholes and the unique characteristics of structures, it is possible to increase safety and the economy of planned decisions for antikarst protection.

The main problem of development in karst terrane is the prediction of karst hazards. The choice of a predictive model of collapse formation in economically developing karst terrane is influenced by both natural and man-made conditions of karst development, and therefore must solve both engineering and economic problems. Among these problems are:

1. The evaluation of the economic efficiency of building in large urban and industrial complexes.
2. The determination of building parameters for residential areas and other enterprises based upon balanced risk (Tolmachev, etc., 1986).
3. The evaluation of karst hazards when choosing sites for development.
4. The comparison of various methods of antikarst protection and the evaluation of their efficiency.
5. The determination of design parameters for antikarst protection.
6. The prediction of man-made impacts upon the predicted parameters of karst development.
7. The prediction of karst subsidence with the goal of providing safety to operating structures.
8. The prediction of the impact of geological environmental pollution from waste disposal sites due to the development of cover collapse sinkholes.
9. The validity of financially insuring engineered structures in karst regions.

At present in engineering karstology, the following models for sinkholes are most often used:

a. Engineering/geological qualitative models which have been widely used from the 1930's to the present.
b. Stochastic models.
c. Deterministic geotechnical rated models

d. Deterministic laboratory models
e. Geophysical models.

When using these models in practice, consideration must be given to the degree of their efficiency. The efficiency is indicated here by the cost/benefit ratio of the minimum acceptable predictive parameters necessary to solve the appropriate engineering/economic problem. The results of the analysis for the efficiency of various models "a-e" with reference to the above-mentioned 9 problems for covered sulphate-carbonate karst where the depth of soluble rocks is greater than 15 m are listed in Table 3.

Weighted predictive models are not an alternative. They would be appropriate for use possibly as a package, more so if the conditions of their application are influenced by the completeness of the initial information available along with its cost and the terms of obtaining it. These models have been approved by the State Venture Antikarst Protection Agency with reference to covered sulphate-carbonate karst. Models of sinkholes are constantly improved with regard to solid engineering/geological situations for specific features of engineering/economic problems.

Table 3. The efficiency of predictable models of sinkholes when solving various engineering and economic problems.

Model	Engineering and economic problems								
	1	2	3	4	5	6	7	8	9
A	+	-	*	-	-	-	-	-	-
B	*	+	*	*	+	-	-	*	+
C	-	*	+	*	*	+	*	*	*
D	-	-	+	+	*	+	*	+	-
E	-	*	+	+	+	*	+	+	*

Notes: + = highly efficient, * = efficient, - = inefficient

Recently the most common problem is the validity of measures to prevent ecological disasters from chemical industries located in hazardous karst terrane arising from the extremely poor antikarst protection of existing facilities and their long-term impact upon the geological environment. Under these conditions, it is necessary to carry out local and facility karst geophysical monitoring of the geological environment with the use of category E Model in combination with Models D, B, and C. Presently, there also exists an acute problem of risk evaluation for insurance of structures in karst terrane. It is self-evident that in this case, Model B is appropriate.

It should be noted that spatial/probabilistic methods of sinkhole prediction are best suited to long-term predictions. It makes no sense to use formulas used in short-term or immediate predictions or probable long-term predictions for small sites because the absolute value of the risk probability turns out to be practically insignificant. In this case the most efficient probabilistic/deterministic method of prediction is based upon considering the probable regularities of the karst process, the mechanism of karst development under definite stochastic man-made influences, and the sufficiently fixed locations of various karst formations.

The use of probabilistic/deterministic methods of karst prediction requires the concurrent application of geophysical methods for the detection of areas of anomalous soil thickness, the physical simulation of sinkholes, and rated methods based on geotechnical models. Probabilistic and probabilistic/deterministic methods of risk evaluation for karst processes, which are well-investigated and proven in practice, can be a good basis for insuring the decision making procedure for existing structures, especially those which have primarily a purely economic value. "Karst risks" for insurance are favorable to both parties of an insurance agreement owing to the following circumstances:

- Karst subsidences, generally sinkholes, have a discrete/local nature in plan of several meters, compared to such natural phenomenon as earthquakes, tsunamis or landslides.

- There are certain symptoms of sinkholes which can be noted in a timely manner in the process of special and objective karst monitoring.

Karst monitoring includes the investigation of a karst area using: geophysical methods aimed at finding karst anomalies in the bedrock, especially at great depths, alarm systems, inspecting for structural deformation, etc. In the detection of spatial symptoms of sinkholes, it is necessary to conduct investigations utilizing laboratory, physical and mathematical models for the evaluation of the time of probable sinkhole formation based upon the detection of karst anomalies and for the probable sinkhole sizes at the base of the foundation (design span of sinkholes). It is the probabilistic evaluation of the temporal and spatial parameters of sinkholes, and the prediction of structural response upon exposure to karst sinkholes that can be the basis for the evaluation of insurance risk. The introduction of insured karst risk will allow a change of the principles of development on karst terrane in Russia.

REFERENCES

Dublyanskaya, G.N., Dublyansky, V.N., 1992, Mapping and division into regions and an engineering-geological estimation of karst territories. Novosibirsk. P. 144 (Russian)

Khomenko, V.P., 1986, Karst piping processes and their prediction. Moscow, Nauka, p. 97 (Russian)

Kuharev, N.M. 1975, Engineering/geological investigations in the field of karst development for building purposes. Moscow:, Stroyizdat, p. 183 (Russian)

Kutepov, V.M., Kozhevnicova, V.N., 1989, Stability of karst terrains. Moscow, Nauka, p. 151 (Russian)

Lyicoshin, A.G., 1968, Karst and hydrotechnical building. Moscow, Stroyizdat, p. 183 (Russian)

Reuter, F., Tolmachev, V., Molek, H., Suderlau, Gl, Chomenko, V., 1990, Bauen und Bergbau in Senkungs – und Erdfallgebieten. Berlin, Akademieverlag, p. 177 (German)

Sorochan, E.A., Troitsky, G.M., Tolmachev, V.V., Khomenko, V.P., Klepicov, S.N., Metelyuk, N.S., Grigoruk, P.D., 1985, Antikarst protection for buildings and structures: Proceedings of the 11th International Conference on Soil Mechanics and Foundation Engineering, San Francisco/12-16 August, Rotterdam-Boston, A.A. Balkema, pp. 2457-2460

Sowers, G.F., 1996, Building on Sinkholes, New York, ASCE Press, p. 202

Tolmachev, V.V., Troitsky, G.M., Chomenko, V.P., 1986, Engineering/building development of karst terrains. Moscow, Stroyizdat, p. 177 (Russian)

Tolmachev, V.V., Reuter, F., 1990, Engineering karstology. Moscow, Nedra, p. 152 (Russian)

Tolmachev, V.V., 1997, Evaluation of antikarst protection efficiency, Proceedings of the 6th Multidisciplinary Conference on sinkholes and the Engineering and Environmental Impacts of Karst, Springfield, MO./6-9 April, Rotterdam-Brookfield, A.A. Balkema, pp. 371-373

Zlobina, V.L., 1986, Underground water impacts on the development of karst-piping processes. Moscow, Nauka, p. 133 (Russian)

Replacing the Beards Creek bridge

STANLEY E. WALKER Haley and Aldrich, Incorporated, Rochester, N.Y., USA

SUSAN L. MATZAT LaBella Associates, P.C., Rochester, N.Y., USA

ABSTRACT

In early March 1994, fresh water (more than 20 mgd) began pouring into the Retsof Salt Mine about 1,100 feet beneath the Beards Creek bridge in western New York. Despite efforts to stop it, the inflow continued until early 1996 when the 6,500-acre mine became completely flooded. The effects of the inflow were twofold; a loss of ground in and above the limestone aquifer about 600 feet above the mine, and the closure of the mine cavity at and spreading radially from the location of the inflow. These effects manifested themselves in the development of two sinkholes immediately upstream of the bridge. Concurrently with the inflow and development of the sinkholes, the bridge began to settle rapidly and was declared unsafe and taken out of service. By March 1996 the bridge had settled more than 15 feet and moved approximately 5 feet upstream toward the sinkholes. To meet the public need, the bridge was replaced, at its previous location.

The authors' firms designed and oversaw the construction of the replacement bridge on an overlapping, fast-track schedule. The approach roadways were restored to a level of serviceability (flood survivability) equivalent to that existing prior to the recent ground subsidence. This required the placement of more than 30,000 tons of new earthfill. The authors' firms also performed close monitoring of displacement and piezometric responses to the fill and structural loadings. This allowed completion and opening of the bridge in less than 10 months while accommodating the anticipated 8 in of construction-phase settlement.

This paper presents summaries of the fast-track design and construction process and of the movements of the bridge and embankments during and following construction.

BACKGROUND

U.S. Route 20A crosses Beards Creek, a tributary of the Genesee River, in the hamlet of Cuylerville near the Town of Geneseo in western New York. (See Figure 1, Site Location Map.)

Starting in March 1994, fresh water (more than 20 mgd) began entering the Retsof Salt Mine about 1,100 feet beneath the bridge crossing the creek. Despite efforts to stop it, the inflow continued until early 1996 when the mine became completely flooded.

The effects of the inflow were twofold; a loss of ground in and above the limestone aquifer about 600 feet above the mine, and the closure of the mine cavity at and spreading radially from the location of the inflow. These effects manifested in the development of two sinkholes and progressive settlement of the land surface (15 ft in the vicinity of the bridge and gradually diminishing to less than 1 ft) over an area of more than 2 square miles surrounding the bridge site. The larger sinkhole is approximately 600 ft. in diameter and 65 ft. deep. The smaller sinkhole, located just upstream of the bridge, is approximately 260 ft. in diameter and 40 ft. deep. The bridge settled 15 ft and moved approximately 5 ft upstream toward the smaller of two sinkholes. The land subsidence imposed significant changes on the channels and floodways of Beards Creek and the nearby Genesee River.

A description and analysis of the geologic and hydraulic changes at the Beards Creek bridge site was presented at the Sixth Multidisciplinary Conference on Sinkholes and the Engineering Impacts of Karst Conference, in Springfield, Missouri, April 1997 (Walker and Matzat 1997).

Bridge Collapse

As the 100-ft long, three-span, reinforced concrete bridge, built in 1936, continued to settle it also shifted laterally toward the creek and the nearby sinkhole. In response to these movements, numerous concentric tension cracks (as wide as 6 in.) with vertical offsets of as much as 24 in. developed in the roadway pavement and land surface to distances of 200 to 300 ft. surrounding the bridge and the sinkhole. These lateral movements of the near-surface soil caused severe buckling of the pavement at the bridge and the thrusting of the west abutment 6 ft. toward the center of the creek. The western pier was also thrust eastward and broken by the shifting of the foundation soils. The deck was more than 16 ft below its original elevation, sagging about 2 ft at mid-length and tilting at least 1 ft downward on its upstream side, toward the sinkhole. Figure 2 depicts the position of the bridge and its foundations relative to their original positions.

Replacement Objectives

The principal objective in replacing the damaged bridge was to re-establish its level of service, including emergency access during floods, as soon as practicable (by Autumn 1997, if possible). Other significant design and construction objectives included: 1) providing a 50-year service life, 2) limiting the acquisition of additional right-of-way, 3) avoiding any expansion of the right-of-way onto an existing historic park between the western approach roadway and the nearby sinkhole, and 4) avoiding disturbance of areas of potential archeological significance (identified within the existing right-of-way during the design).

To meet these objectives, the design called for raising the roadway as much as 18 ft above its sunken level. This raising was needed to allow passage of the 50-year flood beneath the bridge considering anticipated settlements during its 50-year design life. The design also required construction features to not only resist potential future scouring of the bridge's foundations and approaches, but to accommodate a total of as much as 2 ft of settlement anticipated to occur during and following its construction (Matthews, 1996).

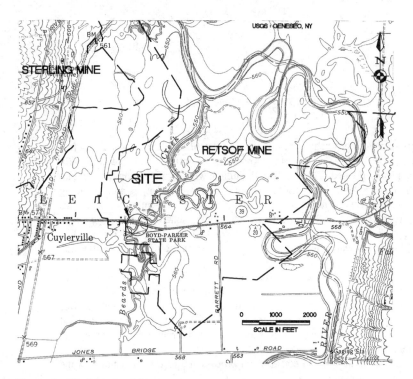

Figure 1: Site Location Map

DESIGN CONFIGURATION

As depicted in Figure 3, the new bridge consists of a 125-ft long single-span, steel girder superstructure with an integral reinforced concrete deck seated on pile-supported concrete stub abutments.

The foundations and abutments of the old bridge were left in-place and the creekbed was regraded with 10 ft or more of stone fill to stabilize the creekbed and buttress the new abutments. Thirty, 40-ft long tapered, concrete-filled, steel piles driven into the fill and natural soils behind the old abutments were used to support the new abutments and to provide further assurance against future scour damage. To compensate for the severely disturbed condition of the subgrade soils, the new approach embankments required special stabilization. The availability of purchasable right-of-way on the north, downstream side of the roadway, allowed conventional flattening of the new embankments to achieve the required stability. Vertical sheet piling (04 to 05 ft long) was driven to depths of about 25 ft and anchored to continuous deadmen buried in the new embankments to retain (to avoid encroachment into the adjacent park land) and to stabilize the upstream sides of the approaches. Sloping, stone riprap was called for where the topography and right-of-way permitted its installation.

To accommodate the anticipated 8 to 16 in. of post-construction settlement and related potential distortion resulting from the installation of the more than 30,000 tons of earthfill, concrete, and steel, the bridge was designed with repositionable, flexible expansion bearings at both ends of the girders, extra clearance between the girders and the abutment backwalls, and flexible, strip-sealed expansion joints at both ends of the deck to allow the deck to move up to 2 in longitudinally.

ANTICIPATED GROUND RESPONSE

Due to its massively disturbed condition, prediction of the ground's response to the proposed construction did not lend itself to conventional analytical approaches. However, the abundance of geologic, hydrologic, geotechnical, and survey data (NYSDOT, 1981, 1994) (Akzo, 1994-1996) gathered during earlier studies relating to the operation of the mine and the design of other bridges and facilities in the in the general vicinity of Beards Creek site and the intensive geologic investigations and subsidence monitoring conducted during the collapse and inundation of the mine provided a substantial empirical database upon which predictions of foundation stability and future ground movements could be made (VanSambeek, 1994,) (Gowan, VanSambeek, and Brekken, 1996).

Relying extensively on this broad database, the authors developed a working stratigraphic model of the bridge site and selected soil strength values and deformation characteristics for use in foundation stability analyses and future ground movement projections. In particular, the New York State Department of Transportation database provided an abundance of shear strength and consolidation characteristics for the near-surface alluvium and the underlying soft lakebed deposits, while the more than two years of bi-weekly subsidence monitoring data from several transects in the immediate vicinity of the bridge provided an exceptional tracking of the site's plastic deformational response to the mine collapse and the related sinkholes. Although, as noted earlier (Walker and Matzat, 1997), the bedrock and deep glacial deposits were subjected to significant changes in piezometric pressures as the deep ground water poured into the mine, the overlying, thick lacustrine clay layers were hydraulically isolated from those pressure changes and were estimated to have yielded less than 1 ft of consolidation related subsidence.

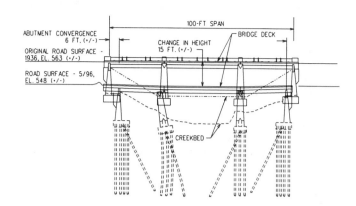

Figure 2: Original and Collapsed Position of Beards Creek Bridge

Interpretation of the voluminous settlement-versus-time data was crucial to the prediction of the on-going effects of plastic deformations of the soils adjacent to the large sinkholes. Although the settlement rates had diminished markedly as the mine filled with water in late-1995, they were still averaging about 0.01 ft every two weeks. If the subsidence were to continue at such an apparently slow rate, the result would be an additional 12.5 ft of settlement over the 50-year design life of the new bridge. However, based on a careful examination and regression analysis of the post-inundation settlement data,

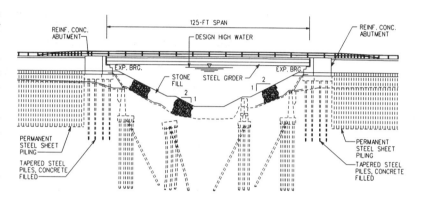

Figure 3: New Bridge Configuration

the data actually indicated less than 1 ft of plastic deformation would likely occur over the next 50 years.

Based on the foregoing considerations the authors concluded that the land beneath the new bridge would likely undergo between 8 to 24 in of settlement during construction and the subsequent 50 years, allocated to approximately 8 in of rapid elastic compression as the fills were placed, 8 in of long-term plastic deformation associated with the on-going adjustment of the upper soil deposits to the presence of the nearby sinkhole, and 0 to 8 in of long-term consolidation of the underlying thick clay layers.

CONSTRUCTION

To have a chance at reopening the road to public traffic by the autumn of 1997, it was imperative that construction be started as early as possible and advanced as rapidly as weather and the ground responses would allow. Accordingly, as soon as the required Corps of Engineers permit to work in the creek was received in mid-September 1996, demolition of the old bridge's deck and piers was started, while contract documents for the initial earthwork were being finalized. The contract for the initial earthwork was awarded in mid-December.

Response Monitoring

To minimize the uncertainty associated with the possible responses of the severely disturbed subgrade soils and yet, allow the earthwork to advance as rapidly as would be prudent, a network of settlement platforms, inclinometers, and piezometers was installed in critical locations before the earthwork was started. The eight settlement platforms were set and secured on the existing subgrade along the roadway centerline on both sides of the creek, with one seated in the creekbed. The four inclinometers and vibrating-wire piezometers were installed, in pairs, with two immediately creekward of the proposed abutments at the centerline of the bridge and the other two about 10 ft south of the abutment wingwalls, toward the sinkhole. The inclinometers extended to 60 ft below the initial subgrade. The piezometers were set about 30 ft below subgrade, where the maximum shear stresses from the new embankments and abutments were expected to develop.

These instruments were calibrated before the initiation of the earthfilling and were read frequently throughout the construction of the approach embankments and the bridge until its opening in October 1997. The data from these readings, along with close surveys of the positions of several critical points on the abutments as they were raised, were carefully evaluated and used to control the progress of the construction.

Design and Construction Progress

The design and construction were advanced in a tandem ("fast-track") sequence to allow the early placement of the lower portions of the heavy earthfill and to promote (and get a measure of) anticipated settlement and pore-pressure responses and to see if the new loads would cause measurable lateral movements in the subgrade soil. Staging of the design and construction provided a logical ("critical path") sequence in which the details of subsequent elements of the bridge and its foundations could be refined and adjusted, if necessary, to accommodate unexpected ground responses.

The earthwork was started (with the oversight and concurrence of NYSDOT engineers) in late December with the placement of crusher-run aggregate fill to raise the northern flanks of the new embankments to the approximate level (El 548, NGVD) of the remnant road grade and with the placement, in early February 1997, of 7 to 10 ft of medium stone fill in the sunken creekbed to stabilize the severely ruptured subgrade and to re-establish the creek's channel. This filling helped to advance the compression of the subgrade and provided a working level well above normal spring run-off creek levels. As depicted and noted on Figure 4, the initial stabilization filling was followed by installation of the embankment-retaining sheet piling along the southern side of the roadway and the raising the roadway embankments to El. 555, about 8 ft above the grade of the sunken roadway. An additional 7 ft of temporary surcharge fill was placed at the abutment locations.

In the meantime, design of the abutments and superstructure was completed and the girders were ordered. In early May, after settlements of about 4 in, the surcharge was removed and the abutment foundation piles were driven and the pile caps cast. By early June, after about 4.5 in. of total subgrade settlement, the abutment backwalls and bridge seats (girder-supporting pedestals) were cast. The embankments were then raised to El. 558.5, about 5.5 ft below the final roadway grade. The elevations and distance between the two abutments were closely monitored. These were found to have settled about 1 in. with no lateral (convergent) movement detected by the control surveys or inclinometer measurements. On 10 July, the four girders were installed. In early August, the embankments were raised to within 1.5 ft of final roadway grade. With the abutments having settled about 2 in since their completion, on 28 August, the bridge's concrete deck was cast and finished. By 6 October, the embankments had been brought to final grade and the roadway was paved (Walker and Divito, 1997).

On 20 October 1997, less than ten months from the start of its construction, the bridge was opened to the public.

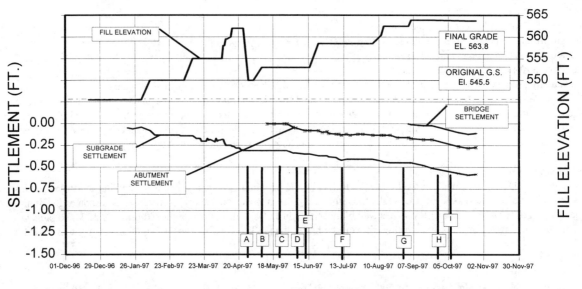

Figure 4: Construction Progress and Settlements

GROUND RESPONSE SUMMARY
During Construction

As shown on Figure 4, the old ground beneath the bridge had settled 7 to 8 in under the weight of the rebuilt roadway and new bridge. However, due to the construction sequencing, the finished deck and roadway had undergone only 2 in of essentially uniform and imperceptible settlement. At no time did the closely monitored instrumentation indicate any excessive pore pressures or instability in the underlying soil.

One Year Later

In late October 1998, the authors revisited the bridge site to observe its condition. The established settlement monitoring points were resurveyed. The survey indicated that the approach embankments and the bridge had undergone approximately 3 in of additional general settlement with a slight (0.1 to 0.4-in) cant downward to the south, toward the sinkhole. It was observed that the flexible bearings on both ends of all four girders were uniformly strained about 0.5 in, indicating that the abutments had converged about 1 in since the original setting of the girders. Follow-up readings of the inclinometers have indicated minor (up to 0.25 in) lateral movements at the ground level (El. 555+/-) adjacent to the abutments toward the creek and to the south, toward the sinkhole since completion of the construction.

PERFORMANCE SUMMARY

During the 10-month-long reconstruction of the Beards Creek bridge, when more than 30,000 tons of earthfill, steel, and concrete were placed within 300 ft of the sinkhole, the ground beneath it yielded up to 8 in of settlement, moved laterally as much as 0.5 in, and demonstrated no detectable signs of foundation instability. During the year following its completion, the bridge and its approach embankments have undergone an additional 3 in of general settlement with some tilt toward the nearby sinkhole and some modest (in the order of 0.5 in) lateral shifting of the abutments toward the creek. Movements of the nature and magnitudes observed were anticipated in the design of the bridge and are well within the projected values. U.S. Route 20A is very heavily traveled, for a rural highway, and the new bridge and reconstructed roadway are performing exceptionally well, considering the site's recent history.

CONCLUSIONS

Prediction of the responses of severely disturbed ground to new loadings associated with the repair or replacement of vital public infrastructure elements damaged by such disturbances presents a significant challenge to design professionals.

An understanding of the course and destiny of the damaging ground disturbance, the geologic setting, and the fundamentals of soil mechanics are essential for predicting the ground's responses to new loads and stresses associated with the stabilization and re-use of the disturbed ground. Documented time-series observations of the ground's failure, even if just during the failure's aftermath, are crucial to establishing a rational basis for predicting future responses of the disturbed ground.

The failure and replacement of the Beards Creek bridge, has provided a useful set of observations, projections, and measured responses to the imposition of significant new earthfill and structure loads on a severely disturbed site.

REFERENCES

Akzo Nobel Salt Inc., 1994-1996. Land Subsidence and Well Water Level Monitoring Data, March 1994 through September 1996, unpublished.

Gowan, S. W., VanSambeek, L. L., and Brekken, G. A., 1996. Anticipated Conditions for the Route 20A Highway and Bridge Reconstruction at Cuylerville, New York, prepared by RE/SPEC, Inc., Rapid City, SD and Alpha Geoscience, Albany, NY.

Matthews, S. L., 1996. Structure Study Report for Replacement of US Route 20A and State Route 39 Bridge Over Beards Creek (BIN 1016090), prepared by LaBella Associates, P.C., Rochester, NY.

Matzat, J. W., 1996. Hydraulic Report for US Route 20A and State Route 39 Bridge Replacement Over Beards Creek, prepared by LaBella Associates, P.C., Rochester, NY

New York State Department of Transportation, 1981. Geotechnical Laboratory Test Results, I-390 Highway Crossing of Genesee River Valley, Livingston County, New York, Albany, NY.

New York State Department of Transportation, 1994. Logs of Test Borings made at Beards Creek Bridge, March-April, 1994, Albany, NY.

VanSambeek. L. L., 1994. Predicted Ground Settlement over the Retsof Mine, prepared by RE/SPEC, Inc., Rapid City, SD.

Walker, S. E., 1996. Geotechnical Evaluation - Replacement of US Route 20A and State Route 39 Bridge Over Beards Creek, prepared by Haley & Aldrich of New York, Rochester, NY.

Walker, S. E. and Divito, R. C., 1997. Geotechnical Monitoring Summary Report, Replacement of U.S. Route 20A and State Route 39 Bridge over Beards Creek BIN 1016090, Town of Leicester, Livingston County, New York, prepared by Haley & Aldrich of New York, Rochester, NY.

Walker, S. E. and Matzat, J. M., 1997. Planning the Replacement of the Beards Creek Bridge, The Engineering Geology and Hydrogeology of Karst Terranes, Beck & Stephenson (eds), pp 373-380. Balkema, Rotterdam. ISBN 90 5410 867 3.

5 Applications of geophysics to karst investigations

Application of electrical resistivity tomography and natural-potential technology to delineate potential sinkhole collapse areas in a covered karst terrane

WANFANG ZHOU, BARRY F. BECK & J. BRAD STEPHENSON P.E. LaMoreaux & Associates, Inc., Oak Ridge, Tenn., USA

ABSTRACT

The potential for sinkhole collapse poses substantial limitations on the development of karst areas, especially where bedrock is covered by unconsolidated material. Sinkholes are likely to form in depressions in the bedrock surface as a result of subterranean erosion of unconsolidated sediment by flowing groundwater. During a site investigation on the Mitchell Plain of southern Indiana, dipole-dipole electrical resistivity tomography was used to identify such depressions, and natural-potential technology was used to detect subsurface areas of localized groundwater recharge. Limestone bedrock is covered by about 9 meters of clayey soil at the site. Following extensive data analysis, two areas of potential collapse were identified. These areas were interpreted as possible collapse hazards because of the coexistence of a substantial depression in the bedrock surface with concentrated groundwater percolation in the epikarst zone. One potential collapse area coincides with an existing sinkhole. The other is coincident with a small valley. Drilling is required to confirm the interpretation in the second area, but the existence of several sinkholes in similar geomorphic settings around the site suggests that this area may be susceptible to sinkhole development.

INTRODUCTION

Karst is a characteristic terrane with distinctive hydrology and landscapes arising from a combination of high rock solubility and well-developed secondary porosity. It usually develops in carbonate rocks (limestone and dolomite)—the focus of this paper—but it may also develop in gypsum, salt, and other soluble rocks. Karst is often characterized by sinkholes, sinking streams, caves, and springs. Most carbonate areas contain solution-enlarged joints, also known as "cutters." Cutters are formed by dissolution of bedrock as groundwater flows through fractures (initial fracture in Fig. 1). These features may be separated by upward-protruding limestone features known as "pinnacles." Carbonate rocks contain various amounts of insoluble materials. When the soluble components are dissolved and removed by flowing groundwater, the insoluble materials are left behind. These insoluble residues accumulate on the sides of pinnacles and in the bottoms of cutters, forming a covered karst terrane (Jennings, 1985). They may

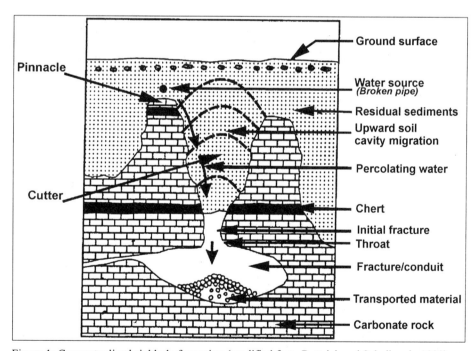

Figure 1: Conceptualized sinkhole formation (modified from Buttrick and Schalkwyk, 1998)

eventually fill the cutters and blanket the rock surface (Fig.1). The bedrock may also be mantled by allochthonous sediments of volcanic, alluvial, or marine origin. Regardless of their source, the sediments cover the limestone and obscure any solution-related features in the underlying bedrock.

Because of the irregular distribution of cutters and pinnacles, it is time-consuming and expensive to define the depth to bedrock with boring data. Such data may even be misleading if data density is inadequate. However, in contrast to other residual soils produced by the mechanical and chemical breakdown of rock from the ground surface downward, the boundary between rock and soil is often relatively sharp in covered karst terranes (Sowers, 1996). Whether residual or transported in origin, the overburden soil has different properties than the underlying carbonate rock. This contrast provides the basis for applying geophysical techniques to characterize the bedrock surface. Electrical resistivity may be the most frequently used method for site investigation in karst areas (Franklin and others, 1981), especially when the overburden is clay-dominated (Cook and Nostrand, 1954). Electrical resistivity tomography (ERT) is an updated version of traditional electrical resistivity methods, where data collection is controlled by an automated instrument (Advanced Geosciences, Inc., 1996) and data are processed using numerical inversion programs (Griffiths and Barker, 1993).

Natural potential (NP)—also known as self potential, spontaneous potential, streaming potential, and SP—can be used to study the electrokinetic effect—i.e., the streaming-potential gradient resulting from subsurface water movement (Ogilvy and others, 1969). Sinkhole collapse is associated with the movement of water and unconsolidated material into the fracture/conduit drainage system underlying a cutter (Fig. 1). The localized groundwater flow generates electrical streaming potential, which is detected by NP method (Erchul and Slifer, 1987). Proper interpretation of NP data allows identification of localized groundwater recharge that may pose a risk of sinkhole collapse. NP technology is the only geophysical method that responds to the *movement* of water rather than the mere *presence* of water.

SITE CONDITIONS

The study site is located on 4.17 Ha (hectares) of undeveloped land and lies on the Mitchell Plain of southern Indiana. Thirty-nine soil borings have shown that the karstic limestones of St. Louis and Salem Formation are overlain by approximately 9 m (meters) of unconsolidated clay. The limestone and the land surface dip approximately 6 m/km toward the west. There are seven sinkholes immediately surrounding the site (A, B, C, D, E, F, and G in Fig. 5). A series of springs and seeps discharge groundwater to a small stream along the base of the slope immediately west and north of the site. Several of the springs discharge from solution-widened joints in the limestone.

MAPPING THE BEDROCK SURFACE WITH ELECTRICAL RESISTIVITY TOMOGRAPHY

Electrical resistivity data were collected along forty-nine transects over an area of approximately 4.20 Ha using dipole-dipole ERT. The electrode spacing was 3 m. The transects were arranged to correspond with a pre-existing soil boring grid. Most of the soil borings were intercepted by at least two perpendicular transects. The lengths of the transects varied from 81 to 249 m. The Sting/Swift electrode system (Advanced Geosciences, Inc.) was used for the data collection.

Apparent resistivity measurements are volume-averaged values affected by all the geologic layers through which the induced electric current flows. These data are processed to generate two-dimensional resistivity models of the subsurface using RES2DINV inversion software (Loke, 1996; Loke and Barker, 1996b). The product of data inversion is a two-dimensional image (a tomograph) showing a distribution of modeled resistivity values. Because a perfect reproduction of the resistivity distribution is impossible, an iterative process is normally used to refine the model until the apparent resistivity distribution matches the raw data as closely as possible. The Root-Mean-Square (RMS) error quantifies the difference between the measured resistivity values and those calculated from the resistivity model. A small RMS value indicates a close match. The average RMS error was 37.1 percent for the transects analyzed during the investigation, with a minimum of 1.8 percent and a maximum of 118.2 percent. Approximately 82 percent of the lines have RMS errors less than 50 percent. (Errors of this magnitude are common when using this technique).

Data from pre-existing soil borings located along the transects were used to locate the limestone/clay boundary in the tomographs because the exact depth of the bedrock surface cannot be determined solely from the tomographs. Even a sharp limestone/clay boundary appears transitional in the processed image. Therefore, a realistic interpretation of the bedrock/overburden boundary cannot be made unless "ground-truth" (boring) data are available for the transect (Griffiths and Barker, 1993). Because the boring data were more localized than the apparent resistivity measurements, incorporation of the boring data into the modeled resistivity profile was not straightforward. Two assumptions had to be made to facilitate the interpretation: (1) the contact between bedrock and soil is laterally continuous and corresponds to a single value of resistivity; and (2) the contact is sharp rather than gradational.

Elevation data from borings located at the intersection of two transects were used to test the reliability of ERT at the site. The difference in the bedrock-surface elevation was calculated from two transects intercepting the same boring. The error distribution is shown in Fig. 2. The elevations interpreted from different transects differ by as much as 10 m. The average difference is 2.4 m. Seventy-four percent of the data points have errors less than 3 m. The causes for large RMS errors and lack of agreement between perpendicular transects are discussed by Zhou and others (1999a). Most importantly, distortions in the measured data can be caused by variations in geology and resistivity adjacent to the transect. For example, cutters/pinnacles that appear on individual tomographs may result from geological features laterally offset from the transect.

The accuracy of the interpretation can be increased by averaging elevations interpreted from intersecting transects. A feature off the line of one transect may be crossed by other transects with different orientations. If site geology is relatively uniform, similar elevations may be obtained from multiple transects, and their average elevations may be considered more reliable. During this investigation, approximately 80 percent of the data points had errors less than 3 m when comparing interpretations from individual transects with the boring data. More than 86 percent of the data points had errors less than 3 m when the average value from

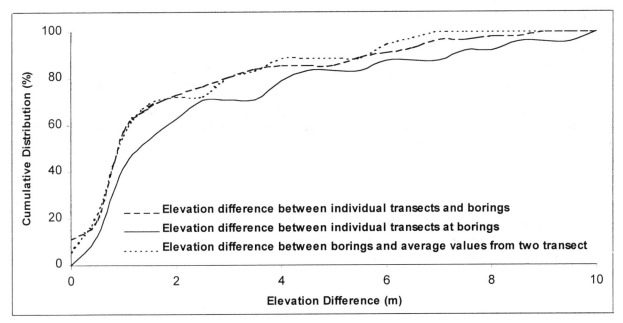

Figure 2: Reliability evaluation of ERT interpretation

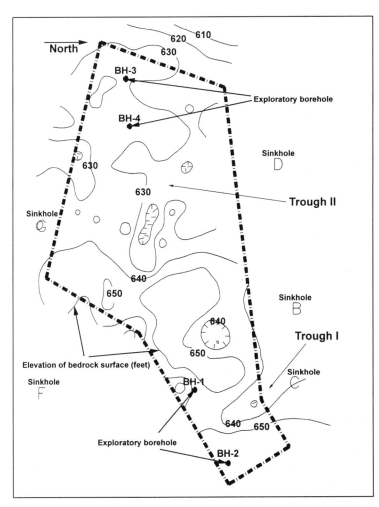

Figure 3: Bedrock-surface interpreted from ERT

intersecting transects was used (Fig. 2), a relatively small but obvious improvement. A bedrock-surface map was generated using the average bedrock elevations from all transect junctions (Fig. 3). Four boreholes were drilled to evaluate the accuracy of the map. The elevation difference between the map and the boring data ranges from 0.2 to 0.6 m, with an average of 0.4 m.

DELINEATING SUBSURFACE AREAS OF LOCALIZED GROUNDWATER RECHARGE WITH NATURAL ELECTRICAL POTENTIAL MEASUREMENTS

NP measurements were made along the ground surface using the voltage-averaging mode of a multimeter with an input impedance of 10 megohms. The reference and roving probes (7.6 and 3.8 centimeters in diameter, respectively) were non-polarizing, copper/copper-sulfate, porous-ceramic electrodes. The multimeter was connected to the reference electrode by 550 m of 18-gauge military communication wire of the type used for induced polarization (IP) surveys. Forty-seven data lines (7.6 meters apart) were measured within an area of approximately 7.6 Ha. NP readings were taken every 7.6 m along each line, except for minor spacing changes near obstacles. The length of the lines varied from 84 m to 244 m. Three base stations were established to provide efficient access to the entire site, including extensive forested areas.

NP measurements are subject to temporal drift. Drift-corrected NP data have been interpreted for karst hydrogeological applications (Lange and Kilty, 1991; Kilty and Lange, 1991). The temporal drift can be corrected using the following equation, assuming that drift is linear between successive base station readings:

$$V_{cj} = V_j - [V_{pi} + (T_j - T_{pi})(V_{ni} - V_{pi})/(T_{ni} - T_{pi}) + V_0],$$

where V_{cj} is the drift-corrected NP value for the measured voltage V_j at time T_j on line j; V_0 is the reference NP value, which is the first reading at the first base station; and V_{pi} and V_{ni} are the previous and subsequent readings at base station i at times T_{pi} and T_{ni}, respectively.

Based on the experiments of Ernstson and Scherer (1986) and analysis of data from the investigation site (Zhou and others, 1999b), the drift-corrected NP can be decomposed into three components: (1) topographic effect (V_j^{te}); (2) residual NP (V_{cj}^{rs}); and (3) noise (α):

$$V_{cj} = V_{cj}^{rs} + V_j^{te} + \alpha.$$

The survey lines were grouped by date, assuming that the ground conditions (temperature, moisture, etc.) did not change significantly during the course of a single day. For each daily group, two lines (with no apparent anomalies) were selected to determine the topographic effect. The drift-corrected NP values were plotted against their corresponding elevations, as shown in Fig. 4. The relationship between NP and elevation was approximately linear, with correlation coefficients ranging from 0.83 to 0.99 and averaging 0.95. The topographic effect for each NP data point was calculated as follows:

$$V_j^{te} = K_j(h_j - h_{0j}),$$

where K_j is the topographic correction factor for line j, which is defined as the NP change per unit elevation increase; h_j is the elevation of the station where V_j is measured; and h_{0j} is the elevation of the first measurement in line j. Data in Fig. 4 revealed that the topographic correction factor varied from -0.38 to -0.75 mV/foot and averaged -0.55, which is within the range reported by other investigators (Aubert and Atangana, 1996; Ernstson and Scherer, 1986).

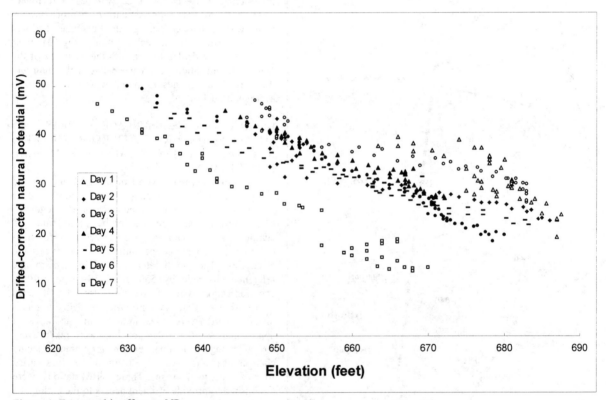

Figure 4: Topographic effect on NP

The residual NP signal was obtained by removing the topographic effect from the drift-corrected NP data and assuming that noise was a negligible component of the data. The residual NP is mapped in Fig. 5, based on the average topographic correction factor and a reference elevation. For profiles with a relatively constant topographic slope, the separation of the residual NP and the topographic effect is similar to correcting the data with a regression on distance because the elevation is generally proportional to distance. However, for profiles measured over varying topography, use of the topographic correction factor produces a much better result than a simple regression on distance. The residual NP component is generally overshadowed by the topographic effect, especially in deep sinkholes. However, removal of the topographic effect reveals the residual NP component, exposing prominent anomalies at the sinkholes and other areas of localized groundwater recharge (Zhou and others, 1998; Zhou and others, 1999b).

The residual NP data represent the effects of local streaming potential. Negative anomalies have been documented in areas with active water infiltration (Lange and Kilty, 1991; Corwin and Hoover, 1979; Stewart and Parker, 1992). The polarity of NP anomalies associated with infiltration in carbonate rocks may be determined at least in part by the pH of the groundwater. It has been suggested that infiltration of neutral water (pH of 7) causes a negative anomaly (Ernstson and Scherer, 1986). Because groundwater in uncontaminated limestone aquifers has a pH of approximately 7, the negative anomalies at the site may be caused by vertical groundwater infiltration.

SUMMARY

Analysis of data in Figs. 3 and 5 reveals two areas with possible sinkhole-collapse risk, where a depression in the limestone surface coexists with localized groundwater recharge. Trough I in Fig. 3 overlaps with Anomaly I in Fig. 5 (Potential Collapse Area 1). Trough II in Fig. 3 overlaps with Anomaly III in Fig. 5 (Potential Collapse Area 2). (The area where Anomaly II is located in Fig. 5 was not covered by the electrical resistivity survey.) Potential Collapse Area 1 coincides with an existing sinkhole (Sinkhole C). This is logical because the bedrock surface at sinkholes is normally lower, and sinkholes generally function as discrete groundwater recharge features in karst terranes. Potential Collapse Area 2 is generally in alignment with a small valley at the land surface. Because the NP anomaly is smaller in magnitude but covers a larger area than that in Potential Collapse Area 1, infiltration is assumed to be more dispersed, and the infiltration rate is assumed to be smaller. Additional drilling is required to confirm the existence of the interpreted depression in the bedrock surface. However, the existence of several sinkholes (e.g., Sinkholes B, D, and E) in similar geomorphic settings around the site suggests that this area may be susceptible to sinkhole development.

The combination of ERT and NP provides useful information for identifying potential sinkhole collapse areas. This case study demonstrates that data from these techniques must be analyzed and interpreted cautiously, even with the aid of boring data. In particular, interpretations are subject to the following limitations.

Dipole-dipole electrical resistivity tomography

(1) Impact of three-dimensional geology: The tomographs are affected by features adjacent to the transects. Interpretation can be improved by using multiple intersecting transects. The use of three-dimensional electrical resistivity methods is even more effective (Loke and Barker, 1996a).

(2) Data quality: Electrical resistivity data are affected by various factors, such as wind, clouds, rain, nearby metallic materials, animals, and even solar flares. The automatic data collection system expedites the data collection process, but it limits the ability to control data quality.

(3) Limitations of electrical methods: The measured apparent resistivity data are average values representing a volume of geologic material. The volume-averaging inherent in resistivity methods tends to obscure small-scale irregularities in the geologic interfaces. The data are more generalized at greater depths.

(4) Non-uniqueness of the modeling results: It is possible for different geological models to produce similar

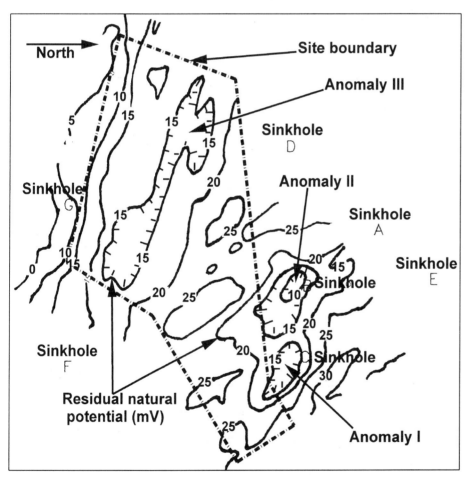

Figure 5: Residual natural potential

profiles of calculated apparent resistivity.
(5) Misinterpretation of soil boring data: Soil borings may record the top of the weathered limestone zone, rather than the top of unweathered bedrock. Some borings may have been stopped by residual chert boulders in the overburden.
(6) Complex geology in karst terranes: Due to complex variations within the bedrock and the unconsolidated sediments, inappropriate interpretation of the tomograph may give misleading results. Isolated, near-surface areas of high resistivity may be caused by concentrations of residual chert, iron oxide nodules, or limestone fragments. An apparent depression in the tomograph may be caused by the presence of a narrow clay- or water-filled fracture or cavity. An apparent pinnacle in the modeled resistivity layers may be caused by the presence of a small air-filled cavity.

Natural-potential technology
(1) Various sources of electrical potential: Natural electrical potential is caused by a variety of discrete physical phenomena acting underground (Zhou and others, 1999b); streaming potential is just one of them.
(2) Electrical interference: NP data may be impacted by various sources of interference, including telluric currents; electrical resistivity variations within the ground; temperature changes in the electrodes; temperature, moisture, and chemical fluctuations in the soil; and currents generated by metallic debris within the survey area.
(3) Inadequate data for component separation: Drift-corrected NP components are not always separable. Their decomposition requires a substantial amount of data.
(4) Averaging the topographic correction factors: Topographic correction factors vary with time and location. Using the average value to calculate a residual NP may obscure small anomalies.

REFERENCES

Advanced Geosciences, Inc., 1996, Sting R1 Instruction Manual (Release 2.0.2). 63p.

Aubert, M. and Atangana, Q.Y., 1996, Self-potential method in hydrogeological exploration of volcanic areas: Ground Water, v.34, no.6, pp1010-1016.

Buttrick, D. and Schalkwyk, van A., 1998, Hazard and risk assessment for sinkhole formation on dolomite land in South Africa: Environmental Geology, v.36, n.1-2, pp170-178.

Cook, K.L. and Van Nostrand, R.G. 1954, Interpretation of resistivity data over filled sinks: Geophysics, v.19, pp761-790.

Corwin, R.F. and Hoover, D.B., 1979, The self-potential method in geothermal exploration: Geophysics, v.44, no.2, pp226-245.

Erchul, R.A. and Slifer, D.W., 1987, The use of spontaneous potential in the detection of groundwater flow patterns and flow rate in karst areas: In: Beck, B.F. (ed.), Second Multidisciplinary Conference on Sinkholes and the Environmental Impacts of Karst, Orlando, Florida, pp217-226.

Ernstson, K. and Scherer, H.U., 1986, Self-potential variations with time and their relation to hydrogeologic and meteorological parameters: Geophysics, v.51, no.10, pp1967-1977.

Franklin, A.G., Patrick, D.M., Butler, D.K., Strohm Jr., W.E., and Hynes-Griffin, M.E., 1981, Foundation Considerations in Siting of Nuclear Facilities in Karst Terrains and Other Areas Susceptible to Ground Collapse: U.S. Army Engineer Waterways Experiment Station, NUREG/CR-2062, 229p.

Griffiths, D.H. and Barker, R.D., 1993, Two-dimensional resistivity imaging and modelling in areas of complex geology: Journal of Applied Geophysics, v.29, pp211-226.

Jennings, J.N., 1985, Karst Geomorphology: Basil Blackwell, Oxford, 293p.

Kilty, K.T. and Lange, A.L., 1991, Electrochemistry of natural potential processes in karst: In: Proceedings of the Third Conference on Hydrogeology, Ecology, Monitoring, and Management of Groundwater in Karst Terranes, U.S. Environmental Protection Agency and National Ground Water Association, December 4-6, Nashville, Tennessee. pp163-177.

Lange, A.L. and Kilty, K.T., 1991, Natural-potential response of karst systems at the ground surface: In: Proceedings of the Third Conference on Hydrogeology, Ecology, Monitoring, and Management of Groundwater in Karst Terranes, U.S. Environmental Protection Agency and National Ground Water Association, December 4-6, Nashville, Tennessee, pp179-197.

Loke, M.H., 1996, Manual for the RES2DINV: Advanced Geosciences, Inc., Austin, Texas.

Loke, M.H. and Barker, R.D., 1996a, Practical techniques for 3D resistivity surveys and data inversion: Geophysical Prospecting, v.44, pp499-523.

Loke, M.H. and Barker, R.D., 1996b, Rapid least-square inversion of apparent resistivity pseudosections by a quasi-Newton method: Geophysical Prospecting, v.44, pp131-52.

Ogilvy, A.A., Ayed, M.A., and Bogoslovsky, V.A., 1969, Geophysical studies of water leakage from reservoirs: Geophysical Prospecting, v.22, no.1, pp36-62.

Sowers, G.F., 1996, Building on Sinkholes: American Society of Civil Engineers, 201p.

Stewart, M. and Parker, J., 1992, Localization and seasonal variation of recharge in a covered karst aquifer system, Florida, USA: In: Back, W., Herman, J.S., and Paloc, H. (eds.), Hydrogeology of Selected Karst Regions, International Association of Hydrogeologists, 13: pp443-460.

Zhou, W.F., Beck, B.F., and Stephenson, J.B., 1998, Identification of vertical ground-water recharge in karst areas with a refinement of the natural potential method using topographic correction factors: In: Jack, J., Barksdale, S., Alverson, M., and Thomas, L. (eds.), Proceedings of Eighth Tennessee Water Resources Symposium, pp2B-1-6.

Zhou, W.F., Beck, B.F., and Stephenson, J.B., 1999a (in press), Defining the bedrock/overburden boundary in covered karst terranes using dipole-dipole electrical tomography: Proceedings of the 1999 Symposium on the Application of Geophysics to Environmental and Engineering Problems (SAGEEP), March 14-18, Oakland, CA.

Zhou, W.F., Beck, B.F., and Stephenson, J.B., 1999b (in press), Investigation of groundwater flow in karst areas using component separation of natural potential measurements: Environmental Geology.

Characterizing karst conditions at a low-level radioactive DOE site

RICHARD C. BENSON & LYNN YUHR Technos, Incorporated, Miami, Fla., USA

ABSTRACT

The St. Louis Airport Site (SLAPS) in eastern Missouri is a former Department of Energy storage area for uranium ore residues. Regional information indicates that the site is underlain by the heavily karstified Mississippian limestones of the Florissant Basin. In order to properly describe geologic and hydrologic conditions at the site, a karst characterization was carried out as part of an Expedited Site Characterization (ESC) investigation.

Key elements of the ESC process, including a core team of senior professionals, a critical review of existing data, and non-invasive measurements, were utilized to develop a conceptual model of karst conditions at SLAPS. Regional data were integrated with site specific boring data to develop an initial geologic model. The review of data indicated a high probability of karst features within the site and a lack of site specific data describing bedrock conditions and the overlying clay confining layer. A field plan was developed to further investigate the confining layer and assess karst development within the bedrock.

The lateral boundary of the 3M clay aquitard is defined by more than 37 existing borings. However, geophysical logs within six existing borings revealed a high degree of lateral and vertical heterogeneity within the unconsolidated materials including the clay confining layer. A buried valley is defined by 33 existing borings, which is supported by microgravity data, time domain electrical sounding data, photo lineaments, and the geomorphic setting. Time domain electrical soundings and microgravity measurements, revealed variations to the top of rock indicating dissolution-enlarged joints. The results of these efforts were integrated into a final conceptual model that provided a understanding of the most probable karst conditions at SLAPS in a cost-effective and efficient manner.

BACKGROUND

The St. Louis Airport site (SLAPS) is a 22-acre site located on the north side of the St. Louis airport (Figure 1). The site was used from 1946 through 1969 for storage of residues from uranium ore processing. Most of these residues were removed and building facilities were demolished and buried on-site. The site was then covered with 1 to 3 feet of clean backfill. SLAPS was selected for use as a demonstration site for the Expedited Site Characterization (ESC) procedures developed by the Department of Energy (DOE).

The ESC approach has incorporated all of the historic and proven approaches to site characterization and combined them with contemporary field techniques to provide an extremely effective means of carrying out the site characterization process. ESC is nothing more or less than an orchestrated and coordinated program of timely site data gathering with rapid turn-around on interpretation and integration of findings. When

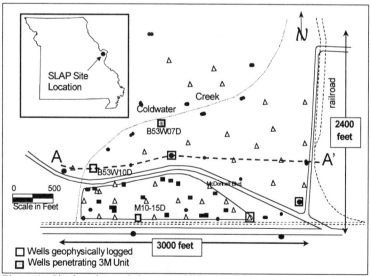

Figure 1. Site location and site map with boring/well locations

we examine the key elements of the ESC process, we see that many of the ESC philosophies have been with us throughout much of the history of engineering geology (Hvorslev, 1949 and Terzaghi and Peck, 1967). Beam, et al (1997) summarize many of the historic aspects which are part of the ESC strategy. The ESC process is discussed further in Benson (1997), Benson, et al (1998) and ASTM PS 85-96.

Four objectives were identified for this ESC demonstration project including:
- Verification of the boundaries of any buried debris;
- Confirmation of the geologic/hydrologic conceptual model, particularly the continuity and thickness of a clay aquitard (referred to as the 3M unit) underlying the site;
- An assessment of karst conditions; and
- Assessment of radioactive contaminant distribution on the surface and with depth.

This paper focuses upon the karst aspects of the SLAP site, including an assessment of the clay aquitard and underlying limestone. Since additional drilling at the site was prohibited, all further geologic and hydrologic investigation of the site had to be carried out using existing data, aerial photography, along with non-invasive methods including surface and borehole geophysics.

SOME KEY ELEMENTS OF THE ESC PROCESS

While all of the components of ESC are critically important, we will focus attention on a few as they relate to the demonstration project at SLAPS.

Core Team

The ESC core technical team consists of 2 to 4 senior experienced professionals with broad expertise in the geosciences and contaminant chemistry. The core team remains on the project from conception through final report, providing technical continuity to the program. The core team are in the field guiding and directly participating in the acquisition of data on a daily basis. The daily results of data acquisition are used to support the on-site decisions concerning the adequacy of the conceptual site model and to guide the next day's work. The many hours on-site provides the core team professionals with an opportunity for first-hand observations, direct interaction with the entire field team which are often critical to understanding site conditions and cannot be obtained in any other way.

A Critical Review of Existing Data

A critical review of existing data serves to identify relevant gaps and errors in the existing data which could then be addressed by subsequent field efforts. The process aids in identifying the data deficiencies as well as guides the strategy and selection of measurements for the field effort.

Developing an Initial Conceptual Model of Site Conditions and a Field Investigation Plan Based Upon Available Data

A conceptual model of site conditions based upon all relevant existing information should be developed prior to starting field work. This model forces us to analyze and integrate the existing data to develop the best possible initial understanding of site conditions. The various sources of data to be obtained will be used to test the hypothesis of the conceptual model, which will be improved as additional information is obtained.

The site characterization process should start with the regional setting, working toward the local setting, and then to the very local boring, sampling, and testing program. This allows us to build our conceptual model in a logical fashion. The focus is upon basics, a keep it simple approach to provide appropriate, adequate, and accurate data, so that borings, piezometers and monitor wells can be more accurately located.

Review of Topographic Maps, Use of Aerial Photography, Development of Site Maps, and On-Site Observations

Some of the oldest, and often lowest cost, but least used site characterization tools are a critical review of topographic maps, aerial photo interpretation and on-site observations. The general character of the setting can be rapidly determined, including topography, drainage patterns, fractures, and structural features, and site specific details identified. In addition, having an experienced professional walk the site and surrounding area can often result in valuable information through simple observations.

Use of Non-Invasive Methods

The primary factor affecting the accuracy of any site characterization effort is the limited number of sample points or borings, resulting in insufficient spatial sampling to adequately characterize the site. Non-invasive geophysical methods allow spatial sampling of geology, hydrology and sometimes contamination to be greatly increased which also improves the probability of detecting geologic anomalies, such as karst.

Interpretation and Integration of Data to Support the Final Conceptual Model.

The process of interpreting each piece of data and then integrating them is often neglected in many site characterization efforts. Multiple sources of data should be used to confirm site specific conditions. When measurements by different methods agree, we can have a high level of confidence in our interpretations. Then when two or more independent sets of data are used to confirm results, our conceptual model is based upon defendable data rather than assumptions and opinions.

APPLYING ESC STRATEGY TO THE SLAPS
A Critical Review of Existing Data

There have been numerous investigations completed at SLAPS by DOE contractors. The study area includes the SLAPS and extends northward just beyond Coldwater Creek (Figure 1). The most recent report for the site is Bechtel's Site Suitability Study Vol. I and II (1994), which was used to provide a summary of existing data. In addition, there are numerous publications by USGS and the Missouri Geological Survey which cover the regional geologic setting.

The existing data in the 1994 report included a wide variety of regional information with 98 references, 12 of which contained more local information. This information was heavily focused on the seismic stability of the area, but included soils, karst, chemistry and flooding information. The balance of the existing data was very site specific and tied to the surficial sampling, and boring or well locations shown in Figure 2. It was this site specific data that was used to create the conceptual cross-section shown in Figure 2.

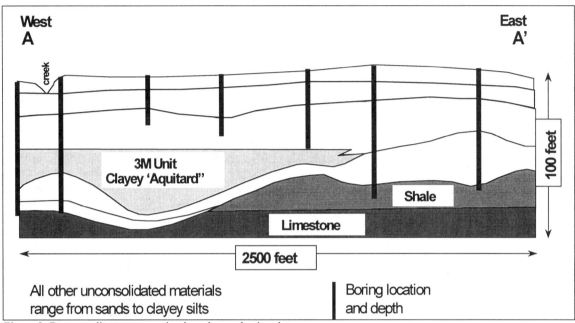

Figure 2. Downgradient cross-section based upon boring data

The geology at SLAPS consists of 50 to 90 feet of unconsolidated materials overlying bedrock (Figure 2). From the surface downward, the geology consists of:
- Unconsolidated silty to clay materials over a sandy gravel. These unconsolidated materials include an aquitard about 20-30 feet thick consisting of clays. This aquitard is referred to as the 3M unit;
- The bedrock at the site consists of deeper Mississippian limestones overlain by Pennsylvanian cyclothems (including shales) on the eastern portion of the site;
- A buried bedrock valley extends north-south through the western portion of the site and is defined by 33 borings.

The State of Missouri is known to have many large sinkholes, caves and spring systems, and the Missouri DNR had raised the issue of potential karst underlying the site. Karst features of concern include dissolution of limestone joints and bedding planes as well as larger cave systems and paleokarst collapse.

Existing assessment of karst conditions were based upon regional literature, the lack of any caves at or near the site in the cave database and only 2 site specific borings which penetrated into weathered rock. The existing assessment of karst suggested that subsidence or collapse is unlikely. In addition, it was felt that the 3M clay unit (which is described as a stiff plastic clay layer) would prevent any possible downward migration of radioactive contaminants into any karst feature and the groundwater. However, maps in existing reports show that the 3M unit does not extend over the eastern third of the site (Figure 2).

At initial glance, the distribution of over 80 borings and wells across the site is impressive (Figure 1). However, only 21 borings even reached top of rock and only two boreholes extended into weathered rock (Bechtel, 1994). Neither borehole extended into rock far enough to obtain the complete profile of unweathered rock. Furthermore, both of the borings into rock were located in the eastern portion of the site. and were not associated with the bedrock valley and greater weathering of limestone which occurs on the western portion of the site. Therefore, an assessment of karst conditions based upon the borehole data is extremely limited in its value; due to the location of the two borings in the eastern part of the site and due to the limited number of boreholes into rock. From existing data we also find that one boring toward the center of the site extended to 94 feet (the top of the limestone) and had 1.1 m^3 of grout take without filling it. This would indicate that there is a potential for karst at the site.

Initial Conceptual Model and Field Investigation Plan

Because of the significant amount of work that had been completed at this site, the initial conceptual model of the site (Figure 2) is more advanced than usual. Even so, there are two questions which remains: 1) what is the extent and homogeneity of the 3M plastic clay unit? and 2) how extensive is the karst development in the underlying rock?

Field Plan - A field plan was developed specifically to address these two issues. The field plan included a review of regional and local karst literature, review of the state cave database, review of topographic maps and aerial photo analysis, outcrop observations and surface and borehole geophysical measurements. These new data were then integrated with the existing data as the conceptual model of site conditions was confirmed or modified. Limitations to the field plan included:
- Additional borings were not allowed due to concerns about possible radioactive contamination;
- Because of budget and time constraints, focus was placed upon obtaining a single line of detailed geophysical data and developing a single E-W, geologic cross-section immediately downgradient of the SLAPS.

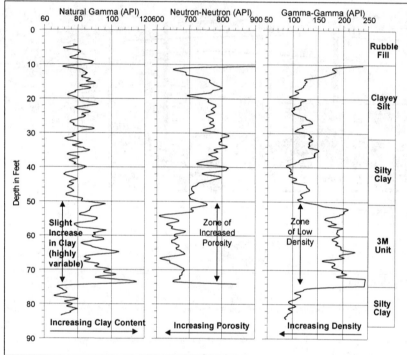

Additional Field Assessment of the 3M Unit

Geophysical logging was carried out in six existing deeper wells cased with stainless steel. The six wells were selected to provide a representative sampling of site conditions both laterally and vertically. Three of these six wells penetrate the 3M unit (Figure 1). The purpose of this logging was to verify the existence and to assess the uniformity of the 3M unit and other unconsolidated materials. The logs include natural gamma (responds to clay content), neutron-neutron (responds to porosity) and gamma-gamma (responds to density). These nuclear logs are capable of measurements through steel casing.

The three geophysical logs shown in Figure 3 are from one well which penetrated the 3M unit. The 3M unit is clearly differentiated by the density and neutron logs (and subtly by the natural gamma log). There appears to be no clear pattern in the geophysical logs from any of the six wells which indicate similar values for a specific lithologic zone, including the 3M unit.

Figure 3. Geophysical logs from well B53W07D.

All of the geophysical logs indicate a moderate degree of variability within each borehole as well as between boreholes. As a result, we can conclude that the unconsolidated materials at SLAPS are not vertical or horizontally homogeneous across the site, including the 3M clay unit.

Assessment of Karst Conditions

To further address the concern of karst features at this site, a wide range of additional data was used including: regional karst literature, aerial photo analysis, site specific boring data, surface geophysical data and most important, regional geomorphology.

The Regional Setting and Its Karst Geomorphology

SLAPS is located upon a main drainage axis of the Florissant Basin, an area of about 29 square miles (Figure 4). The deeper valley seen in the conceptual regional model Figure 5 is a portion of the pre-glacial Florissant Basin which cuts into the Mississippian Limestone. From a geomorphic point of view, one can envision the Florissant Basin as a well developed drainage basin on the order of 100 feet deep in the area of SLAPS. Extensive dissolution and erosion was necessary to remove this amount of rock from the Florissant Basin. The valleys that developed around the perimeter of the basin have developed due to weathering along joints. In addition, the exposed Mississippian-age limestone would have been susceptible to extensive dissolutioning by surface water and shallow ground water.

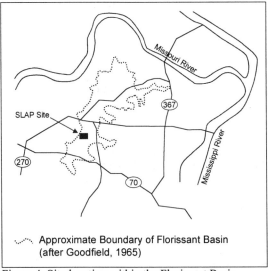
Figure 4. Site location within the Florissant Basin

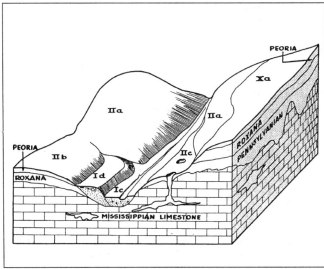
Figure 5. Conceptual model of karst (Rockaway and Lutzen, 1970).

Karst Literature

Rockaway and Lutzen (1970) have done extensive studies on karst in the St. Louis, Missouri area and have developed a conceptual model illustrating the karst features in the area (Figure 5). The overburden soils are silts and clays ranging from 10 to 50 feet thick and the bedrock has an irregular surface. The topography is described as karst (with sinkholes present in the overburden) and there is collapse of cavern roofs and extensive solution features. The assessment of karst by Rockaway and Lutzen (1970) and Lutzen and Rockaway (1989) is primarily based upon sinkhole features observed in topographic maps or aerial photos.

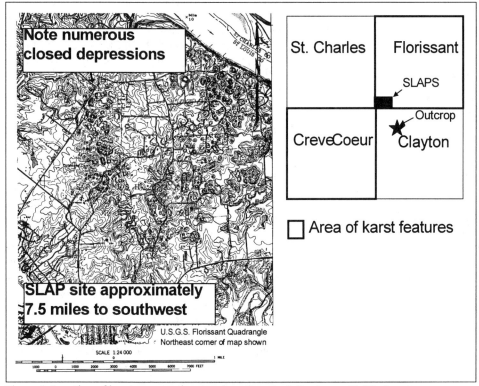
Figure 6. Location of karst areas northeast and southwest of site along with Florissant quadrangle topographic map showing extensive sinkhole development

As of 1990, 5,012 caves have been identified in the State of Missouri according to the Bechtel report (1994). As of 1978, 28 caves had been identified within the corporate limits of the city of St. Louis (Weaver and Johnson, 1980). Many of these caves were used by brewers in the early 1800's because the caves offered natural refrigeration; these are shallow caves which dominantly occur near the Mississippi River.

Cave Database

The assessment of karst by Bechtel (1994) is primarily based upon cave literature from the State of Missouri and St. Louis County. For this ESC effort, a review of the existing literature and the computer records of caves with a representative of the Missouri DNR was completed. There were no records of any caves in the area around or within SLAPS. The fact that no caves are found near the site does not imply that caves are not present, only that they are not accessible or have not been found.

Topographic Maps

The Florissant quadrangle topographic map (USGS, 1954) indicates an area of about 4,000 acres of extensive karst about 7-1/2 miles northeast of SLAPS. This is an area of extensive karst, based upon the localized depressions seen in the topography (Figure 6). Rockaway and Lutzen (1970) also identify a number of karst areas in Creve Coeur quadrangle, about 6 miles to the southwest of the site (Figure 6). Karst features in this area are typically smaller.

Aerial Photo Analysis

Analysis of various aerial photographs in both black and white and color infrared and at different scales suggests the presence of two regional (primary) photo lineaments generally trending northeast, and north-by-northwest. These lineaments are parallel to joints in the rock resulting in valleys seen in the topographic maps of the area (USGS, 1954, St. Charles, Florissant, Creve Coeur, and Clayton). These valleys range from 500 to 1,000 feet long.

Two photo lineaments traverse the SLAP site. The northeast trending lineament is approximately parallel to the main axis of the Florissant Basin. It lies at or near the eastern boundary of the 3M clayey unit and the western extremity of the shales. The northwest lineament lies just a few hundred feet east of the bedrock channel identified in the geologic cross-sections.

Observation of Outcrops

Outcrops of limestones (probably Pennsylvanian age) are found east of the airport (about 2.5 miles from the site) in the southbound lane of Highway 170. These outcrops occur in a road cut and show clear and extensive evidence for dissolution-enlarged joints up to a foot wide and localized dissolution-enlarged cavities of a foot or two in diameter. These outcrops provide a clear indication of karst at an elevation of about 440 feet. The bedrock occurs at an elevation of about 450 feet at SLAPS and appears to be the same age and unit.

Non-Invasive Surface Geophysical Measurements

TDEM Soundings - A time domain electromagnetic (TDEM) sounding measures bulk electrical changes in the subsurface with depth. At SLAPS we expected an electrical contrast between the soils (silts and clays) and the bedrock (limestone). Fifty TDEM sounding measurements were spaced 50 feet apart along a profile line north of McDonnell Blvd. (Figure 1).

The results of the TDEM measurements were modeled as three layers. The interface between the second and third layers is thought to represent the interface between the soils and less weathered limestone. The TDEM data support a bedrock valley centered at approximately Station 600 and a deeper fracture system centered at approximately Station 350 (compare Figures 2 and 7). The TDEM geoelectric profile (Figure 7) indicates a high degree of variability in the top of limestone over the entire site.

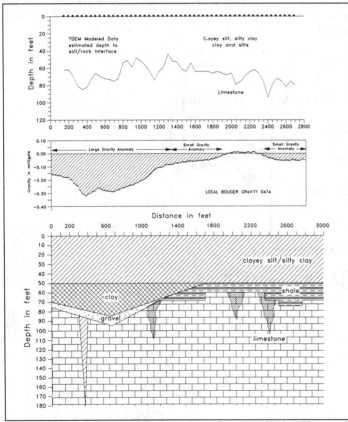

Figure 7. Cross-section based upon surface geophysical data and existing borings

Microgravity Data - Microgravity measurements can be used to detect changes in the density of the underlying soil and rock and can be used to characterize geologic conditions such as bedrock channels, zones of weathered rock, caves, etc. A gravity profile was acquired on the north edge of McDonnell Blvd. and consisted of 281 gravity stations spaced 10 feet apart (Figure 7).

The gravity data indicate one broad area of low values centered at about Station 600 and another narrower area of lower values centered at Station 350 (Figure 7). The broader anomaly can be modeled as a bedrock channel which corresponds with the geologic cross-section for the site (Figure 2) and is also supported by the TDEM data (Figure 7). A second geologic feature was added to the gravity model to account for the narrower anomaly centered at Station 350. A reasonable model fit was obtained by using a vertical sediment filled dissolution-enlarged joint (Figure 7). The two smaller, but broad, gravity anomalies of about 50 to 60 µGals between Stations 1300 to 1900 and Stations 2300 to 2800 can be accounted for by variations in the depth to the top of limestone.

While there is no unique inverse solution to the gravity data, we feel that the fit between the data, the modeled profile (Figure 7), and the initial conceptual geologic model (Figure 2) is based upon reasonable geologic assumptions and provides a reasonable assessment of the subsurface conditions. Overall, the gravity data confirm the bedrock valley on the western half of the site. New information obtained from the gravity measurements included a dissolution-enlarged joint within the buried valley, along with variations in the top of rock to the east.

Interpretation and Integration of Data to Support the Final Conceptual Model of Karst

A final conceptual model of the expected karst conditions at SLAPS is shown in Figure 8. Besides the existing literature and reports, a wide range of independent data were integrated to verify the final conceptual model of karst.

Topographic maps and literature provide the evidence for extensive karst 7 ½ miles northeast (Figure 6) and less extensive karst more than 6 miles southwest of the site. The nearest outcrops clearly show the presence of dissolution joints and cavities.

The buried valley through the site was created by dissolution as part of the Florissant Basin (Figures 4 and 5). The buried valley is defined by 33 existing borings data, microgravity data, TDEM data, photo lineaments, and fits within the geomorphic setting.

During development of the Florissant Basin there was extensive weathering and erosion at the surface of the exposed limestone. The TDEM data and the gravity data both indicate that the top of limestone is highly variable with weathered joints. A major joint in the buried channel was identified by the gravity data.

The lateral boundary of the 3M clay unit is defined by more than 37 existing borings and does not extend over the entire site. The presence of the 3M layer was confirmed by geophysical logs. These logs revealed a high degree of lateral and vertical heterogeneity within the unconsolidated materials including the 3M clay confining layer.

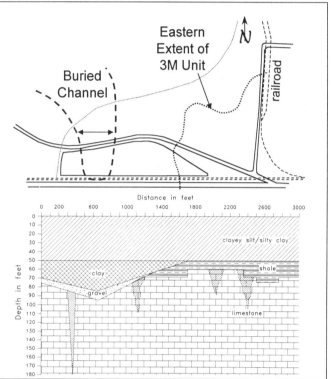

Figure 8. Final conceptual karst model (map and cross-section)

CONCLUSIONS

The karst investigation outlined in this paper, is an example of the application of ESC strategies used to resolve conditions at a site with complex geologic conditions. Figure 8 illustrates how multiple data sets and their integration provide a synergistic improvement of the most probable conceptual model of karst conditions. The final conceptual model has confirmed some of the existing data, added some details, provided an understanding of site geomorphology and identified limitations of the site.

While we have many tools available to help us characterize a site, no single tool will solve all of our problems. We must select an approach using multiple methodologies which are relevant to the geologic setting and the technical problem at hand. The methods of measurement must be selected to fit the scale of measurements and the scale of the expected geologic uncertainties. In all cases, the technical approach to field investigation must be biased to maximize the density and representativeness of spatial data to provide appropriate, adequate and accurate data.

Most critical to the success are the core team of senior experienced hands-on professionals who are sensitive to the issues of geologic and hydrologic uncertainty and possess the skill and persistence to pursue them. A reasonable level of accuracy in site characterization is and has always been an art, mastered both by extensive energies of interest, curiosity and by years of experience. The true standards, guidelines and protocols for correctness and accuracy have been with us for decades. ESC combines these well known and proven concepts along with technological improvements within a strategic framework which, if followed, will dramatically improve results while cutting costs.

The key elements of ESC simply make good common sense, technically and economically. While the process was created to help DOE characterize its waste sites in a more timely and cost effective manner, it is equally applicable to the private sector. With slight modifications, the ESC process can be adapted to a wide range of projects that require site characterization, including karst investigations for both environmental and geotechnical problems.

A properly executed ESC program is a way of doing so at a minimum cost (lower by a factor of 2 to 10), and time (by a factor of 2 to 10 or more), while dramatically improving the data and results (by a factor of 2 to 5) in a conceptual model from which risk assessment, modeling, and remediation can be carried out with confidence.

REFERENCES

American Society for Testing and Materials, 1997. Provisional Standard Guide for Expedited Site Characterization of Hazardous Waste Contaminated Site. ASTM PS 85-96.

Beam, P.; Benson, R.C.; and Hatheway, A.W.; 1997. Lessons learned - A brief history of site characterization. HazWaste World/Superfund XVIII, December 2-4, Washington, DC.

Bechtel National, Inc., 1993. Work Plan - Implementation for the Remedial Investigation / Feasibility Statement for the St. Louis Site, St. Louis, Missouri. August (Bechtel Job No. 14501).

Bechtel National, Inc., 1994. Site suitability study for the St. Louis Airport Site, St. Louis, Missouri.

Benson, R. C., 1997. ESC: Just another new acronym or maybe something meaningful. Environmental & Engineering Geoscience, Vol. III, No. 3, pp. 453-456.

Benson, R.C., Bevolo, A., and Beam, P., 1998. ESC: How it differs from current state of the practice. Proceedings of the Symposium on the Application of Geophysics to Environmental and Engineering Problems, Environmental and Engineering Geophysical Society, Wheat Ridge, Colorado, pp. 531-540.

Hvorslev, J., 1949. Subsurface exploration and sampling of soils for civil engineering purposes. Waterway Experiment Station, Vicksburg, MI, reprinted by ASCE, 1965.

Lutzen, E. E. and J. D. Rockaway, Jr. 1989. Engineering geologic map of St. Louis County, Missouri. Missouri DNR, OFM 89-256 EG, Rolla, MO.

Rockaway, J. D. and Lutzen E. E., 1970. Engineering Geology of the Creve Coeur Quadrangle, St. Louis County, Missouri. Missouri Geological Survey and water Resources, Engineering Geology Series No. 2.

Terzaghi, K. and Peck, R. B., 1967. Soil mechanics in engineering practice, 2nd ed. Wiley, NYC, NY, 729 p.

USGS, 1954. Topographic maps of St. Charles, Florissant, Creve Coeur, and Clayton, Missouri. Photorevised 1968 and 1974.

Weaver, H. Dwight and P. A. Johnson, 1980. Missouri the Cave State. Discovery Enterprises, Jefferson City, Mo.

Microgravity techniques for subsurface investigations of sinkhole collapses and for detection of groundwater flow paths through karst aquifers

NICHOLAS C. CRAWFORD   Center for Cave and Karst Studies, Department of Geography and Geology, Western Kentucky University, Bowling Green, Ky., USA

MICHAEL A. LEWIS   Crawford and Associates, Inc., Bowling Green, Ky., USA

SHAUN A. WINTER   EMPE Inc., Nashville, Tenn., USA

JAMES A. WEBSTER   Geography Department, University of Georgia, Athens, Ga., USA

ABSTRACT

Bouguer gravity can identify locations on the earth's surface that have relatively higher or lower gravity caused by lateral variations in subsurface density. The density contrast of -1.0 to -2.5g/cm^3 between air, water or clay filled cave passages and limestone bedrock can often be detected from the surface by the use of microgravity. Although gravity surveys usually employ a grid pattern, this research has demonstrated the effectiveness of using traverses established perpendicular to linear subsurface features and groundwater flow paths on karst terrain. The directions of these features are determined by a combination of lineament analysis, dye traces and potentiometric surface mapping.

Once a low-gravity anomaly has been confirmed to be a cave, usually by exploratory boring, the general route of the cave passage can be determined by proceeding in a "leap frog" fashion with short parallel traverses. This technique is particularly useful in well-developed karst, where the caves are relatively large and shallow, for installing a monitor or recovery well directly into the karst conduit that drains a hazardous material site. Crawford has drilled over 60 such wells into caves detected by low-gravity anomalies. Several examples are presented in this paper.

Microgravity can also detect voids in the regolith above bedrock and also depth to bedrock. It is therefore a useful tool for identifying potential sites for sinkhole collapses and for investigating existing collapses. This paper discusses microgravity subsurface investigations at recent sinkhole collapses under: a) Interstate 65 at Elizabethtown, Kentucky, b) the main entrance road to Mammoth Cave National Park, and c) a large two-story building in Bowling Green, Kentucky.

INTRODUCTION

Gravity surveys are used to detect variation in the density of subsurface materials. Variations in the earth's gravitational force higher than normal indicate underlying material of higher density while areas of low gravity indicate areas of lower density. In order to detect voids or cavities, very high precision is required. Accurate gravity readings to 10 microgals (1 gal = 1 cm/s^2) are necessary. The earth's normal gravity is 980 gal. A LaCoste and Romberg Model D Microgal Gravity Meter or a Scintrex CG-3M Autograv Microgravity Meter were used for all of the investigations discussed in this paper. Both meters have a 1 microgal sensitivity.

MICROGRAVITY RESEARCH PROCEDURE

A base station is established near the site to be surveyed. Gravity is measured at this base station at approximately a one-hour interval while the meter is being used in order to derive instrument drift. A base station derived instrument drift curve is interpolated to the time of each survey station reading and each station reading is then corrected for instrument drift. Earth tide corrections are made for each gravity reading and differences in elevation between the survey station and the base station are then compensated for using free-air correction procedures. The free-air effect compensates for the decrease in gravity with elevation due to increasing distance from the center of the earth. Free-air gravity indicates that the reference ellipsoid (gravity on a sea level, rotation ellipsoidal model of the earth—a function of latitude) and free-air effect are included in calculating the theoretical gravity. The free-air gravity is modified to obtain simple Bouguer gravity by applying the Bouguer slab effect correction. This correction refers to the attraction of the slab of material, which is caused by variation in density, between the station elevation and sea level. Terrain corrections of the microgravity data are usually not necessary in order to measure relative differences in gravity when measured along a short traverse.

In the karst areas discussed in this paper, the following density values were assumed:

air = 0 g/cm^3	clay = 1.5 g/cm^3
water = 1.0 g/cm^3	limestone = 2.5 g/cm^3

Therefore, density contrasts of -1.0 to -2.5 g/cm^3 exist for any limestone cavity, depending on whether the cavity is filled with air, water, or clay.

Although microgravity subsurface investigations usually consist of measuring gravity at stations established in a grid pattern, Crawford, Webster, and Winter (1989) have demonstrated the effectiveness of using traverses established perpendicular to linear subsurface features and groundwater flow paths in karst terrain.

DETECTION OF SUBSURFACE FEATURES IN KARST TERRAIN

Bouguer gravity can identify locations on the earth's surface that have relatively higher or lower gravity caused by lateral variations in subsurface density. Crawford has used microgravity extensively to locate bedrock caves from the ground surface. The lower density of the air, water, or mud within a cave compared to the surrounding carbonate rock results in a low-gravity anomaly. He has also used microgravity to locate voids in the regolith (unconsolidated material above bedrock) that are potential sinkhole collapses. Since regolith is less dense than limestone bedrock, Bouguer gravity can also identify variations in depth to bedrock. In limestone areas, depth to bedrock is often very irregular with limestone pinnacles that protrude upward and cutters that extend downward. Cutters are V-shaped regolith-filled crevices formed by solution of the limestone by soil water as it percolates down to the karst aquifer. Regolith arches form as regolith spalls into solutionally enlarged voids in the bedrock. In some cases cave streams may then carry off regolith. For these reasons, low-gravity anomalies indicate bedrock caves, voids in the regolith or places where depth to bedrock is abruptly greater and therefore often indicative of places where regolith may be descending into bedrock crevices.

MICROGRAVITY USED TO DETECT BEDROCK CAVES

Several researchers have demonstrated that gravity can be used to detect large bedrock caves (Omnes G. 1976; Kirk K.G. and Werner, E. 1981; Butler, D.K. 1983; Butler, D.K. 1984; Butler, D.K. 1996 and Smith, D.L. and Smith G.L. 1987). Figures 1, 2, 3-1, and 3-2 are examples where relatively large low-gravity anomalies reveal the location of underlying cave passages (Crawford and Webster, 1988 and Crawford, Webster and Winter, 1989). Although some studies had identified low-gravity anomalies that were hypothesized to be caves, few wells had been drilled into anomalies to confirm that they did in fact indicate cave passages previous to research performed by Crawford in 1985. Toxic and explosive vapors rising from contaminated cave passages into homes under Bowling Green, Kentucky in 1984 and 1985 resulted in an intensive effort to find the cave passages over a relatively large area (Crawford, 1989; Crawford, Webster, and Winter, 1989).

Although several geophysical techniques for locating caves under Bowling Green were considered and a few tried, the most successful was microgravity. The best results were obtained by taking microgravity measurements with a LaCoste and Romberg Model D Microgal Gravity Meter at ten-foot intervals along traverses perpendicular to a hypothesized route of a cave stream. The hypothesized route was derived from topographic analysis, knowledge of local hydrogeology, dye traces and a detailed water table map. Voids existing beneath low-gravity anomalies were confirmed by exploratory drilling (Figures 4, 5, and 6). By proceeding in a "leap frog" fashion with short parallel traverses, the route of the cave was established (Figure 4). The precise location of Creason Cave, a paleosection of the large Lost River Cave, was determined in this manner. After confirming its existence with three small-diameter exploratory holes and with a downhole camera, a 30-inch diameter well was drilled into the cave to provide access by cavers for mapping. Crawford has used this microgravity traverse technique on several occasions for locating sites for downgradient monitoring wells or recovery wells into cave streams. For example, lineament analyses with perpendicular microgravity traverses were used to locate a sediment filled cave about 200 feet downgradient from a train derailment site near Lewisburg, Tennessee (Figure 7). Approximately 15,000 gallons of chloroform, a DNAPL, sank into the karst aquifer. Sediment was excavated at a place where the cave roof had collapsed, and a large diameter recovery well was installed to pump chloroform from the aquifer (Crawford and Ulmer, 1994).

MICROGRAVITY USED FOR SINKHOLE COLLAPSE INVESTIGATIONS

Benson, Kaufmann, Yuhr and Martin (1998) have used microgravity to locate and characterize areas susceptible to subsidence and sinkhole collapse along a section of I-70 in Frederick County, Maryland and Crawford has used microgravity to investigate subsurface conditions in the vicinity of sinkhole collapses at several locations. Virtually all sinkhole collapses in karst areas are regolith collapses (Figures 8 and 9). Although bedrock cave roofs do occasionally collapse, these collapses are so rare that they do not constitute a serious threat. However, regolith collapses occur frequently in many karst areas.

Regolith collapses occur due to the formation and the eventual collapse of regolith arches (domes). Urban development in karst areas will often result in an increase in the collapse of regolith arches. Regolith arches form by the downward movement of unconsolidated sediments into voids in the bedrock. In areas where the water table is usually above the regolith-bedrock contact, collapses often occur when the water table drops during droughts or high-volume pumping (Figure 8). Physically, the collapses in these cases are caused by loss of buoyant support for the regolith arches, which span openings in the limestone. Collapses are also caused by spalling of saturated regolith down the opening, enlarging the arch, and eventually causing collapse at the land surface.

Regolith collapses also occur in situations where the water table is usually below the regolith-bedrock contact (Figure 9). Construction and land use changes that concentrate surface runoff in sinkholes, retention basins, ditches, and ponds may locally increase the downward movement of water resulting in the piping of saturated regolith into openings in the limestone (Figure 10). Most of the sinkhole collapses investigated in the Warren County, Kentucky area by the Center for Cave and Karst Studies at Western Kentucky University (Sinkhole Collapse Inventory Vols. 1-4) are of this type (Crawford, Webster, and Veni, 1990). An estimated 70

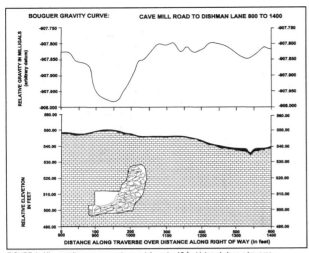

FIGURE 1. Microgravity measurements were taken at a 10-foot interval along a traverse perpendicular to State Trooper Cave in Bowling Green, Kentucky. Microgravity measurements were taken along the proposed route of the Dishman Lane Extension. The route was modified to avoid crossing over the cave at a place where the roof had collapsed.

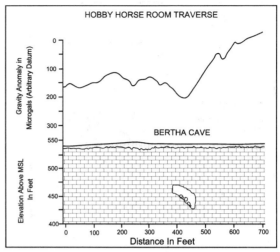

FIGURE 2. Microgravity traverse over Hobby Horse Room in Bertha Cave in Bowling Green, Kentucky. The cave room is over 70 feet below the surface.

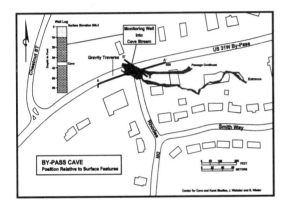

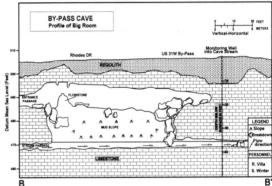

FIGURE 3-2. Profile of Big Room in By-Pass Cave along line B-B'.

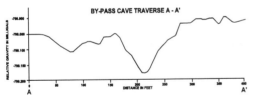

FIGURE 3-1. Low-gravity anomaly along Traverse A-A' over Big Room in By-Pass Cave.

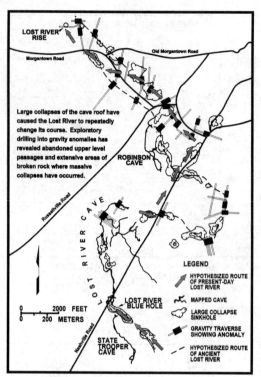

FIGURE 4. Mapped portions of Lost River Cave and hypothesized present-day and ancient routes as determined by microgravity (Crawford, 1986).

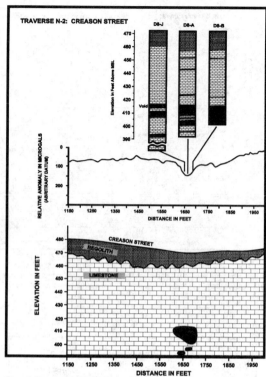

Figure 5. Three borings into this low-gravity anomaly intersected a water-filled cave over 15 feet deep and 50 feet wide.

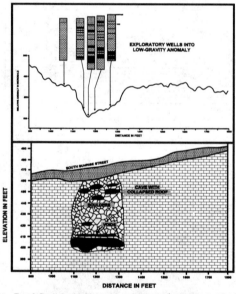

Figure 6. Four borings into this large low-gravity anomaly along South Sunrise Street in Bowling Green intersected numerous voids and boulders indicative of a collapsed bedrock cave. There is no surface expression that might reveal the presence of the collapsed cave. Other borings along the traverse did not intersect voids.

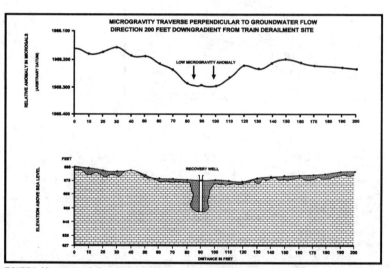

FIGURE 7. Lineament analysis and microgravity traverses were used to locate a sediment-filled cave about 200 feet downgradient from a train derailment site near Lewisburg, Tennessee. Approximately 15,000 gallons of chloroform, a DNAPL, sank into the karst aquifer. Sediment was excavated at a place where the cave roof had collapsed and a large-diameter recovery well installed (Crawford and Ulmer, 1994).

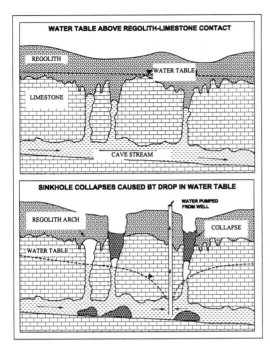
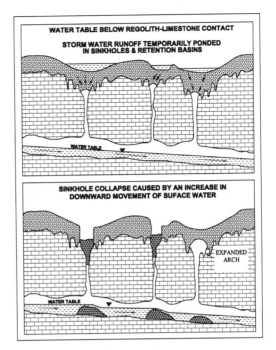

FIGURE 8, LEFT. Sinkhole collapses in areas where the water table is above the regolith-limestone contact are usually caused by a drop in the water table. Regolith arches spanning openings in the bedrock collapse because of the loss of buoyant support and because of downward-moving surface water.

FIGURE 9, RIGHT. Sinkhole collapses in areas where the water table is below the regolith-bedrock contact are usually caused by an increase in the downward movement of surface water. Land use changes and construction activities that concentrate surface water in drains, sinkholes, and impoundments may locally increase downward movement of surface water and induce the collapse of regolith arches.

percent are man-induced collapses of existing regolith arches (Figure 11). Changes in the surface drainage associated with farming, and particularly urban development, are believed to be the primary cause of most collapses.

Microgravity will often reveal the existence of air-filled voids in the regolith above bedrock and therefore identify potential sinkhole collapses. Figure 12 is an example of a microgravity traverse over a regolith void that had collapsed to the surface so that a small hole was actually visible. However, if the regolith voids are beneath the water table, the density contrast between clay (1.5 g/cm^3) and water (1.0 g/cm^3) of only 0.5 g/cm^3 may not be sufficient to permit detection.

Microgravity traverses can be made along foundations previous to building construction to identify low-gravity anomalies. However, depth to bedrock borings into the anomalies are usually needed to establish if they are regolith voids, bedrock caves, or cutters (areas of deep regolith between pinnacles). Microgravity therefore serves as a useful tool for identifying sites that need to be explored by exploratory borings. Although one can model the size and shape of the void responsible for a low-gravity anomaly, and thereby develop a conceptual model, exploratory borings are virtually always necessary to test and improved the model.

Crawford has used microgravity at several sites to investigate subsurface conditions in the vicinity of sinkhole collapses. It provides useful information concerning: a) depth to bedrock, b) extent and shape of the collapse area below the surface, c) location of the crevice or crevices, into which the regolith is collapsing, and d) locations of additional regolith voids in the vicinity of the initial collapse. The following case studies discuss the application of microgravity techniques in greater detail.

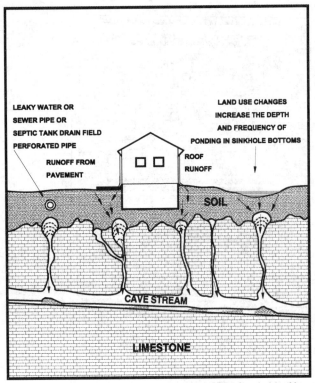

FIGURE 10. Growth of regolith arches toward the surface induced by modification of natural runoff and infiltration conditions.

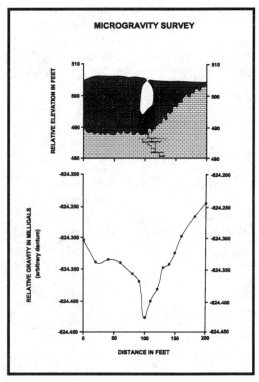

FIGURE 12. Low-gravity anomaly along a traverse over a regolith void that had collapsed all the way to the surface so that a small hole was actually visible.

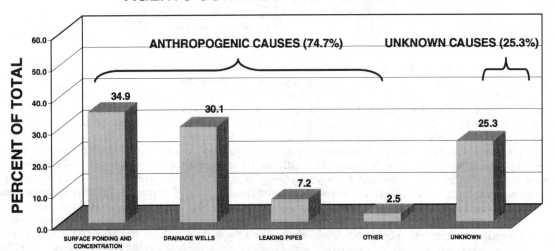

FIGURE 11. An investigation of 80 sinkhole collapses in Bowling Green, Kentucky shows that surface ponding and concentration is the leading cause of sinkhole collapse (Crawford, Webster, and Veni, 1989).

CASE STUDY: SINKHOLE COLLAPSE UNDER INTERSTATE 65

In April 1995 the Kentucky Department of Transportation was notified of a large pothole in the outside lane of I-65 northbound near Elizabethtown, KY. Upon inspection workers discovered a collapsing sinkhole. Loose regolith was removed from the collapse, resulting in an excavation roughly 15 feet broad and 15 feet deep. The hole was then filled with large rock and paved over with asphalt. The site has had numerous sinkhole collapses along drainage ditches and the shoulder of the road dating back to 1978 (Figure 13). The investigation was made to determine: a) locations of additional regolith voids that might be ready to collapse and b) shape of existing collapse and depth to bedrock so that a repair strategy could be formulated that would result in minimal obstruction to traffic flow.

A total of 10 microgravity survey traverses were established perpendicular to the line of sinkhole collapses and parallel to the interstate (Figure 13), and coded A through J. Measurement stations were 5 feet apart and each traverse roughly 12 feet apart.

Traverse A extended for 230 feet along the paved outside shoulder of the northbound lane, centered over the patched sinkhole collapse. The lowest measured gravity occurred over the patched sinkhole collapse (Figure 14). A boring over the anomaly penetrated 14.1 feet of limestone fill used in the sinkhole repair, a 1.5 foot void, and then encountered soft red clay down to 25.1 feet in depth. From 25.1 feet to 30.9 feet the boring intersected stiff red clay and sandy shale.

Traverse B extended for 220 feet along the center of the outside northbound lane, about 2 feet to the west of the asphalt patch used to repair the sinkhole. A small gravity anomaly was detected at approximately the same distance along the anomaly as detected on Traverse A (Figure 14). A boring made into the anomaly intersected a small regolith void and reached bedrock 5.4 feet higher than the boring made along Traverse A. This suggested that the sinkhole collapse occurred directly over a crevice or cutter in the bedrock.

No significant low-gravity anomalies were detected along Traverses C-J. Borings made along these traverses indicated a varying depth to bedrock with evidence of a few small soil voids.

It was concluded that the 1995 sinkhole collapse was a regolith collapse resulting from the raveling of material into a crevice in the underlying limestone bedrock, probably over a period of years. The 1978 collapse along the east drainage ditch increased the movement of runoff water into the subsurface. During large rains storm water sank into this open sinkhole and then flowed laterally across the top of bedrock to sink into nearby bedrock crevices. The piping of regolith into the crevice and upward expansion of regolith voids over time, resulted in the 1994 and 1995 collapses. The vertical fracture crossing under this section of the interstate results in a vulnerability to sinkhole collapses. The fracture is likely associated with the nearby Elizabethtown Fault, which defines a lineament conforming to the existing sinkhole collapses on the site as well as several springs to the west. It was recommended that each of the sinkhole collapses on the site, including the one that had been patched, be thoroughly excavated to bedrock and filled with graded rock, and that the bordering drainage ditches be lined to limit downward infiltration of storm water runoff (Crawford, Tucker and Summerlin, 1995).

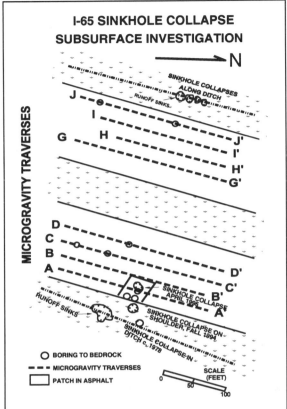

FIGURE 13. Microgravity traverses along the northbound and southbound lanes of I-65 at Elizabethtown, Kentucky.

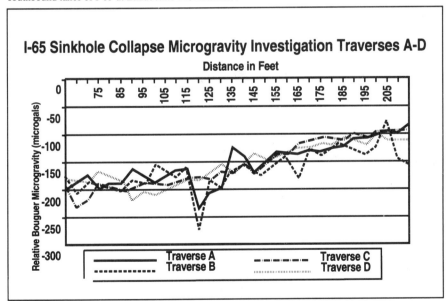

FIGURE 14. Low-gravity anomaly detected in the vicinity of the sinkhole collapse along Traverses A and B.

CASE STUDY: SINKHOLE COLLAPSE UNDER HIGHWAY IN MAMMOTH CAVE NATIONAL PARK

During October of 1995, maintenance workers at Mammoth Cave National Park in Kentucky reported a collapsed area on the east side of the South Entrance Road, one of two main routes to the Park's Visitor Center and Historic Entrance. Surface expression of subsurface activity consisted of a circular collapse about 5 feet in diameter and 5 feet deep along a drainage ditch 30 feet east of the road (Figure 15). Approximately 70 feet east of the road is an older collapse, 25 feet in depth, which includes the entrance to Sloan's Crossing Cave # 1. The site is located on the sandstone caprock near the contact with the underlying limestone, an area where development of vertical shafts tends to be concentrated. A microgravity survey was used to delimit the extent of the existing collapse zone and determine a strategy to protect the road and visitors to the park.

Six traverses were established parallel to the South Entrance Road in the vicinity of the surface collapse. Measurement stations were established along a 5 to 10-foot interval. Figure 15 illustrates the relative location of gravity traverses and surface features at the site.

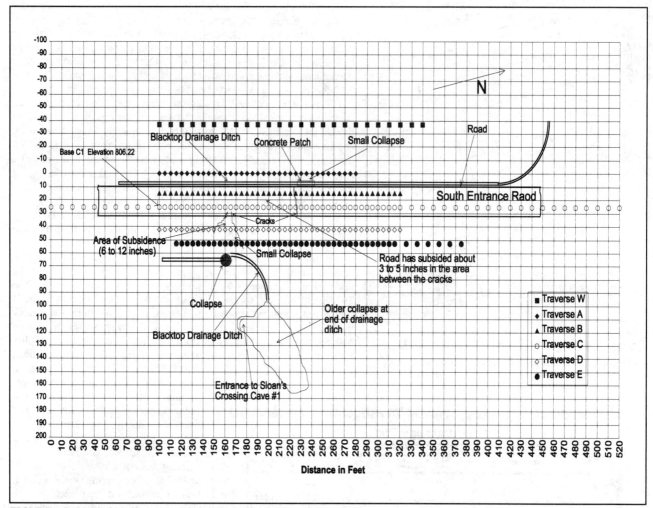

FIGURE 15. Sinkhole collapses and ground subsidence along the South Entrance Road, Mammoth Cave National Park. Microgravity was used to investigate subsurface conditions along Traverses A, B, C, D, E and W.

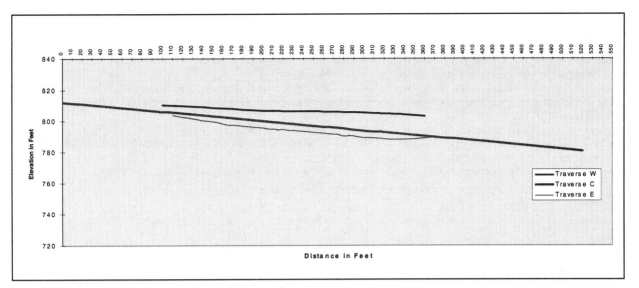

FIGURE 16. Measured elevations along Traverses W, C, and E along the South Entrance Road, Mammoth Cave National Park.

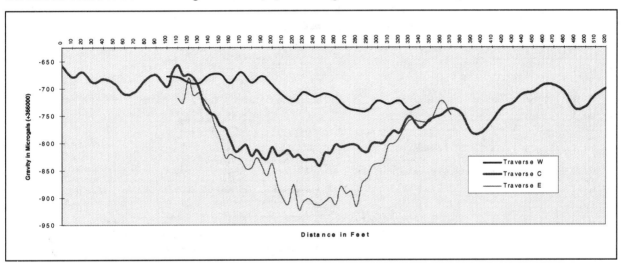

FIGURE 17. Chart showing microgravity results of Traverses W, C, and E measured at site of Mammoth Cave collapse.

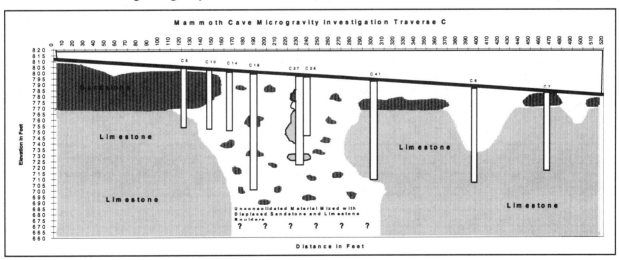

FIGURE 18. Conceptual profile showing subsurface conditions based on exploratory borings and measured gravity along Traverse C.

For the sake of brevity only Traverses W, C, and E are discussed here. Traverses A, B, and D conform to the trend exhibited by the traverses discussed. Figure 16 illustrates measured elevations along Traverses W, C, and E. Elevation along the road (Traverse C) decreases from south to north and from west (Traverse W) to east (Traverse E) perpendicular to the road.

Traverse W extended along the west side of the road, 48 feet west of the road, roughly south-to-north for 240 feet. The data did not indicate any significant low-gravity anomalies along this traverse (Figure 17).

Traverse C extended south to north along the east lane of the South Entrance Road for a total distance of 520 feet. Figure 17 illustrates the detection of a very large low-gravity anomaly approximately 240 feet across with a maximum change of about 150 microgals (Figure 17).

Traverse E, 27 feet east of Traverse C, extended for 270 feet south to north on the east side of the road. The microgravity data show a large low-gravity anomaly similar in shape and extent to Traverse C but with a more dramatic relative change in Bouguer gravity consisting of a 250 microgal drop (Figure 17).

Considered together the microgravity traverses clearly indicate a broad zone of low gravity in the subsurface underlying the Park South Entrance Road and its immediate vicinity, increasing in size and magnitude toward the east. Exploratory drilling was necessary to investigate the subsurface in the vicinity of the large low-gravity anomaly.

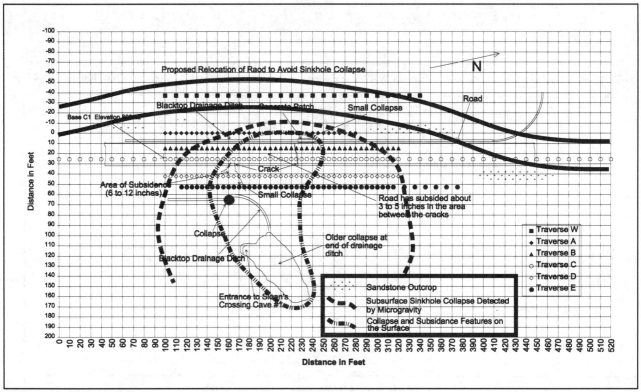

FIGURE 19. Sinkhole collapse area as delineated by microgravity and exploratory borings.

Exploratory wells drilled along Traverse C outside of the zone of the detected anomaly intersected sandstone caprock underlain by competent limestone with small (2-3 foot) clay-filled voids at the contact. (Figure 18). Exploratory borings along Traverse C into the low-gravity anomaly intersected less-dense, unconsolidated material that has filled a large sinkhole created by the collapse of the sandstone caprock into a large vertical shaft complex in the underlying Girkin Limestone. This large cave room was filled with soft unconsolidated material ranging from sand and clay to large limestone and sandstone boulders. Well C19 indicated that the filled vertical shaft was at least 102 feet deep (Figure 18). It was concluded that the recent collapse and subsidence along the Park South Entrance Road was induced by concentrated vertical infiltration of surface runoff along the ditches on both sides of the road. Figure 19 illustrates the lateral extent of the low-gravity/collapse zone as well as proposed relocation of the road atop competent sandstone. Relocation of the road to avoid the sinkhole collapse was cost-prohibitive so the site was mitigated by deep impact compaction followed by a reinforced mat and new pavement (Crawford et al, 1995).

CASE STUDY: SUBSURFACE INVESTIGATION AT INDUSTRIAL SITE ON KARST TERRAIN

Dye traces conducted at an industrial site located on the Lexington Limestone Formation of northern Kentucky indicated the existence of a conduit-flow karst aquifer in the subsurface. In order to help characterize the site and develop an effective monitoring program, microgravity survey techniques were employed to delimit the location of subsurface voids with the ultimate goal of installing a down gradient monitoring well into a cave stream.

Four traverses were made across the site taking measurements at ten-foot intervals with the microgravity meter. At each measuring station the elevation was surveyed. Traverses A and B comprised a perimeter survey around a sinkhole collapse on the site (Figure 20). Traverses C and D extended the range of the survey in an attempt to further delineate the path of the cave stream. Traverse A revealed a small low-gravity anomaly at 130-220ft along the path surveyed. This may represent an increased depth to bedrock, but corresponds well with the occurrence of the large sinkhole on site (Figure 21). Traverse B revealed a low-gravity anomaly from 100 to 200ft along the path surveyed. The anomaly is indicative of the occurrence of a cave, regolith void, or increased depth to bedrock (Figure 22).

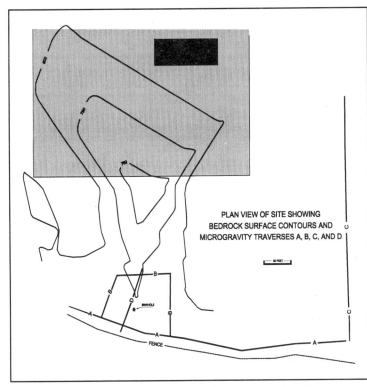

FIGURE 20. Plan view of industrial site showing bedrock surface contours and microgravity survey traverses A, B, C and D.

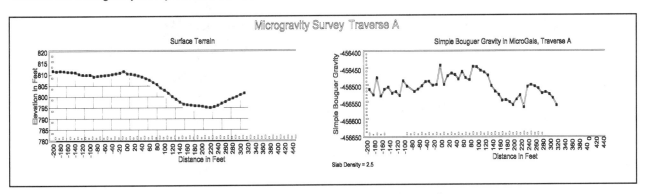

FIGURE 21. Microgravity Traverse A at industrial site, elevation and Bouguer microgravity.

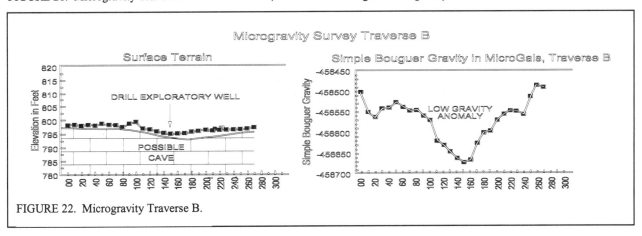

FIGURE 22. Microgravity Traverse B.

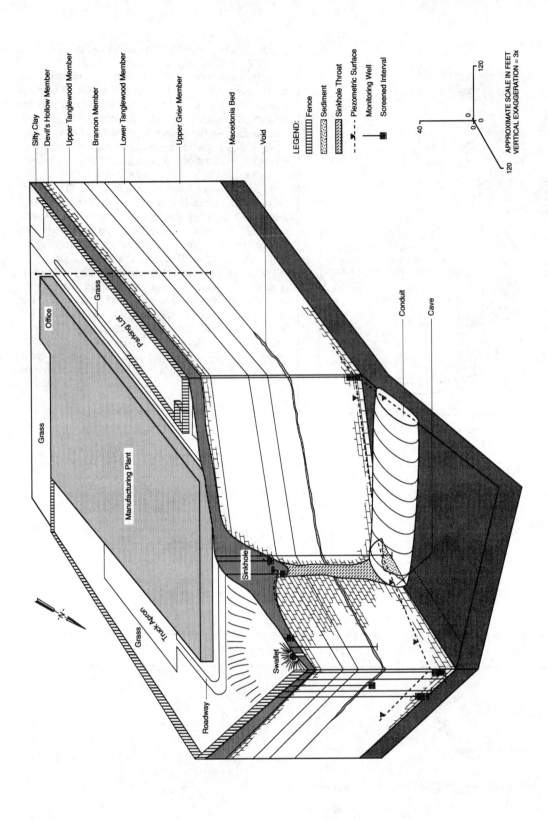

Figure 24. Three dimensional conceptual model illustrating location of cave passage detected by microgravity at industrial site in northern Kentucky. (Figure provided by Dames and Moore, Cincinnatti, OH)

Traverse C produced no low-gravity anomalies and traverse D indicated the same anomaly revealed along Traverse B (Figure 23).

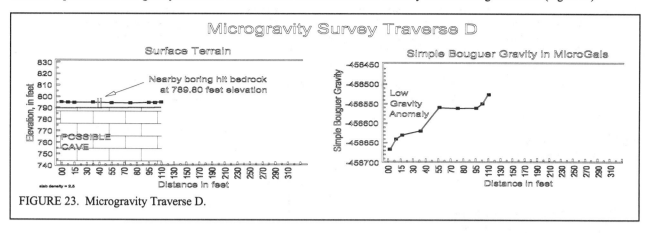

FIGURE 23. Microgravity Traverse D.

Concluding that the low-gravity anomaly confirmed the existence and location of the karst conduit draining the site, it was recommended to the client that the prominent anomaly along Traverse B and D be drilled into to locate the physical cave stream down-gradient from the site. A dye trace should then be performed to further confirm the cave stream as an appropriate monitoring site. Figure 24 shows a three-dimensional conceptual model of the subsurface conditions at the site. The first wells drilled into the low-gravity anomaly B-1, 2,3 and 4 indicated increased depth to bedrock. However, well 10 drilled off the traverse about 100 feet, but in line with the low-gravity anomaly trend revealed by Traverse B and D, intersected an air filled cave passage with a stream at a depth of 117 feet to 130 feet (Figure 24).

CASE STUDY: SINKHOLE COLLAPSE UNDER A TWO-STORY BUILDING IN BOWLING GREEN, KENTUCKY

Visible structural damage to a two-story building in Bowling Green, Kentucky consisted of extensive cracks through the front brick wall of the building extending to the second level (Figure 25). Visible subsidence and ground cracks existed under the outside walls of the structure, and a 6" depression existed in the asphalt parking lot at the rear of the structure. A more extensive investigation from the crawl space under the building revealed cracks in footers and walls supporting the structure, deep cracks in the ground under the structure and collapses under the two central footers and the rear footer. Extent of surface expression of the collapse was approximately 60 feet by 70 feet (Figure 25). Based on preliminary evidence, the site was at risk for a catastrophic collapse. In order to delimit the scale of collapse and determine remedial strategy, a microgravity survey accompanied by exploratory soil borings were conducted.

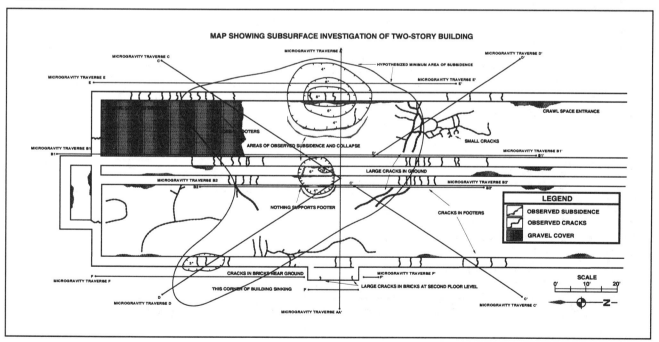

FIGURE 25. Plan view of site illustrating foundation supports of two-story building, location of physical features indicating collapse and/or subsidence, and location of microgravity survey traverses.

A total of 7 microgravity traverses were surveyed under and outside of the building using a LaCoste and Romberg Model D microgravity meter (Figure 25). A 5-foot station interval was used in the crawl space under the building and a 5 to 15-foot interval was used on traverses extending outside of the building. Depth to bedrock soil borings were used to establish depth to bedrock at the front and rear of the building and to calibrate microgravity measurements. Well logs indicate that depth to bedrock varied from 34 feet to greater than 58 feet, the greatest depth achievable by the drill rig (Figure 26).

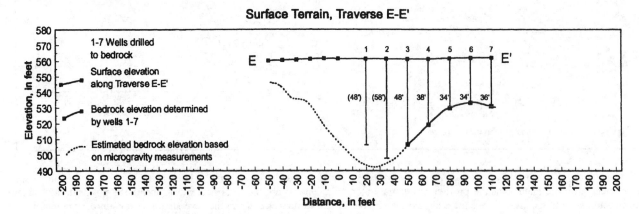

FIGURE 26. Elevation and depth to bedrock measurements along Traverse E- E' of microgravity investigation.

Traverse E-E' extended north to south for 160 feet on the east side of the building (Figure 26). Measured gravity (Figure 27) compared with depth to bedrock at wells 1-7 was used to estimate depth to bedrock along the entire traverse (Figure 26).

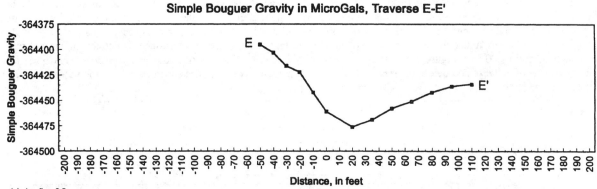

slab density = 2.5
FIGURE 27. Microgravity along Traverse E-E'.

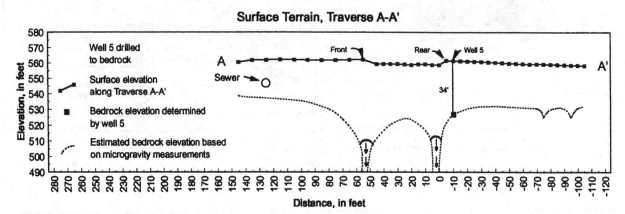

FIGURE 28. Elevation and depth to bedrock measurements along Traverse A- A' of microgravity investigation.

Traverse A-A' was surveyed under and outside of the building, running east to west for 260 feet perpendicular to the building (Figure 28). A profile of measured gravity (Figure 29) was used to estimate depth to bedrock and model anomalies (Figure 28).

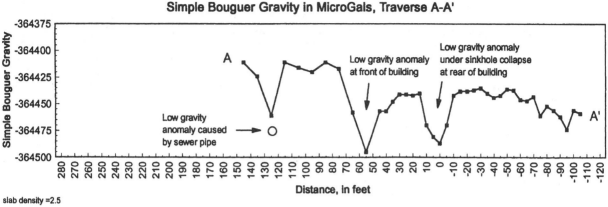

FIGURE 29. Microgravity along traverse A-A'.

The microgravity investigation indicated a very large low-gravity anomaly under the north wing of the two-story building (Figure 30). Borings into the low-gravity anomalies indicated a depth to bedrock in excess of 58 feet. The collapse was mitigated by drilling wells into the low-gravity anomalies and injecting grout under pressure to fill the voids and compact the unconsolidated material under the building.

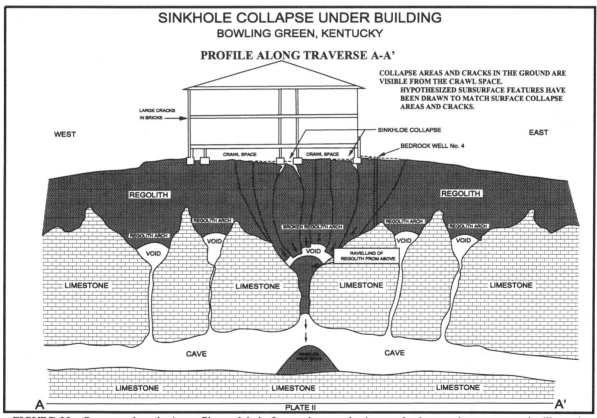

FIGURE 30. Conceptual geologic profile modeled after exploratory borings and microgravity survey results illustrating extent of sinkhole collapse under two-story building in Bowling Green, KY.

CONCLUSIONS

Microgravity has proven to be a useful geophysical tool for investigating subsurface features in karst terrain. It is particularly useful in urban areas where buried pipes, electrical wires, and surface structures often limit the use of some other geophysical techniques. If the caves are relatively large and shallow (less than 100 feet deep), it is particularly useful for locating cave streams for determining groundwater flow direction and for installing monitor or recovery wells directly into the karst conduit that drains a hazardous material site. It is also a useful tool for identifying potential sites for sinkhole collapses and in the investigation of existing collapses. In addition to the case studies discussed in this paper, Crawford and Associates have used microgravity for subsurface investigations at numerous sites, including a proposed site for a dam, along the proposed route for a highway, along the proposed route for a pipeline, for foundation studies for proposed buildings and for karst storm water drainage studies. It has provided very useful subsurface information at the sites investigated. However, it should not be used at all karst sites. If the caves are too small and/or too deep they will not be detected by microgravity. The best results are obtained in well-developed karst with large caves less than 100 feet below the surface.

REFERENCES

Benson, R.C.; Kaufmann, R.D.; Yuhr, L.B. and Martin, D. (1998). Assessment, prediction and remediation of karst conditions on I-70, Frederick, Maryland, 49th Highway Geology Symposium, pp. 1-15.

Butler, D.K. (1983). Microgravimetric and magnetic surveys, Medford Cave Site, Florida, Cavity Detection Research Report 1, Tech, Rep GL-83-1. Vicksburg, MS: Army Engineer Waterways Station, p.92.

Butler, D.K. (1984). Microgravimetric and gravity gradient techniques for detection of subsurface cavities. Geophysics, v. 41, p. 1016-130.

Crawford, N.C. (1989). Toxic and explosive fumes rising from contaminated groundwater flowing through caves under Bowling Green, Kentucky, prepared for City of Bowling Green, Municipal Order No. 85-83 p. 54.

Crawford, N.C. (1995). Microgravity techniques for detection of karst subsurface features, Site Investigations: Geotechnical and Environmental, Proceedings. Twenty-Sixth Ohio River Valley Soils Seminar, Clarksville, IN, pp. 1-24.

Crawford, N., Groves. C., Tucker, R., Forbis, S., Clauson, W., Ericksen, K., and Patrick, K., (1995). Microgravity subsurface investigation at the site of a sinkhole collapse under the South Entrance Road to Mammoth Cave National Park, Kentucky. Report prepared for Mammoth Cave National Park, November 1995.

Crawford, N.C., Tucker, R.B., and Summerlin, M.A. (1995). Microgravity subsurface investigation at the site of a sinkhole collapse under Interstate 65 at Elizabethtown, Kentucky. Prepared for Kentucky Department of Transportation, May 31, 1995.

Crawford, N.C. and Ulmer C.S. (1994). Hydrogeologic investigations of contaminant movement in karst aquifers in the vicinity of a train derailment near Lewisburg, Tennessee. Environmental Geology, V.23, pp. 41-52.

Crawford, N.C. and Webster, J.W. (1988). Microgravity investigation of the proposed Dishman Lane Extension, Bowling Green, Kentucky, report for Law Engineering, Nashville, Tennessee, p. 20.

Crawford, N.C., Webster, J.W., and Veni, G. (1989). Sinkhole collapse problems in Warren County, in Crawford, N.C. (ed.) The Karst Landscape of Warren County, prepared for Bowling Green-Warren County Planning and Zoning Commission, Center for Local Government, Western Kentucky University, Bowling Green, Kentucky, pp.71-115.

Crawford, N.C., Webster, J.W. and Winter, S.A. (1989). Detection of caves from the surface by microgravity followed by exploratory drilling: Lost River Groundwater Basin, Bowling Green, Kentucky, Prepared for the City of Bowling Green, Municipal Order No. 85-83.

Kirk, K.G. (1974). Resistivity and gravity surveys applied to karst research, Proceedings of the 4th Conference on Karst Geology and Hydrology. West Virginia Geological Survey, pp. 61-71.

Omnes, G. (1976). High accuracy gravity applied to the detection of karstic cavities. Karst Hydrogeology, ed. J.S. Tolson and F.L. Doyle UAH, Huntsville, Alabama, USA, International Association of Hydrogeology. V.12 pp. 273-284.

Smith, D.L. and Smith G.L. (1987). Use of vertical gravity gradient analyses to detect near-surface dissolution voids in karst terrains, 2nd Multidisciplinary Conference on Sinkholes and the Environmental Impacts of Karst, Ed. By Beck, B.F. and Wilson, W.L. pp. 205-209.

Two-dimensional resistivity profiling; geophysical weapon of choice in karst terrain for engineering applications

MARK H. DUNSCOMB Schnabel Engineering Associates, West Chester, Pa., USA
ERIC REHWOLDT Schnabel Engineering Associates, Bethesda, Md., USA

ABSTRACT

Geophysics has been used for many years to assist in characterizing the often complex subsurface in karst terrain. However, traditional methods have typically met with limited success due to inherent problems with the use of these techniques in karst terrain which often results in ambiguous information. This has left the geophysicist ever searching for new, more accurate, and cost effective ways to estimate subsurface conditions in carbonate geologies. With the advent of faster computers and inversion model algorithms, the resistivity technique has seen a rebirth in the form of two-dimensional resistivity tomography, particularly due to its usefulness in karst terrain. This method can use up to 150 electrodes in a linear array in conjunction with a computer driven resistivity meter to quickly obtain large amounts of data. A two-dimensional profile can then be generated showing location and depth of measured resistivity values along the electrode array. The measured values are then modeled to remove electrode geometry effects and estimate the true subsurface resistivity conditions.

The two-dimensional resistivity method is shown to have very good resolving abilities in karst terrain to image geologic features such as pinnacled bedrock surfaces, "overhanging" rock ledges, fracture zones, and voids within the rock mass and within the soil overburden. The approximate length of time to perform a survey depends on field conditions and the desired array length. However, an average time for performing a survey with 28 electrodes over average terrain is just over 2 hours. Downloading to a laptop computer and modeling for a cursory evaluation in the field may take 15 to 30 minutes.

Promising future advancements of this technique include faster algorithms for three-dimensional surveys and cross-borehole surveys for increased resolution of near vertical and/or deep features.

INTRODUCTION

Characterization of complex subsurface conditions in karst terrain using geophysical techniques has been implemented for many years to estimate conditions beneath the ground for research and engineering applications. Typically, geophysics is used to help select locations for more invasive investigative techniques such as drilling, or to supplement existing information on subsurface conditions from drilling data and observations of surface features such as sinkholes, springs, fracture trends and lineaments. Features commonly associated with karst geologies that are often the target of geophysical investigations include depth to bedrock and pinnacled bedrock profile conditions, ledge rock conditions, location and size of voids or cavities in soil overburden and rock, saturated "enhanced weathered zones", and the location and size of dry or water-bearing bedrock fracture systems (i.e., joints or faults) and solution pathways.

Many different geophysical methods have been used to characterize karst conditions with varying degrees of success. Most of these methods take advantage of contrasts in earth material properties (density, electrical conductivity, natural potentials, etc.) to locate and quantify karst features. Traditional methods involve recording and filtering data in various forms, and mapping these data to infer the character and location of subsurface features at depth. Some of these methods include electromagnetics (EM), conventional resistivity sounding, microgravity, seismic refraction, induced polarization, and spontaneous potential surveys. These methods may often be considered first for use in karst due to relative ease of use, low cost, and past successes in other geologic terrain. However, these methods typically lack the ability to resolve complex karst features and often lead to ambiguous and/or inconclusive results. Recent developments in ground penetrating radar (GPR) technology have improved this method to provide excellent resolving capabilities; however, conductive, clayey soil overburden typically associated with carbonate geologies often limits the depth of penetration of GPR so that important data at depth is not obtained. As a result, the costs in time and money to perform these often "inconclusive" geophysical studies in karst are not usually viewed as money and time well spent by owners, planners and design engineers alike.

Recent development of relatively fast, two-dimensional resistivity imaging, or profiling, techniques has enabled the engineering geophysicist and geologist to map relatively complex subsurface geologic conditions in karst in greater detail and depth than what could be reasonably assessed using more traditional geophysical techniques. This method involves field measurements made using a computer-controlled data acquisition system connected to a linear array of electrodes inserted in the ground at a constant spacing. Inversion modeling of the resulting data set is carried out using a finite difference or finite element subroutine that provides an approximation of the true earth resistivity distribution. Anomalies within the modeled data may then be used to help identify voids, rock ledges, bedrock fractures and thickness of soil overburden in karst terrain.

This paper provides an overview of the resistivity profiling technique as applied to geotechnical engineering studies in karst terrain. Results of studies performed for existing highway and proposed commercial and industrial development sites in Maryland, Pennsylvania, Virginia and West Virginia are presented and discussed. Typical method limitations and required survey and analysis time are discussed to assist the reader in determining appropriate applications of this very useful geophysical site characterization tool.

HISTORICAL AND THEORETICAL BACKGROUND

The resistivity geophysical method has been utilized in exploration for about a century. Historically, however, it has been slow and difficult to maneuver long cables associated with resistivity data collection through wooded or hilly terrain. With the advent of sophisticated, accurate, and relatively easy-to-handle EM instruments in the 1960's and 1970's, the resistivity technique lost popularity to faster EM methods. The resistivity method has traditionally used four metal stakes (electrodes) which are driven about a foot into the ground. A measurement is obtained and then the electrodes are moved to gain data at different depths or in different locations. Recently, however, computers have made it economically possible to obtain and manipulate large, detailed data sets. Now, many electrodes (generally 20 to 150) can be driven into the ground at a regular interval in a line. A computer guided resistivity meter can automatically take measurements along the "array" of electrodes using four electrodes at a time until all the possible combinations of electrodes within given parameters are used.

Where the site constraints allow, the electrode array can be advanced along the ground in a line by "rolling along" the array (AGI, 1997). This is done when all the measurements in the original array are completed and the data set is extended along the profile line by "flip-flopping" the beginning portion of the array to the end of the array. When all possible measurements have been taken using the new arrangement, the beginning portion of that array is again flip-flopped to the end. In this way, measurements may be obtained indefinitely without leaving gaps in the data. These techniques quickly provide resistivity information with depth and distance along the ground surface, hence two-dimensional resistivity, rather than only resistivity data with depth at one location at a time (one-dimensional resistivity). Furthermore, the data can be stored and downloaded to a computer for display or modeling using a set of partial differential equations that define current flow through the subsurface. Due to the obvious advantages related to the technique, the resistivity method has begun to regain its popularity, especially in the near surface engineering and environmental geophysics industry, and shows great promise for further expansion.

As mentioned above, the resistivity technique uses four electrodes at a time. One pair of electrodes is used to direct electricity into the subsurface. The second pair of electrodes is used to measure the potential drop (voltage) in the earth. The resistance of the ground circuit is calculated using Ohm's Law (Resistance = Voltage ÷ Current). Next, the resistivity is calculated using the electrode geometry and the resistance. Resistivity is a measurement of resistance normalized for distances through the conducting circuit (i.e., the subsurface). The units of resistivity are therefore ohm-length (typically ohm-meters but can be ohm-cm or ohm-ft, etc.). In general, the farther apart the electrodes are placed, the deeper the electricity flows through the subsurface. Therefore, resistivity depth sounding can be conducted by obtaining measurements at successively greater electrode separations. However, the measured resistivity is effected by how the electrodes are placed in the ground with respect to each other (i.e. the electrode geometry). Due to the complexity of carbonate geologies it can become difficult to determine the actual resistivity values of specific subsurface targets. Empirical curves have been developed for specific geologic situations and may be consulted in an effort to determine the actual distribution of resistivity within the subsurface. This is where computer modeling comes into play. Good numerical models have been developed to attempt to solve this problem. This is sometimes called an "inverse" problem. Instead of knowing the characteristics of an object and calculating its resistivity, we know the resistivity and are trying to determine the characteristics (i.e., shape, depth and true resistivity) of the object in the subsurface.

The most commonly used computer modeling programs today for resistivity profiling are generally based on the approach as provided by M. H. Loke and R. D. Barker (1995). These programs are finite-difference or finite element based and use iterative procedures to approach the actual subsurface conditions. The program works by developing a model of the subsurface conditions based on the measured resistivity. The program then calculates a resistivity profile based on the modeled subsurface. If the modeled subsurface comes close to matching the resistivity profile measured in the field, the resistivity profile calculated from the model should be similar to the measured profile. The measured resistivity and calculated resistivity are compared by means of a root mean square (RMS) error. If the RMS error is high, the program then modifies the model and goes through the same process again to try to minimize the RMS error. This process is continued until the model converges to a relatively level RMS error value, or until a specified number of iterations is completed. Most models converge within three to seven iterations. The modeled data can then be exported in an ASCI data format for contouring using standard graphical contouring software to create the final modeled resistivity cross sections. It is important to note that the resulting modeled resistivity profile is theoretically based, and as such does not provide a unique solution to the actual subsurface conditions. Therefore, the model results should be verified by physical field truthing such as drilling. Based on our experience applying this method in karst, as illustrated in the examples that follow, the interpreted model results have closely matched the actual subsurface conditions.

Several different electrode configurations are commonly used in collecting resistivity data. These configurations include dipole-dipole, Schlumberger (or modified Wenner), Wenner, pole-pole, pole-dipole, and square arrays in general order of popularity. Each of these array types can be modeled with the exception of the square array. The dipole-dipole array is most popular because it generally provides the highest precision, provides a reasonable depth of investigation [about 18% of the maximum allowable electrode separation (Roy and Apparao, 1971) depending upon the measured subsurface resistivity values (Edwards, 1977)], and has the greatest sensitivity to vertical resistivity boundaries (Loke, 1998) as are commonly found in karst terrain at pinnacle interfaces, etc. In the setup for the dipole-dipole array, the distance between the two current electrodes is kept small, as is the distance between the two potential electrodes. This distance is kept the same for both sets of electrodes and is called "a". Also, the two sets of electrodes are separated from each other as opposed to standard Wenner or Schlumberger arrangements, where both potential electrodes are placed in between the current electrodes. The distance between the two sets of electrodes is typically a multiple ("n") of "a". In accordance with scientific convention, the two current electrodes are indicated as "A" and "B", and the potential electrodes as "M" and "N". Figure 1 below shows a standard four-electrode dipole-dipole array.

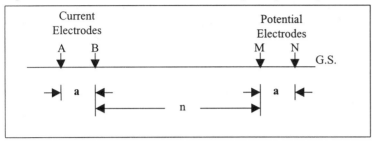

Figure 1: Typical four-electrode dipole-dipole array.

One downfall of the dipole-dipole array, however, is that it does not provide a very high signal for measurement and therefore can produce somewhat noisy data. The Schlumberger array geometry on the other hand generally provides a better signal to noise ratio but does not have quite the resolving power as the dipole-dipole array, especially at depth. The Schlumberger array is also not as sensitive to vertical resistivity boundaries as the dipole-dipole array but is more sensitive to horizontal boundaries. The Wenner array is most sensitive to horizontal resistivity boundaries, but is much less sensitive to vertical boundaries than either the Schlumberger or dipole-dipole arrays. The Wenner array provides good average resistivity measurements with depth, but does not resolve specific subsurface features well. As such, the Wenner array is most often used in obtaining information for purposes such as grounding design. The pole-pole and pole-dipole arrays provide the greatest depth of exploration (about 35% of the maximum allowable electrode spacing depending upon the subsurface conditions) but also provides the worst resolution of the common array configurations available. The square array is sometimes used in situations where there is a very high subsurface resistivity anisotropy (Tsokas, et al., 1997).

DATA INTERPRETATION

Data interpretation of two-dimensional resistivity information is typically done by the geophysicist through manipulation of the measured data by inverse modeling, comparison of the modeled results to the measured data, error estimate analyses, and consideration of potential sources of cultural or natural noise. Modeled results are calibrated by comparing observed anomalies with physical data such as mapped sinkholes, rock outcrops, boring data, aerial photographs, etc. However, there are some specific points to consider and general resistivity anomaly patterns to look for to aid in understanding the data.

In order to understand the two-dimensional resistivity profiles it is important to understand how electricity is conducted through the subsurface. Most earth materials are either good insulators or dielectrics. That is to say, in general, they do not conduct electricity very well. Rather, electricity is conducted through interstitial water by ionic transport. Rock typically has a significantly higher resistivity than soil because it has a much smaller primary porosity, fewer interconnected pore spaces, and is drier than soil. Earth materials, such as clays, that tend to hold more moisture and have a higher concentration of available ions to flow generally conduct electricity better; therefore, have smaller resistivity values. These conditions favor the use of two-dimensional resistivity methods in karst terrain because of the typically high contrast in resistivity values between carbonate rock and typically moist, clayey residual soil overlying it; or karstic features such as voids or very moist deep enhanced weathered zones.

Before looking at the specific resistivity values in a profile, it is important to consider that, as with most geophysical methods, resolution of resistivity data decreases with depth. This is because the number of potentially measured data points decreases with depth as a function of the electrode spacing, type of geometric array used, and the signal to noise ratio. With this in mind, it should be understood that potentially significant features may not be resolved in the resistivity profile, especially as observed near the bottom of the profile. In addition, edge effects at the margins of the model should be considered. Edge effects occur where the finite elements or difference blocks at the edges of the model only have data on one side and exhibit anomalously high or low resistivity values that are not indicative of the true subsurface condition.

Resistivity models generally show the bedrock surface to be pinnacled with a relatively discrete soil/rock interface. Resistivity values of between about 150 to 600 ohm-meters generally delineate the bedrock surface. Because of the generally high contrasts in resistivities among earth material in karst terrain, we have found that contouring the data using logarithmic contours works very well. Additionally, the bedrock interface may lie within a small range of resistivity contour values instead of at one discrete resistivity contour. This range of values can be influenced by the thickness of the enhanced weathered zone overlying the rock, the condition of the bedrock, the thickness of a disintegrated rock zone, if present, and the lack of consistency in defining bedrock in test borings. This is especially true where pinnacled rock is highly jointed.

Voids or solution cavities in carbonate bedrock are generally found to be of two types. The two types include those that are air-filled and empty, and those that are filled with residual soil or water. Isolated zones of anomalously low resistivity measurements within the bedrock mass may indicate a clay-filled void; whereas very high anomalies in the bedrock mass (10,000 to 15,000 ohm-meters or more) may indicate an air-filled void, especially if the observed anomaly lies at or near the bedrock interface. We have observed that these isolated, very high resistivity areas, at or just above the bedrock interface, often correlate with air-filled voids where soil has raveled into a seam or fracture in the bedrock. However, carbonate bedrock tends to exhibit relatively high resistivity values. Therefore, isolated high resistivity anomalies deeper within the rock zone may also indicate very dry, unfractured bedrock.

In addition to subsurface karst features identified by resistivity methods such as voids and pinnacled rock conditions, dry or water-bearing bedrock fractures may also be identified. Narrow zones of lower resistivity within the bedrock mass very often indicate clay- or water-filled fractures or faults.

EXAMPLE RESULTS

To illustrate the results of two-dimensional resistivity profile models in karst terrain and their interpretations, examples of karst features identified at four separate sites are presented below. Dipole-dipole electrode arrays were utilized at each site to maximize penetration depth while maintaining resolution of vertical features. Electrode spacings range between sites from 2 to 5 meters (about 6.5 to 16.4 ft) depending upon the site conditions, anticipated depth to rock, and resolution desired for specific subsurface targets (e.g., fractures and voids).

Bedrock Surface and Fault, Frederick County, Maryland

The resistivity survey conducted at this site was performed at the edge of the paved shoulder of a State highway north of Frederick, Maryland. The highway traverses a line of developing sinkholes that follow regional bedrock structure. The fault identified in the resistivity profile shown in Figure 2 below was traced crossing the highway at this location by aerial photography, and was observed on-strike within a nearby quarry highwall. Additionally, EM data obtained over the profile confirmed the presence of a deep, near vertical, fault-like conductive sheet at the same location.

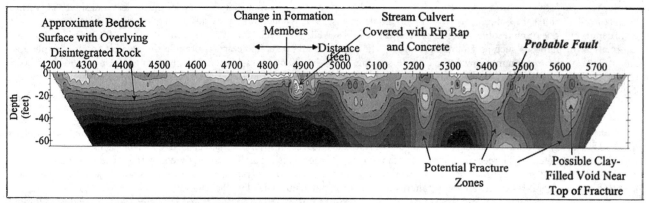

Figure 2: Example of fault zone in modeled resistivity profile in karst.

The fault feature, deep weathered zone, and adjoining bedrock surfaces shown in Figure 2 were later confirmed during drilling and grouting operations conducted to mitigate sinkhole formation within the roadway. Also noteworthy is the difference in karst development between geologic formation members and the presence of cultural noise from a stream culvert crossing beneath the highway.

Disintegrated Rock/Hardrock Interface, Lebanon County, Pennsylvania

Air-track drilling verified depth to bedrock and disintegrated rock conditions within karst from resistivity imaging data at a future power plant site in Pennsylvania. Figure 3 below shows the depth to the bedrock surface interpreted from the resistivity model, the results of the air-track borings performed along the resistivity profile location, and the presence of "overhanging" pinnacles in the bedrock surface.

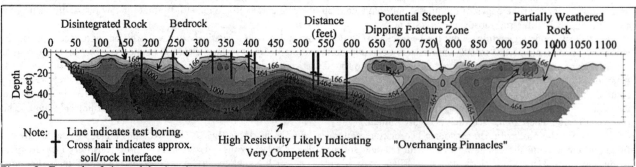

Figure 3: Example of air-track boring data overlain on modeled resistivity profile in karst.

The cross shown on the air-track boring location and depth symbol indicates the approximate bedrock surface interface as recorded from the air-track drill logs. The agreement between the bedrock surface depth estimated from the resistivity model, and the results of the air-track borings are generally very good. However, some variation in rock and disintegrated rock interface depths from the borings is apparent.

Bedrock Cavity and Sinkhole Formation, Pulaski County, Virginia

A remarkable resistivity model profile is shown in Figure 4 that identifies a large bedrock cavity below ground, an opening to the cave at the ground surface, and a large sinkhole developed nearby. The bedrock cavity was well documented in this case and could be accessed to verify its dimensions.

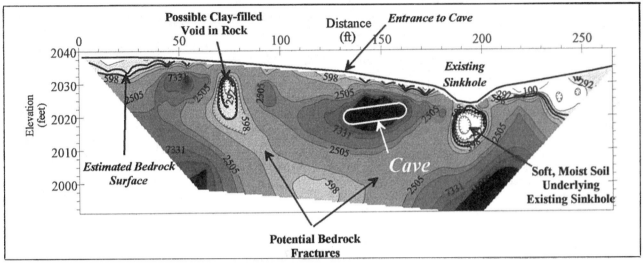

Figure 4: Example of bedrock cavity and adjacent sinkhole developed in karst.

This model profile was utilized for comparison in delineating similar features including additional voids, depth to rock, and rock ledge conditions for a large commercial development site nearby the cave shown in Figure 4. Air-track probes were conducted to confirm the resistivity data for this study and were found to be in excellent agreement with the resistivity data.

Soil Void Formation Over Bedrock Fracture, Mineral County, West Virginia

Large surface sinkhole features were present in the area of the resistivity profile shown in Figure 5 below. A small depression at the ground surface marked by an exposed section of scrap lumber was all that warned of the voided condition located only a few feet below the ground surface.

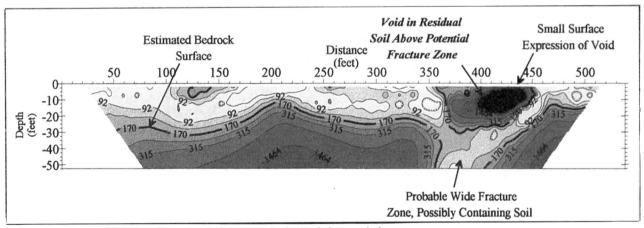

Figure 5: Example of void in soil overburden above major bedrock fracture in karst.

Additional voids were observed to form in an area of concentrated highway stormwater drainage that happened to coincide with a major bedrock fracture. The fracture identified in Figure 5 could be traced from low altitude aerial photography of the site.

CONCLUSIONS

Engineering and environmental geophysicists have struggled for many years in karst terrain because of the inherent difficulties associated with attempting to characterize highly variable subsurface conditions. Two-dimensional resistivity provides a method of imaging the subsurface in karst terrain with accuracy and precision unmatched by other standard geophysical methods. Unlike GPR methods, resistivity is well suited to the normally clayey, low resistive, residual soils derived from carbonate bedrock which often allow for good electrical contact between the electrodes and soil and provide a good medium to direct current into the subsurface. Also, unlike seismic refraction or EM, two-dimensional resistivity has the ability to resolve an undulating, pinnacled bedrock surface as well as highly weathered zones or voids within the rock mass. Although the modeled resistivity profile does not provide a unique solution to actual geologic conditions, physical ground truthing in the field indicates that the resistivity model is very close to being unique, especially where the subsurface conditions are moderately complex. In addition, the resistivity model provides a good estimation of the actual subsurface resistivity conditions without the need for other information related to rock or soil material properties; whereas gravity modeling requires either additional testing of soil and/or rock properties, or gross assumptions with respect to subsurface densities as well as detailed and sometimes time consuming error corrections and data manipulation. Additionally, the two-dimensional resistivity method is relatively quick and measurements can be downloaded to a computer and modeled for a "first cut" look at the data in the field. For example, typical data collection time for a 28 electrode array {135 m (442.8 ft) total length for 5 m (16.4 ft) electrode spacing} in average field conditions may take just over 2 hours (45 minutes for setup, about 1 hour for data collection, and 20 minutes for takedown). Downloading and modeling of that data for a cursory review in the field would take about 20 minutes. Therefore, a line could be conducted and modeled for initial analysis in one morning.

It should be noted that the two-dimensional resistivity technique, like any other, does have limitations. Primary among those limitations in karst terrain is that the method produces a two-dimensional "slice" of the subsurface, and karst features are generally also highly variable in the third dimension. This is amplified by the fact that although the data is modeled and presented two-dimensionally, the measurements include some influences from features in the third dimension and are dependent upon the direction of measurement movement and three-dimensional anisotropy such as bedding planes. Also, although the technique has been shown to provide excellent information on subsurface features, there are resolution limits with respect to potentially significant but small karst features. A feature like this could be, for example, a thin, partially open or clay-filled fracture in the bedrock surface that could potentially become a route for raveling soil and a subsequent sinkhole. Resolution does increase when the electrode spacing is decreased but the trade off is less depth of evaluation.

Areas for future exploration with respect to resistivity include three-dimensional resistivity tomography (Loke and Barker, 1996). Three-dimensional resistivity tomography has the ability to correct for resistivity anisotropy and three-dimensional features in the geology to provide a larger and more accurate image of the subsurface. At present three-dimensional resistivity requires significantly more time to collect and interpret the data and is uneconomical for most sites. However, improvements in computers and model and program algorithms should continue to decrease the time needed to process this type of data. Cross-borehole resistivity in conjunction with induced polarization also could provide a future improvement for locating smaller potentially significant features in karst terrain.

ACKNOWLEDGEMENTS

The authors wish to acknowledge Mr. A. David Martin of the Maryland State Highway Administration who provided ample opportunity to compare modeled resistivity data with other geophysical survey methods, and who allowed our use of data presented in this paper. The authors also wish to acknowledge Mr. Tomasz Labuda (Schnabel Engineering Associates, Bethesda, Maryland) who provided modeled resistivity data for the Mineral County, West Virginia, example site for this paper.

REFERENCES

Advanced Geosciences, Inc., 1997, *Sting R1 Instruction Manual, Release 2.5.5,* AGI, Austin, TX.

Edwards, L.S., 1977, A modified pseudosection for resistivity and IP: GEOPHYSICS, Vol. 42, no. 5 (August 1977); pp. 1020-1036.

Loke, M.H., pers. comm., October 1998; mhloke@pc.jaring.my

Loke, M.H., and Barker, R.D., 1996, Practical techniques for 3D-resistivity surveys and data inversion: Geophysical Prospecting, vol. 44, pp. 499-523.

Loke, M.H., and Barker, R.D., 1995, Least-squares deconvolution of apparent resistivity pseudosections: GEOPHYSICS, Vol. 60, No. 6 (November-December 1995); pp. 1682-1690.

Roy, A. and Apparao, A., 1971, Depth of investigation in direct current methods, GEOPHYSICS, Vol. 36, No. 5, pp. 943-959.

Telford, W.M., Geldart, L.P., and Sheriff, R.E., 1990, *Applied Geophysics,* Second ed., New York, U.S., Cambridge University Press, 770 p.

Tsokas, G.N., Tsourlos, P.E., and Szymanski, J.E., 1997, Square array resistivity anomalies and inhomogeneity ratio calculated by the finite-element method: GEOPHYSICS, Vol. 62, No. 2 (March-April 1997); pp. 426-435.

Karst system characterization utilizing surface geophysical, borehole geophysical and dye tracing techniques

SCOTT GEORGE   QST Environmental, Incorporated, St. Louis, Mo., USA
THOMAS ALEY   Ozark Underground Laboratory, Protem, Mo., USA
ARTHUR LANGE   Karst Geophysics, Incorporated, Golden, Colo., USA

ABSTRACT

A multi-phase characterization study was conducted at an industrial facility located in a karst terrain in central Alabama. The study area covers an area of approximately 1 square mile and contains light non-aqueous phase liquid (LNAPL), coal tar constituents, former disposal lakes, and several landfills. Initial characterization studies identified karst features (pinnacled bedrock, bedrock voids, paleo-sinkholes). Upon evaluation of the initial studies, it was determined that additional analysis was needed regarding the following issues: (1) the interconnection of the underlying "aquifer(s)"; (2) the dominant groundwater flow type (conduit or dispersed); (3) the adequacy of the existing monitoring system to assess groundwater quality at and downgradient of the site; and (4) the potential for conduit flow to allow the rapid offsite migration of constituents.

To address these issues, a study was conducted consisting of: (1) natural potential and electromagnetic surface geophysical surveys; (2) borehole geophysical logging, (natural gamma ray, caliper, conductivity-resistivity; spontaneous potential, borehole fluid logging; and acoustic televiewer); and (3) a comprehensive groundwater tracing study utilizing three dyes (eosine, fluorescein, rhodamine WT).

These studies resulted in an enhanced understanding of the site that allowed the further assessment of potential environmental impacts. The underlying aquifer could be classified as a fully integrated, dispersed flow dominated system with a perennially flooded epikarst. Rapid or long-distance conduit flow was not encountered. This type of system has generally slow groundwater flow rates for karst systems, tends to retain constituents onsite, and is amendable to monitoring with groundwater monitoring wells.

INTRODUCTION

A multi-phase environmental investigation has recently been completed at the subject facility (ESE, 1997). Given the preliminary investigative results indicating releases of constituents (light oils, coal tar, etc.) and the complex karst setting, a program was developed to characterize the karst aquifer system underlying the facility. The investigative program specific to the karst system consisted of surface geophysics, borehole geophysics, and a comprehensive dye tracer study.

As with any environmental investigation of an industrial facility, the principle purpose is to assess the potential impacts to human health and the environment. The groundwater migration pathway is usually one of the primary exposure routes for facility-related constituents to reach potential receptors. Therefore, a full understanding of the aquifer system underlying the facility is required. In karst settings, this understanding is sometimes difficult due to their extreme anisotropic and heterogeneous nature, and the inability to apply standard groundwater flow equations (i.e., Darcy's Law)(Quinlan, 1989). Thus, the understanding of karst aquifer systems requires non-standard investigative techniques.

As will become apparent in the following discussion, karst systems are not all alike. Karst systems are represented by a full range of recharge, flow and storage properties (Smart and Hobbs, 1986). Recharge ranges from concentrated flow directly into a sinkhole to dispersed flow into small fissures and fractures. Flow can range from diffuse to conduit flow. Storage ranges from temporary in conduits to long term storage in the saturated zone. Most karst systems contain components of each of these recharge, flow, and storage types.

HYDROGEOLOGIC SETTING

The location of the study area is presented in Figure 1. The geology and hydrogeology of the study area has been discussed by: Kidd and Shannon (1977); Kidd (1979); Hunter and Moser (1990); Moffett and Moser (1978); Knight (1976); Planert and Pritchett (1989); and Szabo, et al. (1979). The study area is located in the southwestern portion of the Appalachian Valley and Ridge geologic province, which consists of northeast to southwest trending valleys and ridges developed on folded and thrust-faulted sedimentary rocks. The ridges are formed by resistant sandstone and chert beds and the valleys are underlain by less resistant carbonate and shale units. The study area is located in the

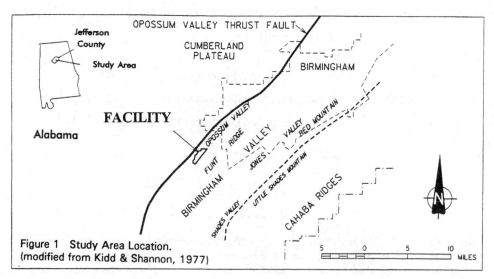

Figure 1   Study Area Location.
(modified from Kidd & Shannon, 1977)

relatively flat Opossum Valley, on the west flank of the northeast plunging Blount Mountain Syncline (Figures 2 and 3). Opossum Valley is bordered on the west by the Opossum Valley thrust fault followed by the Cumberland Plateau (Warrior Basin). Published literature indicates the bedrock in the vicinity of the facility generally dips 20 degrees (°) to the east.

The study area is considered to be a recharge area for the Knox-Shady aquifer group (Planert and Pritchett, 1989). These aquifers are generally carbonates, and the highest yields are from wells that have intercepted interconnected solution cavities. Movement of groundwater is primarily from the higher altitudes adjacent to the ridges to the center of the valleys, but there is also "down-valley" movement in the same direction that the streams flow (Planert and Pritchett, 1989). A large part of aquifer recharge is discharged to streams through seeps and springs. The study area has general groundwater yields from wells of 100 gallons per minute (gpm) or more (Hunter and Moser, 1990). Hunter and Moser (1990) note that extremely variable yields generally occur in areas underlain by extensive cavities. Within the study area, groundwater yields from monitoring wells were highly variable, with one dry well at 200 feet depth and wells above this interval producing abundant water.

The soil cover is generally 5 to 20 feet thick and consists of a clayey residuum derived from weathering of the underlying Ketona Dolomite. This red-brown plastic residuum is composed primarily of clay-sized particles with lesser amounts of silt and occasional sandy zones. The laboratory tests of the material generated vertical hydraulic conductivities in the 10^{-6} to 10^{-8} centimeters per second (cm/sec) range. It is recognized, however, that such materials (particularly over karst systems) contain fractures and macropores that can allow infiltration of water at faster rates than those represented by laboratory permeability tests (Cooley, 1991). Extensive areas of man-emplaced fill (up to 40 feet thick) were encountered, consisting, in a large part, of coal.

Underlying the clayey residuum is the Ketona Dolomite, a massive Cambrian age carbonate unit. The Ketona consists of a light gray to buff, occasionally pink fractured and vuggy dolomite. The lithology consists of a crystalline mudstone (Dunham, 1962) or micrite-dismicrite (Folk, 1959). These terms describe a lime mud [particle size <0.03 millimeter (mm)] without identifiable grains (ooids, peloids) or bioclasts (fossils). The texture includes intraclasts (large particles derived by desiccation breakage or burrow disruption during deposition) and laminates (mm size alternating bands) (Wilson, 1975). Recemented breccia zones where noted, possibly indicating previous or "paleo-karst" conditions. Occasional argillaceous zones were noted. Dissolution is responsible for most of the porosity and permeability in the Ketona Dolomite. The Ketona surface is pinnacled by solutional weathering, with typical top of rock variations of 10 to 20 feet. The bedrock weathering appears to be controlled (in part) by joint and/or fracture zones that roughly parallel the major structural trends of the area (Cooley, 1993).

The dolomite is heavily fractured with fracture angles ranging from horizontal to vertical. Dissolutional enlargement was noted along several vertical fractures and horizontally along bedding planes. The maximum bedrock void height encountered was 18 feet. The Ketona has been noted for fracture zones that have been recemented with calcite and conduits that have been filled with clay and are inactive (Cooley, 1993). Both of these conditions, which restrict or block the flow of groundwater, have been observed in bedrock cores or geophysical logs of the facility.

SURFACE GEOPHYSICAL SURVEY

The natural d.c. electric field in the earth is affected by the flow of groundwater, whose action can be readily detected at the ground surface. Thus, the *natural* field over a groundwater system gives rise to a characteristic pattern of *natural potentials*, or voltages, at the surface. These natural potentials can occur in combination with artificial or *applied potentials* resulting from the accidental or deliberate injection of current into the ground. In the United States, the terms *self-* and *spontaneous potential*, borrowed from downhole terminology, are often applied to the method for measuring natural potentials.

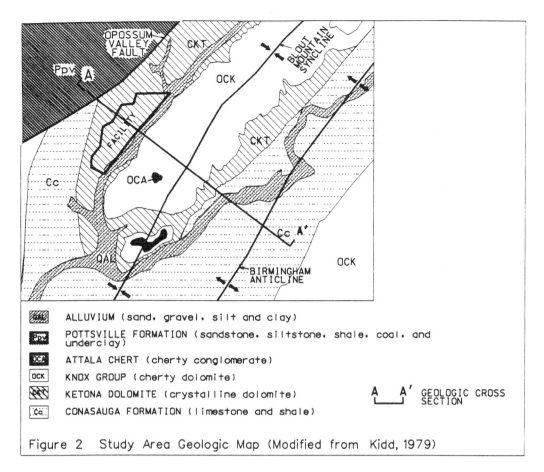

Figure 2 Study Area Geologic Map (Modified from Kidd, 1979)

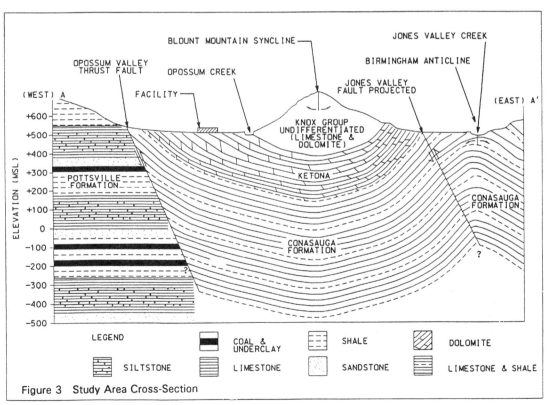

Figure 3 Study Area Cross-Section

Origin of natural potentials

The natural electric currents that occur everywhere in the ground and in the sea come about from chemical reactions around natural and artificial conductors (e.g., metallic mineralization and buried metal), lithologic changes, localized thermal heating and chemical gradients in the subsurface (Semenov, 1980; Sill, 1983). At the ground surface, the most significant effect derives from water moving within soil or rock. This *electrokinetic*, or *streaming* effect arises from the flow of electric charges within a moving fluid, such that the electric potential between upstream and downstream points of a particular flow path is linearly proportional to the driving pressure of the fluid (Ahmad, 1964; Ishido and Mizutani, 1981). The streaming phenomenon occurs not only within granular media, but within fractures as well. The polarity of the potential difference depends on the chemical nature of both the solid and the electrolyte (groundwater). The natural-potential technique has been used extensively for detecting leakage in dams and pipelines and in the search for rising thermal waters (Bogoslovsky and Ogilvy, 1970 and 1972; Corwin and Hoover, 1979).

Within a solution- or fracture conduit or in a cave containing a flowing stream, a potential gradient can develop along the axis of the conduit (Kilty and Lange, 1991). In this case, the upstream end of the channel can become negatively polarized relative to the downstream end or orifice. The normal result is a positive anomaly transverse to the passage axis over the spring, and a negative expression over the point of water entry, such as a sinkhole. Though the observed anomaly shapes take the form of simple positives or negatives, they commonly exhibit compound "M" or "W" shapes, as illustrated by the anomalies over karst springs in Puerto Rico (Figure 4). The natural potential response due to a through-going pathway can be expected to manifest itself on two or more successive lines of a survey grid where these lines are on the order of 100 feet apart and transverse to the flow. Complicating the interpretive recognition of these trends, however, is the observation (based on numerous field occurrences) that the shape and polarity of an anomaly trend can change from line to line, appearing as a simple peak on one line, as a compound W-form on the next, and as an M-figure on a line farther downstream (Figure 5).

In the vicinity of industrial sites, urban centers and cultural features, natural-potential responses are often corrupted by the d.c. or alternating current (a.c.) interference of artificial electrical sources. Cathodic protection in the form of a small d.c. current is commonly applied to metal structures, particularly pipelines, to reduce corrosion. The resulting voltage imparts an exponential signal, or ramp function, to the voltage profile produced by a natural feature. Time-varying electric sources, such as electric trains, generate erratic and excessive electrical "noise" that renders nearby natural-potential measurements almost unreadable without considerable averaging over time (Hoogervorst, 1975).

Field instrumentation and procedures

Typical apparatus employed for the measurement of potentials on the surface include a pair of sealed non-polarizing copper/copper sulfate electrodes connected by a cadmium-bronze wire mounted on a lightweight reel and color-calibrated by distance (Figure 6). Potential differences are logged using a sensitive digital multimeter of very high input impedance (e.g., 1,000 megohms).

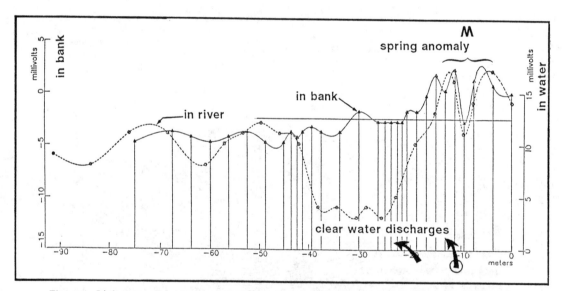

Figure 4 Little Aquas Frias Spring issues from a karst opening in the bank of Rio Grande de Manati in Puerto Rico, giving rise to the M-shaped conduit anomaly in natural-potential profiles recorded both in the bank (solid curve) and in the river (dashed). The introduction of the clear water into the otherwise turbid river resulted in a zone of low potential readings downstream of the spring.

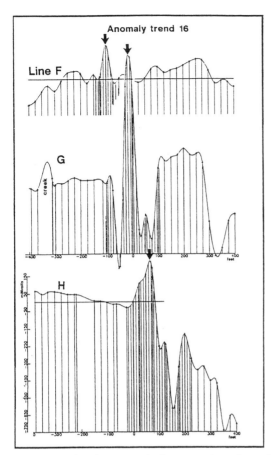

Figure 5 Changing anomaly forms on three lines of Anomaly Trend 16, attributable in part to changes in the configuration of a subterranean karst pathway.

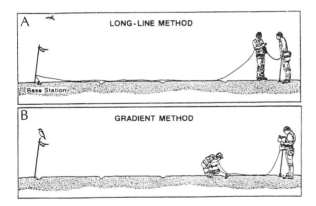

Figure 6 Successive measurements of potential differences along a survey traverse by the long-line method (A) and by the gradient method (B).

In the long-line method, one of the electrodes is planted in soil near the center of the area to be surveyed (base- or reference electrode); while the second of the pair (roving electrode), mounted on a conducting staff, samples voltages in the ground at successive intervals along grid lines of the survey. Readings are taken in shallow holes (of about 4-inch depth) excavated in soil. It is recommended that two or more readings be made at each station in order to insure that the station data recorded are representative. The base location is assigned an arbitrary reference value, usually zero millivolts, to which all other readings are relative. At successive time intervals or at the beginning and end of each traverse, the potential at the base station is re-logged. In the data processing, millivolt changes at the base are distributed by time as a correction over all the stations of a particular traverse. The changes arise from electrode drift and from fluctuations in soil temperature and moisture over the time interval of measurement.

In the gradient method, both electrodes advance successively along a line of the survey, connected by a short length of wire, say 25 feet (see Figure 6). The electrodes should be switched in leap-frog fashion at alternate stations in order to compensate for electrode differences. The individual measurements comprise an electric field gradient, which must be integrated (summed) in order to generate a potential line-profile equivalent to that produced by the long-line method. The long-line method is generally easier to implement and provides more coherent data than the gradient method. However, the subject study area is electrically noisy and the noise level is proportional to the electrode separation. Therefore, it was necessary to employ the gradient method using a 25-foot electrode separation in order to obtain readable data.

Because of the prevalence of buried metal in the form of rails, pipes and scrap on and around the study area, an electromagnetic technique was implemented following completion of the natural-potential data collection. It had been anticipated that many of the mapped potential anomalies could arise from electrochemical action associated with metal not evident at the ground surface; hence, these unnatural sources had to be eliminated from consideration as sites of enhanced permeability.

A Geonics EM-61 device, a time-domain device referred to as a "transient electromagnetic metal (TEM) detection instrument," was utilized to generate a changing magnetic field that induces an electric current to flow in conductive material. An example of the TEM response over a natural potential anomaly on Line H is reproduced in Figure 7 and shows that both positive and negative (high and low) excursions of a compound expression can be attributed to buried metal.

Geophysical Procedures at the Study Area

Nine survey lines, designated A through I, were completed in Area 1, comprising an area downgradient from the main facility but upgradient of Opossum Creek (Figure 8). These lines ranged in length from 1,000 to 4,000 feet and totaled 760 stations and 18,700 feet of traverse. The bulk of the traverses were aligned in a northeasterly direction (N40°E), while one tie line connected Lines A through I at approximate right angles. A common base station (Base D) was established near the center of the survey grid, to which all voltage readings were referred. In the southwestern part of the grid, Lines B and C wrapped around the southern shore of South Pond. It was intended that the lines be regularly spaced at

100 feet; however, intervening railroad tracks, buildings and pavement necessitated irregular separations, locally as great as 190 feet.

In the vicinity of Landfill No. 1 (Area 2), three lines (K, L and M) wrapped around a portion of the perimeter of the landfill. Line separations were between 20 and 50 feet. A local base station was established at the northwest terminus of Line K. Area 2 comprised 144 stations and 3,525 feet of line.

Several possible natural potential anomaly trends, which were partly or totally free of metal and traceable over two or more survey lines, were mapped. A total of six trends in Area 1 and four in Area 2, made up of positive or compound anomalies, were identified from both data profiles and contours. These trends were subsequently drilled in search of high groundwater permeability zones. The voltage anomalies and the associated monitoring wells and observations encountered are presented in Table 1. Each of these monitoring wells encountered at least one void or fractured zone.

The three broad negatives or low zones, designated A, B, and C in Figure 8, were mapped in Area 1. The broad character of these low trends--up to 300 feet in width--suggests a relationship to geologic structure rather than to discrete karst features. Their negative polarities are typical of zones of infiltration, possibly associated with faults and fracture zones. On the other hand, elsewhere in the Valley and Ridge physiographic province, similar lows have been observed over steeply dipping shale beds.

BOREHOLE GEOPHYSICAL SURVEY

The borehole geophysical logging was utilized to characterize the hydrogeology of the subsurface environment with the principal purpose of identifying high groundwater production zones in the carbonate bedrock. The geophysical logging conducted at positive NP anomalies (potential conduits) was used to place well screens at the highest groundwater production zone. The following borehole geophysical logs and primary applications were utilized:
- Caliper log -- void/fracture detection;
- Natural gamma -- lithology determination;
- Single point resistance -- fracture detection;
- Spontaneous potential -- fracture and flow detection;
- EM conductivity and resistivity -- identification of high permeability zones;

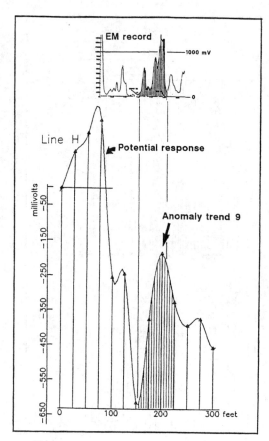

Figure 7 Transient electromagnetic response (EM record) associated with natural-potential anomalies on Line H. The anomaly peak at 200 feet coincides with buried metal detected by the metal detector.

Table 1. NP Anomalies with Associated Monitoring Wells and Observations

Natural Potential Anomaly Trends	Monitoring Wells	Observations
Trend 1, Lines K through M	M-67R on Line K	Lost circulation zone at soil/bedrock interface
Trend 6, Lines K through M	M-1I on Line K	Void at 28-33', utilized as dye introduction point
Trend 9, Lines A through H	M-50R,B,C near Line A	11 foot void at 26 to 37'
	M-69B near Line F	Sandy dolomite and sandstone/fracture at 61'
Trend 11, Lines A through F	M-25D on Line C	Fracture and groundwater production zone at 74-78'
Trend 14, Lines C and D	M-62R,B,C on Line C	Fracture zones at 15-19' and 25-28'
Trend 16, Lines F and G	M-63R,B,C on Line F	Fracture zone at 60-62'
Trend 18, Lines A through E	M-70B on Line E	Paleosinkhole or grike, bedrock at 52' instead of 18'

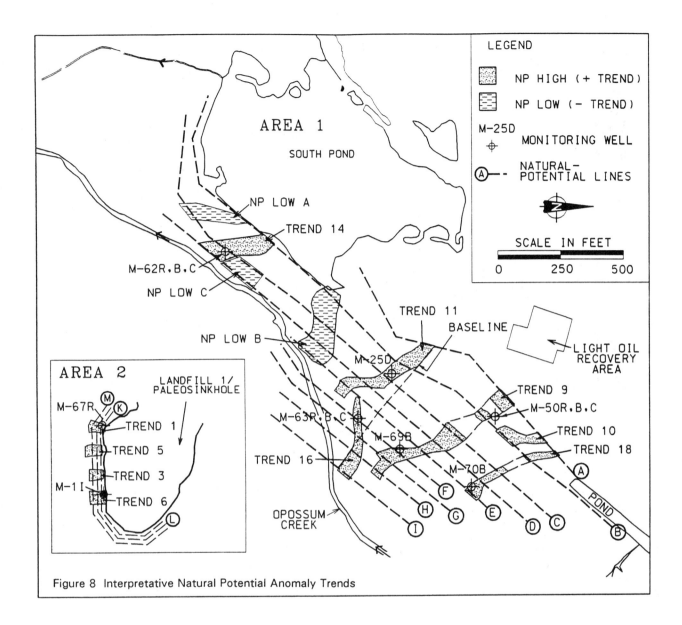

Figure 8 Interpretative Natural Potential Anomaly Trends

- Video inspection -- determination of fracture orientations; and
- Fluid temperature and resistivity -- detection of high groundwater flow zone.

Borehole geophysical logs were run on select existing cased monitoring wells and open holes before the well installation. Not all logs were run on each well (i.e., spontaneous potential logs cannot be run in cased holes). Generally, the deepest monitoring well of a cluster was selected for geophysical logging. Geophysical logging was performed in the open boreholes as soon as practical after the borehole was completed.

The geophysical logging was performed with rental equipment from Colog, Inc., Denver, Colorado. Two of the rig geologists were trained to use the equipment and could readily perform the logging after the well was drilled. Therefore, minimal costs were incurred during intervening periods.

The methodology and expected responses for geophysical logs have been discussed by others [(Keys, 1989), ASTM D5753-95 (1995), and Doveton (1986)] and will not be repeated here. A recent excellent text is "A Practical Guide to Borehole Geophysics in Environmental Investigations" (Keys, 1997). Borehole geophysics provides information regarding the physical properties of the borehole and samples a volume of rock many times larger than the borehole volume. Borehole logs are evaluated together, with each log assisting in interpretation of the others.

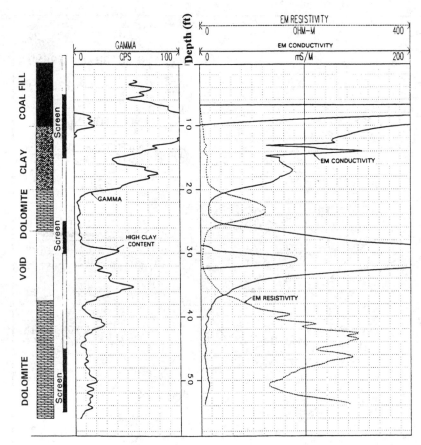

Figure 9 Geophysical Logs for Well Cluster M-50R,B,C

The borehole geophysical logs for monitoring well M-50R,B,C were completed near NP anomaly Trend 9, Line B, and are presented in Figure 9. An 11-foot void was encountered at approximately 26 to 37 feet bgl. The gamma, EM conductivity, and resistivity logs show that the bedrock void was filled with clay. This matches field observations that many of the voids/fractured zones in the bedrock were infilled with clay. Obviously, groundwater and constituent transport characteristics are different with clay-filled voids compared to open conduit systems.

The geophysical log for Well M-25D, completed on NP anomaly Trend 11, Line C is a good example of the responses expected in fractured carbonate rock (Figure 10). These logs show a clear break in the lithology from the clay to dolomite at 10 feet below ground level (bgl). Overall, the logs show a fairly uniform dolomite bedrock without distinct shale or sandstone beds. Two potential well screen intervals were identified: approximately 64 and 76 feet bgl. The geophysical log indications of potential groundwater production at these intervals consist of the following: (1) caliper log–large borehole size or fractures; (2) high EM conductivity and low EM resistivity–high water content; and (3) changes in borehole fluid resistivity–water flow into the borehole. The lower interval was selected for the well screen because it was the larger of the two fracture zones. This was corroborated by the driller who reported greater groundwater flow at this interval.

An acoustic borehole televiewer (ABT) logging was performed in one open borehole (RETEC, 1998). The ABT is an imagining sonde that uses sound waves to scan 360° of the borehole wall. Features such as fractures and bedding planes can be easily identified as an acoustic image. The ABT also includes a magnetometer that senses and references the acoustic signal to magnetic north, allowing the determination of feature orientation. The ABT log showed high angle fractures trending in a northerly direction with dips of 60 to 75°. The ABT log also showed bedding plane orientation with a dip of approximately 25 to 29° to the east-southeast (RETEC, 1998).

DYE TRACER STUDY

Tracer dyes (eosine - Acid Red 87, Color Index 45380; fluorescein - Acid Yellow 73, Color Index 45350; and rhodamine WT - Acid Red 388, no color index number) were introduced into the upper bedrock (the epikarst zone) through epikarstic dye introduction points (EDIPs). For purposes of this study, EDIPs consisted of either: (1) standard 2-inch diameter groundwater monitoring wells screened in voids or high permeability zones or (2) specially designed 4-inch diameter open-bottom wells installed 5 feet into the rock. The very top of carbonate bedrock is often soft and tends to locally reduce infiltration; this was mitigated by extending the EDIP 5 feet into rock.

The epikarst, commonly the upper 30 feet of a carbonate bedrock, commonly contains numerous dissolutionally widened openings and fractures. Lateral permeability within the epikarstic zone is routinely greater than vertical permeability. Flow is usually lateral in the epikarst until arrival at a "drain," with vertical connection to the conduit system in the underlying bedrock. Therefore, dye introduced into the epikarst will move laterally to a drain and then connect with the conduit flow system.

Each EDIP was flow tested with 500 gallons of potable water to ensure good hydraulic connection with the epikarst. To be deemed acceptable, the EDIP had to accept 500 gallons of water at a rate of at least 1, but preferably 5, gallons per minute (gpm). Most of the EDIPs accepted water at 2 to 6 gpm. Approximately 500 to 1,000 gallons of flush water was introduced into each EDIP at a rate of 2 to 6 gpm before dye introduction to ensure the rock surfaces were wet. The eosine and fluorescein arrived at the site as powders and were mixed with water in order to be introduced into the EDIPs in a dissolved form. The rhodamine WT was purchased and used in a liquid form. After the dye

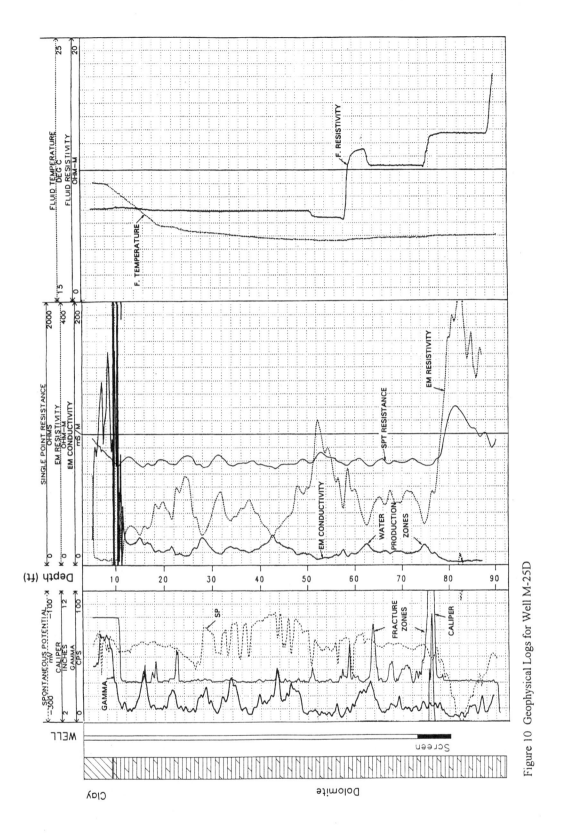

Figure 10 Geophysical Logs for Well M-25D

introduction, an additional 1,500 to 3,000 gallons of flush water was introduced into each EDIP at a rate of approximately 2 to 4 gpm to ensure that the dye reached the conduit flow system.

The dye introduction points were selected to bracket three areas of concern: Landfill No. 1 (LF-1), Light Oil Recovery (LOR) area and the Coal Tar Processing (CTP) area (Figure 11).

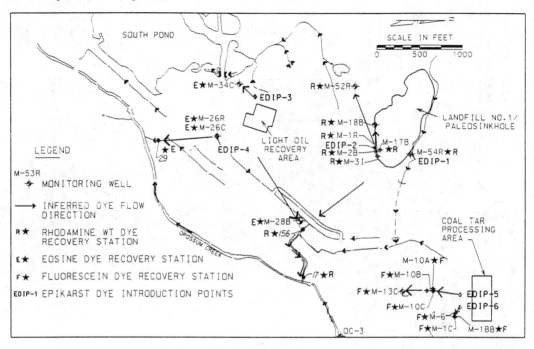

Figure 11 Dye Introduction and Recovery Locations

LF-1 was constructed over a possible paleo-sinkhole composed of alternating clay, sandy clay, and sandstone. LF-1 is at a topographic high. Small upper bedrock voids (3 to 5 feet high) were encountered on the north and south edges of LF-1, and were utilized as EDIPs 1 and 2, respectively. Ten pounds (liquid weight) of rhodamine WT solution was introduced into each of these EDIPs. The solution contained approximately 20 percent dye and 80 percent diluting agent.

The LOR area is located on a broad bedrock high and contained light non-aqueous phase liquid (LNAPL) in the overburden clay and upper bedrock. Eosine was introduced into EDIPs 3 and 4 on the west and east sides of the LOR, respectively. Fifteen pounds (dry weight) of eosine dye mixture was introduced into each of these EDIPs. The mixture was approximately 75 percent dye and 25 percent diluent.

The CTP contained aqueous phase coal tar constituents and potential dense non-aqueous phase liquid (DNAPL). Fifteen pounds (dry weight) of fluorescein dye mixture was introduced into each of two EDIPs (5 and 6) on the downgradient edge of this area. The mixture was approximately 75 percent dye and 25 percent diluent. A 2-foot bedrock void was utilized at EDIP 5.

Each of the existing 146 study area groundwater monitoring wells (with minor exceptions) were sampled for dyes. Each stream and lake at the facility was sampled and 20 dye sampling stations were located along Opossum Creek (the presumed groundwater discharge point). An offsite sanitary sewer line that parallels the creek was sampled. Regional streams and springs were also sampled to determine if a hydraulic connection existed with these distant points. In total, over 200 dye sampling stations were utilized for most of the study.

The duration of the study was 151 days, or 21.6 weeks. Samples were collected weekly. Three weeks of background samples were collected at each station before the dyes were introduced. Based on potentiometric data and preliminary dye tracer results, approximately 74 stations were dropped from the study after 14 weeks because the stations were deemed to be unlikely dye recovery locations.

The detection of dyes was accomplished in two ways: activated carbon samplers and grab water samples. Primary sampling reliance was placed on activated carbon samplers, which are cumulative samplers and permit continuous sampling. Water samples were subjected to analysis at stations where dye was detected in activated carbon samplers or in cases when activated carbon samplers were lost.

The charcoal samplers are packets of fiberglass screening partially filled with approximately 4.25 grams of activated coconut charcoal. The charcoal used in this study was Barnebey and Sutcliff® coconut shell carbon, 6 to 12 mesh, catalog type AC. The samplers are typically about 4 inches long by 2 inches wide. Activated carbon samplers were placed at springs and surface streams in flowing water and firmly anchored in place with wire and weighted in place.

In monitoring wells, the sampler was placed near the middle of the screened interval attached to a dedicated bailer. The bailer was raised about 10 feet in the well and then lowered to the bottom of the well before the charcoal and water samples were collected. This ensured a fresh water sample in the bailer. Monitoring wells were neither pumped nor purged prior to sampling.

Analytical Techniques

After retrieval and transportation to the laboratory, the charcoal samplers were washed under relatively strong jets of water to remove sediment and organic material. Water samples were kept in the dark to prevent or minimize biological or photo-decomposition of any dyes that might be present in the sample. The activated carbon was then eluted for 1 hour with 15 milliliters (mL) of a standard eluting solution which contained 1.45 percent ammonia, 66.50 percent isopropyl alcohol and 32.05 percent water which has been saturated with potassium hydroxide (KOH). Water samples were analyzed directly.

The charcoal elutant and the water samples were analyzed by a Shimadzu® Model RF-5000U Spectrofluorophotometer using a synchronous scan of excitation and emission wavelengths with a 17 nanometer (nm) wavelength separation. Samples were analyzed using a 5 nm excitation slit and a 3 nm emission slit to ensure adequate discrimination between tracer dyes and other fluorescent materials which might have been present. The scanning spectrofluorophotometer provides a plot of fluorescence intensity versus wavelength and generates a characteristic fluorescence emission scan. This is utilized to distinguish specific tracer dyes from background fluorescence.

The typical acceptable emission fluorescence range and reporting levels for the dye utilized in the study are presented in Table 2. The reporting levels are based on three times the method detection limit. The reporting level for the charcoal elutant, due to its cumulative sampling period, was modified, on a per sample basis, to only include results that were 10 times the maximum background levels.

Table 2. Dye Emission Fluorescence Ranges and Reporting Concentrations

Dye	Media	Acceptable Emission Ranges Fluorescence (nm)	Reporting Level (ppb)
Eosine	Elutant	533.0 - 539.6	0.060
Eosine	Water	529.6 - 538.4	0.003
Rhodamine WT	Elutant	561.7 - 568.9	0.465
Rhodamine WT	Water	569.4 - 574.8	0.021
Fluorescein	Elutant	510.7 - 515	0.030
Fluorescein	Water	505.6 - 510.5	0.0015

Three rounds of background data were collected from each sampling station. Background data were critical in subsequent analysis of dye recoveries. In many cases, fluorescence peaks in the acceptable wavelengths and reportable levels were detected before dyes were introduced. Qualitative and quantitative comparisons to emission wavelength scans were utilized to determine dye recovery points. Examples of typical background emission scans are shown in Figures 12a and b, respectively. The broad emission scan of Figure 12a is typical for background groundwater in the study area. A subsequent eosine recovery at this station was easy to discern due to the characteristic emission peak shape (see Figure 12c). The emission scan shown in Figure 12b has the acceptable wavelength for fluorescein dye, but was collected before dye introduction. A subsequent dye recovery at this station was easy to discern with a quantitative comparison (Figure 12d). In general, natural background fluorescence is highest in the green and blue (400 to 500 nm) emission range, and lowest in the orange (~600 nm) emission range (Smart and Laidlaw, 1977).

The locations of the dye recoveries are shown in Figure 11. Dye was recovered at a total of 18 groundwater monitoring wells and three surface water stations (Table 3). The dye recoveries were, in general, located hydraulically downgradient from the dye introduction points. Several dye recoveries, however, were at stations parallel to the potentiometric groundwater contours. No dyes were recovered at Opossum Creek, the principle groundwater discharge point, but dyes were recovered in ditches that flow into the creek. Dyes were recovered at well clusters (wells with screens at different depths in the aquifer) at approximately the same time. Dye recoveries were observed in bedrock wells screened 200 feet below ground level. No dyes were recovered at springs or distant surface water points.

The mean travel rate for the first arrival of tracer dyes was 17.6 feet per day. The mean travel rate for the peak dye concentration was 8.8 feet per day. In general, groundwater travel rates increased with increasing distance traveled. This is reflective of a karst aquifer in which there are preferential flow routes. Typical dye recovery histograms were observed to have relatively broad shapes (Figure 13). This broad recovery pattern, particularly after peak recovery, is indicative of relatively slow groundwater travel rates (USEPA, 1988).

Dye Tracer Results

The dye travel rates presented above are an order of magnitude less than those typically reported for karst aquifers (ASTM, 1995; Ford and Williams, 1989). Most karst groundwater tracing reported in the technical literature has utilized sinkholes, losing streams, or areas near springs for dye introduction. Sinkholes and losing streams usually provide a direct and open connection to a major groundwater transport

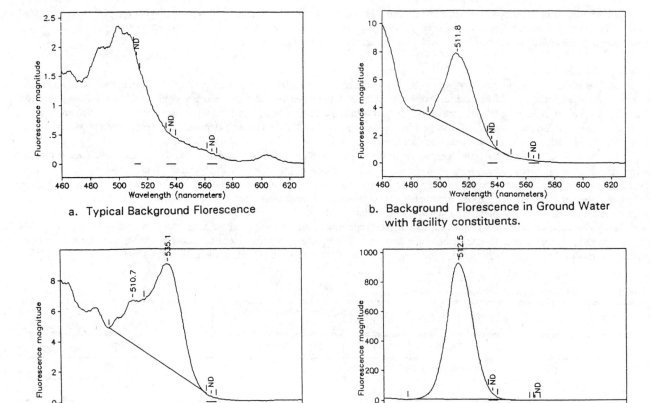

Figure 12 Fluorescence Emission Scans

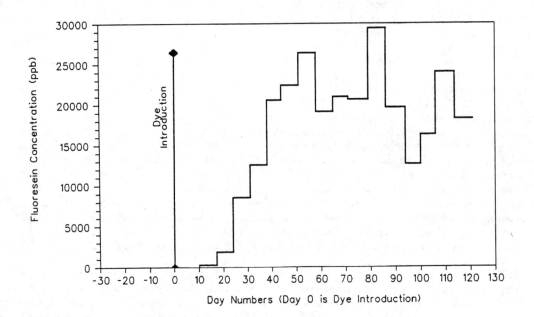

Figure 13 Dye Recovery Histogram

Table 3. Dye Recoveries and Derived Groundwater Travel Rates. Data are Based on Activated Charcoal Samplers

Station Number	StationName	Dye & Relevant EDIP	Peak Conc. (ppb)*	Distance from EDIP (ft)	First Arrival (Days)†	Arrival Peak Conc. (Days)**	Travel Rates	
							First Arrival (ft/day)	Peak Arrival (ft/day)
67	M-54R	RWT/1	87,000	10	1.0	4.5	10.0	2.2
74	M-1R	RWT/2	278	10	18.5	18.5	0.5	0.5
76	M-2B	RWT/2	65,000	60	4.0	60.5	15.0	1.0
79	M-17B	RWT/2	2.68	60	18.5	18.5	3.2	3.2
78	M-3I	RWT/2	2,760	100	4.0	18.5	25.0	5.4
83	M-18B	RWT/2	1.87	260	25.5	53.5	10.2	4.9
85	M-52R	RWT/2	1.31	740	18.5	18.5	40.0	40.0
156	SW	RWT/1 or 2	10.5	--	--	--	--	--
17	SW	RWT/1 or 2	9.52	--	--	--	--	--
129	M-34C	EOS/3	24.8	225	51.0	116.5	4.4	1.9
118	M-26R	EOS/4	11,300	10	1.0	5.0	10.0	2.0
120	M-26C	EOS/4	4,800	10	5.0	81.5	2.0	0.1
149	M-28B	EOS/4	1.07	1,390	19.5	19.5	71.3	71.3
29	SW	EOS/4	2.040	--	--	--	--	--
206	M-6	FL/6	875	60	1.5	90.0	40.0	0.7
165	M-1BB	FL/6	5.51	90	27.5	102.5	3.3	0.9
166	M-1C	FL/6	7.01	150	27.5	41.5	5.5	3.6
181	M-10A	FL/5	93,900	355	13.5	54.5	26.3	6.5
183	M-10C	FL/5	95.4	340	28.0	110.5	12.1	3.1
182	M-10B	FL/5	29,200	345	13.5	82.5	25.6	4.2
196	M-13C	FL/5	1.93	710	54.5	97.0	13.0	7.3
Mean							17.6	8.8

FL = Fluorescein, EOS = Eosine, RWT = Rhodamine WT, SW = Surface Water
* Peak concentration - parts per billion
† Midpoint 1st arrival days after dye introduction.
** Midpoint peak concentration days after dye introduction.

conduit. The hydrologic conditions (including greater epikarst development) may be different in areas near springs than in more remote areas (Aley, 1997). Therefore, the use of sinkholes and losing streams as dye introduction points may not yield results reflective of the hydrologic functioning of the epikarst system as a whole. Areas remote from major groundwater transport conduits may have substantially lower groundwater travel rates. In assessing the potential for constituent migration from industrial facilities, the dyes should be introduced as close as practical to the source areas to obtain data that reflects that part of the karst system. Dye introduced into a sinkhole a quarter mile from a source area can lead to incorrect assessments of the potential risks posed to the environment. In summary, the groundwater tracing results and recommendations reported in the karst hydrology literature are skewed in favor of single traces that typically are not reflective of groundwater flow conditions in many epikarstic systems.

STANDARD METHODS

Although the previous discussion focused on the techniques specially designed to address karst systems, standard investigative methods were also critical to understanding the system. These include: (1) standard groundwater monitoring wells (146 in the study area); (2) seasonal groundwater elevation data (eight events) used to generate maps, statistical tables and vertical gradient profiles; (3) soil/bedrock core lithology descriptions; (4) long duration pumping tests; (5) soil and groundwater constituent distribution maps; (6) historical groundwater constituent concentration trends; and (7) local and regional field reconnaissance. A complete discussion of these investigative methods and results is beyond the scope of this paper. Only a brief summary of results pertinent to the site conceptual model are present.

The study area epikarst groundwater elevations for two seasons [spring (maximum wet period) and the end of summer (maximum dry period)] are shown in Figures 14a and 14b, respectively. The groundwater elevations data clearly show no significant seasonal variations in groundwater flow directions, and indicates groundwater flow to the south-southeast towards Opossum Creek. The groundwater elevations were above the top of bedrock. The vertical groundwater gradient data collected from the well clusters in the study area show upward gradients along Opossum Creek and generally downward gradients away from the creek (Figure 15). This confirms Opossum Creek as a local groundwater discharge point.

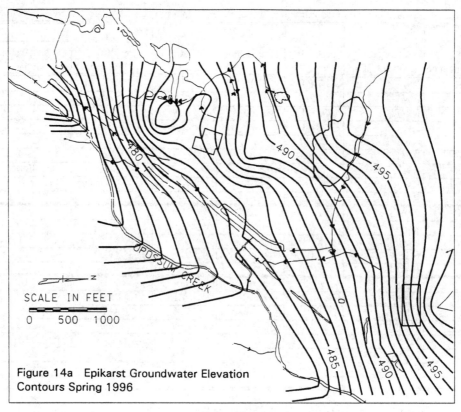

Figure 14a Epikarst Groundwater Elevation Contours Spring 1996

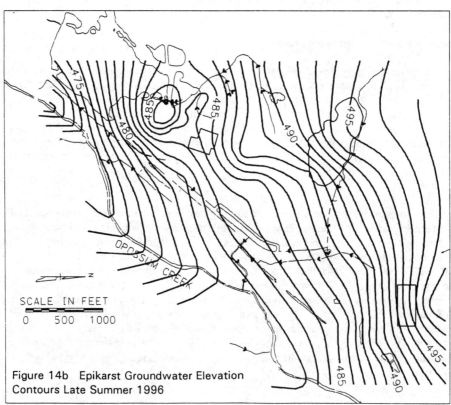

Figure 14b Epikarst Groundwater Elevation Contours Late Summer 1996

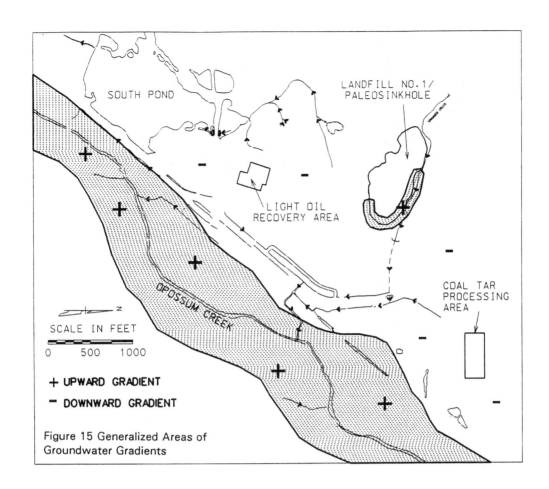

Figure 15 Generalized Areas of Groundwater Gradients

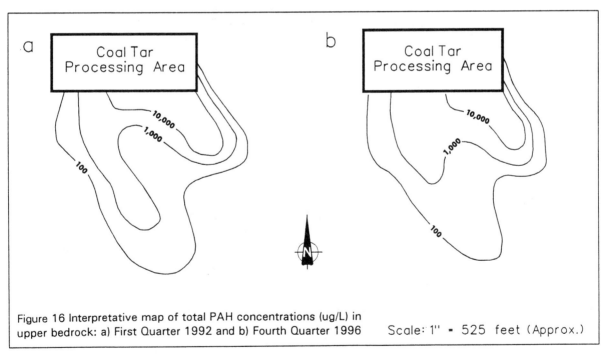

Figure 16 Interpretative map of total PAH concentrations (ug/L) in upper bedrock: a) First Quarter 1992 and b) Fourth Quarter 1996

Scale: 1" = 525 feet (Approx.)

Several aquifer pump tests have been conducted in the study area including a 72-hour pump/recovery test in the epikarst. These tests demonstrated that the upper 200 feet of bedrock and the clay residuum act as one system (i.e., drawdown was immediately observed in residuum and deep bedrock wells after pumping started in the epikarst well). The hydraulic conductivity value reported from this test was 10^{-4} cm/sec.

Historical data were also collected on constituent concentrations in groundwater. This data, collected over several years, indicated that areas of constituent affected groundwater were not expanding (i.e., the plumes have reached a dynamic equilibrium). An example is the relativity constant concentration of polynuclear aromatic hydrocarbons (PAHs) in the upper bedrock adjacent to the Coal Tar Processing Area (Figures 16a and 16b) (RETEC, 1998). It is unlikely that equilibrium conditions would exist in a fully mature open karst system.

The usefulness of field reconnaissance in karst system characterization cannot be over emphasized. The field work extended for several miles beyond the facility and included detailed surveys of likely groundwater discharge points for seeps and springs. A description of the regional hydrogeology was critical in understanding the local flow system. Soil and bedrock lithology descriptions were previously discussed in the Hydrogeologic Setting section.

SITE CONCEPTUAL MODEL

The results of the surface geophysical survey, borehole geophysical survey, and the dye tracer study were integrated with the groundwater elevation data, lithology descriptions, aquifer pump test results, constituent distribution maps, historical data, and reconnaissance observations to refine the existing conceptual model of the aquifer system.

The epikarst in the study area is perennially saturated. Therefore, seasonal variations in groundwater flow directions (as observed in some periodically flooded epikarst systems) are not expected. This was confirmed with the seasonal groundwater elevation maps. The system is fully interconnected, with dyes arriving at both surface water points and wells screened 200 feet into bedrock. Given the extensive fracturing and high number of small dissolutional voids observed in bedrock cores and the dye tracer results (which included a high number of recoveries at wells, relativity slow travel rates, and broad recovery histograms), an aquifer system dominated by dispersed flow is inferred.

In carbonate aquifers, dispersed flow is defined as an intermediate between conduit and diffuse flow end members (Smart & Hobbs, 1986). Dispersed flow refers to water movement through fissures (< inch), fractures (~ inches) and small conduits. It does not appear that the system under study has developed a mature conduit drainage system. Dispersed aquifers tend to have less spatial variability than aquifers dominated by conduit flow. Although on a small scale (inches to feet), a high degree of variability in aquifer parameters can be expected, on a more regional scale (hundreds to thousands of feet), a dispersed flow aquifer will behave more like a typical "diffuse" flow aquifer. The dispersed flow system allows the use of groundwater elevation contours to infer flow directions and the use of wells to monitor groundwater quality.

The subsurface consists of four interconnected units: (1) coal fill, (2) clay residuum, (3) upper 30 feet of bedrock (epikarst), and (4) lower (>30 feet) bedrock (Figure 17). The physical properties and groundwater transport characteristics of these units are different. Due to their position at the release points and higher capacity to retain constituents, the coal fill and clay residuum can be considered reservoirs that release constituents slowly to the underlying bedrock. The ability of the upper bedrock to allow rapid groundwater and constituent transport varies. In areas where rock has experienced only minor dissolution and/or considerable clay or calcite infilling of fractures/voids (closed system), the permeability is low and transport rates will be slow. Eventually, constituents may make their way to more open parts of the bedrock system and travel at higher rates (open system).

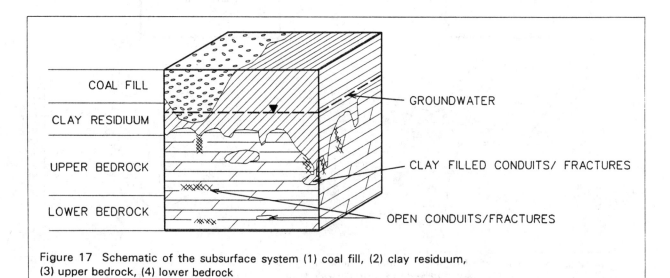

Figure 17 Schematic of the subsurface system (1) coal fill, (2) clay residuum, (3) upper bedrock, (4) lower bedrock

Once the dissolved phase constituents reach the open bedrock system, the environment changes. Although higher travel rates can be expected, the physio-chemical processes (dilution, oxygenation, biodegradation) acting on these constituents increase. The degradation rates within this part of the system may match the reservoirs ability to supply constituents. Obviously, this will vary with source area configuration and constituents of concern. Sulfate is an example of a conservative constituent that may travel long distances in both open and closed parts of the system with little attenuation. Low levels of benzene, however, may rapidly degrade in an open conduit system with the increase in oxygen and microbial activity. Therefore, a source area may remain indefinitely without expansion of the bedrock groundwater constituent plume. Historical studies conducted on groundwater plumes within the study area support the concept of a dynamic equilibrium or steady state condition of the constituent affected groundwater areas.

REFERENCES

Ahmad, M.U. 1964. A laboratory study of streaming potentials. Geophysical Prospecting, v. 12, p 49-64.

Aley, Thomas. 1997. Groundwater Tracing in the Epikarst. Proceedings of the Sixth Multi-Disciplinary Conference on Sinkholes and the Engineering and Environmental Impacts of Karst, Springfield, Missouri, p 207-211.

American Society for Testing and Materials (ASTM). 1995. Standard guide for design of groundwater monitoring systems in karst and fractured-rock aquifers. ASTM Standard D5717-95, 17p.

ASTM. 1995. Standard Guide for Planning and Conducting Borehole Geophysical Logging, Designation: D5753-95, 8p.

Bogoslovsky, V.A. and A.A. Ogilvy. 1970. Natural potential anomalies as a quantitative index of the rate of seepage from water reservoirs. Geophysical Prospecting, v. 18, p 261-268.

Bogoslovsky, V.A. and A.A. Ogilvy. 1972. The Study of Streaming Potentials on Fissured Media Models. Geophysical Prospecting, v. 20, p 109-117.

Cooley, Tony. 1991. Approaches to Hydrogeologic Assessment and Remediation of Hydrocarbon Contamination in Clay-Covered Karsts with Shallow Water Tables. In the Proceedings of the Third Conference on Hydrogeolgy, Ecology, Monitoring, and Management of Groundwater in Karst Terrains.

Cooley, Tony. 1993. Engineering Characteristics of the Pinnacled Surface of the Ketona Dolomite in the Birmingham Valley of Alabama. In: Applied Karst Geology, Beck (ed.) Balkema, Rotterdam, p207-213.

Corwin, R. and D. Hoover. 1979. The self-potential method in geothermal exploration. Geophysics, v. 44 (2), p 226-245.

Dakhnov, V.N. 1962. Geophysical well logging. Colorado School of Mines, Quarterly, v. 57 (2), 445p.

Doveton, J.H. 1986. Log Analysis of Subsurface Geology, John Wiley and Sons, New York, p150-157.

Dunham, R.J. 1962. Classification of carbonate rocks according to depositional texture. In: Classification of Carbonate Rocks. Am. Ass. Petrol. Geol. Mem. 1, p108-121.

Environmental Science & Engineering, Inc. (ESE). 1997. Draft Phase I and II RCRA Facility Investigation and Groundwater Quality Assessment Report. Koppers Industries, Inc., Woodward Coke Facility, Dolomite, Alabama.

Folk, R.W. 1959. Practical petrographic classification of limestones. Am. Ass. Petrol. Geologist Bull. 43, p1-38.

Ford, Derek C. and Paul W. Williams. 1989. Karst Geomorphology and Hydrology. Unwin Hyman, Boston. 601p.

Hoogervorst, G.H.T.C. 1975. Fundamental noise affecting signal-to-noise ration of resistivity surveys. Geophysical Prospecting, v. 23 (2), p 380-390.

Hunter, J.A., and Paul H. Moser. 1990. Groundwater Availability in Jefferson County, Alabama, Special Map 224 Geological Survey of Alabama, 68p.

Ishido, T. and H. Mizutani. 1981. Experimental and theoretical basis of electrokinetic phenomena in rock-water systems and its applications to geophysics. Journal of Geophysical Research, v. 86 (B3), p 1763-1775.

Keys, Scott W. 1997. A Practical Guide to Borehole Geophysics in Environmental Investigations. CRC Press, 176p.

Keys, Scott W. 1989. Borehole Geophysics Applied to Groundwater Investigations. National Water Well Associations, 313p.

Kidd, J.T. 1979. Structural geology of Jefferson County, Alabama. Alabama Geological Survey, Atlas 15, 89p.

Kidd and Shannon. 1977. Preliminary aerial geologic maps of the Valley and Ridge province, Jefferson County, Alabama Geological Survey Atlas 10, 41 p.

Kilty, K.T. and A.L. Lange. 1991. The electrochemistry of karst systems at the ground surface. National Water Well Association, Third Conference on Hydrogeology, Ecology, Monitoring and Management of Ground Water in Karst Terrains: Proceedings, p 163-177.

Knight, Alfred L. 1976. Water Availability, Jefferson County, Alabama: Geological Survey Map 167, 31 p.

Moffett, T.B., and Paul H. Moser. 1978. Groundwater Resources of the Birmingham and Cahaba Valleys of Jefferson County, Circular 103, Alabama, Geological Survey of Alabama, 78p.

Planert, M., and Jane L. Pritchett. 1989. Geohydrology and susceptibility of Major Aquifers to Surface Contamination in Alabama; Area 4. U.S. Geological Survey Water Resources Investigation Report 88-4133, 31p.

Quarto, R. & D. Schiavone. 1996. Detection of cavities by the self-potential method. First Break, v. 14 (11), p 419-431.

Quinlan, James F. 1989. Groundwater Monitoring in Karst Terrains: Recommended Protocols and Implicit Assumptions. USEPA 600/X-89/050, 79p.

RETEC, 1998. Focused Groundwater Investigation Report. Woodward Tar Facility, Dolomite Alabama.

Sabatini, D.A. and Austin, T.A. 1991. Characteristics of Rhodamine WT and Fluorescein as Adsorbing Groundwater Tracers. Ground Water, Vol. 29:3, p 341-349.

Semenov, A.S. 1980. Elektoreazvedka Metodom Estestvennogo Elektricheskogo Pola [Electrical Prospecting with the Method of the Natural Electric Field]. Izlatelstvo "Nedra", Leningrad. 3d Ed., 448p. [In Russian]

Sill, W.R. 1983. Self-potential modeling from primary flows. Geophysics, v. 48 (1), p 76-86.

Smart, P.L. and Laidlaw, I.M.S. 1977. An Evaluation of Some Fluorescent Dyes for Water Tracing. Water Resources Research, Vol. 13:1. p 17-33.

Smart, P.L. and Hobbs, S.L. 1986. Characterization of Carbonate Aquifers: A Conceptual Base. In the proceedings of the Environmental Problems in Karst Terrains and Their Solutions Conference, October 28-30, 1986, Bowling Green, Kentucky, National Water Well Association, p1-14.

Smart, P.L. 1984. A Review of the Toxicity of Twelve Fluorescent Dyes Used for Water Tracing. National Speleological Society Bulletin, Vol. 46. p 21-34.

Szabo, W.M. et al. 1979. Engineering Geology of Jefferson County, Alabama. Atlas 14 Geological Survey of Alabama.

U.S. Environmental Protection Agency. 1988. Application of Dye Tracing Techniques for Determining Solute-Transport Characteristics of Groundwater in Karst Terrains. EPA/904/6-88/001, 103p.

Wilson, J.L. 1975. Carbonate Facies in Geologic History. Springer-Verlag, New York, 471p.

Integrated geophysical surveys applied to karstic studies

ROB MCDONALD & NICHOLAS RUSSILL TerraDat, Cardiff, UK

ROB DAVIES TerraDat, Sandown Village, Vic., Australia

ABSTRACT

The investigation of features associated with karst environments is an important application of non-invasive geophysical methods. Although the physical principles underlying individual methods are well established, the merits of integrating different techniques in order to reduce model ambiguity are explored. We present case histories that illustrate how combinations of gravity, resistivity tomography and seismic refraction surveys enable improved model reliability of structures that would not generally be resolved by single-discipline approaches.

Microgravity methods in particular are generally employed to identify karstic features relevant to engineering work. While localised gravity 'low' anomalies are indicative of mass deficiency in the sub-surface we show that it is essential to carry out follow-up targeted seismic and/or resistivity tomography surveys to establish whether the anomaly is caused by a rockhead depression/sinkhole or by a sub-rockhead cavity. In addition, seismic surveys are shown to be useful for resolving ambiguities in interpreting drilling results from a targeted solution feature.

INTRODUCTION

Of the various methods that have been applied to the investigation of limestone solution features several have provided notable successes and are now routinely in use, i.e., resistivity tomography, microgravity, electromagnetic ground conductivity mapping and ground radar. Limestone environments tend to provide favourable conditions for the use of geophysical methods because there are relatively few material types present and they have very distinct physical properties. Limestone is very dense and very resistive to the flow of electricity, typical clay infill has a lower density but conducts electricity very well, and air voids have zero density and block the passage of electricity.

It is through combination of the survey methods that the nature of identified anomalous ground conditions can be interpreted, e.g., an area of anomalous low density may be due to the presence of an isolated air-filled void or a larger zone of clay filled fractures - this ambiguity can be resolved by resistivity tomography which would display anomalous high or low resistivities respectively.

Ground radar is a method that can provide high resolution information about the subsurface but its successful use is highly site dependent. Clay materials at or near surface severely attenuate the radar signal which prevents sufficient depth penetration. Highly fractured or broken rock can produce too many reflecting signals which detriments reliable interpretation. Best results are achieved when the radar antenna can be placed directly onto the rock surface and moved steadily in short stepped increments along a particular traverse line. The presence of an irregular surface and scrub causes the antenna to lift off the rock causing a great reduction in data quality as much of the transmitted signal is reflected at the surface.

Resistivity has a longer history in mapping buried karstic features than any other geophysical method. It offers advantages of cost, speed of acquisition and ease of processing over gravity measurements. Recent advances in acquisition have resulted from computer-controlled multi-channel meters and have been complimented by vast improvements in interpretation software (e.g. Loke, 1996). The presence of infill sediments or water greatly improves the detectability of solution features due to the enhanced local conductivity within more resistive host limestone bedrock. The theoretical resolution limitations that affect the use of resistivity may be reduced in some instances by more dominant secondary effects such as drying-out halos and fracturing associated with the voids which could produce measurable anomalies due to the greater volume of affected material.

Theoretically the sensitivity of resistivity measurements increases with decreasing potential electrode separation and so the optimum array for mapping buried voids is the pole-dipole array (Lowry and Shive, 1990). However, practical considerations and processing techniques that iterate through model pseudo-sections leads us to favour the dipole-dipole array which also serves well for mapping the large horizontal boundaries associated with soil, rock and water table layers.

Microgravity surveys are widely employed for the detection of solution features and the increasing availability of high precision semi-automated instruments that can read to accuracies better than 5 microgals has enabled targeting of relatively small features. It is vitally important that all corrections are applied to maintain the level of acquired data accuracy which includes removal of all local terrain effects which can often be more significant than measured anomalies where topography is irregular. Observation of a significant and localised gravity 'low' anomaly in a limestone terrain indicates the presence of solution features or a buried rockhead depression given that terrain effects have been reliably removed from the dataset. Where targeted seismic or resistivity tomography data show that there is no bedrock depression at the site of the gravity anomaly then the likely presence of a sub-rockhead solution feature has been identified. If the rockhead geometry derived from the follow-up resistivity and/or seismic data suggests a possible sinkhole then the gravity results can be modelled to determine whether there is evidence to suggest the presence of an underlying cavity.

Large sites requiring investigation can prove too laterally extensive for early use of combined resistivity and gravity methods. In such instances it is often useful to carry out a reconnaissance electromagnetic (ground conductivity) survey in an attempt to identify anomalous areas of the site to best target more detailed geophysical work. Various equipment and operational parameters can be set to investigate an appropriate depth range and features such as infilled sinkholes, cavities and fractures may cause measurable anomalies depending on the local conditions, depth of burial etc. While it could not be guaranteed or expected that all features would be identified using such a rapid reconnaissance approach it is often possible to locate the most significant shallow structures which can help to identify potential problem areas and it is certainly more beneficial than a random scatter of boreholes which are typically very unlikely to hit isolated karstic features.

Seismic refraction methods can prove very useful for proving the presence and topography of rockhead. While resistivity tomography can often be used to map limestone rockhead as well as investigating sub-rockhead structures there are some instances where the electrical contrast between sediments and saturated/fractured bedrock containing sediment infill or conductive water can be insufficient for clear identification of rockhead. Seismic refraction methods provide a reliable means for profiling rockhead without significant ambiguity because the velocity of limestone bedrock is generally beyond that possible for even very compacted unconsolidated sediments. While seismic refraction is not best suited for investigating solution features within bedrock it can prove to be a very useful tool for solving the main potential ambiguity affecting gravity surveys, i.e., whether an anomaly is due to a sub-rockhead cavity or a localised undulation in bedrock topography.

CASE EXAMPLES

Geophysical Investigation in Eire

Results from a recent geophysical survey carried out in a limestone environment in Eire are presented here to illustrate the useful combination of resistivity tomography and microgravity techniques. The survey was carried out to target solution features and fracture zones that would represent preferential flow paths within bedrock. A representative selection of the results are shown in Figure 1. The high degree of correlation between the survey methods is evident and the position of anomalous bedrock structures would have been identified by both survey methods individually. However, while the individual methods enable location of features they do not enable certain identification of their physical nature for the following reasons: (i) the microgravity survey has identified the presence of localised mass deficiencies but whether they are due to rockhead undulations or sub-rockhead solution features is not certain, and (ii) the resistivity tomography data reveals the presence and cross-sectional geometry of anomalously conductive zones within bedrock but it is not certain whether they are due to local variations in permeability or due to solution of bedrock.

Correlation of the resistivity and gravity data enables examination of the physical characteristics of any identified anomalous features as well as providing geometrical control for modelling the gravity data. It is apparent from the results presented in Figure 1 that each of the main low resistivity anomalies is also associated with a mass deficiency identified by the gravity survey. If these low resistivity zones were caused by localised water ingress where there is no significant solution of bedrock then there would not be any associated negative gravity anomalies of the observed amplitude.

Given that an anomalous feature has been identified as being due to bedrock solution by reconnaissance geophysical work, it is then possible to carry out a more detailed investigation to attempt to constrain its physical characteristics in three dimensions. Orthogonal aligned resistivity traverses (or true 3D survey) can provide the geometrical constraint on gravity modelling necessary to enable reliable estimation of the density contrast of the solution feature with competent bedrock. An alternative approach would be to carry out a detailed gravity survey on a grid to enable calculation of the total mass deficiency and plan geometry of the solution feature or fracture zone. Incorporating the cross-sectional information derived from the original resistivity tomography data would again enable reliable calculation of the density contrast. The density contrast information could enable estimation of the void/bedrock ratio within the anomalous zone given assumption about the likely infill material whether water, sediment or air. Evidence about likely infill material can be derived from knowledge of local water levels, sediment transport and from measured resistivity values and groundwater conductivities.

Geophysical Survey At Pembroke Dock, Wales

Results from a geophysical survey carried out to target possible solution features beneath a developed site are presented in Figure 2. The case example was selected to illustrate the sometimes useful and often essential combination of a seismic refraction survey with microgravity data. A significant gravity anomaly was identified by the microgravity survey carried out over the development site. The thick surface asphalt cover and presence of underlying coarse grained fill materials made it difficult to carry out a resistivity

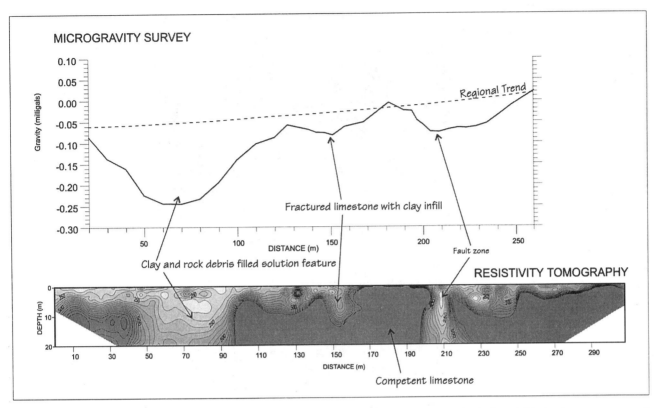

Figure 1: Correlation of microgravity and resistivity tomography at a site in Eire. Features including sinkholes fracture zones and faults are clearly identified and were confirmed with targeted boreholes.

tomography survey to investigate the large gravity anomaly. The asphalt renders it difficult to gain sufficient electrical contact between the electrodes and the subsurface and the electrical contrast between limestone and coarse fill materials is not as clearly defined as the seismic contrast between loose materials and bedrock. Boreholes targeted on the gravity anomaly suggested that bedrock continued to depths in excess of 16m and in nearby places only to 10m which suggests that either some of the holes passed down through narrow voids/fractures or some of the holes hit large boulders at shallower depth. In order to resolve this apparent ambiguity a seismic survey was carried out across the main identified gravity anomaly which revealed that average deepest bedrock depths are apparently about 10m below surface and the conclusion was that some of the boreholes had indeed passed down through narrow local voids/fractures.

CONCLUSIONS

Geophysical methods offer a number of alternative approaches to the detection and mapping of buried karst features. Microgravity in particular is a very powerful tool for establishing the presence of karst features which often give rise to measurable negative gravity anomalies. If only microgravity data is considered it can prove difficult to distinguish between markedly different possibilities and it is therefore best to use an additional geophysical method to constrain the interpretation. A key constraining factor is knowledge of the depth to bedrock which can be derived using resistivity tomography or seismic methods and input to the gravity modelling process.

Under favourable circumstances resistivity tomography can be used to map karst features directly in cross-section, however, it is recommended that anomalous zones are targeted using microgravity to identify whether they relate to fracture zones of voids. Without additional gravity data conclusive interpretation can be difficult, particularly in areas that have variable soil cover and/or laterally inhomogeneous rock conductivity. Resistivity tomography can be used to map the soil cover for input to the gravity modelling process and to identify the origin and approximate geometrical form of anomalous gravity zones.

Seismic refraction surveys are generally unable to map sub-rockhead solution features, however, the method provides a powerful means of accurately mapping soil depths which is particularly useful for input to the gravity modelling process in areas where the soil/rock density contrast is not expressed as a significant resistivity contrast.

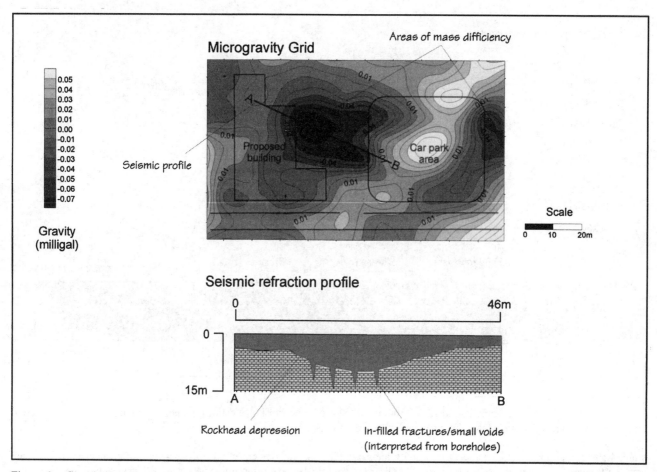

Figure 2: Correlation of microgravity and seismic refraction results at a site in South Wales, UK. An area of identified mass deficiency is shown by the seismic results to be due to a bedrock depression/sinkhole feature. The results from the seismic survey also suggested that some boreholes had penetrated narrow void/fractures within the bedrock depression.

REFERENCES

Loke, M.H. and Barker, R.D. 1996. Rapid least-squares inversion of apparent resistivity pseudosections. Geophysics, 60, 1682-1690.

Lowry, T and Shive, P.N. 1990. An evaluation of Bristow's method for the detection of subsurface cavities. Geophysics, 55, 514-520.

A case study of the reliability of multi-electrode earth resistivity testing for geotechnical investigations in karst terrains

M.J.S.ROTH, J.R.MACKEY & C.MACKEY Lafayette College, Easton, Pa., USA
J.E.NYQUIST Temple University, Philadelphia, Pa., USA

ABSTRACT

The thinly mantled karst areas of northeastern Pennsylvania have been associated with a number of structural failures caused by sinkholes. Geotechnical investigations in these areas have primarily relied upon intrusive probe tests, either borings or air-track drilling (Thomas and Roth, 1997). The silty clay soils overlying the carbonate bedrock have limited the use of non-intrusive ground penetrating radar and electromagnetic methods; however, multi-electrode earth resistivity testing may be a suitable method for locating subsurface features associated with sinkhole formation in these areas.

The case study presented in this paper concerns a site in Northampton County where sinkholes have been occurring at an accelerated rate due to the influence of construction (Chen and Roth, 1997). The bedrock at the site is located between a meter and approximately 10 meters below the surface. The soil is primarily a silty clay. Seventy resistivity lines using a variety of electrode and line spacings were conducted at the site. Using the data obtained from the earth resistivity tests, 17 borings were located to assess the accuracy of the resistivity results.

Comparisons between boring and resistivity results demonstrate that the resistivity tests provide good information concerning the depth to the irregular surface of the bedrock. However, use of the test to locate subsurface anomalies associated with sinkhole formation and to determine their size will require additional research. The paper presents the results of the study and provides recommendations for geotechnical investigation in silty clay soils overlying shallow carbonate bedrock.

INTRODUCTION

Regions of thinly mantled karst in northeastern Pennsylvania have been associated with structural failures due to sinkholes. At present, geotechnical investigations in the area typically use intrusive tests, such as borings or percussion (air-track) drilling (Thomas and Roth, 1997). Unfortunately, these intrusive probes do not provide sufficient information concerning variations in depth to bedrock and they have a low probability of locating subsurface anomalies that could result in sinkholes. The presence of clay soils overlying the carbonate rock of the area has limited the applicability of non-intrusive methods such as ground penetrating radar (GPR) and electromagnetic methods. Micro-gravity measurements provide valuable data concerning anomalies, but are time consuming and expensive for most sites. Also, in areas of shallow bedrock with an irregular surface, seismic reflection methods do not provide reliable data.

Multi-electrode resistivity testing has recently been introduced as a viable method for locating subsurface features associated with sinkholes. However, this method has not been well-documented as a tool for geotechnical site investigation. Questions regarding the necessary number of tests, electrode spacing, and reliability of results still exist. This paper presents a case study involving a site in Northampton County, Pennsylvania, which is experiencing accelerated sinkhole formation due to recent development (Chen and Roth, 1997). Data from 70 resistivity tests at varied spacings were collected at the site over a three-month period. Seventeen borings were drilled at selected locations for correlation with resistivity test results. This paper includes a description of the site geology, the testing performed, and typical results from both resistivity tests and borings. A discussion of the correlation between borings and resistivity profiles is presented and an assessment of the reliability of resistivity testing is given. Finally, recommendations concerning the use of resistivity testing in geotechnical investigations are presented.

BACKGROUND
Site description

The case study site is a portion of an athletic field complex owned by Lafayette College in eastern Pennsylvania. The 0.36 km^2 (90-acre) complex was developed into athletic fields during the 1970's. Development of the site consisted of extensive soil grading to provide level playing fields and drainage. Sinkhole formation at the complex has occurred more rapidly since site development. Sinkhole occurrences at the site were studied by Chen and Roth (1997). This study found that sinkhole density for the complex prior to and after development was approximately 10 sinkholes/km² and 45 sinkholes/km², respectively. Currently, approximately 15 sinkholes form at the complex each year.

The study site is in the northwest corner of the athletic complex. The area is essentially level with a small surface slope towards a drainage culvert. The site contains a visible line of sinkholes at its northern boundary at a bearing of approximately N 60° E.

Site geology

The study site is underlain by the Epler formation. This formation is composed of interbedded limestone and dolomite (USGS, 1967). The bedrock surface for the area is known to be extremely irregular, characterized by pinnacles and solution enlarged fissures (Thomas and Roth, 1997). Strike is N 60° E and dip is to the southeast at approximately 45 degrees. Several nearby outcrops confirmed the bedding and dip. Site soils are characterized as silty clays.

Site investigation

Investigation for the purpose of this case study included GPR, multi-electrode resistivity testing, and traditional borings. GPR equipment at frequencies of 100 MHz and 250 MHz was tried at the site. The penetration depth for both frequencies was approximately one meter or less and the data obtained revealed no significant information concerning the subsurface.

Multi-electrode resistivity testing was conducted with a 28-electrode system using a dipole-dipole electrode array. Initial testing at the site consisted of 27 lines spaced on an approximately 5-meter grid. The electrode spacing for these lines was 3 meters. The grid was approximately parallel and perpendicular to strike. The results of these tests were reviewed and six anomalies were selected as locations for additional testing. Tests at these anomaly locations were run in tighter grids and cross patterns using an electrode spacing of one or two meters.

Resistivity profiles were created through an inversion process and were used for characterization of the subsurface. The inversion process is discussed in deGroot-Hedlin and Constable (1990) and Loke and Barker (1996), among others. The profiles illustrate trends in resistivity that may be interpreted to represent subsurface features of interest. Resistivity profiles typically showed an irregular surface at the contrast between low and high resistivity materials. Subsurface anomalies of high and low resistivity were also observed.

Based on the results of the resistivity tests, boring locations were selected to confirm the preliminary subsurface interpretations. Borings were located where earth resistivity tests indicated possible anomalies, e.g., water- or air-filled voids, fractures, and regions of high over low resistivity; 17 borings were drilled for correlation purposes. Test borings in the soil materials were conducted using a hollow-stem auger. Split spoon samples were taken at 1.5 meter intervals to characterize the soils. A rock core barrel was used to obtain samples in the bedrock. Where possible, at least 1.5 meters of core was taken at the bottom of each boring. If an anomaly below the bedrock surface was indicated by the resistivity testing, rock cores were taken to at least five feet below the location of the suspected anomaly.

Soil specimens were visually classified as orange-tan clayey silt to silty clay with varying amounts of rock fragments. The rock specimens recovered were limestone. The rock cores exhibited fractures and varying degrees of weathering. Fractures were noted at a range of angles, with the majority being at approximately 40 degrees. Air, soil, and water-filled voids were discovered during this portion of the investigation.

Groundwater was encountered during the drilling, but the depth to groundwater was highly variable. Even though there was minimal rainfall during both the resistivity testing and the drilling operations at the test site, water was encountered at depths varying from one meter to more than 10 meters. Water levels may have been influenced by the method of coring, but it appears as if there is some perched groundwater at the site. Future work is required to characterize the groundwater conditions.

INTERPRETATION OF RESISTIVITY RESULTS

Subsurface features of interest at the site included depth to bedrock and the location, size, and type of voids. Boring results were overlaid onto resistivity profiles to aid in the interpretation of the resistivity test results. The following sections discuss interpretation of depth to bedrock and void locations.

Depth to bedrock

Figure 1 illustrates a typical resistivity profile and associated boring data. The bedrock surface correlates well with resistivity values of approximately 600 Ohm-m (Ω-m) and greater. Because resistivity values are significantly affected by the presence of moisture, different resistivity values may be obtained at different times of the year for the same material. The effect of this anticipated change in resistivity values was observed at the test site. Data obtained during the site investigation during the summer of 1998 was compared with data obtained during January of 1998. Resistivity values were significantly lower during the winter, but the overall trends in the data observed during the winter and summer were essentially the same. Because of the variability of the results with moisture, a specific value of resistivity should not be selected to determine the depth to bedrock. The value for a set of tests conducted at approximately the same time should be selected with reference to boring data.

For most of the resistivity tests conducted at the site with electrode spacings of three meters or larger, resistivity pseudosections indicated a decrease in resistivity at depths greater than approximately 10 meters. There is no indication from the boring data that this decrease in resistivity is associated with changes in the subsurface conditions. It is the authors' interpretation that these results are due

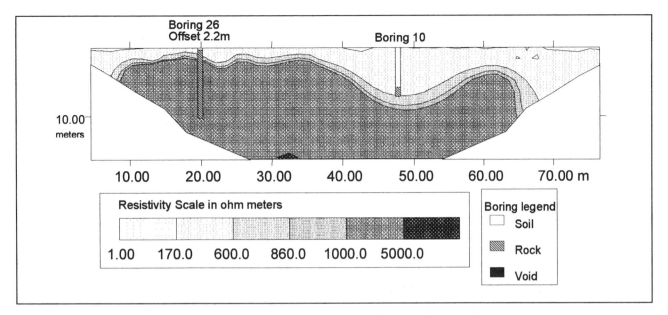

Figure 1. Typical resistivity profile (#29) with boring data

to three-dimensional effects. As the pairs of electrodes being tested are at greater and greater distances, the current introduced into the subsurface is more likely to find a path of less resistivity. Only in areas where the bedrock is of high quality (i.e., with very few open fractures or joints), would the profiles show a continuing increase in resistivity with depth (Roth, Mackey, and Nyquist, 1999).

Void locations

Interpretation of the resistivity data to locate voids proved to be difficult at this site. Three-dimensional effects played a significant role in the difficulties observed. Specifically, if a void is insulated from the low resistivity surface materials by the surrounding bedrock, the presence of the void will not be detected by the test because the current will find a less resistive path. However, if a low resistivity connection exists between the surface materials and the void, e.g., a water or soil filled fracture, the void may be detected. Figures 2 and 3 illustrate this problem. These resistivity lines were run perpendicular to each other and crossed at the location of Boring 1 (the location of a 0.2 meter high void that contained wet soil). The resistivities observed at the location of the void are very different. This is most likely due to the orientation of a soil or rock filled fracture intersecting the void causing low resistivities in one direction. The void was apparently insulated from the current in the other direction.

RELIABILITY OF EARTH RESISTIVITY TESTS

Based on the results of the case study, the reliability of resistivity tests in karst geology overlain by clay soils is dependent on a number of factors. Orientation of the line, electrode spacing, and line spacing all impact the accuracy of test results. The following sections describe the effects of these factors on the reliability of multi-electrode resistivity results.

Line orientation

As described above, line orientation proved to be critical to the results of this study with respect to locating voids. Resistivity results were influenced by three-dimensional effects related to the irregular bedrock surface and the joints and fractures in the rock. Voids which were located during the subsurface sampling program were often observed as anomalies in one line orientation but were not observed in a perpendicular orientation. Figures 2 and 3 illustrate this at the location of Boring 1; Figures 4 and 5 illustrate this at the location of Boring 26a. Interpretations of the results for the depth to bedrock were not significantly affected by line orientation.

Influence of probe spacing

A tradeoff exists between probe spacing and depth of the profile, and between probe spacing and detail of the profile. As electrode spacing increases, the depth of the profile increases, while the detail of the profile decreases. The opposite is also true. Therefore, the scope of the investigation will dictate the choice of electrode spacing. For geotechnical investigations, the depths of interest are typically less than 15 meters. Therefore the electrode spacings used in this project ranged from one to four meters. This resulted in an approximate depth of testing from five to about 20 meters.

Figures 2 and 6 show the same resistivity line at spacings of 1m and 2m respectively. The 1m line is centered over the 2m line. A comparison of the two figures reveals that interpretation of depth to bedrock is consistent at both probe spacings. The interpretation of the variability of the bedrock surface is also relatively unaffected by the change in probe spacing. This suggests that for a test aimed at defining the bedrock surface, larger electrode spacings will be adequate. Based on a comparison of each line to Boring B1,

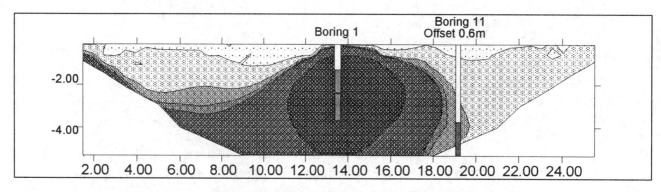

Figure 2. Resistivity profile (#61) conducted perpendicular to strike with electrodes at 1 meter spacing. (For resistivity scale and boring legend, refer to Figure 1.)

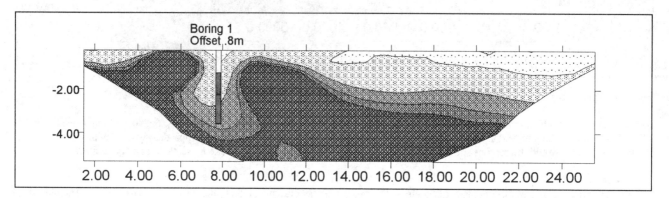

Figure 3. Resistivity profile (#53) conducted parallel to strike with electrodes at 1 meter spacing. (For resistivity scale and boring legend, refer to Figure 1.)

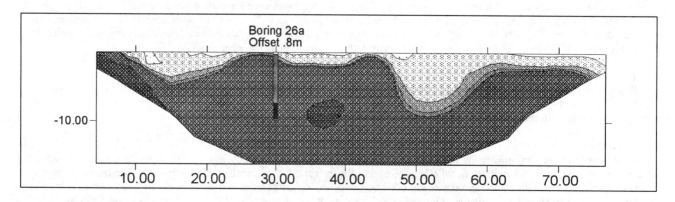

Figure 4. Resistivity profile (#25) conducted perpendicular to strike with electrodes at 3 meter spacing. (For resistivity scale and boring legend, refer to Figure 1.)

both tests indicate a possible anomaly. However, neither test was able to accurately determine the size or depth of the anomaly encountered.

Line spacing

Distance between adjacent resistivity lines is also a significant factor in the reliability of the method. Both 2 and 5m offsets were used in this study. Figures 7, 8, and 9 illustrate the effect of offsetting. Boring 3 encountered a 1.2 meter high void at a depth of 3.3 meters. The line shown in Figure 7 was conducted directly over this void. The lines shown in Figures 8 and 9 are offset by approximately 2 and 4 meters, respectively. The void is clearly illustrated by the resistivity line shown in Figure 7; Figure 8 still indicates a possible void; Figure 9, at an offset of 4.2 meters, does not indicate a possible void.

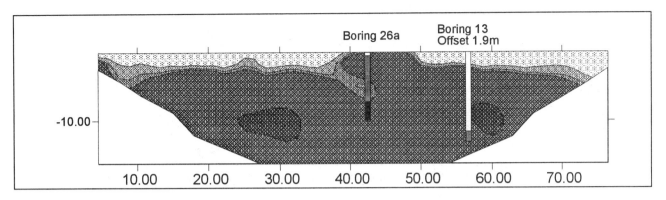

Figure 5. Resistivity profile (#74) conducted parallel to strike with electrodes at 3 meter spacing. (For resistivity scale and boring legend, refer to Figure 1.)

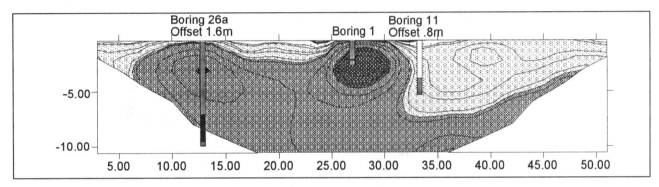

Figure 6. Resistivity profile (#62) conducted perpendicular to strike with electrodes at 2 meter spacing. (For resistivity scale and boring legend, refer to Figure 1.)

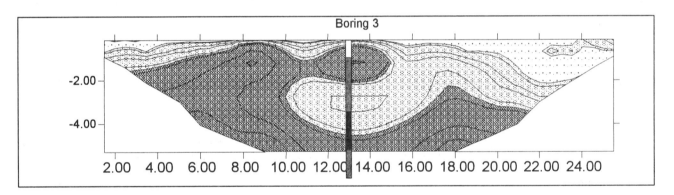

Figure 7. Resistivity profile (#44) located directly over 1.2 meter void. (For resistivity scale and boring legend, refer to Figure 1.)

In general, offsets greater than approximately 4 meters from a known void did not indicate the presence of the void. In all cases, the areal extent of the void is not yet known. Further testing is required to refine interpretation of these results.

RECOMMENDATIONS FOR GEOTECHNICAL INVESTIGATIONS

Multi-electrode earth resistivity testing appears to be a valuable tool for geotechnical exploration in areas of karst overlain by clay soils. By using intrusive methods to confirm the interpretation of the results, this test can predict depth to bedrock and determine trends in the bedrock surface. Electrode spacing can be selected based on the requirements of the project. However, the reliability of this method is still in question with regard to locating and determining the size of possible voids. Three-dimensional variability and the effects of line orientation both have significant influence on the results and require further study. The use of three-dimensional resistivity tests may overcome some of these effects.

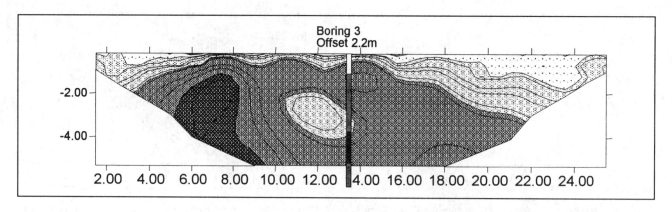

Figure 8. Resistivity profile (#43) located 2.2 meters west of 1.2 meter high void. (For resistivity scale and boring legend, refer to Figure 1.)

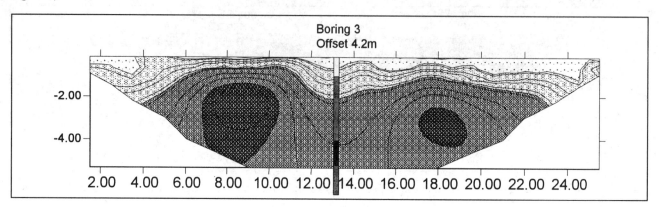

Figure 9. Resistivity profile (#42) located 4.2 meters west of 1.2 meter high void. (For resistivity scale and boring legend, refer to Figure 1.)

ACKNOWLEDGEMENTS

This study was supported in part by the National Science Foundation under Grant No. CMS-9734899. This support is gratefully acknowledged.

REFERENCES

Chen, L., and Roth, M.J.S. (1997). "Sinkhole case study: Athletic fields, Lafayette College, Easton, Pennsylvania," inThe Engineering Geology and Hydrogeology of Karst Terranes, B. Beck and B. Stephenson, eds., A.A. Balkema, Rotterdam, 37-40.

deGroot-Hedlin, C. and Constable, S.C. (1990). "Occam's inversion to generate smooth, two-dimensional models from magnetotelluric data," Geophysics, 55, 1613-1624.

Loke, M.H., and Barker, R.D. (1996). "Rapid least-squares inversion of apparent resistivity pseudosections by a quasi-Newton method," Geophysical Prospecting, 44, 131-152.

Roth, M.J.S., Mackey, J.R., and Nyquist, J.E. (1999). "A case study of the use of earth resistivity in thinly mantled karst," SAGEEP.

Thomas, B., and Roth, M.J.S. (1997). "Site characterization for sinkholes in Pennsylvania and New Jersey," in The Engineering Geology and Hydrogeology of Karst Terranes, B. Beck and B. Stephenson, eds., A.A. Balkema, Rotterdam, 281-286.

USGS (1967). "Geologic Map of the Easton Quadrangle, New Jersey-Pennsylvania," U.S. Geological Survey, Washington, D.C.

Remediation of leakage of an earth-filled dam and reservoir by geophysically-directed grouting, Washington, Missouri

DAVID G.TAYLOR   Strata Services, Incorporated, St. Charles, Mo., USA
ARTHUR L.LANGE   Karst Geophysics, Incorporated, Golden, Colo., USA

ABSTRACT

Prior to remediation measures taken in November 1997, the water level of a private reservoir owned by WWAB Enterprises of Washington, Missouri had been falling at a rate of approximately 2½ centimeters (one-inch) per day during late summer and fall. The reservoir, covering approximately twelve acres, is retained by a 10.4-meter (34-foot) high, 183-meter (600-foot) long earth-filled dam founded on Jefferson City Dolomite of Ordovician age. The estimated leakage of approximately 12.6 liters/second (200 gallons/minute) resurfaced in the form of several springs up to 335 meters (1100 feet) downstream from the dam.

Initial attempts at remediation were made to seal off at depth a swallow hole in the basin area approximately 90 meters (300 feet) upstream of the dam. A drill hole was advanced down the center of the swallow hole which penetrated solutioned dolomite and voids. Cementaceous grout slurry totaling 18.3 cubic meters (647 cubic feet) was injected via the drill hole into the swallow-hole feature. This procedure appeared to control the lake-water exfiltration at the time.

The reoccurrence of leakage during the summer of 1996 when the lake level was at 3.7 meters (12 feet) below normal pool prompted the implementation of a geophysical natural-potential survey along the dam crest and its slopes during August 1996. The resulting six d.c. electrical anomalies delineated a pattern of interconnecting channeled fluid flow beneath the dam and its abutments. The two principal electrical anomalies, along the crest of the dam, were bracketed by five exploration/grout holes located on 6.1-meter (20-foot) intervals. One of the holes penetrated a solutioned zone having strong hydraulic conductivity between 12.96 meters (42.5) and 13.48 meters (44.2 feet) depth, within the underlying dolomite. The discovery hole and two adjoining holes penetrated the dissolved zone, where they lost 9.8 liters/second (150 gallons/minute) of circulation, showed a depressed piezometric level and strong downhole acoustic emission anomalies. Dye injected into these holes communicated visually with the downstream rises.

A total of 17.4 cubic meters (614 cubic feet) of cementaceous grout slurry was injected through the three grout holes and into the dissolved limestone. Within hours, the downstream leakage and springs significantly reduced in flow. This remediation resulted in complete stabilization of the reservoir to normal pool elevation. The amount of lake-water leakage terminated through the dissolved zone is estimated at 12.6 liters/seconds (200 gpm).

INTRODUCTION

Water level in a private reservoir in Washington, Missouri, was dropping, due to leakage, at a rate of approximately 2½ centimeters (one-inch) per day prior to November 1997. The reservoir is owned by WWAB Enterprises and is situated 6 kilometers (3.7 miles) southwest of downtown Washington—a city on the Missouri River approximately 65 kilometers (40 miles) west of St. Louis (Figure 1). The reservoir is retained by an earth-filled dam and covers approximately 5 hectares (12 acres). The dam measures 10.4 meters (34 feet) in height and 183 meters (600 feet) in length and is founded on Jefferson City Dolomite of Ordovician age. Water loss, estimated at approximately 12.6 liters/second (200 gallons/minute), reappeared at springs between the foot of the dam and 1100 feet downstream.

INITIAL REMEDIATION ATTEMPTS

In September 1995, Strata Services, Inc. attempted to seal off at depth a 1.2-meter (4-foot) diameter swallow hole in the basin area approximately 90 meters (300 feet) upstream from the dam (Figure 2). A drill hole advancing down the center of the swallow hole penetrated solutioned dolomite and voids beneath a 6.1-meter (20-foot) deposit of soft clayey soil. A total of 3.82 cubic meters (135 cubic feet) of sand-cement slurry was poured into the annular space around a 10.16-centimeter (4-inch) casing, which thoroughly filled the exposed sinkhole void. After allowing an overnight set period, the casing was reentered with a 12.3-centimeter (3.75-inch) bit and drilling penetrated broken solutioned rock and voids to a depth of 11 meters (36 feet) where it encountered competent carbonate rock. Dye placed in the bore hole was chased with water at a rate of approximately 12.6 liters/second

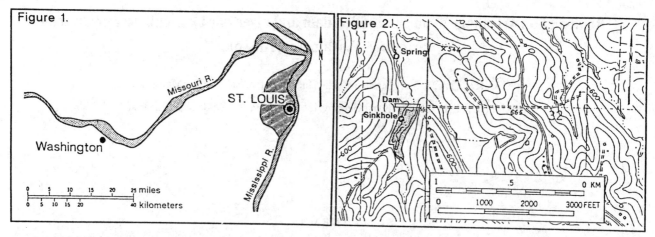

Figure 1. Location of Washington, Missouri on the Missouri River.

Figure 2. Location of the WWAB Reservoir, showing the swallow hole (Sinkhole) to which the initial remediation efforts were directed. Prior to remediation, dye-tracing demonstrated communication between the swallow hole and seeps downstream of the dam and the depicted spring.

(200gallons/minute) and reappeared after six hours at various donstream spring locations including a suspect spring 335 meters (1100 feet) downstream of the dam.

A neat cement grout slurry having a volume of 14.5 cubic meters (512 cubic feet) was injected through the casing and into the underlying solutioned dolomite, thereby sealing the swallow hole. This procedure appeared to control the lake water exfiltration at the time.

SUCCESSIVE REMEDIATION MEASURES

Leakage reoccurred during Summer 1996 with lake level 3.7 meters (12 feet) below spillway level, whereupon exploration was redirected to the dam in an attempt to intercept likely solution conduits within the foundation rock. Prior to test-drilling, a geophysical survey was planned in order to detect and map possible conduits from the dam surface and abutments.

Geophysical investigations: The natural-potential method

The geophysical method adopted was that of natural potential (NP), first utilized in the early 1800's for detecting conductive minerals in the ground (Fox, 1830; Rust, 1938). For groundwater applications the technique utilizes distortions of the d.c. natural electric field in the ground brought about by the *electrokinetic effect* of localized groundwater flow paths (Ishido & Mizutani, 1981). The effect results from the convective flow of electric charges within a moving fluid that gives rise to a return conductive current, producing by Ohm's Law a voltage referred to as the *streaming potential*. The streaming phenomenon occurs not only within granular materials but also within fractures (Bogoslovsky & Ogilvy, 1972). The natural-potential technique has been utilized extensively for detecting leakage in dams, reservoirs and pipelines, and in the search for rising geothermal waters as well (Bogoslovsky & Ogilvy, 1970; Corwin & Hoover, 1979). Within the past decade the method has been shown to produce significant voltage anomalies over caverns and subterranean streams in karst. Details on the theory of the natural electric field in karst are presented by Kilty & Lange (1991), and examples of NP anomalies over caves and underground streams are described in the companion paper by Lange & Kilty (1991). Figure 3 depicts characteristic anomaly types observed over caves and springs.

Natural potentials at the surface are measured using a pair of non-polarizing electrodes connected by a distance-calibrated wire to a high-impedance multimeter mounted on a reel. One electrode, (the *reference-*, or *base-*) electrode is planted in the ground at a central location while the second (*roving electrode*) is moved along a grid of previously flagged traverses covering the area of interest. The roving electrode is inserted into shallow holes sufficiently deep to contact soil moisture. At any one particular station, the measured voltage (in millivolts) thus represents the difference of potential between the remote station and the reference electrode. The resulting voltage readings are then logged along with the time and site descriptions.

The data processing software corrects for voltage drifts in the ground and in the electrodes that occur during the day. The software generates graphic profiles of NP along the individual lines of the grid. The interpreter then attempts to correlate characteristic features of the plotted profiles from line-to-line. The resulting anomalies can also be transferred to a plan map of the site and anomaly trends mapped. The interpreter may also generate a contour map based on amplitudes of the potential distribution; however, anomaly trends are normally more readily linked by viewing stacked profiles, where anomaly shapes as well as amplitudes are evident.

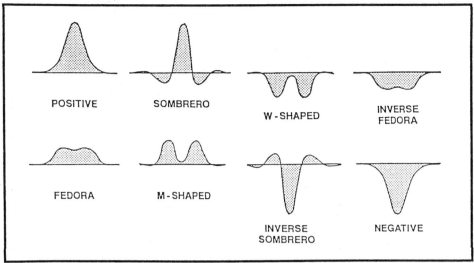

Figure 3. Typical NP anomaly types commonly observed over caverns and karst springs. The changing shapes are attributed to changes in the geometry of the voids as well as variations in flow conditions from, say, a tubeful situation, to one of a free-flowing underground stream or pond.

The natural-potential grid

A grid composed of six survey lines, designated A through F, were laid out in an east/west orientation along the dam, parallel to its axis. Line A paralleled the north edge of the dam crest, eight feet north of its midline designated as an east/west baseline. Lines A, B and C along the face of the dam were 9.15 meters (30 feet) apart, while Line E in wooded ground immediately beyond the toe of the dam was 44.2 meters (145 feet) north of the midline of the dam crest. Line F followed a high-water line along the south face of the dam, roughly 8.8 meters (29 feet) south of the baseline. The traverses varied in length from 221 meters (725 feet)(Line F) to 259 meters (850 feet) (Line E). The six-line grid is shown in idealized form in Figure 4. NP readings were made along the lines at stations spaced 3.81 meters (12.5 feet) apart and closer, where detail was needed. One common base station was established at the approximate mid-point of the baseline on the dam crest.

The geophysical survey was conducted by The Geophysics Group on the dates 30 and 31 August 1996. During the two-day study, 422 NP stations were logged, not counting re-occupations of the base station at the beginning and end of each line segment. The coverage totaled 1364 meters (4475 feet) of line.

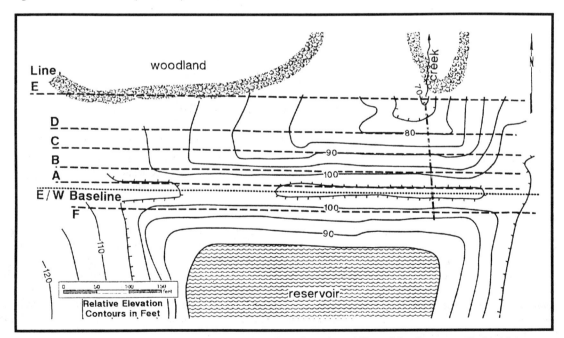

Figure 4. The natural-potential grid of six traverses and a baseline along the midline of the dam crest. Under high-water conditions the lake would normally discharge to the creek north of the dam via the drain shown by the double-dashed line.

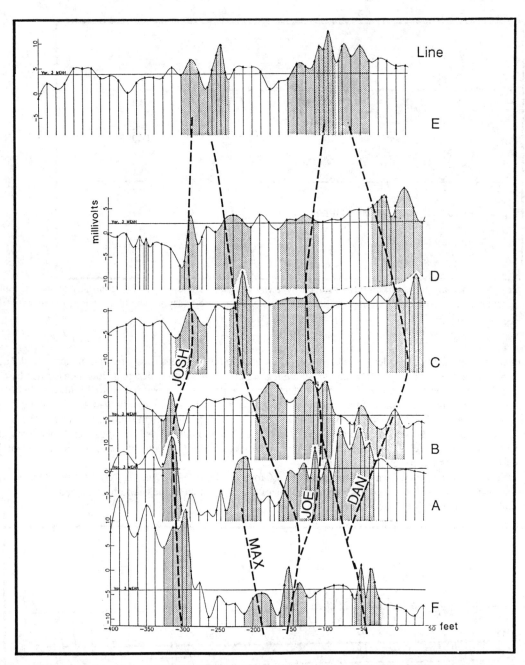

Figure 5. The western portions of six natural-potential profiles in their proper north/south relationship. Deduced anomalies are shaded and connected by dashed lines to depict the interpreted anomaly trends theoretically associated with underlying groundwater flowpaths. The anomaly trends have been designated arbitrarily for participants in the survey, including the dog Max.

Results of the natural-potential survey

An example of the manner in which individual NP anomalies could be traced from line-to-line, using both their shapes and amplitudes is provided in Figure 5. Anomaly trends were designated by names of the individuals participating in the survey. Although some of the linkages appear tenuous or ambiguous, others are more evident; for example, the pronounced sombrero-like anomaly "Josh," observed near the west side of the grid--between 6.1 and 12.2 meters (20 and 40 feet) beyond the edge of fill at the surface. "Josh" can be traced across all six lines of the grid. This anomaly trend is likely the result of a narrow roundabout water pathway within the west abutment of the dam. Anomalies recognized from the NP profiles are translated into a system of anomaly trends in plan view in Figure 6. Their interwinding pattern resembles a network of anastomose solution tubes commonly found along bedding planes in nearly horizontal carbonate rocks.

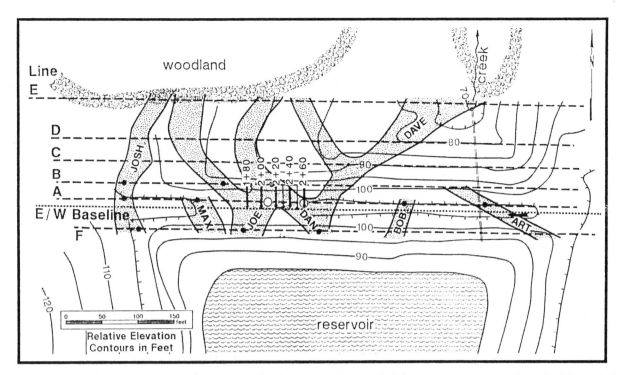

Figure 6. NP anomaly trends across the dam and woodland (Line E), showing their interpreted interconnections. Such interwinding pathways are typical of anastomotic groundwater flow along bedding plane openings in a near-horizontal karst environment. Recommended drilling sites are shown by solid circles and two open circles—the latter being the favored drilling targets. The actual locations drilled are shown by arrows.

Twelve of the most prominent anomaly peaks were selected as recommended drilling sites, of which two near the midpoint of the dam on Line A appeared most promising.

Remediation procedures

A remediation effort on the dam was undertaken by Strata Services in November 1997. The two prominent electrical anomalies along Line A—"Dan" and "Joe"—were bracketed by five exploration holes 6.1 meters (20 feet) apart drilled to a depth of 19.82 meters (65 feet) along the midline of the dam eight feet south of Line A (Figure 7). One of the holes (#2+40) penetrated a solutioned zone within the dolomite, having strong hydraulic conductivity at depths between 12.96 and 13.48 meters (42.5 and 44.2 feet), immediately below the top of bedrock. Casing was set at 13.29 meters (43.8 feet) and drilling resumed; however, all circulation was lost right out of the end of the casing for the duration of the hole. Subsequently two holes were drilled at offsets of ten feet on either side of #2+40. The three of these holes that penetrated the main part of the solutioned zone all lost 9.5 liters/second (150 gallons/minute) of circulation, showed a depressed piezometric level and produced strong downhole acoustic emission anomalies (Figure 8). Dye injected into these holes was detected visually at the downstream rises.

Two of the three holes that penetrated the zone of maximum flow coincided with the western half of NP anomaly "Dan", projected onto the midline of the dam; however, the lesser narrow anomaly "Joe" was not directly drilled. Judging from the northwesterly orientation of anomaly trend "Dan" (Figure 5), NP Line A crossed the feature at an angle between 40 and 50°; thus broadening the apparent stream width one and a half times. In retrospect, drilling on the anomaly's western (left) peak, rather than its medial low would have tapped directly into the zone of high permeability.

A total of 3.62 cubic meters (128 cubic feet) of neat grout slurry and a subsequent 13.76 cubic meters (486 cubic feet) of sand-cement slurry was injected in hole 2+40. As grouting progressed, water levels in adjoining holes 2+20 through 2+60 rose to the surface, or near-surface, indicating excellent subsurface lateral grout movement and sealing of underlying void space. Grout pressures at the pump manifold slowly rose to a final maximum of 3.4 atmospheres (50 pounds/inch^2).

The total weight of grout materials injected was 17,142 kilograms (37,800 pounds) of well-graded river sand and 12,834 kilograms (28,300 pounds) of Portland Type 1 cement, representing approximately 29.9 tonnes (33 tons) of solids. This remediation resulted in complete stabilization of the reservoir to normal pool elevation. The amount of lake-water leakage terminated through the dissolved zone is estimated at 12.6 liters/second (200 gallons/minute).

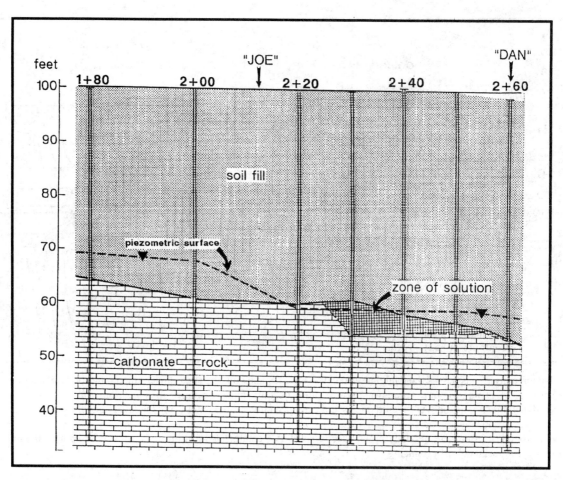

Figure 7. Vertical cross-section along a portion of NP Line A showing the solutioned zone intercepted by four drill holes. High NP amplitudes occurred over the four holes on the left in a region where Anomalies "Dan" and "Joe" merged. Anomaly "Joe" was not specifically drilled. Because Line A ran along the edge of the dam crest, drill holes were displaced to the midline of the dam, eight feet south of Line A.

REFERENCES

Bogoslovsky, V.A., and Ogilvy, A.A., 1970. Natural potential anomalies as quantitative index of the rate of seepage from water reservoirs: Geophysical Prospecting, v. 18, 261-268.

Bogoslovsky, V.A., and Ogilvy, A.A., 1972. The study of streaming potentials on fissured media models: Geophysical Prospecting, v. 20 (1), p.109-117.

Fox, R.W., 1830. On the electro-magnetic properties of metalliferous veins in the mines of Cornwall: Royal Society of London, Philosophical Transactions, v. 2, p. 411.

Ishido, T., and Mizutani, H., 1981. Experimental and theoretical basis of electrokinetic phenomena in rock-water systems and its applications to geophysics.: Journal of Geophysical Research, v. 86 (B3), p. 1763-1775.

Kilty, K.T., and Lange, A.L., 1991, Electrochemistry of natural potential processes in karst: Proceedings of the Third Conference on Hydrogeology, Ecology, Monitoring, and Management of Ground Water in Karst Terranes, Nashville, Tennessee, December 4-6, 1991, p. 163-177.

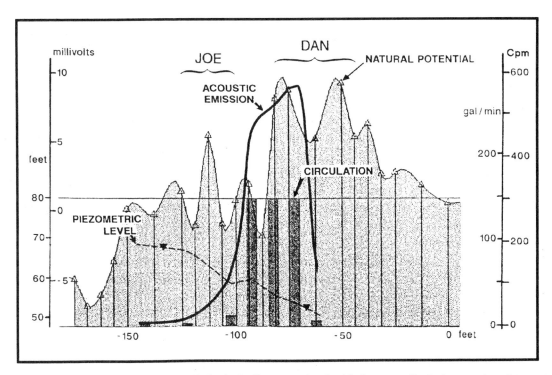

Figure 8. Superimposed graphs of the measured physical effects associated with the zone of solution openings intercepted during drilling. The original NP profile for a portion of Line A (along the dam crest) is stippled. The piezometric surface as logged in the holes (measured in feet of relative elevation) is shown by the dashed line. Volume of fluid accepted by each hole is depicted in the bar graph designated "Circulation." Downhole acoustic emission levels are traced in counts/minute by the solid curve. Drill holes were sited on the midline of the dam, eight feet south of the anomalies on Line A.

Lange, A.L., and Kilty, K.T., 1991, Electrochemistry of natural potential processes in karst: Proceedings of the Third Conference on Hydrogeology, Ecology, Monitoring, and Management of Ground Water in Karst Terranes, Nashville, Tennessee, December 4-6, 1991, p. 179-196.

Rust, W.M., Jr., 1938, A historical review of electrical prospecting methods: Geophysics, v. 3 (1), p. 1-6.

6 Governmental role in karst areas: Regulations and education

Education about and management of sinkholes in karst areas: Initial efforts in Lebanon County, Pennsylvania

E. DRANNON BUSKIRK JR Harrisburg Area Community College, Pa., USA
MICHAEL D. PAVELEK II & RACHEL STRASZ Greater Lebanon Refuse Authority, Pa., USA

ABSTRACT

Thousands of sinkholes and landscape depressions occur in a karst belt that crosses Lebanon County in south-central Pennsylvania. Despite frequent land subsidence, road collapses, stream disappearances, and other karst happenings, sinkhole education and management have only recently been addressed. During the fall, 1997 through spring, 1998 period, a group of five organizations conducted a project for preventing and abating pollution associated with sinkholes.

This highly successful project accomplished the following educational achievements: (1) after advertising in local newspapers, provided landowners of sinkholes, businesspersons, and others with information on groundwater and pollution relationships, drainage control, erosion prevention, and sinkhole closure; (2) conducted a sinkholes management workshop; (3) developed and gave a presentation entitled, "Sinkholes in Lebanon County" to various community groups; and (4) co-sponsored seminars on sinkholes and related topics through association with Harrisburg Area Community College.

Sinkhole management measures developed during the project were of the following types: (1) information on the locations of sinkholes, suitable for incorporation into GIS databases and other uses; (2) a Risk Priority Assessment Checklist for classifying sinkholes; and (3) screening of selected chemicals in private well waters, with the findings coordinated with a conservation agency and landowners of sinkholes. An additional project achievement from the educational and management aspects was enhanced cooperation among agencies that deal with water resource protection, sinkhole management, and related subjects.

Successes, failures, and lessons learned, such as the importance of joint sponsorship of activities, complete this report on the sinkhole project.

INTRODUCTION

Lebanon County in south-central Pennsylvania, USA, has a long history of karst-related phenomena. In the past year some memorable events included the development of a sinkhole that swallowed a stream and subsidence beneath an apartment complex that forced 11 families from their homes. The Borough of Palmyra has an especially impressive record of collapsed roads, cars of an automobile dealership lost underground, a house partially collapsed into a sinkhole, and rushing waters heard beneath lawns and streets following storm events.

Karst geology, exacerbated by building practices and stormwater drainage, is the source of this havoc and humor. Hundreds of sinkholes and thousands of landscape depressions (i.e., including sinkholes forming or filling) occur within a band of carbonate bedrock, that crosses the county (Berg, T.M. and others, 1980).

Residents and community leaders are increasingly concerned about the potential for groundwater contamination through trash or surface drainage into sinkholes. During the fall, 1997 through spring, 1998 period, five organizations in the area conducted a project for preventing water resource degradation from, and providing education about, pollution associated with sinkholes. The organizations were the Greater Lebanon Refuse Authority, Harrisburg Area Community College, Lebanon County Conservation District, Lebanon County Planning Department, and Swatara Creek Watershed Association.

The project had the following goals: identification of sinkholes, sinkhole users, and landowners; conveying information about sinkholes, water quality, and waste disposal practices to both responsible parties and the general public; interagency cooperation and resource sharing; and inspection of sinkholes and properties, with corrective actions taken as appropriate, mainly through emphasis upon voluntary participation. To achieve the goals, this project had both education and management components.

EDUCATION

Sinkhole and karst-related education takes many forms as practiced in nearly a dozen states in the central and eastern parts of the country. Among these educational activities are workshops sponsored by the Virginia Karst Project of the VA Department of Conservation and Recreation, workshops in 8 states from Alabama to Minnesota, training for facilitators, and educational materials for school teachers produced by Project Underground (Zokaites, 1998). In the Middle Atlantic area, educational efforts include promotion of awareness about limestone geology, education on engineering, regulatory aspects, and management aspects of sinkholes, assistance for municipal officials, and a reference collection of karst-related documents maintained by North Jersey Resource Conservation and Development (Drewes, 1998). The Conservation District of Clinton County, PA, has developed, and is distributing, a video tape showing sinkhole closure methods. A widely-read agricultural newspaper in Lebanon County and neighboring areas published a two-part series on sinkholes as threats to farmers (Andrews, 1994).

In Lebanon County education is recognized as a key to enlisting the support of the general public, businesspersons, and landowners in protecting sinkholes as potential portals to groundwater contamination. The project conducted the following activities:

- Advertisements in local newspapers;
- Distribution of information pamphlets, fact sheets, video tapes, and other educational materials;
- Development and delivery of a sinkhole presentation to community groups;
- Sponsorship and cooperation on workshops and seminars.

In local newspapers, the project displayed advertisements inviting sinkhole owners to participate in sinkhole evaluation and education activities. Persons who responded were given sinkhole inspections and were provided with background and technical information. The closure of a sinkhole was video recorded, with the videotape used on a television news broadcast.

Existing information on the potential effects of sinkhole trash on water quality, surface drainage, erosion control around sinkholes, and sinkhole closure methods were given to landowners, respondents, and participants at meetings, as appropriate. With the cooperation of landowners, photographs were taken of representative sinkholes to illustrate groundwater protection for educational purposes.

A presentation entitled; "Sinkholes in Lebanon County" was given to the Lebanon County Builders' Association, Lebanon Chapter League of Women Voters, a local Lions Club, and other community groups. Contractors and developers were provided information on the negative effects of construction activities for sinkholes, with alternatives for mitigation. The Builders' Association and the Conservation District were supplied with video tapes on techniques for closing sinkholes without causing surface collapses or groundwater contamination.

The project sponsored a very successful day-long workshop on sinkhole management. This workshop, including a field trip, provided information on regional groundwater quality, stormwater drainage, sinkhole closure methods, governmental regulations, and reference resources. The workshop was attended by over 50 persons representing approximately 30 organizations, including municipalities, state agencies, consultants, universities, landowners, and professionals from neighboring states. A second scheduled workshop on groundwater remediation, although inexpensive and staffed with experienced people, was canceled due to low pre-registration.

The project also sponsored seminars, including a discussion on sinkhole density as a planning tool for land development. These seminars, sponsored jointly with Harrisburg Area Community College, were interactive experiences in which invited experts from the college and community, often from various disciplines, reacted to the seminar presentations so as to enhance the discussions.

Persons participating in the various educational activities and other citizens requesting information were provided existing literature, mainly publications from federal and state agencies. This literature ranged in scope from single pages telling how pollutants in aquifers can affect both local and distant water supplies, to booklets giving detailed aspects of stormwater management and sinkhole closure.

Other related educational efforts were conducted by participating organizations. The Conservation District provided additional information on wellhead protection, manure control, and farmstead water management. Harrisburg Area Community College conducted a "3R's Series" of seminars, workshops, and symposia on resource reduction, reuse, and recycling, as part of continuing efforts to enhance public awareness of practical solutions to resource problems. Through the Web site of the Pennsylvania Department of Environmental Protection, the activities of the award-winning Greater Lebanon Refuse Authority are shown. These activities include ongoing dissemination of information about sinkholes and enforcement actions on commercial or large-scale violations of solid waste regulations involving sinkholes.

MANAGEMENT

Concurrent with educational objectives, the project sought to contribute to improved sinkhole management in the county. These efforts were of the following types:

- Documentation of sinkhole locations;
- Coordination of information on groundwater quality with the locations of wells and sinkholes;
- Development of a checklist for sinkhole classification and evaluation.

Although anecdotal evidence existed about the abundance of sinkholes in the county, the extent and locations were unknown to project staff. The project initially intended to use aerial infrared photography and other remote sensing imagery to identify sinkholes. It was subsequently discovered that many sinkhole and landform depression locations had already been recorded by the PA Geological Survey and were entered onto 7.5-minute topographic maps (Kochanov, 1988). The project acquired these maps and provided copies, and lists of sinkhole locations according to latitude and longitude, to the Lebanon County Planning Department and the Lebanon County Conservation District. This information is now readily available for many uses, such as incorporation into Geographical Information System (GIS) databases for evaluation of building permit applications, monitoring of land use practices, and coordination with property ownership and tax maps.

The project screened groundwater data on private wells in Lebanon County for potential aquifer contamination. These water sources had been tested for selected chemical parameters in 1991. Although no groundwater pollution was found associated directly with known sinkholes, data showed three groundwater locations to have nitrate concentrations approximately 2.5 times above the maximum recommended levels. This information was referred to the Conservation District for follow up with the landowners as to water supply risk, appropriate pollution control approaches, and potential remediation practices. No new sinkholes used for waste disposal were identified through this water quality screening effort.

The project team developed a Risk Priority Assessment Checklist to be used in the evaluation of sinkholes, wells, household wastewater systems, and fertilizer or pesticide practices. This checklist contains criteria for scales of best-to-poor ratings. The criteria for the status of sinkholes include the following descriptors: size of sinkhole; drainage area around sinkhole; position of sinkhole in relation to potential sources of water contamination; and the maintenance condition of the sinkhole. Representative aspects of the checklist are displayed in Figure 1. The checklist was made available to sinkhole owners, and to agencies such as the Conservation District, Greater Lebanon Refuse Authority, and others to provide guidance in assessing the potential extent of sinkhole problems relating to water quality management. In conjunction with the Risk Priority Assessment Checklist the Conservation District and associated agencies use Pennsylvania Farm*A*Syst materials for evaluating and managing water resources of farms in an environmentally sensitive manner (Penn State Cooperative Extension, 1997).

Sinkhole Status	4 Best	3 Good	2 Fair	1 Poor	Score
Size of Sinkhole	<1 foot in diameter	1<2 feet in diameter	2<3 feet in diameter	3+ feet in diameter	
Drainage area into sinkhole	1-2 acres	<3-5 acres	>5-15 acres	16+ acres	
Position of the sinkhole in relation to possible sources of contamination	Upslope from all sources of contamination, and all surface water is diverted away from the sinkhole.	Downhill from sources of contamination and surface water runoff is diverted from the opening.	------------------	Downslope from most sources of contamination, and some surface water runoff reaches the sinkhole.	
How has the sinkhole been maintained? *	An approved treatment plan has been implemented	Some sort of preventative action has been taken.	Nothing has been done to the sinkhole.	The sinkhole is filled with debris and garbage.	
Water Well Condition					
Type of well	Off site private or municipal source (therefore, site conditions or management practices do not affect well quality.	Drilled well	Driven well	Dug well (this has the potential for the maximum pollutants to enter into aquifers).	
Position of well in relation to potential sources of contamination? **	Upslope from all sources of contamination, and all surface water is diverted away from well.	Downhill from sources of contamination, and surface water runoff is diverted from casing.	Downslope from most sources of contamination, and some surface water runoff reaches well.	Well is located in a depression. Ponding occurs around the well during storm events.	
Distance between well and potential sources of contamination	More than 300 feet from a contamination source.	150-300 feet from a contamination source.	75-150 feet from a contamination source.	Less than 75 feet from a contamination source.	

* A "poor" evaluation to this question indicates a definite threat to the groundwater, and the need for immediate action.
**A "poor" evaluation to this question suggests a definite threat to drinking water, and constitutes a need for immediate action.

WORKSHEET AVERAGE = Σ Scores ÷ number of features. Score = average value for management, for comparison with the following values: 4.0 -3.6, best management; 3.5-2.6, good management; 2.5-1.6, fair management; 1.5-1.0, poor management.

Figure 1: Risk Priority Assessment Checklist for Sinkholes and Water Wells, illustrating two of four assessment categories.

SUCCESSES, FAILURES AND LESSONS LEARNED

The sinkhole management workshop was a major success. The large workshop had a wide array of participants, including representatives from all local municipalities, except one in Lebanon County, that have sinkhole problems. Letters from several private citizens, engineering firms, and out-of-state organizations expressed appreciation for the information that was provided, particularly for materials relating to sinkhole closures. A second scheduled workshop on water resources remediation was organized, but was canceled because of projected low attendance. Possible reasons for the failure were the absence of co-sponsors for the second workshop, and a schedule conflict. Several organizations held a water resources clean-up seminar during the same week.

During the project period, several private citizens with newly formed sinkholes requested three types of information: technical instructions for closing sinkholes, liability responsibilities of sinkhole owners, and sources of financial assistance for sinkhole closures. Helpful insights from authorities on the last two items were sparse, especially opportunities for financial assistance. In Pennsylvania, public moneys for sinkhole closures are available only if the sinkholes threaten public water supplies or other crucial community resources. In Iowa sinkholes apparently may be dealt with through a farmland conservation reserve program, but survey findings showed that landowners were unlikely to participate because they needed the land for economic viability and for other reasons (Huber, 1990). In past years in Lancaster County, PA, farmers in the Pequea – Mill Creek area had a special education program and free repairs of sinkholes under Section 319 Nonpoint Source funding (Andrews, 1994).

Several municipalities attending the sinkhole workshop subsequently requested public agency assistance in removing waste materials from sinkholes within their jurisdictions. One sinkhole, containing potentially harmful materials, was inspected, cleared, re-inspected, and closed. This occurrence demonstrated that municipalities within Lebanon County have an agency for enforcement of regulations designed to prevent groundwater contamination by sinkhole waste materials.

Some project events of the "3R's Series," including a seminar on sinkhole densities, had relatively low attendance. This pattern suggested two potential design drawbacks: an inappropriate schedule (i.e., a mid-afternoon hour was poorly supported, although the academic and business communities indicated initial enthusiasm for the time slot) and the need for additional sponsoring organizations to bolster publicity, networking, and attendance at gatherings.

The high level of interest and positive feedback from civic organizations that received the presentation entitled "Sinkholes of Lebanon County" were significant. These reactions stimulated the Greater Lebanon Refuse Authority to continue offering the program as a public education service.

Additional benefits of the educational and management achievements of the project have been to enhance working arrangements among local agencies charged with water resource protection and related issues, and to engender increased confidence among the public for these agencies in Lebanon County. Letters of appreciation and oral comments attest to this conclusion.

ACKNOWLEDGEMENTS

Activities were achieved through much time and expertise given by participating organizations and individuals. Documents for the workshops were provided by several organizations, including the United States Geological Survey, the Pennsylvania Department of Environmental Protection, and the Pennsylvania Geological Survey. The sinkhole project was partially funded by the Citizen Education Fund of the League of Women Voters of Pennsylvania under a grant from the Pennsylvania Department of Environmental Protection. Other principal resources were provided by the Greater Lebanon Refuse Authority, Lebanon County Conservation District, and the Math, Science and Allied Health Division of Harrisburg Area Community College. The Greater Lebanon Refuse Authority conducted overall project direction. The manuscript for this report was processed by Anna Wilson-Borges of Harrisburg Area Community College.

REFERENCES

Andrews, Andy, November 16 & December 3, 1994, Sinkholes potential environmental threats – strain farmers' patience, money (part 1 of sinkhole series) and There are solutions to sinkhole problems (part 2 of sinkhole series), Lancaster Farming, volume 40, nos. 3&4, 3+pp.

Berg, T.M., Edmunds, W.E., Geyer, A.R., and others, 1980, Geologic map of Pennsylvania, Pennsylvania Geological Survey, 4th Series map, scale 1: 280,000, 3 sheets.

Drewes, Donna, November 1998, North Jersey Resource Conservation and Development, personal communication.

Huber, Gary, April 1990, Landowner perceptions of sinkholes and groundwater contamination, Journal of Soil and Water Conservation, pp. 323-327.

Kochanov, William E., 1988, Sinkholes and karst-related features of Lebanon County, Pennsylvania, open file report 88-02, Bureau of Topographic and Geologic Survey, Department of Environmental Resources.

Penn State Cooperative Extension, 1997, Pennsylvania Farm*A*Syst, packet of worksheets and related materials, The Pennsylvania State University publ. 2M497PS, 26 pp.

Zokaites, Carol, October 1998, Project Underground, personal communication.

Irish methodologies for karst aquifer protection

DONAL DALY Geological Survey of Ireland, Beggars Bush, Dublin, Ireland
DAVID DREW Geography Department, Trinity College, Dublin, Ireland

ABSTRACT

Carboniferous limestones are the most extensive rock type in Ireland. They have a range of hydrogeological characteristics from fissure flow to wholly conduit flow dominated systems. Most of the limestones are overlain by variable thicknesses and types of Quaternary sediments.

The need to produce a national groundwater protection scheme that would (a) be relatively uncomplicated and therefore usable by non-specialists, (b) cover the full range of hydrogeological settings present in Ireland, and (c) show all the groundwater protection zones on one integrated map, has meant that the scheme is not based solely on the distinctive character, uncertainty and problems posed by karst aquifers. However, karst is included as an important element.

The national groundwater protection scheme is based on the concepts of groundwater contamination risk and risk management and integrates the four elements of risk – hazard, vulnerability, groundwater value and groundwater protection responses for potentially polluting activities. The mapping of groundwater vulnerability and groundwater value (in terms of aquifer category) is the basis of the land surface zoning component of the groundwater resources protection scheme. Recharge through karst features such as swallow holes and sinking streams is taken into account in vulnerability mapping, and areas around these features are classed as 'extremely' vulnerable. In producing aquifer maps, the distinction is drawn between limestones with and without a high degree of karstification. In the groundwater protection responses, the greatest degree of restriction is on developments on extremely vulnerable, regionally important karst aquifers, as this is the highest risk resource protection zone. At present, no special allowance is made for karst in delineating source protection zones, apart from specifying areas in the vicinity of karst features as 'extremely' vulnerable and categorising all the zone of contribution around wells and springs in areas with a high degree of karstification as an inner protection area.

It is recognised that the above methodology is not sufficiently discriminatory always to be able to identify fully the character of karst hydrogeological systems on a small scale. Therefore a second approach designed to complement the land surface zoning technique is presently being developed. This will assess the degree to which a carbonate aquifer functions karstically by calculating indices of karstification. The indices will then be combined with the data from the land surface zoning to yield more robust estimates of overall vulnerability.

INTRODUCTION

Recognition of the distinctive character of karsts and the specific problems posed in terms of protecting karstic groundwater is now widespread. For example, in Europe the Co-operation in Science and Technology programme COST65 (COST Action 65 1995) was dedicated to exchanging information on karst water protection practices on a pan-European basis, whilst its successor COST Action 620 is charged with developing karst-specific protection criteria.

Amongst the major problems typical of karst areas are:
♦ uncertainty concerning the nature of the karst system in a particular region (recharge-flow-discharge);
♦ the great spatial variations in karst aquifer properties;
♦ the great temporal variations in flow patterns.

Despite the recognition of the uniqueness of the karst hydrogeological regime, only limited progress has been made towards karst-specific methodologies of vulnerability assessment. The Swiss EPIK method is the only wholly karst oriented method available (Doerfliger 1997), but it was developed with the karsts typical of Switzerland in mind. (EPIK is an acronym for Epikarst, Protective cover, Infiltration conditions and Karstic network.) Elsewhere in Europe adaptations of existing vulnerability methods such as DRASTIC and SINTACS have been used in karst regions.

Karst specific management codes and karst specific legislation has probably been developed in the USA to a greater extent than elsewhere. Such developments have occurred at Federal, State and local levels and have been largely documented in the Sinkhole Symposia; for example, in contributions to the fifth multidisciplinary conference on sinkholes *Karst Geohazards* (Beck 1995).

This paper summarises the approach now used in Ireland and outlines a second approach which is being developed at present.

BACKGROUND TO GROUNDWATER PROTECTION IN IRELAND

In Ireland, limestones (of Mississippian age) underlie almost half of the land surface and up to 80% of the more economically significant lowlands. While they are the most important aquifers in Ireland overall, they vary in aquifer category from regionally important to poor. They are variable in their hydrogeological characteristics: they have permeability systems in a continuum between fine fractures and large conduits; permeability usually decreases with depth; the degree of solution/karstification is highly variable, yet solution has impacted on all the limestones to some degree (and therefore all can be said to be karstified). Conduit flow is a major component of groundwater flow in some circumstances but not in others where virtually all flow is in fractures. The hydraulic properties vary depending on the limestone lithology, degree of bedding, structural history and position within regional groundwater flow systems. However, knowledge of the extent of karstification is imperfect and the situation is further complicated by the fact that most of the limestones are overlain by variable and often unknown thicknesses of Quaternary sediments.

A national groundwater protection scheme (DoELG/EPA/GSI, in press) is being used as a means of assisting public authorities to meet their statutory responsibilities for the protection and conservation of groundwater resources, and of providing geological and hydrogeological information for the regulatory process, so that developments can be located and controlled in an environmentally acceptable way. As the protection scheme is being applied by regulators who have no specialist training in hydrogeology, it was decided that (a) the scheme must be relatively simple, (b) the hydrogeological principles on which the scheme is based must cover the full range of hydrogeological settings, and (c) the groundwater protection zones for an area must be capable of being shown on one integrated map. As all bedrock units are classed as aquifers in Ireland, the groundwater protection scheme covers the total land surface. The only subsoil classed as an aquifer is sand/gravel.

In view of the importance of karstification and karst aquifers in Ireland, the protection scheme specifically includes both. However, karst is incorporated to only a limited extent; limited because of the lack of available data, and also because the scheme is for use by non-experts and so must be both simple and comprehensive.

SUMMARY OF PROTECTION SCHEME

The national groundwater protection scheme is based on the concepts of groundwater contamination risk and risk management and integrates the four elements of risk – hazard, vulnerability, groundwater value and groundwater protection responses for potentially polluting activities.

Land surface zoning provides the general framework for the groundwater protection scheme. The outcome is a map, which divides any chosen area into a number of groundwater protection zones according to the degree of protection required. There are three main hydrogeological elements to land surface zoning:

♦ Division of the entire land surface according to the vulnerability of the underlying groundwater to contamination. This requires the production of a vulnerability map showing up to four vulnerability categories.
♦ Delineation of protection areas surrounding groundwater sources (usually public supply sources); these are termed source protection areas.
♦ Delineation of areas according to the value of the groundwater resources or aquifer category; these are termed resource protection areas.

These three elements are integrated together to give maps showing groundwater protection zones; source protection zones and resource protection zones.

The location and management of potentially polluting activities in each groundwater protection zone is by means of a groundwater protection response matrix for each activity or group of activities, which describes (i) the degree of acceptability of each activity, (ii) the conditions to be applied and, in some instances, (iii) the investigations that may be necessary prior to decision-making.

While the two components – maps showing the zones and the codes of practice – are different, they are incorporated together and closely interlinked in the scheme.

The protection scheme covers the full range of hydrogeological settings present in Ireland, varying from sand/gravel with intergranular flow to limestones with karstic flow. All four elements – vulnerability, source protection areas, aquifer categorisation and groundwater protection responses – take account of karst.

VULNERABILITY MAPS

Vulnerability is a term used to represent the intrinsic geological and hydrogeological characteristics that determine the ease with which groundwater may be contaminated by human activities (Daly and Warren, 1998).

The factors determining the vulnerability assessment process in Ireland are as follows:
♦ The target at risk is taken as the first groundwater encountered in bedrock and sand/gravel aquifers (i.e. the top of the aquifer where all the aquifer is saturated or the water table in the aquifer where there is an unsaturated zone). Therefore, the target is not the water issuing from a spring or abstracted from wells, water in the discharge zone of aquifers or water in tills.
♦ The bedrock in Ireland is not taken to be a significant factor in attenuating pollutants or in defining vulnerability. This is because once contaminants enter the bedrock there is little attenuation other than by dilution (usually relatively limited) when

groundwater is reached, owing to the fissure permeability that characterises Irish bedrock. Therefore, the presence of an unsaturated zone in bedrock is assumed not to be a significant factor.
- As a consequence of the above points, the reference location that is of interest in defining the vulnerability is the top of the bedrock, in the case of bedrock aquifers, and the water table in the case of unconfined sand/gravel aquifers.
- In Ireland, point sources have caused most of the contamination problems to date. Consequently, in mapping the vulnerability, it is assumed that contamination more frequently originates below the soil (topsoil) zone and therefore that the release point of contaminants into the geological environment is 1-2 m below ground level (bgl). (However, where the release point is at the ground surface from diffuse sources, this is taken into account in the groundwater protection responses.)
- Therefore in mapping the vulnerability of bedrock aquifers, the pathway for contaminants is from 1-2m bgl to the top of the bedrock. Vulnerability is assessed on the basis of the (largely) vertical transport of contaminants to the top of the bedrock.
- In mapping the vulnerability of bedrock aquifers (including karst aquifers), the elements considered are (a) the thickness and permeability of the subsoils and (b) the type of recharge (diffuse or point, the latter occurring via surface karst features).
- Therefore, the only specifically karst factor taken into account in mapping vulnerability is recharge through karst features such as swallow holes, sinking streams and collapse features. Groundwater is classed as 'extremely' vulnerable within 30 m of karstic features (including along the area of loss of losing or sinking streams) and within 10 m on either side of losing streams upflow of the area of loss. The distances can be varied depending on the circumstances - for instance, they are increased where overland surface runoff is likely.
- The vulnerability of groundwater is assessed according to four classes: extreme (E), high (H), moderate (M) and low (L). Further details are given in Table 1.

Table 1   Vulnerability Mapping Guidelines

Vulnerability Rating	Hydrogeological Conditions				
	Subsoil Permeability (Type) and Thickness			Unsaturated Zone	Karst Features
	high permeability (*sand/gravel*)	moderate permeability (e.g. *sandy subsoil*)	low permeability (e.g. *clayey subsoil, clay, peat*)	(*sand/gravel* aquifers <u>only</u>)	(<30 m radius)
Extreme (E)	0 - 3.0 m	0 - 3.0 m	0 - 3.0 m	0 - 3.0 m	–
High (H)	>3.0 m	3.0 - 10.0 m	3.0 - 5.0 m	>3.0 m	N/A
Moderate (M)	N/A	>10.0 m	5.0 - 10.0	N/A	N/A
Low (L)	N/A	N/A	>10.0 m	N/A	N/A
Notes: i) N/A = not applicable. ii) Precise permeability values cannot be given at present. iii) Release point of contaminants is assumed to be 1-2 m below ground surface.					

GROUNDWATER SOURCE PROTECTION ZONES

Two source protection areas (SPAs) are delineated around each public supply well and spring: an Inner Protection Area (SI), defined by a 100-day travel time within the aquifer from any point below the water table to the source, and an Outer Protection Area (SO), encompassing the source catchment area or zone of contribution (ZOC). The Inner Protection Area aims to protect the source against the effects of human activities that might have a rapid effect on the source, especially microbiological pollution. The Outer Protection Area covers the total source catchment area, or zone of contribution to a well or spring. The SPAs are combined with the vulnerability map to give the groundwater source protection zones, as shown in Table 2.

The boundaries of the SPAs are based on the horizontal flow of water to the source, whereas the vertical movement of water and contaminants to the bedrock aquifer is taken into account in the vulnerability rating.

There is no difference in the general principles used in delineating source protection zones in karst and non-karst areas. However, karstification is taken into account, firstly because karst features are classed as 'extremely' vulnerable areas and secondly, all the area of karst limestone in the ZOC is classed as an inner protection area (SI), as travel times karst sources are generally rapid. (Numerical modelling is not used to delineate the SI area for karst sources.)

GROUNDWATER RESOURCE PROTECTION ZONES

For any region, the area outside the source protection zones is classified into one of three basic categories depending mainly on their permeability, areal extent, and storage capacity. Each category is further split into two or three sub-categories, according to the type and variability of permeability:
- Regionally Important Aquifers (R): These are subdivided into 'Rg' (sand/gravel aquifers), 'Rk' (karst limestone aquifers) and 'Rf' (aquifers with fissure flow).

- Locally Important Aquifers (L): These are subdivided into 'Lk' (karst aquifers), 'Lm' (Generally Moderately Productive), 'Ll' (Productive only in Local Zones) and 'Lg' (Sand/Gravel).
- Poor Aquifers (P): These are sub-divided into 'Pl' (Generally Unproductive except for Local Zones) and 'Pu' (Generally Unproductive).

Groundwater protection zone are produced by combining the vulnerability and the aquifer categories (see Table 2). Each protection zone has a code which represents both the groundwater value (use for public supply or aquifer category) and the groundwater vulnerability. Thus, for any site or area, a groundwater protection zone category exists which influences the degree of protection required for the groundwater beneath that site.

There are gradations in the degree of karstification in Ireland from slight to intensive. Where karstification is slight, the limestones are similar to fissured rocks and are classed as Rf, although some karst features may occur. Aquifers in which karst features are more significant are classed as Rk. Within the range represented by Rk, two sub-types are distinguished, termed Rk^c and Rk^d.

Rk^c are those aquifers in which the degree of karstification limits the potential to develop groundwater. They have a high 'flashy' groundwater throughput, but a large proportion of flow is concentrated in large conduits. Numerical modelling using conventional programs is not usually applicable, well yields are variable with a high proportion having low or minimal yields, large springs are present, storage is low, locating areas of high permeability is difficult and therefore groundwater development using bored wells can be problematical.

Rk^d aquifers are those in which flow is more diffuse, storage is higher, there are many high yielding wells, and development of bored wells is less difficult. These areas also have caves and large springs, but the springs have a more regular flow.

While all three regionally important limestone aquifers – Rf, Rk^d and Rk^c – are shown on the aquifer maps (where all are present), only Rf and Rk aquifers are shown on the groundwater protection zone map. It was concluded that the distinction between the two types of karst was for the purpose of indicating groundwater development potential, and that the difference would not affect the groundwater protection responses.

Table 2 Matrix of Groundwater Protection Zones

VULNERABILITY RATING	SOURCE PROTECTION		RESOURCE PROTECTION Aquifer Category						
			Regionally Imp.		Locally Imp.		Poor Aquifers		
	Inner	Outer	Rk	Rf/Rg	Lk/Lm/Lg	Ll	Pl	Pu	
Extreme (E)	SI/E	SO/E	Rk/E	Rf/E	Lm/E	Ll/E	Pl/E	Pu/E	↓
High (H)	SI/H	SO/H	Rk/H	Rf/H	Lm/H	Ll/H	Pl/H	Pu/H	↓
Moderate (M)	SI/M	SO/M	Rk/M	Rf/M	Lm/M	Ll/M	Pl/M	Pu/M	↓
Low (L)	SI/L	SO/L	Rk/L	Rf/L	Lm/L	Ll/L	Pl/L	Pu/L	↓
	→	→	→	→	→	→	→	→	

In Table 2, the arrows on the matrix indicate directions of decreasing risk: the arrows from top to bottom indicate decreasing likelihood of contamination (decreasing vulnerability), and the arrows from left to right indicate decreasing consequences of contamination (decreasing value of the target).

GROUNDWATER PROTECTION RESPONSES

Risk management for potentially polluting activities within the groundwater protection scheme is by means of groundwater protection responses for each activity. Groundwater protection responses set out recommended responses to development, showing (a) whether such a development is likely to be acceptable, (b) what further investigations may be necessary, and (c) what planning or licensing conditions may be necessary. They take account on the elements of risk, i.e. the hazard posed by the proposed action, the groundwater vulnerability, the location relative to wells and springs, and the aquifer category. Therefore, karstification is taken into account in deciding on the responses. Restrictions on most developments are greater in karst areas than elsewhere, except where the vulnerability is classed as low or moderate.

Four levels of response (R) to the risk of a potentially polluting activity are used:

R1 Acceptable subject to normal good practice.

$R2^{a,b,c,\ldots}$ Acceptable in principle, subject to conditions in note a,b,c, etc. (The number and content of the notes may vary depending on the zone and the activity).

$R3^{m,n,o,\ldots}$ Not acceptable in principle; some exceptions may be allowed subject to the conditions in note m,n,o, etc.

R4 Not acceptable

Table 3 shows a groundwater protection response matrix where the hazard is landspreading of organic wastes. Response matrices have also been prepared for landfills and on-site wastewater treatment systems.

Table 3 Response Categories for Landspreading Activities

VULNERABILITY RATING	SOURCE PROTECTION		RESOURCE PROTECTION Aquifer Category					
			Regionally Important (R)		Locally Important (L)		Poor Aquifers (P)	
	Inner	Outer	Rk	Rf/Rg	Lm/Lg	Ll	Pl	Pu
Extreme (E)	R4	R4	R3²	R3²	R3¹	R3¹	R3¹	R3¹
High (H)	R4	R2¹	R1	R1	R1	R1	R1	R1
Moderate (M)	R3³	R2¹	R1	R1	R1	R1	R1	R1
Low (L)	R3³	R2¹	R1	R1	R1	R1	R1	R1

(from DoELG/EPA/GSI, in press)

In Table 3, the response $R3^1$, for example, indicates that the proposal is not generally acceptable in this hydrogeological setting, unless there is a consistent minimum thickness of 1 m of soil and subsoil. Response $R3^2$ is similar, except that a consistent minimum thickness of 2 m of soil and subsoil is required over bedrock. This greater thickness is in response to the greater value of regionally important aquifers relative to locally important and poor aquifers, and to the faster velocities in regionally important bedrock (particularly limestone) aquifers. This latter point is important because one of the main hazards from landspreading of untreated wastes is provided by faecal bacteria and viruses.

The following general groundwater protection response relates specifically to landspreading on karst:
In karst areas farmers must take account of karst features such as swallow holes, caves, streams connected to karst systems, etc. Karst features can occur in any limestone rock type irrespective of aquifer category. Landspreading within 30m of karst features is not permitted.

FURTHER DEVELOPMENT OF THE IRISH METHOD OF VULNERABILITY ASSESSMENT

The Irish method of determining vulnerability is similar to the Swiss EPIK method referred to previously, but it omits the epikarst factor and takes only limited account of the nature of the karst system, as the target is the watertable rather than the discharge point. The method is essentially *causally* based, assessing factors (e.g. protective cover, recharge type) that are likely to increase or decrease pollutant attenuation and/or transit time. However, the extent to which these parameters actually operate as assumed is uncertain. For example, macro-pores may increase the permeability of a 'protective' cover, whilst dolines may operate as concentrated infiltration routes, or if clay infilled, as low permeability features. Also, layering methods are data-demanding and data of adequate quality are not always available. Thirdly, it is not obvious what degree of mapping resolution is required to deliver a vulnerability map of a specified reliability.

It is proposed to develop a second-pass method of vulnerability assessment, complementary to the existing method and designed to be applied in cases where detailed discrimination is essential. This attempts to establish an index of degree of karstification for an area. It is therefore an *effect* based method, directly measuring the degree to which karstification has taken place and the degree to which the hydrogeological system functions karstically.

The indicators used to determine an index of karstification are detailed in Drew, Burke and Daly (1996). It is envisaged that the indicators will be combined into three groups with a single, composite index being ascribed to each group. An overall Index of Karstification (Ki) will then be derived by combining the three sub-indices.

The groups of indicators will be:
- Degree of development of surface karst landforms (e.g. dolines, karren)
  (non-existent to well developed and abundant)
- Characteristics of the hydrological system (e.g. recharge mode, degree of cave development, storage)
  (diffuse flow systems to sink-conduit dominated systems)
- Spring outflow characteristics (e.g. chemograph-hydrograph form, system memory)
  (Time invariant flow, and water quality to flashy characteristics/low system memory)

In theory, a low Ki index for an area should correspond with a low vulnerability index as determined via the existing methodology (protective cover + recharge type) for a particular area. The eventual use of such a combined method should generate a more robust methodology for the determination of karstic groundwater vulnerability.

CONCLUSIONS

The Irish methodology for groundwater protection is a pragmatic approach intended to be both easy to comprehend and to implement. It is based on land surface zoning, which indicates the general degree of risk posed by hazards to groundwater, and groundwater protection responses for different hazards, with the levels of response depending on the degree of risk. The singularities of karst hydrogeology are incorporated into the scheme to some extent. It is recognised however, that reliance on a Protective Cover-Infiltration Mode vulnerability zoning method and the relatively simple subdivision of carbonate aquifers into three categories (Rf, Rk^c, Rk^d) is restricted both in terms of spatial resolution and in terms of its ability to identify complex karst systems. A complementary karst index method is being developed to address the above mentioned deficiencies and to take account of the often site-specificity of karst locations.

ACKNOWLEDGEMENTS

The influence of discussions with Catherine Coxon, Paul Johnston and Bruce Misstear, TCD; Gerry Carty and Margaret Keegan, EPA; and colleagues in the GSI Groundwater Section, in particular Geoff Wright, Jenny Deakin, Bob Aldwell and William P. Warren, is gratefully acknowledged.

This paper is published with the permission of Dr. P. McArdle, Director, Geological Survey of Ireland.

REFERENCES

Beck B.F. (Editor), 1995. Karst Geohazards. Engineering and environmental problems in karst terrane. Proceedings of the 5th multidisciplinary conference on sinkholes and the environmental impacts of karsts. A.A. Balkema, Rotterdam, 581p.

COST-Action 65, 1995. Karst groundwater protection. Final report - European. Commission, Report EUR 16547 EN, Brussels-Luxembourg, 246 p.

Daly, D. and Warren, W.P. 1998. Mapping groundwater vulnerability: the Irish perspective. In Robins, N.S. (Ed.) Groundwater Pollution, Aquifer Recharge and Vulnerability. Geological Society Special Publication No. 130, 179-190.

DoELG/EPA/GSI, in press. Groundwater protection schemes. Department of Environment and Local Government, Environmental Protection Agency and Geological Survey of Ireland.

Doerfliger N. and Zwahlen F., 1997. EPIK: A new method for outlining of protection areas in karstic environment. In Gunay, G. and Johnson, I. (Eds.), Karst waters Environmental Impacts. Balkema, Rotterdam, p117-124.

Drew D.P., Burke, A. M. and Daly, D. 1996. Assessing the extent and degree of karstification in Ireland. In Rozkowski A. (Ed.) Proceedings of International Conference on Karst Fractured aquifers - Vulnerability and Sustainability, Katowice-Ustron, Poland *1996*, p 37-47.

Maryland's zone of dewatering influence law for limestone quarries

MOLLY K. GARY   MD Department of the Environment, Baltimore, Md., USA

ABSTRACT

In 1991 the Maryland General Assembly passed legislation requiring the Maryland Department of Environment (MDE) to establish zones of dewatering influence (ZOI) around limestone quarries in karst terrane and to administer a program requiring the mining company to mitigate or compensate affected property owners within the zone. The mining industry challenged the constitutionality of the statute and was successful in obtaining injunctive relief preventing MDE from establishing any ZOI. The statute was ultimately upheld and in December 1994 the Department began the implementation phase. The law applies to limestone quarries in four counties (Baltimore, Carroll, Frederick, and Washington) where the quarries have an appropriation permit to dewater in order to facilitate mining. Within the ZOI, the mining company is responsible for replacement of water supplies (domestic, municipal and industrial) which fail due to declining water levels and to repair damage to real property occurring as a result of sudden land subsidence as determined by MDE. In the instance of a water supply failure the quarry must replace a failed water supply unless they can prove by clear and convincing evidence that they were not the cause of the failure. In the instance of a sinkhole the Department will investigate the proximate cause of the sinkhole and make a determination whether quarry dewatering caused the sinkhole. Regulations are currently being promulgated to define the quarries' responsibilities within the ZOI. A team approach is used in defining the ZOI. A plan is organized by MDE staff geologists and hydrologists to identify the data needed to determine the ZOI for each site. There is no set method for defining a ZOI, the geology and hydrology of each site is unique and presents a special challenge as each ZOI is initiated. Information is collected from field geology and mapping, dye tracing, collection of existing published data and cultural information, stream gauging and all other pertinent information. The information gathered is used to prepare a written report outlining the data collected and the support for defining the zone. Maps delineating the zone are prepared using digitized information from Geographic Information System (GIS) and Computer Aided Drafting (AUTOCAD). Once a ZOI has been determined for a quarry, the mining company has an opportunity to review and comment on the proposed ZOI. The company can either accept the zone as defined by MDE or appeal the location. As each ZOI is completed it is made a part of the surface mine permit via Departmental modification. Conditions specific to the ZOI are added to the permit requiring the permittee to provide proof of liability insurance and to provide a water supply replacement plan within an established amount of time from issuance of the modification. As of September 1998, ZOI's have been established for six (6) of the existing eighteen (18) quarries covered by this law. An additional four (4) ZOI are in various stages of completion and one (1) new quarry has proposed a ZOI as a part of their permit application. All new quarry applications and modifications to existing quarries will require the mining company to propose a ZOI for the quarry to be reviewed by the Department.

INTRODUCTION

In 1991 the Maryland General Assembly passed legislation requiring the Maryland Department of Environment (MDE) to establish zones of dewatering influence (ZOI) around limestone quarries in karst terrane and to administer a program requiring the mining company to mitigate or compensate for damage due to quarrying within the zone. The law applies to limestone quarry operations in four counties where the quarries are dewatered to facilitate mining. Other limestone areas of the State are excluded because based on topography they require minimal dewatering to facilitate mining. Non-limestone quarry operations and sand and gravel pits are also excluded from the law. Regulations are currently being promulgated which will assist MDE in implementing the law. These regulations have been through several drafts; industry input and input from local citizens groups was solicited during the writing and revision of the regulations.

Since the mid-1980's rural areas of the State have been rapidly developing, reflecting urban sprawl outside the Baltimore-Washington business corridors. The increase in development and road building increases the demand for aggregate. Existing quarries have expanded to meet this demand, while at the same time they are being encroached on by urban sprawl. The limestone formations mined for aggregate are commonly the primary aquifer used for local water supplies. Sinkhole activity is common in portions of Baltimore, Carroll, Frederick and Washington counties. All citizens are concerned about the adequacy of their water supplies and protection of their property. With the prospect of new quarries opening and existing ones expanding, citizens sought protection from the perceived increase in sinkhole occurrence and affect on water supplies resulting from an increase in mining activity.

The Surface Mining Act (§15-801 through 834 Environment Article Annotated Code of Maryland) was amended in 1991 to protect property owners from well failures and unexpected damage to structures and real property arising from the occurrence of sinkholes caused by quarry dewatering. The law directs MDE to establish zones of dewatering influence around mining operations in Baltimore, Carroll, Frederick and Washington counties. The mining company is responsible for replacement of water supplies (domestic, municipal, and industrial) which fail due to declining water levels and to repair damage to real property as a result of sudden land subsidence (sinkholes) determined by MDE to be caused by quarry activity. This report will describe the elements of the law and proposed regulations and MDE's implementation of the law.

ELEMENTS OF ENVIRONMENT ARTICLE §15-812 And §15-813 ANNOTATED CODE OF MARYLAND
Legislative Intent, Criteria for definition of ZOI
" The General assembly finds that in certain regions of the State dewatering of surface mines located in karst terrain may significantly interfere with water supply wells and may cause in some instances sudden subsidence of land, known as sinkholes. Dewatering in karst terrain may result in property damage to landowners in a definable zone of dewatering influence around a surface mine."
[Environment Article §15-812 (a)]

MDE must define the ZOI for any mining operation that has a Water Appropriation permit for pit dewatering and a Surface Mining Permit. The law specifies that

"the areal extent of the zone of dewatering influence shall be based, as appropriate, on local topography, watersheds, aquifer limits, and other hydrogeologic factors, including the occurrence of natural fractures, cracks, crevices, lineaments, igneous dikes, changes in rock type, and variations in water-bearing characteristics of formations
[Environment Article §15-813(b)(2)].

Once the ZOI has been established as a condition of the surface mining permit, the mining company is subject to the requirements described below.

Loss of water supply
In the event of a well failure, the mining company is presumed liable if the well is located within the ZOI. The mining company must replace the water supply, unless they can prove based upon clear and convincing evidence that the proximate cause of the water supply loss is not the result of quarry dewatering. In the case of a domestic well, the replacement well, retrofitted well or water supply must meet the minimum yield requirements established in Maryland regulations (COMAR 26.04.04.07). A municipal, industrial, commercial, institutional or farming water supply is considered replaced adequately if a new or retrofitted well or other alternative water supply is capable of yielding water equal to the volume used or needed by the property owner before the disruption of the supply.
[Environment Article §15-813 (c)]

Sinkhole damage
In the event of real or personal property damage caused by sudden land subsidence, MDE must determine if the proximate cause of the land subsidence was the result of quarry dewatering. To determine the proximate cause of the sinkhole, MDE will conduct an investigation evaluating the geologic features of the sinkhole, topography, surrounding land uses, weather data, and site specific conditions. The mining company also has the opportunity to provide MDE with information concerning the cause of the sinkhole.

An example of the difficulties in determining the proximate cause of a sinkhole is demonstrated by the investigation of the catastrophic sinkhole that occurred near Westminster, MD on March 31, 1994. The sinkhole opened up in the middle of a State road and claimed a man's life, after he drove into it in the dark. Within 1 mile of the sinkhole, are two municipal water supply wells and an active quarry operation. The sinkhole occurred in an area mapped on the geologic map as metabasalt. The sinkhole measured approximately 26½ feet by 20 feet, and was 15 feet deep according to the author and other MDE personnel who examined the sinkhole on the day it opened. The sinkhole was quickly filled and the road repaired leaving little time for investigation. The sinkhole was re-opened in October 1994 to allow an engineered repair of the roadway. The re-excavation of the sinkhole revealed an isolated pinnacle of limestone in the center of the roadway alignment.

In cooperation with the Maryland State Highway Administration, a dye trace was conducted under the direction of Dr. James Quinlan of Quinlan and Associates, Inc. The purpose of the study was to determine if there was a hydraulic connection between the sinkhole and the quarry or other pumping locations. Sampling stations were placed throughout the surrounding valley and in a nearby quarry, which is down gradient from the injection site and sinkhole. There was no dye recovered in the sample sites either upgradient or downgradient of the injection site. Therefore, there is no conclusive evidence that quarry dewatering was the cause for the sinkhole.

The sinkhole fatality resulted in a negligence and wrongful death suit brought by the family against the mining company, the State Highway Administration and the Maryland Department of the Environment. While the ZOI does not address personal injury only property damage, the plaintiffs sought to prove that the Route 31 sinkhole was within the ZOI. Since the sinkhole precluded the ZOI, MDE had not made a determination as to whether the sinkhole formed within the ZOI. The plaintiffs believed it to be beneficial to their case to demonstrate that the sinkhole was within the ZOI. While the matter was not adjudicated, some of the difficulties the plaintiffs were having in developing evidence of a hydrologic connection between the sinkhole and the quarry serves as guidance to MDE in the establishment of future ZOIs. Historical water level data is a key element in determining impacts due to dewatering and the importance of verifying the accuracy of data used became very clear. When using monitoring well data to help determine the ZOI it is very important that there are sufficient properly placed monitoring points around the site. A discontinuity in the monitoring network raises questions about the reliability of using that information to define the ZOI when a data gap remains. It was also noted that the ZOI should be able to be readily identifiable in the field, the use of GPS and other mapping techniques may help with this verification. The suit was settled out of court with all parties involved, if the case had gone to trial the items previously identified may have posed significant evidentiary hurdles to overcome.

In a case where MDE's investigation into a sinkhole determines that the sinkhole was the result of quarry dewatering, the mining company must pay monetary compensation to the affected property owner, or repair the damage caused by the sudden subsidence. The property is considered repaired if it is restored to its condition before the subsidence of the surface of the land. If the property is not capable of being restored to a pre-subsidence condition, the mining company must monetarily compensate the owner. The amount to be compensated is the difference between the fair market value of the property before subsidence and the fair market value following subsidence. The mining company and the property owner may also agree on monetary compensation or other mitigation in lieu of restoration.

Regulations COMAR 26.21.01.03 (proposed)

The regulations will further define the responsibility of the Department in defining the ZOI and the responsibility of the company after establishment of the ZOI. The regulations will allow the Department to require the permittee to submit additional information that may be beneficial in determining the ZOI. Such information includes monitoring well data, boring logs, information on known sinkhole occurrences, aquifer pump tests and other information deemed necessary by the Department. The regulations will also provide for specific permit conditions dealing with the ZOI. Such conditions may include the requirement for the company to establish a monitoring well network in the vicinity of the quarry. The company may be required to measure stream flow and precipitation on a continuing basis in the vicinity of the quarry. The permittee may be required to report all sinkholes that occur within the ZOI to the Department. The permittee will be required to maintain liability insurance and provide a water supply replacement plan to the Department. These two items will be required within a specified timeframe after the modification of the surface mine permit.

The regulations will lay out specific actions to be taken by the permittee when a sinkhole opens up within the ZOI and when a water supply fails. The regulations will most certainly provide for public notice for persons affected by the ZOI and provide for a public informational hearing.

Exceptions

Compensation, restoration, or mitigation does not apply to improvements that are made to real property within a zone of influence following the decision of the Department to issue a new surface mine permit, or; improvements that are made to real property following the establishment of a zone of dewatering influence as a condition of an existing permit. These exceptions are outlined in, Environment Article §15-813 (I) Annotated Code of Maryland.

Contested case hearing

Once MDE has defined the ZOI for a quarry the zone is presented to the mining company for comment. Upon approval by the Department of the location of the ZOI, a public informational hearing is conducted to allow review of the zone by the public. A mining company or person with legal standing to sue can challenge the final decision of the Department on the location of the ZOI through a contested case hearing. If the aggrieved party has standing and can show that there is a substantial likelihood of prevailing on the merits of the case, a temporary stay may be granted. A stay may only be granted if the delay would not harm public health or safety or cause environmental harm to natural resources.

DEFINING THE ZONE OF INFLUENCE FOR A QUARRY

The ZOI will encompass an area of responsibility for the quarry in which they are accountable for water supply failures and sinkholes that are shown to be the result of quarry dewatering. Research for the ZOI includes an investigation of existing data and permit files within MDE, published data, and geologic literature. Local governments, the State Highway Administration and soil conservation districts are contacted to compile information they have, which would prove useful to the study. Interviews with local citizens may provide valuable data regarding historical sinkhole occurrences and past water supply problems.

A team approach is used in outlining the strategy and methods to be used in defining a particular zone. Based on the relevance and detail provided in the available published information, a determination is made of additional information required to designate an accurate ZOI. A dye trace study may be designed; local streams gauged and well levels measured. Extensive fieldwork is done to verify existing sinkholes, local geology and important topographic features. Streams, springs and seeps are identified for potential use as monitoring points in a dye trace and for the hydrologic information they provide. All of the information is compiled into a report describing the proposed ZOI and information used in its' determination.

Maps showing the limits of the zone of influence are prepared using digitized information from GIS and Computer Aided Drafting (AUTOCAD) systems. At a minimum the maps will show property lines, topography, geology, quarry permit limits, well locations, and the proposed ZOI. The map may also show introduction and sampling points for a dye trace.

Prior to the ZOI being presented to the mining company it is reviewed by staff from the Maryland Geological Survey and other professional staff within the Department of the Environment. After the ZOI is reviewed in-house it is presented to the mining company for their review and comment. This allows the mining company to discuss any concerns with the Department and present any information they may have not previously shared.

A public meeting is held in the vicinity of the quarry for each proposed ZOI. All property owners within and adjacent to the proposed ZOI are notified in writing of the hearing date, time and location and a map depicting the proposed ZOI is provided to them. After the public meeting and when all supplemental information is taken into account the ZOI is made a part of the Surface Mine Permit via Departmental Modification and special conditions are added to the permit pertaining to responsibilities of the permittee.

Figure 1: Typical collapse sinkhole used as a dye introduction point. Frederick County, Maryland 1998

Each time a Surface Mine permit is modified or when new information is obtained the ZOI may be modified at the discretion of the Department. In instances where there is a monitoring well network set up for a quarry information may be provided by the ongoing data collection that provides important information on the affect of the quarry dewatering on the surrounding area. Pump tests required for other permits might provide valuable information on drawdown of the water table. As the quarry deepens or expands laterally there may be additional information that can be gathered that was not previously available.

IMPLEMENTATION TO DATE

As of December 1998, six (6) ZOI have been established, one in Baltimore county, one in Carroll county, one in Frederick county and three in Washington county. Four (4) additional ZOI are in various stages of completion, these are located in Baltimore, Carroll and Frederick counties. As the ZOI law has been implemented the method of determining zones has evolved from mostly literature search to extensive fieldwork and data collection including dye tracing. Department personnel have received training in dye trace technique and have been certified on the Shimadzu 5301 spectrofluorophotometer recently purchased by the Mining Program. To date the quarry operators have been increasingly willing to work with the Department in supplying geologic and hydrologic information and assisting with dye trace studies by supplying water trucks to flush dye and providing easy access to sample points.

While there is frequently extensive literature available for quarry sites, field investigation allows MDE to look at the specifics of a site with an eye toward dewatering impacts. GPS systems have been used to more accurately locate pertinent features as opposed to digitizing features from existing maps. The use of Arcview as a GIS application has aided in the compilation of cultural information saving hours of digitizing from tax maps. The increasing cooperation between the Department and industry has made all of the field research undertaken easier and has provided otherwise unavailable information to the Department. The team approach to defining ZOI is beneficial particularly in light of the extensive ZOI research required and the increasing responsibilities of Department personnel.

As a requirement of the modification of the surface mine permit for the ZOI, each site must provide a water supply replacement plan to the Department. This plan is reviewed and approved after which it becomes a part of the surface mine permit. The water supply replacement plan outlines the measures to be taken by the quarry operator in the instance of a water supply failure within the ZOI. This helps ensure a speedy supply of potable water to the affected property owner while the cause of the loss of water supply is investigated. It is important to ensure that the loss in water supply is not the result of a pump failure or other mechanical problem. While quarries are presumed liable for water supply loss within the ZOI, it is advisable that a quarry investigate the cause since drought conditions can cause temporary water supply failures, old wells can collapse and there are always opportunists looking to take advantage of a situation.

Not long after implementation of a particular ZOI an affected property owner within the ZOI notified the Department that their well no longer had ample capacity to supply water to their home. The quarry operator in question was notified and after determination that there was no mechanical problem with the well, a new well was drilled for the property owner in a timely fashion and at the quarries expense.

CONCLUSIONS

The Maryland Zone of Influence law provides property owners in the vicinity of quarries a mechanism to seek compensation for well failures and property damage that occurs as a result of quarry dewatering. This law provides an additional level of protection to those property owners within the ZOI. No rights or privileges are taken away from the property owner by this law. Each site will have distinct and unique geologic and hydrologic factors that will influence determination of the ZOI. The ZOI determinations will be ongoing as further data is collected and evaluated.

The comments made in this paper have been made to provide general information within a professional educational forum and not to provide specific advice to readers of the paper. If any of the material presented in this paper is applicable specifically to the reader, the reader should seek legal counsel to evaluate his specific facts in relation to possible legal action.

STATUTE
Environment Article §15-812-813 Annotated Code of Maryland.

Regulating construction of manure storage systems in sinkhole prone areas of Minnesota

DAVID B. WALL   Minnesota Pollution Control Agency, St. Paul, Minn., USA

ABSTRACT

Construction of large liquid manure storage systems has greatly increased during the past decade in the karst region of southeastern Minnesota. Soil subsidence under a liquid manure storage system could breach the integrity of the liner, causing either a catastrophic release of manure to ground water or a slow undetected manure seepage problem. The probability of soil subsidence varies greatly across this region of the state. Construction of new liquid manure storage systems in higher risk areas for sinkhole formation creates heightened concerns about water quality protection. To minimize the risks of siting new manure storage systems in the karst region, the Minnesota Pollution Control Agency has developed and implemented a policy to evaluate relative risks of soil subsidence prior to approving feedlot construction permits. Permitting decisions depend on the results of a site-specific karst investigation, the proposed volume of manure to be stored and the type of liner proposed for the storage system. Precautionary measures required of some livestock producers have included one or more of the following: locating the feedlot in a less vulnerable area, using less permeable liner materials, and rerouting roof runoff waters.

NEW MANURE STORAGE SYSTEMS IN THE KARST REGION

Trends

The portion of Minnesota most susceptible to sinkhole formation is limited mostly to nine counties in the southeastern part of the state. Livestock agriculture is a dominant land-use in these counties, with regional livestock numbers exceeding 500,000 cattle and 600,000 hogs. While total livestock numbers have not changed significantly during the past decade, there has been a trend of consolidating livestock into larger operations. In this part of the state, the number of permits issued for feedlots expanding to over 300 animal units (e.g. 750 hogs, 300 beef cattle or 215 dairy cows) has increased from less than 10 per year in 1990 to over 100 per year in 1998. Most of the new larger facilities in this region are swine and dairy operations. To store the manure produced at these facilities, the most common practices are to construct reinforced concrete (R/C) structures below total confinement barns for liquid swine manure and earthen basins lined with cohesive soils for liquid dairy manure. The size of R/C structures to hold manure for over 1000 hogs is usually 0.2 to 0.5 acres, holding 350,000 to over a million gallons. It is common practice to construct two to four R/C structures at one location. Manure storage basins for the new larger dairy operations are commonly one to four acres, holding between two to ten million gallons. Both structures are typically designed to hold manure produced throughout one year, or more.

Construction Practices

Since 1991, R/C structures over 500,000 gallons and all earthen basins have been required to be prepared by or under supervision of a registered professional engineer and meet construction requirements limiting seepage and liner damage (see Minnesota Pollution Control Agency, 1997; Minnesota Pollution Control Agency, 1998). The plans and construction reports are reviewed by professional engineers at the Minnesota Pollution Control Agency (MPCA) prior to construction. Despite the precautionary measures to minimize seepage from manure storage systems, it is recognized that standard construction of R/C structures and cohesive soil liners will not prevent all liquids from seeping from the manure storage system. The seepage rates will vary with soil type, liner construction, pit/basin age, manure properties, climate, management and maintenance. In certain sensitive areas it has become more common in recent years to overlay cohesive soil liners with a synthetic liner or geosynthetic liner. Some R/C structures have been constructed using liquid seals, more commonly know as "waterstops," to further reduce the amount of seepage liquids.

Benefits to Water Quality

Livestock agriculture has some water quality benefits in the karst region that help to offset some of the risks to water quality. Manure applied to land planted to row crops can reduce soil erosion. Hay-land and pasture associated with cattle operations results in very little soil erosion and pesticide transport in this region of steeply sloping soils.

The trends to construct new larger feedlot facilities and the associated liquid manure storage systems can result in further protection of surface water quality. Liquid manure storage structures increase management flexibility, making it easier to apply at proper rates and to avoid winter-time manure application. Also, a majority of the older, smaller feedlot facilities are located adjacent to streams and do not have containment of manure or manure-contaminated runoff. The new larger facilities all have total containment of manure with no discharge into surface waters from rainfall and snowmelt. As new larger facilities replace the smaller feedlots there is less manure runoff into streams. Also, the liquid manure at the new facilities is usually injected below the soil surface and is less subject to surface runoff compared to the typical soil surface spreading practices of many older feedlot facilities.

Risks to Water Quality

While there are a number of water quality benefits associated with liquid manure storage systems, there are also several heightened risks. One possible risk is the failure of the walls of the manure storage system to hold the manure with a resulting river of manure flowing down a valley and into a stream. This has not been known to occur in Minnesota, likely due in part to engineering review and regulation of construction activities. What has occurred in Minnesota are basin overflows and intentional discharges from manure storage structures. Enforcement of such violations has increased substantially during recent years in an effort to curb blatant violations and mismanagement.

Three potential water quality risks associated with liquid manure storage systems in the karst region include: 1) seepage of contaminants through the liner and underlying soil to fractured bedrock and subsequently to ground water; 2) soil subsidence below the structure which breaches the integrity of the concrete, geosynthetic or soil liner, causing a slow and perhaps undetectable leaking of manure from the storage system to ground water; and 3) a large sinkhole forming below a manure storage system leading to a rapid flow of manure into ground water or causing a collapse in a basin sidewall and a pouring out of manure onto the ground surface.

Manure entering ground water will discharge into streams within a period of time ranging from hours to decades depending on the site-specific hydrogeology. The karst region of Minnesota maintains a large number of high quality trout streams. A rapid discharge of a large quantity of manure into a stream will destroy the aquatic life for a stretch of the stream and also result in increased nutrient loading into the receiving waters of the Mississippi River system. Manure which flows in the ground water for a longer period before discharging into streams will be more diluted and may not destroy aquatic life, but will threaten drinking water supplies as it travels toward the stream, and contribute to stream pollution upon discharge.

Risks associated with slow seepage through the liner are reduced somewhat by Minnesota requirements for a minimum ten-foot separation distance between the bottom of standard reinforced concrete and earthen manure storage structures and underlying bedrock. If a composite liner or other nearly impermeable liner system is used, then the required minimum separation distance from bedrock is five feet. Requirements to minimize the risks associated with soil subsidence as new liquid manure storage systems are constructed in the karst region is the primary subject of the remainder of this paper.

SOIL SUBSIDENCE RISKS AT MANURE STORAGE SITES

Learning experiences from sinkholes forming under municipal wastewater treatment ponds

Between 1974 and 1992, sinkholes opened below three of the twenty-two municipal wastewater treatment ponds in Minnesota's karst region. Sinkholes developed in Altura's ponds in 1974 during construction and in 1976 when it first filled to capacity (Alexander and Book, 1984). A sinkhole developed in a Lewiston pond in 1991 after eighteen years of use (Jannik et al., 1992). Several sinkholes developed in a Bellchester pond in 1992 after twenty-two years of use (Alexander et al., 1993). The amounts of partially treated wastewater draining into sinkholes at the three respective sites was 3.7, 2.3, and 7.7 million gallons. The ponds were constructed of earthen materials with a designed seepage rate not to exceed 3500 gallons per acre per day. Several sinkholes are located within about a mile from all three sites, yet no sinkholes have been identified within a quarter of a mile from the sites.

These failures clearly demonstrate the potential for sinkholes to develop in southeastern Minnesota when large quantities of liquids are stored in sinkhole prone areas with minimal barriers between the liquid and underlying materials. Similar problems could develop when storing liquid manure on top of permeable liner materials. However, there are several notable differences between these failed municipal wastewater treatment systems and manure storage systems currently being constructed. The maximum allowable design seepage rate for manure storage systems is 500 gallons/acre/day, seven times less than the old municipal wastewater ponds. These design seepage rates assume that the ponds remain full and they do not account for seepage reductions caused by the physical, chemical and biological sealing which takes place at the manure/soil interface. In addition, the size of even the largest manure storage systems is smaller than the municipal ponds. These differences between the failed municipal systems and manure storage structures are worth recognizing, but they are not great enough to warrant complete disregard of the risks associated with siting liquid manure storage systems in sinkhole prone areas. It is also important to note that the contaminant concentrations in manure are often over 100 times greater than municipal wastewater pond liquids, and thus the environmental consequences of a catastrophic manure release could be much worse than municipal pond failures.

Use of Sinkhole mapping and research efforts

Sinkhole mapping and research completed during the past two decades has made it easier to determine the relative soil subsidence risks when siting new liquid manure storage systems in Southeastern Minnesota. Sinkhole probability maps have been completed for three counties (Dalgleish and Alexander, 1984; Alexander and Maki, 1988; Witthuhn and Alexander, 1995) and additional hydrogeologic investigation has been conducted in the other karst areas. The probability of sinkhole formation has been found to vary tremendously across the region. Some areas have in excess of 50 sinkholes per square mile and other areas have no sinkholes. Often high density clusters of sinkholes are adjacent to areas with scattered individual sinkholes. Bedrock composition, topographic position in the landscape and thickness of glacial materials over bedrock have all been found to affect the likelihood of sinkhole formation.

Most sinkholes in southeastern Minnesota appear where there is less than 50 feet of surficial cover over carbonate and sandstone bedrock. The proximity of nearby sinkholes remain the single best predictor of new sinkhole development (Witthuhn and Alexander, 1995). Magdalene and Alexander (1995) concluded that on the scale of several kilometers, new sinkholes in Winona County tend to develop in the areas of existing sinkholes, especially near newly developed sinkholes. The risk of soil subsidence has generally been found to increase in areas of ponded or intermittently flowing water, and in areas with indications of more extensive karstification, including areas with disappearing streams, caves, dry valleys, springs and solution cavities.

REGULATORY POLICY TO MINIMIZE RISKS
Overview

The rapid increase in the construction of large liquid manure storage structures in southeastern Minnesota, coupled with experiences of sinkhole development in three municipal wastewater treatment ponds, prompted the MPCA (the state regulatory agency for feedlot activities) to consider measures to minimize risks associated with construction of liquid or semi-solid manure storage structures in sinkhole prone areas.

Beginning in 1995, the MPCA has worked to develop and implement a policy that will reduce environmental risks associated with construction of liquid manure storage systems in the karst region, yet maintain the feasibility of constructing manure storage systems throughout much of this area. Guidelines were developed so that a general indication of environmental risk can be readily evaluated in karst regions and precautionary measures can be taken. The information used to evaluate the potential for sinkhole formation, and, in general, how this information is used in making permitting decisions, is described on the following pages. Specific permitting decisions are made on an individual case-by-case basis after considering numerous factors. The intent of the guidelines is to allow the producers and their technical advisors to understand sinkhole risk considerations early in the planning and site selection process, prior to substantial investment of time and money.

Listed below are three steps which producers are required to take when considering construction of a liquid or semi-solid manure storage system in areas where sinkholes could potentially form.

Step 1 - Conduct site investigation for sinkholes and other karst features.
Step 2 - Submit site investigation to state and/or county officials so that the karst risk factor may be determined.
Step 3 - Determine manure storage system options and requirements.

Step 1- Site Investigation for Sinkholes and Karst Features

A site specific investigation is used to gather information needed to evaluate the risks of soil subsidence at a proposed manure storage site. The following is required for the site investigation:

- Sinkhole Maps - A copy of any published sinkhole location and/or probability maps showing the area within a few miles of the proposed facility. If a sinkhole map shows the proposed manure storage site location to be in an area designated as "low" or "no" probability, then the other steps for the site investigation need not be completed.
- Field Inspection - a map of the proposed site showing the location of all small and large depressions in the landscape. At a minimum, all land within a 700 foot radius of the potential manure storage structure location must be closely inspected. The best period of time to conduct this investigation is when crop-cover, leaf cover, and snow-cover are minimal.
- Sinkhole/depression Characteristics - a description of the following for all sinkholes and potential sinkholes identified in steps 1 and 2: a) whether the sinkhole is currently open or has been filled; b) decade when formed, if known; c) position on landscape; d) depression diameter and depth, and e) other possible explanations which may explain the hole or depression.
- Other karst features - a description of other notable potential karst features, located within 1 mile of the proposed facility, including disappearing streams, caves, dry valleys, springs or solution cavities.
- Soil borings or soil trench information - The minimum soil boring depth must be to a point 10 feet below the bottom of the manure storage system. The karst risk factor (step 2) will be determined by assuming that the bedrock elevation is at the bottom of the shallowest boring. Deep soil borings which extend beyond the minimum required depth are optional and can be used to demonstrate a lower sinkhole risk potential. A minimum of four borings are required for the first one-half acre of storage system surface area. A minimum of two additional borings shall be taken for each additional one-half acre of storage structure surface area. If the borings indicate an uneven bedrock surface or highly variable soil conditions, additional borings will be required.
- Other Potential Diagnostic Work - The MPCA may require other work as deemed necessary by agency staff, possibly including: deeper borings to determine the characteristics of underlying bedrock, ground penetrating radar or other geophysical investigations to better diagnose subsurface conditions, trenching, or other karst investigative techniques.

The following additional information is needed for liquid manure storage structures proposed in counties where a sinkhole location/probability map has not been prepared:

- Soils Maps and Aerial Photos - topographic maps, soil survey maps and aerial photos of all land within a one mile radius of the site. All known open and filled sinkholes must be highlighted on these maps. Closed depressions identified on topographic maps are to be identified and inspected.
- Land owner interviews - a list of all long-term residents (living in area at least 15 years) and land owners in the area who were interviewed and asked about the location of existing and filled sinkholes located within a 1 mile radius of the proposed facility. All sinkholes or potential sinkholes (open or filled) are to be identified on a map or photo of the site.
- Well Logs - Geologic information from well logs within a 2 mile radius of the proposed site location

Step 2. Determination of Karst Risk Factor

Information obtained under Step 1 is submitted to the MPCA or delegated county authority so that a karst risk factor for the site under consideration may be estimated. The karst risk factor is determined from available sinkhole probability map information, along with site specific soils, landscape function, geology, and sinkhole information. Karst experts from other organizations may be consulted during the review of more complex cases. The following site specific information is considered when determining the karst risk factor:

a) density of sinkholes;
b) the topographic and geologic setting which sinkholes are found;
c) patterns and characteristics of nearby sinkhole formation;
d) type and condition of first encountered bedrock;
e) depth to bedrock;
f) soil and subsoil types; and
g) identification of other karst features (e.g. disappearing streams, blind valleys, dry valleys, caves, springs, and karst features observed in exposed bedrock along roadways).

Sinkhole characteristics roughly representing various karst risk categories are listed below. While these general descriptions largely refer to proximity to sinkholes and sinkhole densities, the other site specific variables noted above are often evaluated for proposed sites in order to determine the most fitting risk category. The following descriptions are only intended to serve as general guidelines.

- No Risk - Areas where the first encountered bedrock is not subject to sinkhole formation.
- Low Risk - Areas underlain by carbonate bedrock, but in which very few sinkholes are found. No known sinkholes exist within a 1 mile radius of the proposed site, and the soils and geologic information indicate that there is minimal risk of sinkhole formation at the site under consideration.
- Moderately Low Risk - No sinkholes or buried sinkholes are known within a 1/2 mile radius of the proposed site. However, widely scattered sinkholes have been identified in the area and the depth to bedrock is less than about 50 feet.
- Moderate Risk - No sinkholes or buried sinkholes are known within a 1/4 mile radius of the site. However, there are scattered sinkholes (e.g. 2 - 5 sinkholes in a 1 mile radius of proposed site) and/or other geologic factors that make the area susceptible to sinkhole formation.
- Moderately High Risk - Similar sinkhole densities as high risk zones, but the soils and other information about karst features indicate that the specific site of construction has a lower sinkhole risk than the high risk category.
- High Risk - There is typically either 1 sinkhole or buried sinkhole within a 1/4 mile radius or 2-4 sinkholes or buried sinkholes within a 1/2 mile radius and the soils and karst feature information indicates minimal protection.
- Very High Risk - Sinkholes are common in the area, but sinkhole densities are less than in the extremely high risk areas (e.g. 2 to 4 sinkholes in a 1/4 mile radius or 5 or more sinkholes within a 1/2 mile radius).
- Extremely High Risk - Sinkholes are the dominant landform, with typical sinkhole densities exceeding about 4 sinkholes in a 1/4 mile radius from any point.

Step 3. Determine Manure storage system options and requirements

MPCA recommendations are that the proposed liquid or semi-solid manure storage systems be:

a) located as far as possible from topographic lows, depressions or ravines;
b) located as far as possible from existing or historically filled sinkholes;
c) located in an area with the greatest thickness of fine-textured soils;
d) constructed with very low seepage liners;

e) constructed so as to minimize the amount of rainfall and roof runoff water infiltrating soils in the area of the manure storage system;
f) not constructed when very large volume manure storage systems are proposed in high risk karst areas;
g) not constructed when soil excavation reveals indications of historic or potential future sinkhole formation.

After the sinkhole risk factor and the combined storage capacity of all structures on site has been determined, Table 1 is used as a general guideline for identifying recommended options for manure storage structures and associated liners. The options for manure storage are intended to be guidelines only. Best professional judgment is used when determining allowable manure storage system options. Consideration is given when a new manure storage structure is designed to correct existing surface or ground water pollution problems without a significant expansion in operation size. For example, at existing operations, it can be better for the environment to have a new liquid containment structure built in a sinkhole prone area than to have direct feedlot runoff into streams or the continued use of an old structure that was constructed using less stringent standards. Other considerations include: maximum manure volume to be stored in any single manure storage structure, site history and management, planned contingency efforts, and specific properties of cohesive soils.

Table 1. General guidelines for manure storage system options in different karst risk zones. The letters A-G correspond with letters in the table. For example, a five million gallon storage structure proposed in a moderate karst risk area could be constructed using options D, E, F, or G. Design capacity considers the combined storage capacity of all manure storage structures on the property.

A. Cohesive soil liner designed/constructed to seep no more than 0.018" per day when full (of water) and a design thickness of 2 feet or greater.
B. Reinforced concrete structure constructed in accordance with MPCA standard requirements.
C. Cohesive soil liner designed/constructed to seep no more than 0.012" per day when full (of water) and with a liner thickness of 3 feet or greater.
D. Cohesive soil liner designed/constructed to seep no more than 0.0089" per day when full (of water) and a thickness of 4 ft or greater.
E. Composite liner system or upgraded concrete liner. A composite liner system consists of a combination of compacted clay covered by an approvable geomembrane or geosynthetic liner. For concrete, an upgraded system includes a steel reinforced floor and a waterstop or water sealant in all construction joints and control joints.
F. Above ground storage system.
G. Solid manure handling systems only.

Karst Risk	Design capacity in millions of gallons						
	≤ 0.25	0.25-0.5	0.5 - 1	1 - 2	2 - 4	4-8	≥ 8
No Risk or Low risk	A-G	A-G	A-G	A-G	A-G	A-G	A-G
Moderately low risk	A-G	A-G	A-G	A-G	B-G	C-G	C-G
Moderate risk	A-G	A-G	A-G	B-G	C-G	D-G	E-G
Moderately high risk	A-G	A-G	B-G	C-G	D-G	E-G	G
High risk	C-G	C-G	D-G	E-G	F-G	G	G
Very high risk	E-G	E-G	F-G	G	G	G	G
Extremely high risk	E-G	G	G	G	G	G	G

POLICY IMPLEMENTATION

The requirements described above gradually developed from 1995 to 1997 as individual feedlot permit applications were reviewed for karst risks. Since 1997, the review process has become more standardized, and has resulted in the construction of numerous manure storage system liners which combine two-feet of compacted cohesive soils and a 40 to 60 mil synthetic liner. Some producers have had to locate their new barns in a lower risk area than originally proposed.

There are currently many environmental protection demands surrounding feedlots, including enforcement of intentional manure discharges, open lot runoff problems, land application of manure issues, engineering review for new sites, hydrogen sulfide and other air emission issues, feedlot abandonment concerns, manure storage system construction problems, livestock access to public waters, manure stockpile runoff, silage liquids runoff, dead animal disposal, old and poorly lined manure storage systems, and other problems stemming from mismanagement of manure. The intent of the MPCA feedlot program is to allocate limited staff resources in a manner which balances addressing the issues which are causing immediate environmental problems, with the need for taking preventative measures to minimize catastrophic problems in the future. Regulating the siting and designs of new manure storage systems in areas prone to sinkholes should help to reduce the chances of catastrophic problems from occurring in the future. It is likely that the regulations and guidelines will continue to be refined as more is learned about karst processes and the effects of storing liquid manure in Southeastern Minnesota.

ACKNOWLEDGMENTS

The author gratefully acknowledges Jeff Green of the Minnesota Department of Natural Resources and Dr. E. Calvin Alexander, Jr. of the University of Minnesota for sharing their experiences and expertise of karst processes during development of MPCA karst guidelines.

REFERENCES

Alexander, E. C. Jr., and Book, P.R., 1984. Altura Minnesota lagoon collapses. in Beck, B.F. (ed.), Proc. First Multidisciplinary Conf. on Sinkholes, Balkema, Boston. p. 311-318.

Alexander, E.C., Jr., Brogerg, J.S., Kehren, A.R., Graziani, M.M. and Turri, W.L., 1993. Bellchester Minnesota lagoon collapse. in Beck, B.F. (ed.), Applied Karst Geology. Balkema, Rotterdam. p. 63 to 72.

Alexander, E. C. Jr. and Maki, G.L., 1988. Sinkholes and sinkhole probability, Plate 7 of Balaban, N.H. (ed.), Geologic Atlas Olmsted County, County Atlas Series C-3. Minnesota Geological Survey.

Dalgleish, J..B. and Alexander, E.C., Jr., 1984. Sinkholes and sinkhole probability, Plate 5 of Balaban, N. H. and Olsen, B.M. (eds.), Geologic Atlas Winona County, County Atlas Series C-2. Minnesota Geological Survey.

Magdalene, S. and Alexander, E.C., Jr., 1995. Sinkhole distribution in Winona County Revisited. in Beck B.F. (ed.) Karst GeoHazards. Balkema, Rotterdam. p. 43-51.

Minnesota Pollution Control Agency. 1998. MPCA Guidelines for design of cohesive soil liners for manure storage structures. Feb. 1998. draft. 28 pp.

Minnesota Pollution Control Agency. 1997. Guidelines for concrete manure storage structures. Dec. 1997. 14 pp.

Witthuhn, K. M. and Alexander, E.C. Jr. 1995. Sinkholes and sinkhole probability, Plate 8 Geologic Atlas Fillmore County, County Atlas Series C-8. Minnesota Department of Natural Resources and Minnesota Geological Survey.

7 Dye tracing and the delineation of karst groundwater basins

Karst groundwater basin delineation, Fort Knox, Kentucky

DENNIS P. CONNAIR & SCOTT A. ENGEL Dames and Moore, Cincinnati, Ohio, USA
BRIAN S. MURRAY Science Applications International Corporation, Oak Ridge, Tenn., USA

ABSTRACT

Evaluation of karst groundwater quality concerns at Fort Knox Kentucky have required the development of a sitewide karst groundwater flow model and basin delineation investigation. The karst aquifer underlying the base is developed within approximately 60 m of the St. Louis Limestone and is bounded on three sides by baseline streams. The underlying Salem Limestone acts as a regional aquitard and provides a lower limit to karst aquifer development. The study area covers over 130 km^2 and contains over 200 inventoried karst features. As a part of this investigation, innovative multiple dye trace events were conducted throughout the study area using up to six dyes per event with a total of 8 dyes used to conduct 14 dye traces during three seasonal events. Dye trace results, structural and topographic controls, spring characteristics, and normalized base flow were used to establish groundwater basin limits and boundary zones and to develop a conceptual sitewide groundwater flow model. The findings of this investigation will be used to assess the groundwater contaminant contribution from source areas in individual basins, develop an effective groundwater monitoring program, and guide future groundwater management strategies.

INTRODUCTION

This paper presents a case study of karst groundwater basin delineation conducted for the U.S. Army Armor Center and Fort Knox at Fort Knox, Kentucky, the findings of which suggest a conceptual model of groundwater flow that may be useful in evaluating karst aquifers in similar settings elsewhere.

The Fort Knox Military Reservation occupies 422 km^2 (109,270 acres) in north-central Kentucky, approximately 50 km (30 miles) south of Louisville and 30 km (18 miles) north of Elizabethtown (Figure 1). Fort Knox has been in use by the military since 1918 and is an active U.S. Army Armor Center with the principal mission of basic combat training and advanced individual training in armored vehicles. Various operations at Fort Knox involve handling and storage of hazardous materials, which require the facility to maintain a permit under the Hazardous and Solid Waste Amendments of 1984 and to meet Resource Conservation and Recovery Act regulations.

The primary objective of ongoing sitewide karst groundwater studies at Fort Knox is to establish individual groundwater basin limits and to characterize water quality so that appropriate groundwater management strategies can be implemented. Fort Knox is working diligently with state and federal regulators to identify cost-effective approaches to assess the environmental impact of historical and current waste disposal practices on groundwater at the facility. The agreed upon sitewide approach provides a broad perspective of the potential influences of solid waste management units (SWMUs) and areas of concern (AOCs) on the water quality of individual springs. Working at the sitewide scale allows for consideration of all potential sources of impact within an individual spring basin and accounts for fluctuations in groundwater quality during monitoring activities due to seasonal changes, divergent flow, and other large scale variables.

The specific objective of basin delineation is to identify contaminant source areas on the base that have the potential to impact individual springs. Most of the identified SWMUs and AOCs are located in and around the cantonment area of the base. Being located on a topographic high, the cantonment area also serves as the headwater region for many of the sitewide groundwater basins. Therefore, it is critical to establish the limits of those basins in order to develop an effective groundwater monitoring program. The results will be used in conjunction with planned rainfall-response studies to identify appropriate monitoring points and sample collection times for sitewide groundwater quality assessment and monitoring and for site-specific SWMU and AOC investigations.

GEOLOGIC SETTING

The cantonment area of Fort Knox is located on an upland sinkhole plain. Bounded by Otter Creek to the west, Salt River and Mill Creek to the east, and the Ohio River valley to the north, this upland constitutes the study area for this investigation (Figure 1). The upland area is underlain by the Mississippian St. Louis Limestone, whereas the steep valley walls of the baseline streams are cut through to the underlying Mississippian Salem Limestone, Harrodsburg Limestone, and Borden Formation. The St. Louis Limestone, which is particularly susceptible to karstification resulting in typical surficial karst features and limited surface stream development, can be as much as 70 m (230 ft) thick in the area (Swadley 1963; Withington and Sable 1969; Kepferle 1977; Kepferle and Sable 1977) with a thick, typical limestone residuum soil. The Salem Limestone is 24 to 40 m (80 to 130 ft) thick, characterized as a mixture of limestone, shale, and dolomite, and acts as a regional aquitard, limiting karst development to the St. Louis Limestone.

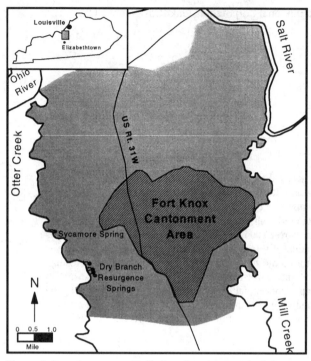

Figure 1. Location map with limits of study area shaded gray and primary base flow resurgence spring labeled.

Primary resurgent base flow springs, Sycamore Spring and the Dry Branch resurgence springs are located at the St. Louis Limestone/Salem Limestone contact along Otter Creek on the western boundary of the study area (Figure 1).

METHODS

The sitewide karst groundwater assessment was based on work started in 1994 for the site specific investigation of one SWMU. Since then, the assessment activities have followed an approach similar to that outlined in ASTM D5717-95 (1995), which involves the development of a conceptual model followed by subsequent testing and refinement of the model using an array of investigation tools. The tools employed for this study include an extensive (and ongoing) spring survey and geologic reconnaissance (Dames & Moore, 1996), qualitative and semiquantitative dye tracing (Dames & Moore and SAIC, 1998), and an evaluation of normalized base flow after Quinlan and Ray, 1995.

Survey activities covered all of the upland study area and baseline streams except where access is limited in artillery impact zones. Separate dye tracing events were conducted in dry and wet season conditions with injections into open sinkholes, sinking streams, existing monitoring wells, and drilled injection points. Up to six dyes were injected per tracing event to maximize use of field time monitoring the large, and in places, remote study area. The use of multiple dyes required innovative analytical methods to segregate and quantify the concentration of dyes at each resurgence point. Continuous monitoring with automatic water samplers was used in some locations to help verify detections by tracking breakthrough curves of multiple dyes. A total of eight different dyes were used to conduct 14 traces in three events.

During each trace, parameters such as temperature, pH, specific conductance, and flow rate were measured on a daily to weekly basis for all springs. The flow rate data were used to identify representative base flow conditions for normalized base flow evaluation. Normalized base flow, typically used to evaluate the size and hydraulic character of surface drainage basins, can be applied with caution to karst drainage basins (Quinlan and Ray, 1995). The method works under the assumption that basins in a similar physical setting and climate will have a similar base flow discharge per unit drainage area (typically expressed as cubic feet/second/square mile of basin area [cfsm], or liters/second/km^2 [Lsk]).

FINDINGS

Dye tracing results suggest that karst groundwater flow, on a sitewide basis, is controlled directly or indirectly by local stratigraphy, geologic structure, and changes in baseline stream level during the recent geologic past. These controls define the elements of the conceptual model for sitewide groundwater flow. Figure 2 illustrates the application of this model to idealized cross sections through the two dominant groundwater basins on site, Dry Branch Basin and Sycamore Spring Basin highlighted on Figure 3 with hachure marks and identified as basins #1 and #2, respectively.

The role of stratigraphy in the control of sitewide groundwater flow was easily recognized from the spring survey reconnaissance data. The survey found most major and many minor springs to occur at the St. Louis Limestone/Salem Limestone contact. The relatively insoluble character of the Salem Limestone, especially with respect to the highly soluble St. Louis Limestone, limits conduit development; and therefore, the vertical development of the karst aquifer. The Salem Limestone and underlying formations are recognized to act as a regional aquitard, restricting the downward flow of water and forcing groundwater to migrate horizontally along its contact with the St. Louis Limestone. On Figure 2 this is depicted by the lack of significant flow conduits into and through the Salem Limestone and underlying formations and by the presence of resurgent springs at the contact.

The structural inclination of the St. Louis Limestone/Salem Limestone contact is generally subdued with a regional dip of approximately 0.5 percent from east to west. Structural anomalies along the contact include a dome uplift in the northwest portion of the study area (Muldraugh Gas Storage Field) with an adjacent shallow syncline that is inferred to plunge south from the Ohio River escarpment and turns west to cross Otter Creek. There are very few control points to define the trend of the synclinal axis, but existing data indicate that its lowest point intersects the ground surface near Sycamore Spring on Otter Creek. These structural controls have contributed to groundwater flow and resulting conduit development from east to west toward Sycamore Spring with secondary resurgent points at other locations along the contact structure. On Figure 2, structural control is illustrated by the tendency for conduits to follow the contact and parallel bedding planes within the St. Louis. Conduit development along the structural features of the St. Louis Limestone/Salem Limestone contact is exemplified in the dye trace flow patterns illustrated on Figure 3. Dyes injected in the eastern portion of the cantonment routinely flowed to resurgent springs 4 or more km to the west rather than resurfacing in springs 0.5 km away along the eastern escarpment.

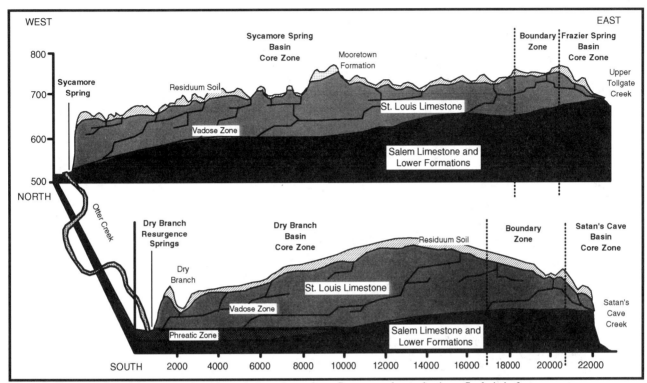

Figure 2. Conceptual model cross-sections through the primary base flow groundwater basins. Scale is in feet.

Hydrologic baseline appears to be a controlling localized influence for groundwater flow to springs along the northern and eastern escarpment. These springs are characterized by a low or absent base flow, a corresponding small basin area, and a flashy response to storm events. For these springs, the incisement into the underlying formations by nearby baseline streams generates a steep, localized hydraulic gradient that negates the westward structural control.

The southwestern portion of the study area is dominated by the Dry Branch Basin and its resurgent springs along Otter Creek. These springs are topographically higher than the downstream Sycamore Spring and are approximately 20 feet above the inferred St. Louis Limestone/Salem Limestone contact. The position of these springs appears to be controlled by the localized base level for Otter Creek. Several of the Dry Branch resurgent springs are artesian and attest to the phreatic storage of groundwater between the base of Otter Creek and the underlying St. Louis Limestone/Salem Limestone contact

Higher-elevation springs and caves occur along Otter Creek downstream from the Dry Branch Basin resurgence. These springs likely represent historical base levels resurgences which developed similar to the Dry Branch resurgent springs when those reaches of Otter Creek were at higher elevations. The remanent springs currently function as overflow routes for the Sycamore Basin and/or drain a small, localized basin area.

GROUNDWATER BASIN DELINEATION

Base flow spring groundwater basins delineated within the study area are illustrated on Figure 3. Basin boundaries are based on the conceptual model, structural controls, dye tracings, an other assessment activities conducted to date. Two basins, Sycamore Spring Basin (#2) and Dry Branch Basin (#1), dominate the cantonment area of the base and account for almost 80% of measured sitewide groundwater discharge.

Sycamore Spring is an artesian spring on the floodplain of Otter Creek which serves as the primary resurgence for the Sycamore Spring Basin. Except for several localized epikarst springs in the upper basin and several overflow springs in the lower basin, Sycamore Spring is the only point of groundwater resurgence for the Basin. Observed discharge rates range from 0.03 to 0.6 m^3/s (1 to 20 cfs) and eight of 14 traces emerged at Sycamore Spring with sharp breakthrough curves and high flow rates ranging from 0.17 to 1.04 m/s (2.6 to 16.3 ft/min). The basin has a peculiar shape, starting in the cantonment and trending north before turning southwest toward Otter Creek. This bending form may be a reflection of the underlying structural syncline. The well-developed surficial topography and apparent subsurface conduit system in the north-trending segment encompasses several sinking streams whose surface flow runs south. It is likely that a paleovalley or high level conduit system developed on this north-south trend and was pirated by the younger, structurally-controlled Sycamore Spring Basin.

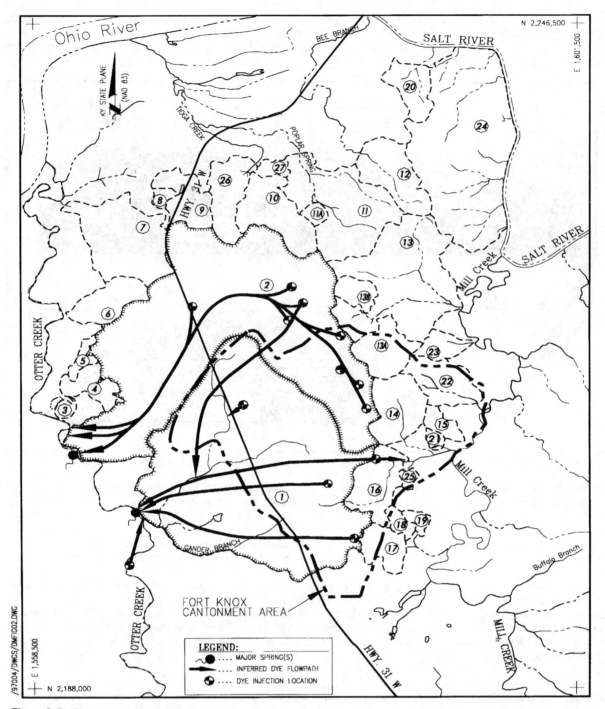

Figure 3. Projected groundwater basins in the Fort Knox area.

Dry Branch Basin discharges to Otter Creek from a series of springs including a blue hole, several streambed artesian springs, and bank discharge springs. Discharge rates are difficult to estimate due to the location of the springs and the mixing of extrabasinal groundwater originating from a sinking reach of Otter Creek outside the study area. Combined base flow from the springs is estimated at 0.04 m^3/s (1.5 cfs). Dry Branch Basin has several well-developed surface streams that appear to interact with epikarst and shallow conduits by gaining and losing reaches and which ultimately emerge in the Dry Branch resurgent springs. The surface streams also function as overflow routes during high flow events. A second, deeper flow path from the eastern headwater reaches of the basin is indicated by dye trace results (Figure 2). Recharge over the western portion of the basin is likely captured by the shallow conduits that stair-step through the bedding planes of the St. Louis Limestone. Dyes from six traces emerged in the Dry Branch Basin resurgent springs but complex breakthrough curves and slower flow rates suggest an anastomosing path for the shallow component of flow.

Normalized base flow values calculated for the basins depicted on Figure 3 range from 1.1 to 2.0 Lsk (0.1 to 0.18 cfsm) with a sitewide value of 1.5 Lsk (0.14 cfsm). These values suggest the karst aquifer relies on autogenic recharge with minimal exogenic input.

CONCLUSIONS

The delineation of groundwater basins in a karst setting is possible using a carefully constrained evaluation method. For the Fort Knox area, the incorporation of information gathered on the geologic setting and the identification of resurgence springs by groundwater dye tracing allowed for identification of the dominant elements controlling groundwater flow and direction. This information was further used to construct a unifying conceptual model that guided the placement of groundwater basin boundaries. Normalized base flow values calculated for each basin could then be used to assess individual basin contribution to the sitewide groundwater and to further constrain basin limits. The findings of this investigation will allow an assessment of the relative contribution to groundwater contamination from SWMUs and AOCs located within each basin, suggest how and where to monitor for groundwater quality, and guide future groundwater management strategies. Future assessment work planned for the site includes rainfall-response studies using quantitative tracing and spring gauging and refinement of groundwater basin boundaries.

ACKNOWLEDGEMENTS

The authors would like to acknowledge the support of the Fort Knox Directorate of Base Operations Support provided by Al Freeland and Donnie McGar under Director Colonel Philip Jones. At the U.S. Army Corps of Engineers, Nashville District, Walter Green and Tom Lerner secured the necessary funding and provided overall guidance of the assessment program. Dave Harmon of the University of Kentucky Water Resources Research Institute provided invaluable feedback and guidance on behalf of the Kentucky Department for Environmental Protection. Dr. Charlie Riggs and Dave Cika of Sverdrup Environmental, St. Louis, MO provided program and project support on much of the work indirectly related to the sitewide assessment. Stu Edwards of Dames & Moore, Cincinnati, OH provided technical guidance and review for all work products relating to the sitewide assessment, and Bill Eckhoff of Dames & Moore provided field management of drilling and other related assessment tasks. And finally, Dr. Nick Crawford and Rick Tucker of Crawford and Associates, Inc. provided vastly experienced perspective and innovative solutions to the challenges of maximizing our dye tracing results within available time, budget, and access limitations.

REFERENCES

American Society for Testing and Materials (ASTM), 1995. Standard guide for design of ground-water monitoring systems in karst and fractured-rock aquifers. ASTM D5717-95.

Dames & Moore, 1996. Spring survey and initial background spring monitoring report, Fort Knox groundwater study. In cooperation with Sverdrup Environmental for USACE Nashville District, IDT Contract No. DAC62-93-D-0028.

Dames & Moore and SAIC, 1998. Draft sitewide karst groundwater assessment report, Phase I activities, U.S. Army Armor Center and Fort Knox at Fort Knox, Kentucky. For USACE Nashville District, IDT Contract No. DACA62-94-D-0029.

Kepferle, R.C., 1977. Geologic map of the Vine Grove Quadrangle, Kentucky. U.S. Geological Survey, Geologic Quadrangle Map GQ-645; 1:24,000 scale.

Kepferle, R.C. and E.G. Sable, 1977. Geologic map of the Fort Knox Quadrangle, North-Central Kentucky. U.S. Geological Survey, Geologic Quadrangle Map GQ-1375; 1:24,000 scale.

Quinlin, J.F., and J.A. Ray, 1995. Normalized base-flow discharge of groundwater basins: a useful parameter for estimating recharge area of springs and for recognizing drainage anomalies in karst terrains. Proceedings of the 5th Multidisciplinary Conference on Sinkholes and Environmental Impacts of Karst, April 1995, Gatlinburg Tennessee. Balkema, Rotterdam.

Swadley, W.C., 1963. Geologic Map of the Flaherty Quadrangle, Kentucky. U.S. Geological Survey, Geologic Quadrangle Map GQ-299; 1:24,000 scale.

Withington, C.F., and E.G. Sable, 1969. Geologic Map of the Rock Haven and Laconia Quadrangles, Kentucky-Indiana. U.S. Geological Survey, Geologic Quadrangle Map GQ-780; 1:24,000 scale.

Dye study tracks historical pathway of VOC-bearing industrial waste water from failed pond at metals coating facility

LOIS D.GEORGE & GHEORGHE M.PONTA P.E.LaMoreaux & Associates, Inc., Tuscaloosa, Ala., USA

ABSTRACT

Nearly ten years after initiating a ground-water assessment and documenting a stable mass of VOC-bearing ground water in the upper karst aquifer, the Post-Closure Care regulatory process required a dye study to trace ground-water movement. The regulatory impetus was the location of a major spring, regionally downgradient from the facility, used as a municipal drinking water supply.

A well drilled strategically in the area of highest concentrations of VOCs of the stable plume was used as the point of injection of fluorescein. Charcoal packets and water samples were collected from 29 locations (on-site monitoring wells, monitoring wells and productions wells at an adjacent facility, surface water sites, and the major spring – a diffuse flow, conduit spring with three points of emission). The bedrock fracture system in the vicinity of the facility is not well developed or well connected. The distances from the injection point to monitoring/sampling locations ranged from 45.7 meters (150 feet) to 4.9 kilometers (3.05 miles). The duration of the test, including pre-injection monitoring, was 118 days.

A cocktail of potable water (7,182 liters or 1900 gallons) preceded the injection of the dye solution, and a chaser of about 4,158 liters (1100 gallons) followed the injection of fluorescein. The initial movement of the dye was a combination of the head created by the injection of drinking water and dye, and the non-pumping conditions at the adjacent facility. Within less than 24 hours, the imposed head dissipated and water levels in the injection well returned to pre-injection levels.

The head imposed or induced with injection resulted in an almost unique occurrence that pushed water out to the northwest fractures, in pathways other than normal. The only other time that those northwest pathways were probably used, temporarily completing the connection between the facility and the major spring, was when the pond failed, and a slug of VOC-bearing water was induced into the ground-water system.

HYDROGEOLOGIC SETTING AND CONTAMINANT CHARACTERIZATION

The geology at the site is typical of the terrain associated with limestone bedrock. At the site, the irregular surface of the limestone is overlain by a mantle of clay-rich, unconsolidated material (overburden). Ground water in the limestone aquifer, occurs under semi-confined to confined conditions. Variations in the direction of ground-water movement at the site are controlled by pumping of production wells at the adjacent manufacturing facility, immediately west of the site.

Movement of ground water containing volatile organic compounds (VOCs) occurs within fractures/bedding in the upper portion of the bedrock, and locally within the chert-rubble zone at the bedrock-overburden contact. Measured hydraulic conductivities in bedrock, observations that the abundance of fractures and ground water decrease with depth, and analyses of ground water from deep wells demonstrate the presence of little, if any, VOCs at depths exceeding 27 meters (90 feet) below land surface. The horizontal and vertical distribution of Tetrachloroethylene (PCE) in ground water has been assessed and monitored using the wells installed at the site and adjacent to the facility. Monitoring of ground water since 1989 indicates that the mass of VOC-bearing groundwater has been spatially stable.

PCE in ground water at the facility is most probably the result of an historical (1977) failure of a wastewater pond, which was near the southwestern corner of the facility. The bedrock pinnacle, which underlies the northern edge of the former pond, provided a potential pathway for PCE to reach the chert-rubble zone at the bedrock-overburden contact (Figure 1). PCE has been confirmed in the chert-rubble zone in this area (south of the pinnacle) at some of the highest concentrations identified in the sub-soil at the site. The chert-rubble zone in this area is within the zone of fluctuation of the potentiometric surface as a result of pumping at the adjacent manufacturing facility and/or seasonal fluctuations in the water surface.

Ground-water monitoring and the distribution of PCE in soil south of the bedrock pinnacle suggest that this compound is sorbed to soils near the top of bedrock or sorbed to soils in fractures near the top of bedrock. No evidence of Dense Non-Aqueous Phase Liquids (DNAPL) has been identified in soil or ground water at the site. Fluctuation of the water surface, as a result of pumping at the adjacent facility and recharge at the bedrock pinnacle, provides a flushing mechanism which desorbs some PCE from soil, supplies dissolved PCE to ground water, and maintains the spatially stable VOC mass.

The entire course of the nearest perennial stream was observed by canoe within a 1.6 km (1-mile) radius of the facility. No springs were observed along this portion of the creek, nor were any springs observed along the entire 10.5 km (6.5-mile) course, which was traversed. The stream has been sampled at locations upstream and downstream of the facility. No VOCs have been detected.

The system of fractures (joints) in the vicinity of the site consists of two sets, which trend toward the northeast and northwest. Both sets of fractures occur throughout the area. However, the northwest-trending set is dominant, as evidenced by the alignment of the stream valley, and alignment of karst depressions in the area.

A major spring (Spring), 4.9 kilometers (3.05 miles) northwest of the site, is a municipal water supply. The Spring is considered a diffuse flow conduit spring with emissions visible at three nearby locations. The estimated 7-day Q_2 (low flow) is 34 million liters (9 million gallons) per day (mgd) and average flow is about 158.76 million liters per day (42 mgd). The computed variability in flow is 200 percent. Water is withdrawn by the treatment plant by a 24-inch intake pipe approximately 3 meters (10 feet) from the largest flowing of the springs. The plant is permitted by the state environmental agency as a surface water filtration plant. The recharge area of the Spring is 218 km^2 (84 square miles). The quality of water at the spring is influenced by many potential sources of inorganic and organic constituents.

THE INJECTION WELL

A recovery well, designated RW-1, was installed to extract ground water with elevated concentrations of PCE and degradation products and to prevent the migration of PCE-containing ground water from the source area. The well was also installed for use as the injection well for this dye study.

A borehole for the well was advanced through the overburden (residuum) from land surface to the top of bedrock. To aid in the characterization of hydrogeologic conditions, soil samples in the overburden were collected continuously using a split-spoon sampler, and bedrock was drilled by air rotary method as opposed to previously specified coring. The air rotary method was used, after reconsideration of coring using drilling water and/or drilling mud, so as to enable easy recognition of water-bearing zones and to eliminate the potential for the introduction of drilling fluids into the ground-water system. Lithologic descriptions based on returns (drill cuttings), rig behavior and penetration rate, and geophysical logging allowed the identification of fracture zones at 7-8.84 m (23 to 29 feet), 12.80 to 15 m (42 to 49 feet, 16.76 to 17.37 m (55 to 57 feet), 23.45 to 23.77 m (77 to 78 feet) and, 25.60 to 26.21 m (84 to 86 feet) below land surface. The total depth drilled was 29.38 m (96.38 feet) below land surface. The borehole was drilled to 8 inches in diameter to facilitate installation of 4-inch diameter PVC casing and screen for the recovery well. The well was screened from 18.60 to 29.26 m (61 to 96 feet) below land surface.

THE DYE TEST

The injection of dye was timed to be dependent on the hydrologic conditions in the vicinity of the injection well. As tracing the potential connectivity from the site to the Spring was the primary focus of the study requirements, hydrologic conditions conducive to dye migration off site should occur at, or just prior to, dye injection. Ground-water withdrawals at the adjacent manufacturing facility occur on an as-needed basis during the operation of the plant and affect the movement of ground water at the site. The pumps are not typically operating during weekends and holidays. To simulate conditions that would be most conducive to off-site ground-water movement, the dye was injected on December 22, 1996, after a period of 2 non-operation (non-pumping) days at the manufacturing facility.

Natural or man-made background fluorescence of the ground water was monitored prior to injection of the dye (background concentration). Passive dye detectors (charcoal packets) were placed at selected locations, including wells, surface water, springs and storm sewers 18 days prior to injection of the dye at well RW-1. Some potential sources for background fluorescence are detergents, bathroom cleaners, pigments for inks and dyes, antifreeze, industrial wastes, naturally-occurring mineral fluorescence, and residual dye from previous studies. The first set of charcoal packets for the background portion of the dye was installed on December 4 through 6, 1996. The detectors were removed for evaluation at 9 days and 2 days prior to injection of the dye, (on December 13 and December 20, respectively).

On December 22, 1996, between 15:16 hours and 15:31 hours, 2.27 kg (5 pounds) of fluorescein (powdered fluorescein mixed with a total of 26.46 liters or seven gallons of distilled water) were injected in well RW-1 at the site. A drop hose was installed in the injection well (RW-1) such that the bottom opening was within the screened interval of the well and approximately 3 m (10 feet) below the top of bedrock. A slug of potable water (7,182 liters or 1900 gallons) preceded the injection of the dye-containing mixture (dye and water) via two drop hoses. Care was taken to avoid contamination of clothing and the area around the point of injection. A chaser, of approximately 4,158 liters (1100 gallons), followed the injection of fluorescein.

Water-level changes before injection (during the slug) at RW-1 are presented in Figure 2. At the beginning of the slug, the water level rose about 4.27m (14 feet) above the initial level and then declined, as the water infiltrated, to 2.1 m (7 feet) above the initial level (about 3,780 liters or 1,000 gallons were added at this time). Thence the water level stabilized for the rest of the injection.

No significant change in water levels in monitoring wells was measured during the injection process. At wells MW-1S, MW-2, and MW-3, water levels rose slightly in the range of 0.030 m (0.10 feet), and at MW-1D, a slight decline in water level was measured (less than 0.015 m or 0.05 feet). These wells are located between 45.70 to 61.00 m (150 to 200 feet) from RW-1 (Figure 1). The water level at RW-1 was measured at 0.018 m (0.06 feet) above the pre-injection level at 13:00 hours on December 23, 1996, about 22 hours after the dye injection.

DYE TEST RESULTS

The data from a single dye trace generally reflect conditions for that particular test, and especially for that particular discharge. The shape and magnitude of dye-recovery curves are influenced, primarily, by the amount of dye injected, the velocity and magnitude of flow, the mixing characteristic within the flow system, the sampling interval, and whether the discharge was diluted by non-dyed waters (i.e., other waters from the recharge area).

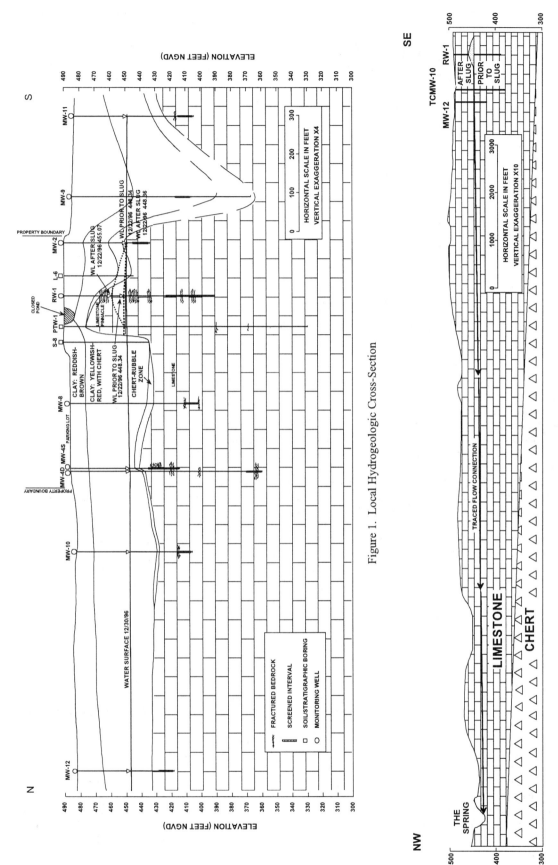

Figure 1. Local Hydrogeologic Cross-Section

Figure 3. Regional Hydrogeologic Cross-Section

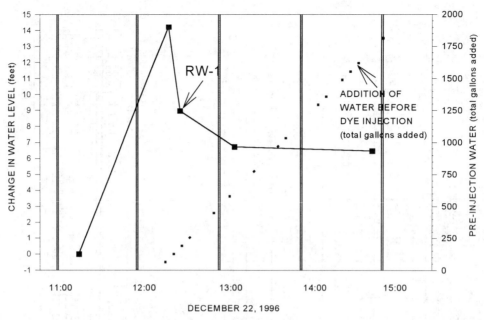

Figure 2. Water-Level change at RW-1 before the injection

Fluorescein was detected in the elutant from all four charcoal packets (SP-1, SP-2, SP-3, SB-1) at the Spring. Figure 3 is a generalized hydrogeologic cross section showing the site and the Spring and shows the pre- and post-injection water levels at the site. Time of travel (time to leading edge) was a maximum of 48 hours (maximum hours based on time charcoal packets were in place) after injection (on December 24, 1996), with the peak dye concentration being detected on December, 30, 1996. Dye detections at the springs are shown on Figure 4.

The peak dye concentration of elutant from charcoal packets from SP-1 is 16.20 parts per billion (ppb), and was detected in the packet collected 4 days (96 hours) after dye injection. At SP-1 the last charcoal packet where the dye was detected, was recovered on January 13, 1997, 528 hours after dye injection, with a concentration of 2.93 ppb. The peak dye concentration of the dye in charcoal packets from SP-2 is 104 ppb, and was detected 4 days (96 hours) after injection. The highest concentration of the dye detected in SP-3 is 233 ppb, and was encountered 4 days (96 hours) after injection. The highest concentration of the dye detected in SB-1 is 146 ppb, and was encountered 4 days (96 hours) after injection. The highest concentration of dye (Figure 4) was detected in SP-3 (maximum reading was 233 ppb) followed by SP-2 (104 ppb) and SP-1 (16.20 ppb). At SP-2, SP-3, and SB-1, the dye was detected in all the charcoal packets recovered during the 100 days period.

The results suggest that the ground water is traveling between the site and spring through a fracture system. However, because of differing concentrations detected and times of arrival of peak concentrations at the 3 springs, ground-water movement through the fracture system and the discharge at the springs is affected by conditions throughout the recharge area (218 km^2 or 84 square miles) for the Spring. The travel time (time to leading edge) of the dye from RW-1 (injection well) to the Spring is a maximum 48 hours. The dye traveled between the injection well and the spring in a maximum 48 hours, a distance of approximately 4,900 meters (16,100 feet) (flow velocity of 102 m/hour (335 feet/hour)). The difference between the water level elevation in RW-1 and spring elevation is approximately 9.14 m (30 feet), corresponding to a gradient of the water surface of about 0.1 degree, toward the spring (Figure 3).

The response of the system and the movement of the dye is a combined result of water injected at RW-1 before and after dye injection (head of water in RW-1, Figure 1 and 2), and non-pumping conditions (natural condition) at adjacent manufacturing site. The injected water resulted in an imposed artificial head at the injection well. The water and dye moved out of the injection well through the north-west trending fracture. Within less than 24 hours, the imposed head dissipated and water levels in RW-1 returned to pre-injection levels and fluctuations. The imposed or induced head resulted in an almost unique occurrence, which pushed water out to the northwest fractures, other than normal flow paths. The only other time that those northwest pathways were probably used was in 1977 when the pond failed, and a slug of water was induced into the ground-water system.

On December 29, the pumps were turned on at the adjacent manufacturing (normal conditions), and resulted in appearance of low concentrations of dye in the pumping wells. Under "normal conditions" (pumps on at the manufacturing facility, no artificially induced head in RW-1), significant concentration of dye remained in the well RW-1 indicating the well documented, low conductivity of the system. The last charcoal packet collected in RW-1 on March 31, 1997 still had a concentration of 1,430,000 ppb of dye, and the water sample had 288,000 ppb of dye (approx. 9% of the dye).

The slug flow produced by the injection procedure demonstrated connection between the two locations; but further demonstrated that the connection only functioned for a limited time in response to increasing the head at the well by 2.13 to 4.27 m (7 to 14 feet). After the water levels recovered from the injection, water was drawn to the pumping wells at the adjacent manufacturing facility. Three months after injection, significant dye remained in RW-1, and under "normal conditions" dye had not reached any on-site monitoring wells, except MW-3. This is an

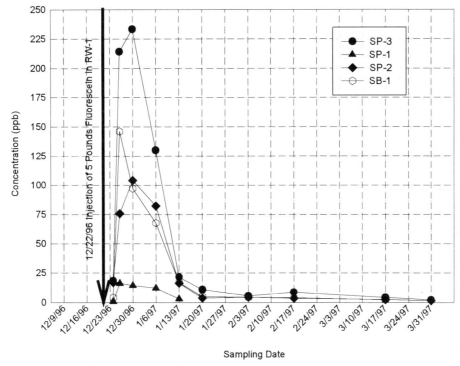

Figure 4.: Fluorescein Concentration in Charcoal packets at spring

expected result of the low gradient and slow movement of ground water under natural conditions.

At the site, the dye was detected only in MW-3 in concentrations from 1.43 ppb (March 3, 1997) to 0.53 ppb (March 17, 1997). The dye was not detected in the other wells sampled and tested. The fluorescein was not detected in other wells most probably because the other wells were not influenced by the slug induced at RW-1, as the bedrock at the site is predominantly of low hydraulic conductivity. When stressed the system exhibits a northwestern anisotropy due to the abundance of similarly oriented fractures. Under normal conditions and under conditions of limited stress (pumping at the adjacent manufacturing facility) interconnection of the northeast and northwest oriented fracture systems provides a weak diffuse flow system for the movement of ground water and movement of PCE bearing ground water.

The MW-3 well is 45.72 m (150 feet) (flow velocity based on computation, 0.027 m/hour or 0.09 feet/hour) southwest of RW-1. The dye detected in this well indicates that the well has a hydraulic connection to RW-1 and that the movement of the dye and ground water was influenced initially by the induced head at RW-1 and subsequently water withdrawal at the manufacturing facility.

At the manufacturing facility the dye was detected in four wells (two pumping wells, and two monitoring well). The pumping wells (TCMW-2 and TCMW-4) are reportedly 91.44 m (300 feet) deep. TCMW-2 is a pumping well 107 m (350 feet) west of RW-1. Water samples only were collected during the dye study. Dye was detected for the first time in the water sample collected on March 3, 1997, arrival time of the dye (time to leading edge) was 71 days after injection, or 1,032 hours (flow velocity 0.104 m/hour or 0.34 feet/hour). The peak dye concentration dye was detected on March 31, 1997, 100 days after injection, or 2,400 hours. TCMW-3 is a deep monitoring well (depth is 67 m or 220 feet below top of casing) 76.20 m or 250 feet northwest of RW-1. Charcoal packets were used in well TCMW-3 during the dye test study. Dye was detected for the first time (time to leading edge) in elutant from the charcoal packet collected on December 30, 1996, arrival time of the dye being 8 days after injection or 192 hours (flow velocity o.40 m/hour or 1.3 feet/hour). The peak concentration of the dye was encountered on February 17, 1997, 57 days after injection.

TCMW-4 is a pumping well situated 110 m (360 feet) southwest of RW-1. Water samples only were collected during the dye test study. Dye was detected for the first time in the water sample collected on January 6, 1997, arrival time (time of leading edge) of the dye being 15 days after injection, or 360 hours (flow velocity 0.3 m or 1.0 feet/hour). The peak concentration of the dye was encountered on January 20, 1997, 29 days after injection, or 696 hours, which correspond to the lowest water level elevation in TCMW-4, during the dye study. TCMW-8 is a monitoring well situated 53 m (170 feet) northwest of RW-1. Charcoal packets were used in well TCMW-8 during the dye test study. Dye was detected only in the charcoal packet collected on January 20, 1997, arrival time of the dye being 29 days after injection or 696 hours (flow velocity 0.07 m or 0.24 feet/hour).

The appearance of the dye in the wells at the adjacent facility is a result of the water withdrawal during pumping at TCMW-2 and TCMW-4. Dye was not encountered at the facility when the pumps were not operating.

At the Spring the dye was detected only in the charcoal packets. Dye was not detected in any water samples collected at the springs during the 100 day test. The passive samplers continuously absorb the tracer dye passing through them; thus they are cumulative samplers. The concentration of any dye in a sampler is, therefore, greater than the concentration in the water. Fluorescein dye is preferentially sorbed on activated charcoal when the charcoal is present in water containing fluorescein. The distribution coefficient (Kd - ratio of concentration of fluorescein in charcoal to that in water) is not available. However, based on the results at RW-1 the concentration of dye in the elutant is 3 to 60 times greater than the concentration in water.

The concentration of fluorescein in the elutant is about 3.5 times less than that in charcoal, because 15 ml of elutant are used to extract fluorescein from 4.5 grams of charcoal. As a consequence the charcoal in packets from RW-1 should have contained 10 to 210 times more fluorescein than that in the water.

The very high concentrations of fluorescein in water from RW-1 cause saturation of the charcoal, and the apparent Kd's may be too low - but would represent a minimum for Kd. In addition, wide variations in the concentration of fluorescein in water from RW-1 indicate that fluorescein may be stratified within the well bore. If we use the data for fluorescein packets from SP-3 (Spring) we can determine a maximum value for Kd, as follows: The detection limit of fluorescein in water is 0.0005 ppb and the reporting limits is 0.0015 ppb. Using the reporting limit as the concentration, when fluorescein is not detected; and concentrations reported for the elutant from charcoal packages in SP-3 at 2.03 to 21.4 ppb range; the ratio of fluorescein in elutant to that in water has a range of about 1,350 to 14,500. Obviously the Kd for the charcoal/water exchange is 3.5 times higher than that for the elutant/water or about 4,700 to 50,000.

According to Ozark Underground Laboratory the concentration factor or Kd of dye on charcoal is time dependent and demonstrated to be several orders of magnitude after only one week and as demonstrated by the above. The system may or may not be at chemical equilibrium after one week. Under natural flowing conditions the system probably does not attain equilibrium, because the concentration of fluorescein changes continuously as a slug-flow passes through the spring. This confirms the relationships seen during the test, where elutant concentrations at the spring sites range vary from the 21 to 2 ppb and water analysis resulted in below detection. Therefore, it is likely that no dye would have been detected in water samples from the spring.

WATER LEVELS

Water-level fluctuations are related to the amount of precipitation and the water withdrawal at the adjacent manufacturing facility (the rate and duration of pumping at each well is not known). The highest ground-water levels across the site were measured between December 20, 1996, to December 29, 1996, when the pumps where off at the manufacturing facility and, on March 17, 1997, two weeks after a precipitation event, indicating the lag between the rain, infiltration, water-level response in the ground-water system.

Water-level fluctuations, over a period of 112 days (December 9, 1996 - March 31, 1997), at the site range from about 0.625 m (2 feet) at the beginning of the test, and in most of cases are in the range of less than 0.15 m (0.5 feet), illustrating that the water surface fluctuations are not dramatic at the site. The water level change across the area covered by monitoring wells is only about 0.60 m (2 feet) indicating that the ground-water surface is nearly horizontal.

Under natural conditions the ground-water gradient is nearly horizontal and ground-water velocities are low. Ground-water withdrawals at the adjacent manufacturing facility locally affect the direction of ground-water movement and gradient locally.

By contrast to the subtle conditions at the site in responding to climatological events, surface water levels and the pool at the Springs exhibit rapid responses due to precipitation and ground water recharge. The pool level rose over 1.25 m (4 feet) after the rainfall event. These conditions demonstrate the influence of precipitation over the recharge area of the springs and illustrate the contrast in hydraulic conditions between the well developed karst at the springs and the poorly connected system at the site.

VOCs CONCENTRATIONS AT THE SPRING

During the dye test, samples were collected for VOC analysis from selected monitoring sites. The PCE concentrations detected at the springs were consistently within a range from 0.28 to 0.66 ppb, at levels, 10 times below the Maximum Contaminant Level (MCL) of 5.0 ppb. The concentrations detected were fairly constant over the period of testing and could indicate a constant source from some location within the 218 km^2 (84 square mile) area of recharge. In contrast, the dye detected at the spring reflects movement of a slug or pulse through the system as a result of the imposed artificial head.

In addition, in comparing the water quality characteristics of ground water at the site and the springs, the ratio of dye concentrations in water, PCE concentrations and constituents from city water are not consistent. The dye concentration in elutant from charcoal packets changed more than 2 orders of magnitude as the dye moved through the spring area. When detected, PCE varied only 0.2 ppb in the spring discharge. Therefore, the data indicates connection between the site and the springs during slug movement, other sources of PCE, and very limited migration of dye and city water since recovery of water levels after injection.

CONCLUSIONS

Under induced artificial head, affected by the injection of water and dye, a hydraulic connection between RW-1 and the Spring was temporarily realized. A slug or pulse of dye-containing water was pushed rapidly through the system via pathways not active under normal conditions.

The water-level data collected before injection, during the injection of city water and the fluorescein dye, and during the remainder of the dye test period are consistent with and confirm monitoring data collected since initiation of investigations at the site. The regional direction of ground-water movement is to the northwest and the ground-water surface at the site is nearly flat, exhibiting a low gradient resulting in low velocity unless disturbed or influenced by outside influences (i.e., ground-water withdrawal at the adjacent manufacturing facility or slug flow).

The VOC's detected at the Spring remained fairly constant, at generally 0.6 ppb or less (well below the MCL of 5.0 ppb), over the period tested, with the exception of samples collected during the flood event (no VOC's detected). The dilution factor for dye in the elutant is over one million, and the highest concentration of PCE ever detected at site was about 1,200 ppb. Therefore, if PCE can survive the trip from the site to the Spring, the maximum concentration would have been 0.12 parts per trillion. This data in conjunction with the dye results suggests the potential for different sources for the two constituents (dye and VOC's). The PCE at the Spring cannot be directly or solely attributed to the site due to many other potential sources identified (previous dye studies) and sources unknown throughout the 84 square mile recharge area.

ACKNOWLEDGEMENTS

The authors would like to acknowledge our client, anonymously, whose site was investigated and who funded the project; and the project team at P.E. LaMoreaux & Associates, Inc. (PELA) that contributed to the collection, interpretation, and presentation of information.

REFERENCES

Driscoll, Fletcher G., Principal Author and Editor, 1987, Groundwater and Wells (2nd ed): Johnson Division, St. Paul, MN, 1089 p.

LaMoreaux, P. E. and Associates, 1989-1997, Consultants Reports

Ozark Underground Laboratory, 1996, Procedures and Criteria. Analysis of Fluorescein, Eosine, Rhodamine WT, Sulforhodamine B, and Pyranine Dyes in water and Charcoal samples.

Quinlan, James F., 1990, Special Problems of Ground-Water Monitoring in Karst Terranes: in D.M. Nielsen and A.I. Johnson, eds. "Groundwater and Vadose Zone Monitoring," American Society for Testing and Materials STP 1053, Philadelphia), p. 275-304.

Quinlan, James F., 1990, Standard Operating Procedures for Fluorometric Analysis of Rhodamine WT and Fluorescein Dyes: Appendix B, Revision #4 (by Howard Trussell).

Quinlan, James F., and Ewers, Ralph O., 1986, Reliable Monitoring in Karst Terranes: It Can Be Done, But Not By An EPA-Approved Method: Ground Water Monitoring Review, Vol. 6, No. 1, p. 4-6.

Quinlan, James F., Ewers, Ralph O., and Field, Malcom S., 1988, How to use Ground-Water Tracing to "Prove" that Leakage of Harmful Materials from a Site in a Karst Terrance will not occur: Abstract, Proceedings of the Second Conference on Environmental Problems in Karst Terranes and Their Solutions Conference, National Water Well Association, Dublin, Ohio, p. 289.

U.S. Environmental Protection Agency, 1986, Test methods for evaluating solid waste, physical/chemical methods, SW-846 (3rd ed.): U.S. Government Printing Office.

U.S. Environmental Protection Agency, 1988, Application of Dye-Tracing Techniques for Determining Solute-Transport Characteristics of Ground Water in Karst Terranes (PB92-231356): U.S. Environmental Protection Agency, Region IV, EPA Report 904/6-88-001.

U.S. Environmental Protection Agency, 1991, Standard operating procedures and quality assurance manual (ESDSOPQAM): U.S. Environmental Protection Agency, Region IV, Environmental Services Division, Athens, GA.

U.S. Environmental Protection Agency, 1992, RCRA ground-water monitoring, draft technical guidance: EPA Report 530-R-93-001 (PB93-139350).

A method for correction of variable background fluorescence in filter fluorometry

S.R. LANE & C.C. SMART Department of Geography, The University of Western Ontario, London, Ont., Canada

ABSTRACT

Semi-quantitative and quantitative tracing requires reliable time-concentration data to define tracer breakthrough and recovery. Fluorescent tracer tests should be designed for minimum dye mass injections while maintaining quantitative detectability above background fluorescence levels. In filter fluorometry this becomes difficult when i) background fluorescence is large and ii) background is variable. In these cases the signal of the dye can easily be misinterpreted in concentration compromising data reliability, or in the worst cases become lost within background noise and natural variability.

Grab sample analysis and continuous flow filter fluorometry revealed a diurnal signal in 'green' background fluorescence in surface streams. The fluctuating background resulted in low uranine concentrations being indeterminate without a correction method capable of handling variable background. 'Blue' background fluorescence was linearly correlated to green background fluorescence. Simultaneous recording of green and blue background allows the blue background to be used as a proxy for green background. This method for determination and removal of a high and variable green background allows an accurate and more reliable determination of uranine breakthrough and recovery.

1. INTRODUCTION

Water tracing is widely used in karst terrain but also has applicability to surface investigations. The determination of discharge in surface river systems is an example of a surface application of water tracing techniques. Prior to this study a continuous flow dilution gauging system using fluorescent dyes, and upgraded Turner model 10 series filter fluorometers (Smart et al., 1998) was developed and evaluated. This technique measures discharge through reliable time-concentration data which define tracer breakthrough curve and recovery, with potential for both surface and sub-surface application.

Semi-quantitative and quantitative tracing requires reliable time-concentration data to define tracer breakthrough and recovery. The definition of a coherent breakthrough curve also lends credibility to a qualitative trace, especially where natural and artificial background are present. Spectacular breakthrough curves leave little doubt about the result of a trace but high concentrations of fluorescent dyes are undesirable for aesthetic, budgetary and environmental reasons. However, minimalist tracing poses the risk of losing the dye or misinterpreting a spurious peak especially if background fluorescence is high or variable.

Background fluorescence can be large relative to the fluorescence of a low concentration of dye especially in the green region of the spectrum, and increasing towards shorter wavelengths. If the background signal also fluctuates then identifying the tracer, let alone the determination of dye concentration becomes difficult with a filter fluorometer. Variable green background combined with conventional subtraction of pre-monitory background may result in negative readings of dye concentration. Smart and Smart (1991) proposed a method for correction of green background using a blue background signal. The principle is that green background fluorescence arises largely as the tail of the high background signal in the blue. It therefore is possible to correct for background interference with green dyes by measuring the blue background and applying a suitable multiplier to calculate expected green background fluorescence. Field testing of the principles and viability of this background correction technique in filter fluorometry is the purpose of this paper.

2. TRACER TECHNIQUE

Natural background influences are studied for two days in a small surface stream. Uranine (acid yellow 73, C.A.S. number 518-47-8) and rhodamine WT (acid red 388, C.A.S. number 37299-86-8). Both fluorescent dyes are safe, inexpensive, and easily detectable in the presence of each other, although both also exhibit undesirable survivability characteristics. Neither dye can be considered perfect, but both are commonly used fluorescent tracers selected for many tracer applications. Rhodamine WT has shown a lower than expected recovery in stream environments (Bencala et al., 1982). This loss is believed to be due to adsorption onto clay mineral surfaces (Shiau, Sabatini and Harwell, 1993). Uranine suffers from photodegradation and high natural background levels in the green region of the spectrum (Smart and Laidlaw, 1977).

Sampling and analytical techniques have associated errors. Sample handling may be a significant source of error without rigorous care in handling protocol, specifically while sampling in the field where controls are less rigorous than in the laboratory. Even within highly controlled spectrofluorometric laboratories, inter-laboratory comparisons of identical samples result in very different breakthrough curves (Behrens et al., 1992). Continuous flow fluorometry, as both a sampling and analytical strategy, eliminates the possibility of these random contamination errors by eliminating the need for operator handling during sampling and analysis (Smart et al., 1998). Stream flow is directed into the fluorometer and measurements are completed as flow passes through the detection chamber. High frequency continuous flow measurement results in very precise, contamination free measurements. Continuous flow fluorometry is best combined with grab sampling to ensure 'Quality Assurance/Quality Control' by maintaining a chain of custody with grab samples.

3. METHOD

Field work for this study was undertaken at Medway Creek, a small surface stream flowing through glacial clay and till deposits in London, Ontario Canada. The stream flows over cobble pools and riffle bed with extensive clay drapes. Some rhodamine WT adsorptive loss onto these clay deposits was expected (Bencala et al., 1982, Shiau, Sabatini and Harwell, 1993). A stage monitoring station operated by the Upper Thames River Conservation Authority was the base of operations, providing easily accessed power and stage based discharge figures for dye recovery estimates.

Calibration was completed using 100 ppb anthracene, 10 ppb rhodamine WT and 10 ppb uranine for the blue, red and green filter sets respectively (see Table 1 for filter configurations). Anthracene was adequate as a blue standard but was found difficult to dissolve in water at concentrations above 100 ppb. Calibration solutions were mixed in distilled water. Filter fluorometers were configured in grab sample mode when calibrated for grab sample background analyses in the laboratory and in continuous flow mode when calibrated for tracing studies in the field (Smart et al., 1998).

Table 1: Filter and lamp setup for the three turner model-10 filter fluorometers

Component	Configuration Nomenclature: Turner Designs part number (10-), Kodak Wratten gel filter (KW) or Corning Colour Specification number (CS)		
	Blue	Green	Red
Lamp	10-049 near UV, 310-390nm	10-045 blue, 400-600nm	10-046 clear quartz
Reference Filter	clear glass	10-053 535nm sharpcut lowpass, KW16	10-053 535nm square sharpcut lowpass
Primary Filter(s)	360nm interference, CS7-7	10-062 455nm sharpcut lowpass, KW3 10-061 390-500 bandpass, KW47B	10-103 550nm interference
Secondary Filter(s)	435nm interference, KW98	10-058 325-700 bandpass, CS4-97 10-059 410nm sharpcut lowpass, KW2a 10-060 510m, sharpcut, KW12	10-058 325-700nm bandpass, CS4-97 10-052 570nm sharpcut lowpass, CS3-66

Automatic samples for background studies were collected from October 30/98 at 0600 hours to November 01/98 at 0600 hours at 30 minute intervals. Additionally, background samples were collected every 2-3 days from November 5/98 to December 02/98 at 1800 hours at the same location each day. Samples were gathered with an American Sigma 6201 automatic water sampler or manually for the daily samples and then returned to the laboratory at the University of Western Ontario and analyzed on a Turner model 10 series filter fluorometer in grab sample configuration and also on a PTI-QM1 spectrofluorometer. It proved more difficult with two dimensional scanning on the spectrofluorometer to detect small differences in background fluorescence. It would have likely required full 3d scans, followed by mathematical integration in order to detect these small differences as reliably. In contrast, the automatic integration of filter fluorometry more readily distinguished small differences in background fluorescence. The weakness of filter fluorometry is it's inability to quantify fluorescence in anything but an integrative manner.

A series of controlled tracer tests were conducted over 1997-98. The focus of this paper is the first trace conducted on November 30/98, hereafter referred to as trace98-11-30/01. Trace98-11-30/01 reported here was undertaken to test the blue background correction procedure. For trace98-11-30/01 1.180g uranine and 1.057g rhodamine WT were injected simultaneously in liquid form from the stream bank. The dye injection point was approximately 40m upstream of the sampling point, with dye travel times of 35.3 minutes to first arrival and 43.5 minutes to breakthrough curve peak. The injection distance and riffle beds ensured complete lateral mixing of the dye cloud from injection to sampling point.

All dye tracing experiments were performed during comparable low stream discharge levels ranging from 90-160L/sec (source: Upper Thames River Conservation Authority) and during overcast days to minimize any uranine photodegradation. Additionally, a thick vegetation cover on both stream banks provided shade to much of the stream, further reducing any potential of uranine photodegradation. Abundant carbonate exists in the well buffered Medway stream waters of pH~8 eliminating any pH effects on uranine. Without moving into a karst system, the site provided straightforward conditions for testing the conservatism of uranine behavior and possible adsorptive loss of rhodamine WT during the traces.

Continuous flow fluorometry, with readings taken every 0.2 seconds, combined with automatic grab sampling every 5 minutes were the chosen sampling techniques. Three Turner model 10 series filter fluorometers were set up in continuous flow mode. Each fluorometer was configured with one of the following filter and lamp sets; blue to read background, green for uranine and red for rhodamine WT. Filter and lamp combinations are outlined in Table 1. The filter fluorometers were parallel linked with lengths of garden hose and stream flow pumped at a rate of ~22.7 L/min into each fluorometer from an identical sampling location 30cm above the stream bed. This setup ensured that each fluorometer detected identical stream conditions at any given moment in time.

Each filter fluorometer was modified for automatic datalogger range switching with appropriate electronic relay systems and Campbell CR-10 datalogger programming (Smart *et al*, 1998). In addition to automatic range switching, programming included; noise reduction through rapid averaging, voltage to ppb conversion, 1 second data recording and automatic breakthrough curve integration for discharge calculation. Datalogger programming also included recording of water temperature for real-time rhodamine WT temperature compensation, using a copper/constantan thermocouple temperature probe. The fluorescence of rhodamine dyes has significant temperature coefficients of ~-2-3% per °C (Käss, 1992). Rhodamine WT temperature compensation and cross fluorescence corrections for rhodamine WT and uranine were determined empirically and post-processed. Cross fluorescence calculations were based upon corrections provided by Smart *et al* (1998) and Käss (1992). Additional post-processing included standardizing injected dye masses to 1.0g to facilitate dye recovery comparisons.

4. RESULTS

All traces completed with uranine and rhodamine WT had the following characteristics (e.g. Figure 1); i) at the tail end of the trace uranine concentration was unexpectedly well above or below green background fluorescence levels with pre-monitory green fluorescent background removed, ii) Upper Thames River Conservation Authority discharge figures, when used to calculate dye recovery, imply a significant loss of both uranine and rhodamine WT, iii) rhodamine WT recovery is considerably less than uranine recovery, and iv) the tail ends of the breakthrough curves cross over with uranine standardized concentration falling below that of rhodamine WT. The results of trace98-11-30/01 (Figure 1) exhibits each of the above characteristics with a uranine breakthrough tail falling below green background fluorescence. Trace98-11-30/01 is the focus of the remainder of the paper.

Both rhodamine WT (49.6% recovery) and uranine (80.8% recovery) are showing non-conservative behavior. Lower than expected uranine recoveries and a uranine concentration falling below background levels at the tail end of the trace suggest a fall in green background levels during the course of the trace. Green background must be examined before any reasonable comparisons can be made between uranine and rhodamine WT breakthrough curves and recovery estimates.

Standard subtraction of average pre-monitory background is inadequate as this resulted in calculated uranine concentrations well below background near the end of the trace (Figure 1). An alternative method of green background correction might be to extrapolate pre-monitory background through the entire trace. This is only possible if i) all dye is rapidly recovered and ii) background trends are linear. These conditions were not true for trace98-11-30/01 as green background appeared to fall and then rise during the course of a day (Figure 2).

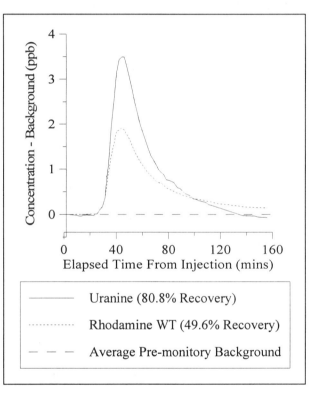

Figure 1: Mass standardized breakthrough curve results for uranine and rhodamine WT, trace98-11-30/01. Average pre-monitory background subtracted (green background ppb equivalent uranine, red background ppb equivalent rhodamine WT). Dye recovery based upon Upper Thames Conservation River Authority discharge numbers.

Figure 2 shows the results of the 24 hr background fluorescence study. Blue background is the highest, followed by the green and then red background. Both the blue and green background fluorescent signals have a diurnal signal, whereas red background exhibits a steady but surprisingly irregular background signal. The diurnal signal of the blue and green background fluorescence has an inverse relationship with incoming solar radiation when compared to the results from a pyranometer operating during the experiment. Many organics fluoresce in the blue, and to a lesser extent in the green, wavelength ranges (Guibault, 1990). These fluorescent materials include introduced organics, e.g. gasoline, and natural organics, e.g. fulvic and humic acids (Mobed *et al.*, 1996). An interaction between organics and incoming solar radiation could cause a fall then rise in green and blue background fluorescence over the course of a day, as in Figure 2.

The similar patterns in blue and green background fluorescence in Figure 2 also suggests a coherent cross fluorescence relationship between green and blue backgrounds (Smart and Smart, 1991). This relationship is supported by a linear regression of green and blue fluorescence in Figure 3 (green background=blue background*0.0171). A linear regression of blue and green background fluorescence over the longer time scale of a month shows that the relationship still holds, shown in Figure 4 (green background=blue background*0.0173), but that there are additional complexities beyond the scope of this paper suggested by significant residuals. Green and blue background fluorescence fluctuations over an hour observation results in a similar multiplier (green background=blue background*0.0167) but at this scale background noise dominates over any linear relationship. Green and blue background fluorescence show a predictable interdependence based upon linear correlations over different time scales.

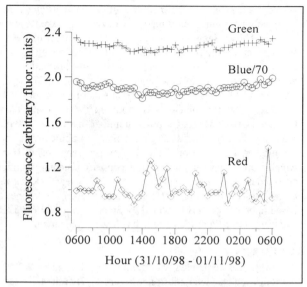

Figure 2: 24 hour observation of Medway background fluorescence in the blue, green and red. Automatic grab sample analysis. Note: blue/70 for plotting convenience.

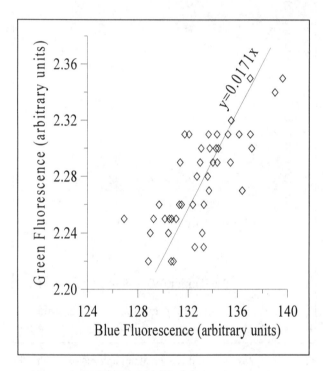

Figure 3: 24 hour observation of Medway background fluorescence green vs. blue. Regression statistics: 2 tailed, df=47, rcrit=+/-0.278, robs=0.839, residuals random.

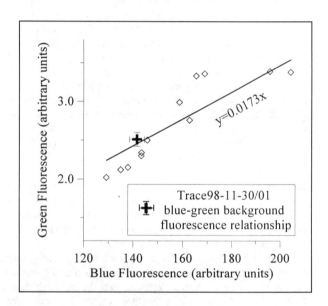

Figure 4: 05/11/98-02/12/98 observation of Medway background fluorescence green vs. blue. Regression statistics: 2 tailed, df=10, rcrit=+/-0.576, robs=0.888, residuals non-random.

The blue-green background fluorescence relationship is also suggested by the fluorescence spectra of a number of different fluorescent backgrounds. Figure 5 shows a number of spectrofluorometer scans of different waters. The high organic content of the leachate and road runoff is reflected in a high blue background fluorescence, relative to the longer wavelength regions of the spectrum. Calculated green/blue background fluorescence ratios decrease as the expected level of organics of the waters increase suggesting an organic specific relationship and particulate scattering influences (Table 2). Prior to a study using this technique of proxy background correction the multiplier must be determined for the study site in question.

Table 2: Green/blue fluorescence ratios for a variety of backgrounds. Tap and distilled water not expected to contain significant organics, but do contain particulate matter increasing scattering background, and the resulting ratio. Blue and green regions of the spectrum defined by 424-491nm and 491-575nm respectively.

Sample	Green/Blue Fluorescence Ratio
Landfill Leachate	0.512
Road Runoff	0.558
Medway Stream	0.660
Johns Flowstone	0.733
Tap Water	0.977
Distilled Water	1.077

Green background was determined using blue background readings and an applied multiplier of 0.0167, determined prior to trace98-11-30/01 (see Figure 4). The trend of the blue background and background corrected uranine breakthrough curve is shown in Figure 6. With real-time background correction uranine recovery estimates increase to 90.0% from 80.8% for trace98-11-30/01. Additionally, the tail end of the uranine breakthrough curve no longer falls below the background green fluorescence level. This green background correction technique successfully improves the quantitative detectability of low uranine concentrations where i) background fluorescence is large relative to dye fluorescent signal and ii) when fluctuating background fluorescence levels give unexpected and inconsistent results.

With green background properly subtracted uranine shows recovery and breakthrough curve detail closer to that expected for a conservative tracer (Figure 6). The corrected uranine peak remains significantly higher than the rhodamine peak, suggesting a significant dye loss of during trace98-11-30/01 reflected in a calculated recovery of 49.6%. This loss is supported by the findings of Bencala *et al.* (1982) where recoveries were as low as 45% of that expected in a mountain stream environment. Low rhodamine WT recovery and a breakthrough curve tail rising above the uranine tail suggest that at least some rhodamine is being reversibly lost by adsorption onto stream bed clays and re-released after the dye cloud passes. Further study is needed for confirmation of controls on rhodamine WT loss and desorption (Sabatini, *et al*, 1993).

CONCLUSIONS

This study has shown that dynamic background correction in the green is possible using blue background fluorescence and a multiplier as a proxy for green background. Background fluorescence can be a problem in the shorter wavelength regions of the spectrum but this technique may in many cases be able to compensate for background effects increasing the quantitative detectability in filter fluorometry. By extension, green background should be able to compensate in the same manner for fluctuations in blue background under the breakthrough curve of a blue fluorescent dye but with expected lower precision from a higher background and noise levels in the blue region of the spectrum.

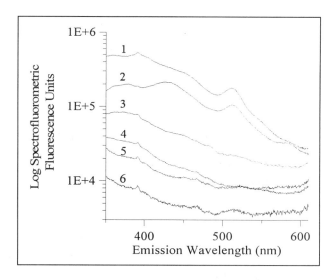

Figure 5: Background spectra, synchronous scans delta lambda = 20nm, slits 5nm (on PTI QM-1 spectrofluorometer)

1) Keele Valley Landfill Leachate, Toronto, Ontario, 12/08/98
2) Road Runoff, London, Ontario 7/10/98
3) Medway Stream, London, Ontario 31/10/98
4) John's Flowstone, Scott Hollow Cave, W.Va. 15/5/98
5) Tap Water
6) Low Grade Distilled Water

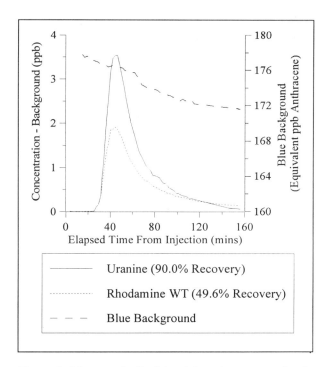

Figure 6: Mass standardized breakthrough curve results for uranine and rhodamine WT, Trace98-11-30/01. Background subtracted (uranine: blue background correction, rhodamine wt: pre-trace average subtracted), blue background readings used for green correction included (right axes).

One of the greatest problems in contemporary fluorometric dye tracing is the effects of natural background fluorescence. To some extent, the nature of the spectrofluorometric scan has overcome the problem of unquantifiable background under a trace. Unfortunately, this sensitive analysis still experiences the random handling errors associated with the potentially, insensitive technique of grab sample collection and analysis. Whatever the sampling or analytical technique employed, for reliable time-concentration data in fluorometric dye tracing, background fluorescence must always be carefully considered.

ACKNOWLEDGEMENTS

Field and Laboratory work have been generously funded by the Natural Sciences and Engineering Research Council of Canada, and the University of Western Ontario. Acknowledgments go to S. Worthington (McMaster University), S. Lane (University of Akron), K. Karaunartne (University of Western Ontario) and M. Helsten (Upper Thames River Conservation Authority) who have helped in many ways with this study.

REFERENCES

Bencala, K.E., Rathbun, R.E., Jackman, A.P., Kennedy, V.C., Zellweger, G.W., and Avanzino, R.J., 1977. Rhodamine WT losses in a mountain stream environment, *Water Resources Bulletin*, vol. 19, no. 6, p 943-950.

Behrens, H., Benischke, R., Reichert, B. and Zupan, M., 1992. Results with Fluorescent Tracers. In Behrens, H., R., Benischke, (Eds.) *Investigations with natural and artificial tracers in the karst aquifer of the Lurbach System (Peggau-Tanneben-Semriach, Austria), Steirische Beitraege zur Hydrogeologie*, vol. 43, p. 81-90

Guibault, G.G. 1990. *Practical Fluorescence, 2nd Edition*, Marcel Dekker, Inc., New York, 812 pp.

Käss, W., 1992. *Geohydrologische Markierungstechnik*, Gebrüder Borntraeger, Stuttgart, 519 pp.

Mobed, J.J., Hemmingsen, S.L., Autry, J.L. and Mcgown, L.B., 1996. Fluorescence characterization of IHSS humic substances: total luminescence spectra with absorbance correction, *Environmental Science and Technology*, vol. 30, no. 10, p. 3061-3065.

Shaiu, B.J., Sabatini, D.A. and Harwell, J.H., 1993. Influence of rhodamine WT properties on sorption and transport in subsuface media, *Ground Water*, vol. 31, no. 6, p. 913-920.

Smart, P.L. and Laidlaw, 1977. An evaluation of some fluorescent dyes for water tracing, *Water Resources Research*, vol. 13, p. 15-33.

Smart, C.C. and Smart, P.L., 1991. Correction of background interference and cross-fluorescence in filter fluorometric analysis of water tracer dyes. In Quilan, J.F. and A. Stanley (Eds.) *Proceedings of the Third Conference on the Hydrogeology, Ecology, Monitoring and Management of Ground Water in Karst Terranes, National Ground Water Association*, Dublin, Ohio, p. 475-491.

Smart, C.C., Zabo, L., Alexander, C., and Worthington, S., 1998. Some advances in fluorometric techniques for water tracing, *Environmental Monitoring and Assessment*, vol. 53, no. 2, p. 305-320.

Non-linear curve-fitting analysis as a tool for identifying and quantifying multiple fluorescence tracer dyes present in samples analyzed with a spectrofluorophotometer and collected as part of a dye tracer study of groundwater flow

RICHARD B.TUCKER   Crawford and Associates, Incorporated, Bowling Green, Ky., USA

NICHOLAS C.CRAWFORD   Center for Cave and Karst Studies, Department of Geography and Geology, Western Kentucky University, Bowling Green, Ky., USA

ABSTRACT

Non-linear curve-fitting analysis computer software can be used to segregate the emission spectra from a synchronously scanning spectrofluorophotometer data stream into the individual spectra created by each fluorescent molecule present in the sample. Synchronous scanning of a fluorescent molecule on a spectrofluorophotometer generally produces a data stream that can be modeled with a Gaussian curve. By specifying that the data stream is to be modeled with Gaussian curves of varying heights and widths, a data stream generated by the synchronous scanning of a sample containing multiple fluorescent molecules can be broken down into the individual spectra making up the data stream. This is the case even when the emission spectra of several molecules overlap. Once the data stream has been broken down into its constituent spectra, the individual molecules present in the sample can be identified by the wavelength of the peak locations of the individual Gaussian curves. The non-linear curve-fitting analysis software can then perform integration on each Gaussian curve to determine the area bounded by the curve. Through the use of calibration curves derived from similar analysis performed on laboratory standards, it is possible to calculate the concentration of each fluorescent molecule found to be present in the sample.

Non-linear curve-fitting analysis software provides two statistics-based methods for checking the correctness of the generated model. The first method pertains to how accurately the individual Gaussian curves define the original data stream. The second method provides a complete statistical analysis of each Gaussian curve determining to what extent a curve fits into the complete model. In order for a model to be considered acceptable, it must meet certain criteria under both methods.

Crawford and Associates has used non-linear curve-fitting analysis in order to analyze samples containing as many as seven dyes. Non-linear curve-fitting analysis has proven to be an essential tool for analyzing samples from dye tracer investigations at sites requiring multiple dye injections into the same or adjacent drainage basins.

INTRODUCTION

Fluorescent dye tracing is the tool of choice for characterizing groundwater flow in karst terranes. Over the past few years, demanding new standards have been placed upon the field of dye tracing by both regulatory agencies and environmental consulting firms. These new standards have forced traditional dye tracing methods, which often relied upon subjective analytical techniques, to be replaced with techniques more soundly rooted in scientific principles. At the same time, major improvements have been made in the instruments used to analyze samples for fluorescent molecules. Today, the instrument of choice for analyzing samples for fluorescent dyes is the synchronously scanning spectrofluorophotometer. This instrument is capable of analyzing for virtually an unlimited number of fluorescent dyes, without the need for special filters for each dye. This new capability, along with improved dye tracing protocols, quickly generated interest in conducting simultaneous dye tracer investigations in which numerous dyes could be injected at the same time. Such investigations provide considerable cost and time savings. Non-linear, curve-fitting software makes it possible to perform these types of investigations.

ANALYTICAL PROCEDURES FOR FLUORESCENT DYES
Charcoal Preparation

1. Charcoal dye receptors are washed under a high-speed jet of tap water to remove as much sediment as possible.

2. A typed label containing the site location name, sample number, and date of collection are stapled to each receptor.

3. The receptors are placed in an oven and dried for 12 hours at 120° C.

4. 0.5 grams of charcoal is weighed and placed into a disposable plastic container labeled with a sample identification number.

5. The remainder of the charcoal is returned to its original zip-lock bag and, along with the cotton receptor, stored until the dye trace investigation is complete.

6. 5.0 ml of Smart solution (an eluent consisting of 1-propanol 100% assay, distilled water, and ammonium hydroxide 28-30% assay mixed at a ratio of 5:3:2) is added to the charcoal and the plastic container is capped.

7. After 30 minutes, all of the elutant is transferred into a KIMAX®51 borosilicate glass test tubes that is then sealed with a polypropylene cap.

8. The sample trays are then placed in a constant temperature bath, covered to prevent photochemical decay and allowed to equilibrate to the 30° C analysis temperature.

9. The test tube is then placed in the Shimadzu RF 5301PC spectrofluorophotometer for analysis by synchronous scanning.

Charcoal Analysis

Analysis on a scanning spectrofluorophotometer provides the lowest detection limits and most reliable dye analysis. For a typical analysis for Tinopal CBS-X Direct Yellow 96, Fluorescein, Rhodamine WT, Eosine, FD&C Red 3, D&C Red 28 or Sulphorhodamine B, a synchronous scan is performed where the excitation and emission monochromators are kept at a fixed wavelength separation during the scan. The emission spectra from the synchronous scan is displayed on the monitor and plotted on a laser printer (Figure 1). The printout has the file number, sample identifier, job name, collection date, and scanning parameters at the bottom of the page. If the scan indicates positive results for fluorescent dye, a second printout is made utilizing spectrum integration and calibration curves stored in the computer to determine the concentration of the dye in question (Figure 2). If the emission spectra from two or more dyes overlap (Figure 3), then the spectra for each dye are separated by use of a non-linear curve-fitting computer program specifically designed for spectral separation. Spectrum integration and calibration curves stored in the computer are then used to determine the concentration of each individual dye present in the sample. For a sample where the concentration is approaching the quenching threshold, serial dilutions are made until it can be scanned without quenching occurring.

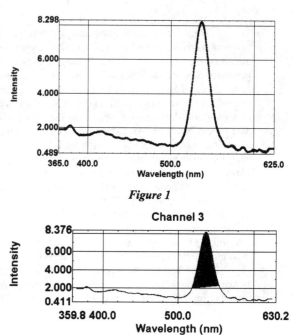

Figure 1

Figure 2

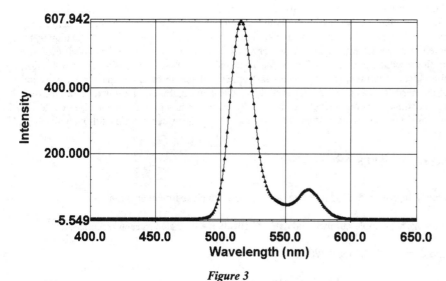

Figure 3

Water Sample Preparation

1. Water samples are removed from locked refrigerator.

2. A typed laboratory identification tag containing the project name, site location number and collection date is taped to each sample bottle.

3. A 3.0 ml aliquot is withdrawn from the bottle using a disposable polyethylene pipette and placed into a KIMAX®51 borosilicate glass test tube which is then sealed with a polypropylene cap.

4. The sample trays are then placed in a constant temperature bath, covered to prevent photochemical decay and allowed to equilibrate to the 30° C analysis temperature.

5. The test tube is then placed in the Shimadzu RF 5301PC spectrofluorophotometer for analysis by synchronous scanning.

Water Sample Analysis

The analysis for water samples is performed by synchronous scanning on the spectrofluorophotometer. The synchronous scanning technique is similar to the analysis for eluted charcoal samples with the scanning parameters adjusted to compensate for shifts in the excitation and emission maximum wavelengths, as well as differences in the Stoke's shift caused by the differences in pH and polarity of water (as compared to elutant). The emission spectrum from the synchronous scan is displayed on the monitor and plotted on a laser printer. The printout has the sample identifier, job name, collection date, and scanning parameters at the bottom of the page. If the scan indicates positive results for fluorescent dye, a second printout is made utilizing spectrum integration and calibration curves stored in the computer to determine the concentration of the dye in question. If the emission spectra from two or more dyes overlap, then the spectra for each dye is differentiated by use of a non-linear curve-fitting computer program specifically designed for spectral separation. Spectrum integration and calibration curves stored in the computer are then used to determine the concentration of each individual dye present in the sample. For a sample where the concentration is approaching the quenching threshold, serial dilutions will be made until it can be scanned without quenching occurring.

NON-LINEAR, CURVE-FITTING ANALYSIS PROCEDURES

Baseline Processing

Baselines are subtracted utilizing a non-parametric algorithm that employs a Gaussian weighting function. This algorithm can fit any shape baseline when defined by a sufficient number of points. This method is particularly effective when processing spectral data that extends in range from the ultraviolet through the visible portion of the spectrum (Figure 4). When processing spectral data within a narrow range of the spectrum, the non-parametric algorithm functions, in effect, the same as a linear baseline subtraction.

Initial Peak Placement

Initial peak placement is performed by the software. Peak placement locations are determined through the use of data smoothing and baseline subtraction algorithms that are performed on a copy of the data set. No alterations are made to the actual data set. Once the data set has been modified, the non-linear, curve-fitting software places peaks at locations where detection of local extrema occurred (Figure 5). After the initial peak

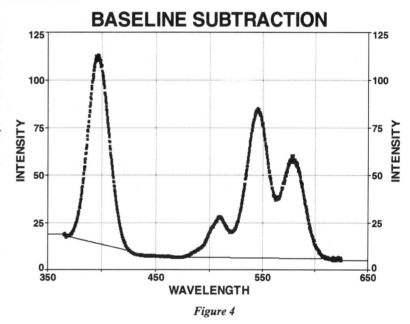

Figure 4

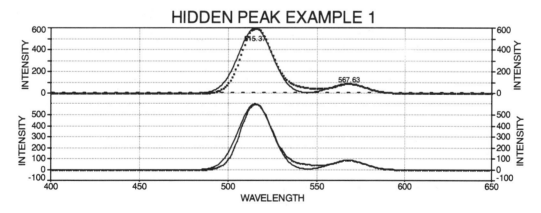

Figure 5

placement, the software performs a Residuals Analysis in order to locate areas under the original curve that are not defined with the current model (Figure 6).

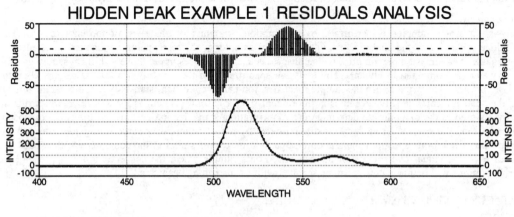

Figure 6

The software gives the user the option of letting the software add additional peaks, automatically, to compensate for the discrepancy; or the user may elect to add additional peaks based upon their visual perception of where peaks may be needed (Figure 7).

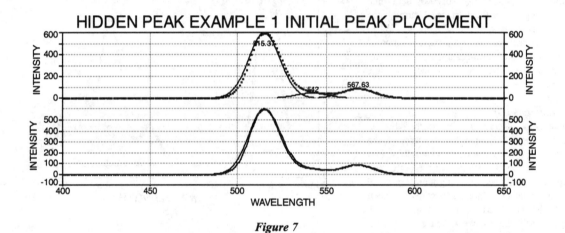

Figure 7

After adding a peak, a second Residuals Analysis is performed (Figure 8). Additional Residuals Analyses are performed each time any modification is made to the peaks in the model. Typically, the user should allow the software to add additional peaks, removing any peaks that

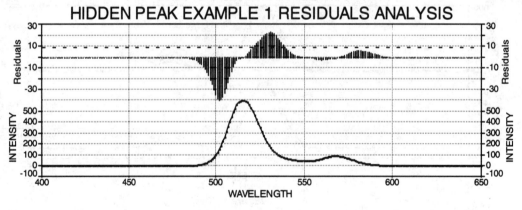

Figure 8

are perceived to be placed incorrectly. This method helps to insure the best initial peak placement, with both the software and the user providing input.

Non-Linear Peak Fitting

The algorithm used by the software to adjust the peaks from their initial placement is the Levenburg-Marquardt minimization algorithm. Like all non-linear minimization algorithms, a model that fits the data better prior to use of the algorithm will produce end results that are more representative of the true nature of the data set.

Once the fitting routine is completed, the software allows the user to either make further modifications to the model, or to review the fitted data. Two types of review are provided by the software, graphical and numerical.

Graphical Review

Graphical review allows the user to see graphically how well the model represents the original data set (Figure 9). The graphical review also provides the user with information regarding the numeric quality of the fit through 3 statistical calculations: 1) the coefficient of determination (r^2), 2) the fit standard error (SE), and 3) the F-statistic.

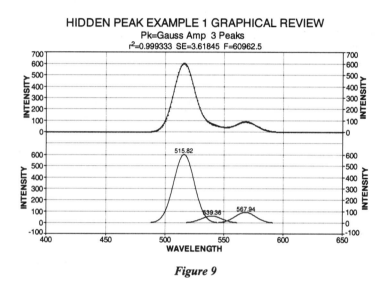

Figure 9

Numeric Review

Numeric review allows the user to view a complete analytical summary of the fit including the following items:

1. Fitted parameters including goodness of fit and fitted parameter values
2. Measured peak characteristics and analytical areas
3. Parameter statistics for each peak
4. Analysis of Variance
5. Details of the fit such as convergence state, iterations, and fit settings

By reviewing these values, the user can determine whether the fit should be accepted or if additional modifications to the model are necessary.

NON-LINEAR, CURVE-FITTING ANALYSIS USED TO IDENTIFY EOSINE DYE IN A SAMPLE CONTAINING THREE DYES WITH OVERLAPPING EMISSION SPECTRA

A dye tracer investigation was performed at a low level radioactive waste burial ground at the K-25 Gaseous Diffusion Plant inside the Oak Ridge Reservation. Four dyes were injected into monitoring wells in and adjacent to the burial ground. Three of the dyes were detected on activated charcoal dye receptors recovered from two springs adjacent to each other (hereafter referred to as Spring 1 and Spring 2). At both locations, Fluorescein and Sulphorhodamine B dye were easily identified in the scans for the samples due to the localized maximums produced by each dye (Figures 10 and 11). The Eosine dye did not produce a localized maximum and could only be detected through the used of non-linear, curve-fitting analysis. By week 17 of the investigation, Eosine produced a localized maximum in the data stream for the scan of the activated charcoal dye receptor recovered from Spring 1 (Figure 12). Eosine never produced a localized maximum on any sample recovered from Spring 2 throughout the duration of the investigation, even though its presence was apparent in the scan for week 18 (Figure 13).

By utilizing non-linear, curve-fitting analysis, dye was detected at a location that would have been classified as non-detect using conventional analysis only. It was also possible to detect the dye at Spring 1 nearly 6 weeks earlier than with conventional analysis.

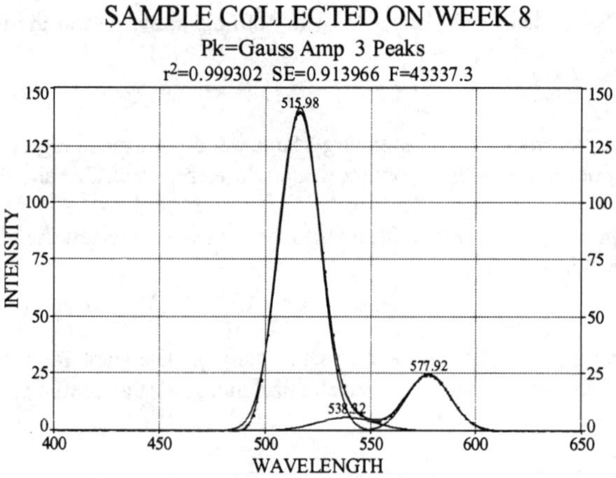

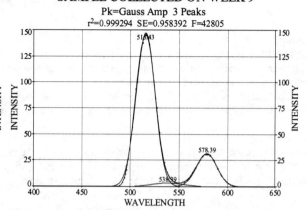

Figure 10. First Eosine detection at Spring 1 occurred during week 8. Eosine did not produce a localized maximum in the data stream.

Figure 11. First Eosine detection at Spring 2 occurred during week 9. Eosine did not produce a localized maximum in the data stream.

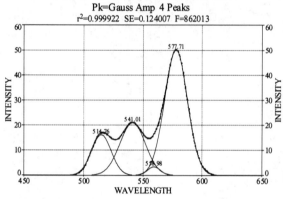
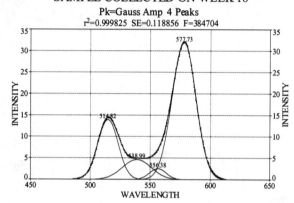

Figure 12. Eosine produces a localized maximum at Spring 1 during week 17.

Figure 13. Eosine obviously present at Spring 2 during week 18, but does not produce a localized maximum.

CONCLUSIONS

Non-linear, curve-fitting analysis has proven to be an effective tool in multiple tracer investigations and in situations where background dye levels were sufficient to interfere with detection of planned tracer dyes. Dye tracer investigations have been performed where as many as 7 dyes have been injected simultaneously. The cost and timesaving made possible by performing multiple tracer tests makes non-linear, curve-fitting analysis a cost-effective tool as well.

SOURCES

Crawford and Associates, Inc. Karst Groundwater Investigation Research Procedures, 1997.

Crawford and Associates, Inc. *Laboratory Report*, K-25 Gaseous Diffusion Plant Dye Tracer Investigation, 1996.

Rundel, Robert, Ph.D. *Technical Guide*, PeakFit Non-Linear Curve-Fitting Software Manual, 1991.

8 Environmental hydrogeology in karst terrane

Shallow lateral DNAPL migration within slightly dipping limestone, southwestern Kentucky

MARK JANCIN & WALTER F. EBAUGH Nittany Geoscience/Chester Engineers, State College, Pa., USA

ABSTRACT

DNAPL migration from source areas has mostly been governed by gravitative movement along beds dipping ~1° NW that show diffuse-flow, semi-confined characteristics. This shallow, phreatic migration has occurred within relatively fractured and vuggy beds of the Mississippian St. Louis Limestone at depths beneath ground of 12-15 m, over lateral distances to 230 m. Near source areas, vertical product migration is occasionally deeper than 21 m beneath ground. Migration along the top-of-bedrock (TOB -- generally at depths of 1.5-4.6 m) appears to be restricted to distances of less than 46 m, and therefore is unlikely to foster significant offsite migration. However, opportunely located TOB lows can provide routes for free-product entry, into the main lateral-migration beds, that are not located beneath source areas. Preferred migration along layer diplines appears to reflect: (1) bedding-plane partings; (2) interconnected, layer-bounded joints; (3) minor solutional enhancement of the above two features, especially the first; (4) sufficiently large and interconnected pore spaces, dissolution pits, and vugs; and (5) any possible solution conduits (for which there is no direct evidence, but some local occurrences are hinted at by site hydraulic characteristics). Local deflection of migrating free product away from the bedding dipline can be expected where (1) NW-trending bedding, joint, or vug-network pathways are poorly developed or absent in proximity to such available pathways trending along other directions, or (2) clay-filled depressions in the TOB surface, deeper than about 12 m beneath ground, act as barriers to downdip migration in the uppermost bedrock. Where such critically deep TOB lows might be filled with relatively permeable sands, lateral DNAPL migration from the uppermost bedrock into the soil mantle could occur along the updip slope of the TOB low. The directions of groundwater flow and transport (to the W) and DNAPL migration (to the NW) are divergent, which probably led to a plan-view aqueous plume that, over decades, had advanced toward the NW and W while tapering toward the migration limit to the NW.

INTRODUCTION

The subject industrial facility in southwestern Kentucky has been operating for over 80 years. (The site identity and location are proprietary and are not specified in this paper.) Earlier product and waste handling allowed dense nonaqueous phase liquids (DNAPL) to enter the subsurface, where it has migrated both vertically and laterally. Soluble aqueous organic constituents derived from the DNAPL are present in groundwater beneath the site. Chieruzzi et al. (1995) have described pumping tests at this same site, as well as a successful pumping system for hydraulic control and DNAPL recovery within the upper part of the limestone bedrock. Groundwater largely behaves in a diffuse, approximately Darcian manner (equivalent homogeneous medium) on the scale of roughly 152-305 m.

The present study involved the development of a conceptual site model for contaminant fate and transport, and was partly motivated by the presence of a karst spring, 3 km from the site, which serves as the sole-source drinking supply for the local community. The inferred, predominant direction of flow is based on integrating site water levels (under both natural and pumping conditions) with information on regional directions of fracture traces, joints, and cave passages. Based on free-product occurrences in wells and groundwater quality analyses, DNAPL has migrated along the dipline of the slightly inclined beds within perched pathway networks of bedding partings, joints, and vugs, and in some areas has probably utilized intergranular porosity.

GEOLOGIC FRAMEWORK OF THE KARST

The site is located within the Interior Lowlands Province of the United States. Within this physiographic province it falls in the Mississippian Plateau, which super-regionally is called the Pennyroyal Plateau or Plain, underlain by limestone bedrock. The Pennyroyal Plain has large tracts which are characterized by sinkholes (closed surface depressions) and conduit-influenced subsurface drainage. However, within the site itself, none of these karst features have been identified.

The Pennyroyal Plain in this region is underlain by the Ste. Genevieve Limestone and the St. Louis Limestone. Both of these formations are of Mississippian age, with the younger Ste. Genevieve overlying the older St. Louis. At the site, generally the topmost

1.5 m of bedrock is the Ste. Genevieve. Bedrock cores from the northern site area show the Ste. Genevieve Formation comprises oolitic limestone or vuggy, clastic limestone containing bryozoans and brachiopods. The St. Louis cores show fine to medium-grained limestone, often fossiliferous and cherty, varying from vuggy to relatively compact (Chieruzzi et al., 1995). Super-regionally, the site is near the troughline of the very gently northward plunging, very low amplitude structural basin that exposes the Western Kentucky Coal Field in the northern part of western Kentucky. Bedrock-layer dip angles in the general site area apparently vary between about 1/2° and 1°, but the strata regionally undulate and the dip direction varies. The best estimate of the site-specific bedding orientation is a northeast strike, with a dip of 1° to the northwest.

Regional baselevel streams trend roughly north-south, with flow to the south. These two streams are separated by an east-west distance of roughly 24 km, within which the site is located. These streams feed the Red River to the south of the site, which in turn discharges to the Cumberland River in the area of Clarksville, Tennessee. All of these streams follow meandering paths incised into the limestone. Regional topography tends to be very level to gently rolling, with locally varying degrees of sinkhole development. Seasonal and wet-weather karst swamps and lakes are common. Karst springs, karst windows, and swallets are developed to varying degrees across this region of low relief. Many surface depressions in the region tend to pond on an intermittent basis, but otherwise do not act as swallets. The site is relatively flat due to regrading of the land, with ground elevations of 166-169 m -- gentle hills surrounding the site stand ~7-11 m higher. The site water table generally is about 1.5 m below the ground surface and between 6 and 9 m above the top of bedrock.

Borings from the north part of the site show the soil mantle generally has a topmost thin layer of loess (silt) and alluvial soils (clay, silt, and fine sand mixes) overlying red, cherty clay that is residual from the underlying limestone (Chieruzzi et al., 1995). Monitoring well logs and borings show the soil-mantle thickness varies from 7-16 m with a typical thickness of ~8 m -- the thickness variations generally reflect the relief on the underlying top-of-bedrock (TOB) surface. Discontinuous sand lenses to stringers locally occur in the soil mantle, typically varying from 0.3 to 1 m in thickness as sinuous deposits. These sands probably were deposited along former paths of minor creeks that pass through the site. Locally these sands are directly on the TOB. The TOB lows do not systematically coincide with relatively thicker sand deposits -- the location and thickness-distribution of the sands is not governed by the varying depths to the TOB. Maximum TOB relief over the northern site is ~7 m, with the steepest local slopes (along the same TOB low) varying from ~45° to 24°. Northwest of the site boundary along well nest M-22 the TOB relief is 6 m, implying very steep to subvertical local TOB slopes. Figure 1 shows a cross section through the northwestern edge of the site. This section illustrates part of the TOB surface and site water table. The site and surrounding area comprise subsoil karst with a shallow water table and perennially saturated TOB.

DIRECTIONS OF STRUCTURAL ELEMENTS AND CAVES

The overall plan-view configuration of the northern site TOB is a combination of fairly equant highs and lows mixed with varyingly more elongate highs and lows. For the more elongate highs and lows the preferred trends are (in order of decreasing frequency): (1) NW; (2) WNW to E-W; (3) NE; and (4) N-S to NNW. There are no deep TOB slots persisting across the site.

Part of a fracture-trace (photolinear) map of the northern site and adjacent areas has been presented in Chieruzzi et al. (1995, their fig. 2). Two main sets of fracture traces were identified (with significant dispersion about these trends): (1) within 15° of N-S; and (2) within 10° of E-W. Subsequently, two boreholes were drilled along two of these fracture traces with the expectation that such locations might prove useful for pumping-test wells because of relatively enhanced bedrock permeability due to the presence of close fractures or solution conduits. Short-term pumping tests at these two locations provided very low sustainable yields (significantly <1 gpm). 3-m spacing to Geoprobe borings across these two photolinears did not show any correlation to TOB lows (Chieruzzi et al., 1995).

No site-specific or nearby data on joint orientations and spacing are available. Site bedrock well logs (continuous cores) make occasional reference to subvertical fractures or veins and, much more frequently, to subhorizontal (presumably layer parallel) fractures, but no orientations on the former are available. Of course, there is a very strong inherent geometric bias against vertical wells intersecting subvertical joints that are spaced more widely than the wellbore. It is virtually certain that the site bedrock is pervasively jointed. The bedrock cores show the joints encountered are generally quite tight with very few gaps due to solution.

The nearest available generalizations of joint orientations we found are from Kemmerly et al. (1987) who studied limestones near Clarksville, Tennessee. Their three generalized, systematic joint-set strikes are: (1) N 20-30° W (subparallel to the local bedding dipline); (2) N 20-40° E; and (3) N 70-80° E (subparallel to the local bedding strike). Senior author Phillip R. Kemmerly (personal communication, 1996) noted that this third joint set is preferentially solutioned in the limestones he has examined in the Clarksville area.

Through Dougherty (1985) and the Western Kentucky Speleological Society (WKSS) (Mylroie, 1981, 1982 & 1983 combined, 1984; unpublished information and personal communications, 1996) information has been accessed on 32 known caves in the county containing the site. The WKSS is dedicated to the exploration, mapping, and documentation of western Kentucky caves. The nearest caves to the site are about 10 km to the northeast, near both banks of a baselevel stream. Within this distance six known caves are located, including the second longest that is entirely beneath the county (mapped at 3168 m). The longest cave that is entirely within the county is Carter Cave (mapped at ~4570+ m). The WKSS recently completed a dye trace of a stream in Carter Cave that showed a hydrologic connection to nearby Glover's Cave (which extends across the county boundary to the west). Glover's presently is mapped at 3660+ m -- the Carter-Glover's cave system is one of the longest yet known in western Kentucky.

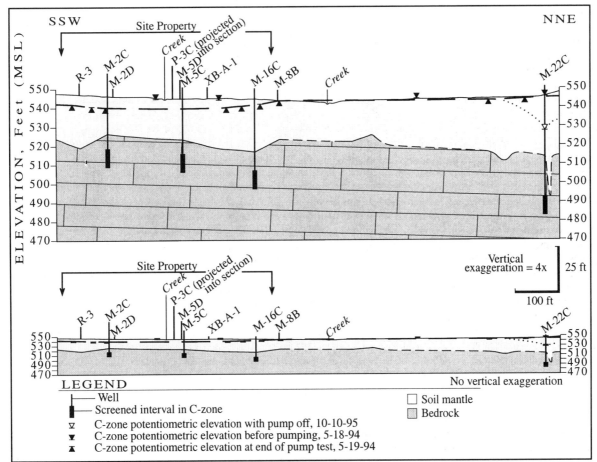

Figure 1: Cross section through the northwest edge of the site in southwestern Kentucky. Upper section shows a vertical exaggeration of 4X; lower section is identical but shows no vertical exaggeration.

Review of available cave maps and descriptions suggests significant joint control of the cave passages, as inferred from segments that are conspicuously straight. Most meandering passages have shorter, straight component segments that are linked into an overall sinuous configuration. Most of the caves have map-view patterns that are angulate to sinuous single conduits; occasionally the passages splay into a branchwork form. The frequency of passage orientations for the entire county is the same as that generalized for the six caves nearest the site. In order of decreasing frequency these trends are: (1) E-W to WNW to ENE; (2) N-S to NNW; (3) NE; and (4) NW. Dispersion about these orientations occurs. These preferred passage orientations each correspond to either a fracture-trace set, a Clarksville-area joint set (Kemmerly et al., 1987), or both. Overall, passage cross-sectional dimensions are highly variable, but there is a strong tendency for width-to-height ratios to exceed 3:1. This indicates that solution is most pronounced along bedding, although solution within beds has been guided by joints.

With some exceptions, known caves are throughways for diverted surface streams that resurge along the banks of the major baselevel streams. Most of these caves occur within 3 km to either side of these baselevel streams. These streams meander but have overall N-S trends with flow to the south. The cave passages tend to have dominant E-W (to WNW to ENE) trends, subparallel to the tributary surface streams they capture. These caves are acting as subsurface, dendritic feeders to the baselevel streams. Those parts of the regional karst that are farther than roughly 3 km from these baselevel streams appear to have very limited development of accessible caves.

At least two reasons can be suggested for this regional cave distribution: (1) farther from the baselevel streams the water table may generally be shallower, leaving relatively little vertical room in any vadose-zone bedrock for large cave development and human access; and (2) the geologic origin of the caves is tied to a regional fluviokarst setting, and over geologic time the larger conduit networks tend to become increasingly well developed in directions headward along their tributary-stream drainage basins. With one baselevel stream located about 6 km east of the site, and the other about 16 km west, the headward expansion and integration of such cave networks has not had sufficient time to reach the location of the site. However, interconnected proto-conduits that are far too small for human penetration can exert strong influence on the direction and rate of groundwater flow, and dye tracing (Ray, 1997) in a nearby, apparently cave-free groundwater basin suggested a relatively low capacity, but well integrated, conduit-flow system in that area.

GROUNDWATER HYDROLOGY

Monitoring wells are screened within four depth zones at the site which have been designated as the following four aquifer zones: A-zone wells are screened within the surficial soil unit from 1.5 to 4.6-m depths; B-zone wells are screened within the surficial soil unit directly on top of the bedrock from 6 to 9-m depths, +/- 1.5 m; C-zone wells are screened within the upper bedrock aquifer zone from 12 to 15-m depths, +/- 3 m; and D-zone wells are screened within the lower penetrated bedrock aquifer zone from 18 to 21-m depths, +/- 4.6 m. The C-zone screens correspond to the uppermost vuggy and relatively highly fractured bedrock; D-zone screens correspond to a lower vuggy and relatively highly fractured bedrock zone. Generally, these two vuggy zones are separated by a relatively compact zone of bedrock of 3 to 6-m thickness which occurs at depths of 15 to 18 m. Groundwater in the C and D-zones behaves in a confined to leaky-confined manner. 3-m screens on site wells permit sustainable yields of less than 2 gpm, indicating low permeability to the bedrock in the northern part of the site. No large solution voids have been encountered in site wells.

Site groundwater recharge appears to be distributed (diffuse). No focused recharge points are located within or near the site. Head-level data sets generally show a fairly well-defined water table and concordant trends to potentiometric contours in all four aquifer zones. However, horizontal hydraulic gradients vary directionally, in time and location, from N to NW to W to SW. There is no present justification for projecting these gradient directions very far from the site. Under background, non-pumping site conditions well M-22C has shown head-level oscillations of almost 6 m over time (Figure 1), while C-zone levels within the northern site generally have oscillated <0.6 m (this same generalization applies to the other three aquifer zones). Since there is no offsite pumping, it is possible this well nest is influenced by a nearby solution conduit. Elsewhere along the northern site, a locally steep gradient configuration sometimes reverses direction, which also raises the possibility of localized conduit influence.

Pumping tests have been conducted at the site toward developing an effective hydraulic gradient control system and DNAPL recovery system. Starting in 1995, hydraulic control has been maintained through sustained pumping of 0.9 gpm from bedrock recovery well P-3C, screened in the C-zone. During a 32-hr pumping test in 1994, a rate of 1.4 gpm in P-3C was sufficient to generate an approximate 116 to 166-m radius of influence within the C-zone bedrock aquifer. Figure 1 shows a cross-sectional view through the flank of this drawdown cone. B-zone drawdown during this same pumping test (which directly stressed the C-zone) was centered at well M-5B, about 91 m WNW of pumped well P-3C. This WNW-trending spatial offset between the centers of C and B-zone drawdown is parallel to the long axis of the C-zone head-level contours during this same test (Chieruzzi et al., 1995, their fig. 4) -- this latter configuration maintained during pumping through 1996. These observations suggest enhanced C-zone horizontal conductivity along this WNW trend, as well as enhanced vertical hydraulic communication between the B and C-zones in the vicinity of the M-5 well nest.

C-zone time-drawdown data from the 32-hr pumping test showed a very good fit to a Hantush leaky aquifer solution (Chieruzzi et al., 1995). This fit to an equivalent homogeneous (porous) medium, Darcian-flow solution suggests the predominance of diffuse flow over the C-zone area that experienced drawdown. At the same time, a very low hydraulic gradient between pumped well P-3C and an observation well located 61 m to the SW is consistent with the observation that the latter well showed a blowout (freeflow) during the drilling of the former well. These observations suggest the possibility of conduit connection between these two wells.

Ray (1997) has discussed the role of dye tracing, low-flow stream gaging, and unit-discharge analysis for accurate delineation and understanding of several groundwater basins in Kentucky. One of the basins he described is located near, but is separate from, the basin containing the subject industrial site. Ray emphasized the importance of intermittent, overflow conduit systems in these and other basins within the Mississippian carbonate aquifer of central and southern Kentucky. The small basin (31 km^2) dye traced near the site has no known caves, but dye injection at swallets and sinkholes showed conduit-influenced, fast flow (290 to 1280 m/day) to the water-supply spring in the area, which is perennial. This study by Ray and the Kentucky Division of Water showed that this water-supply spring drains a basin separate from that containing the site.

CONCEPTUAL SITE MODEL

Bedrock cores from the northern part of the site show varying degrees of development of solution pits and vugs, bedding-parallel partings or fractures, and subvertical fractures. No large solution voids have been encountered, although a 46-cm-tall void was noted in one well at a depth of 11 m beneath ground. Smaller layer-bounded voids to about 5 cm in height are occasionally noted in many well logs. To aid in further conceptualizing the likely nature of onsite bedrock pores and pathways, bedrock exposures in area streambanks and quarries were examined. These exposures showed that selective solution generally appears more focused along bedding than along steeply dipping joints, and that many of the subvertical joints are layer-bounded.

The site model for groundwater flow forms the core for prediction of the extent of the organic aqueous constituents derived from the DNAPL. Flow will be biased in directions of decreasing head but controlled by available pathways. In order of decreasing importance these pathways are: (1) bedding-controlled features including partings and degrees of perching due to the presence of insoluble chert layers and relatively tight limestone lithologies; (2) interconnected joints, sometimes layer-delimited and enhancing conductivity parallel to bedding; (3) minor solutional enhancement of the above two fracture types, especially the first; (4) relatively large and interconnected pore spaces, dissolution pits, and vugs; and (5) any possible solution conduits (for which there is no direct evidence but, as discussed above, some local occurrences are hinted at by site hydraulic characteristics). Based on pumping tests, flow appears to be approximately diffuse on the scale of roughly 152-305 m. Such a Darcian block may be bounded by local zones of enhanced fracture or solution-conduit permeability -- such remains to be characterized. Any laterally persistent conduits will exert strong influence on aqueous transport. It is unlikely the bedrock aquifers surrounding the site are isotropic; therefore, one should not assume groundwater flow will strictly be directed along maximum hydraulic gradients.

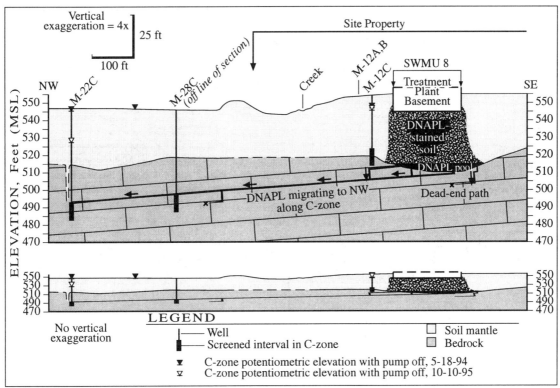

Figure 2: Cross section through the north part of the site in southwestern Kentucky. The section is approximately parallel to the dip direction of the limestone. Upper section shows a vertical exaggeration of 4X; lower section is identical but shows no vertical exaggeration.

Based on available information concerning directions of structural elements and caves, and site drawdown, the preferred pathways for aqueous transport are likely to be within roughly 10° of E-W and toward the west. Flow to the SW, subparallel to the strike of bedding, is also a possibility that cannot be ruled out at present. Poorly constrained gradient data suggest the possibility of localized transport toward the east. Since typical C-zone head potential in the northern part of the site is about 166 m in elevation, it is possible the site groundwater regionally discharges at this, or a lower, elevation along small (non-baselevel) streams to the west or southwest of the site. Groundwater impacted by site operations is not expected to reach the water-supply spring located about 3 km to the west.

Organic constituents derived from solution of the DNAPL are naphthalene, polynuclear aromatic hydrocarbons, phenols, and benzene-toluene-xylenes. We totaled the concentrations of all these constituents from sampled C and D-zone wells and developed isoconcentration maps for the northern site C and D-zone aquifers -- these maps show the known distribution of aqueous constituents in the upper bedrock. On these same maps we located DNAPL occurrences in C and D-zone wells, along with likely potential surface DNAPL source areas. These maps fostered the following conclusions: (1) C and D-zone total concentrations fall off rather sharply to the S, SW, and W of the plant process area; (2) D-zone concentrations diminish rapidly toward the NW, while much higher C-zone concentrations persist in this direction; and (3) isoconcentration contours reflect the prevailing flow direction to only a limited degree -- to a much greater extent, higher constituent concentrations reflect nearby free product. Offsite DNAPL has been found only in two C-zone wells (including the most source-distant well, M-22C, located 128 m NW of the northwestern site boundary). Free product locally is present in any or all of the four aquifer zones within site boundaries (occasionally deeper than 21 m beneath ground), near the surface source areas.

DNAPL migration from source areas has mostly been governed by gravitative movement along beds dipping ~1° NW. This shallow, phreatic migration has occurred within the C-zone at depths beneath ground of 12 to 15 m, over lateral distances to 230 m. Figure 2 shows a dip section through DNAPL-bearing well M-22C and the inferred primary source area (treatment plant basement).

The DNAPL entered the C-zone along a TOB low beneath the treatment plant basement. Migration along the TOB (generally at depths of 1.5-4.6 m) appears to be restricted to distances of less than 46 m, and therefore is unlikely to cause significant offsite migration. However, opportunely located TOB lows appear to have provided routes for free-product entry, into the main lateral-migration beds, that are not always located beneath source areas. The same bedrock pathways that are exploited by the flowing groundwater will tend to be utilized by the DNAPL but, unlike the groundwater flow, the DNAPL migration is largely driven by gravity (negative buoyancy). Chieruzzi et al. (1995) showed the observed pores in some site limestones are large enough to explain the occurrence of DNAPL without having to invoke the existence of conduits (with a calculated pore size similar to those in medium to coarse sands required for DNAPL entry).

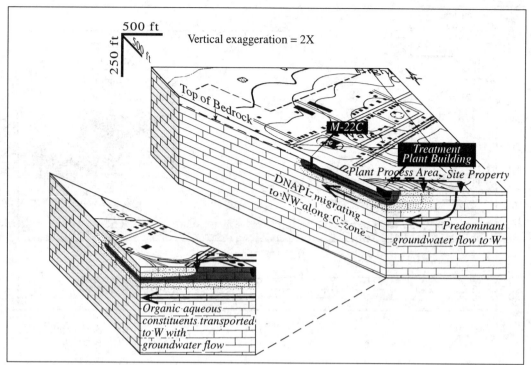

Figure 3: Pull-apart block diagram summarizing the hypothetical models of groundwater flow, constituent transport, and DNAPL migration for the site in southwestern Kentucky. A vertical exaggeration of 2X is shown.

Local deflection of migrating free product away from the bedding dipline can be expected where (1) NW-trending bedding, joint, or pore-network pathways are poorly developed or absent in proximity to such available pathways trending along other directions, or (2) clay-filled depressions in the TOB surface, deeper than about 12 m beneath ground, act as barriers to downdip migration in the uppermost bedrock. Where such critically deep TOB lows might be filled with relatively permeable sands, lateral DNAPL migration from the uppermost bedrock into the soil mantle could occur along the updip slope of the TOB low.

DNAPL migration across the bedrock layering is possible and expected to be guided by: (1) steeply dipping pathways provided by zones of fracture concentration or widened fractures; (2) solution or collapse pits (as observed in some of the area caves); (3) local occurrences of critically large pores or vugs having vertical interconnectedness; and (4) possibly, facies transition zones within the bedrock. However, only two (onsite) D-zone wells contain DNAPL, indicating migration across the bottom of the C-zone is rather uncommon. As well, the DNAPL in one of these wells may have entered this aquifer zone along the flank of an opportunely located TOB low, rather than by routing through the base of the C-zone.

Figure 3 illustrates the divergent directions of groundwater flow and transport (to the W) and DNAPL migration (to the NW). Although the density of offsite monitoring wells does not permit thorough testing of the following idea, it is expected that transverse DNAPL migration to the NW has fostered a plan-view aqueous plume which, over decades, had expanded toward the NW and W while tapering toward the migration limit to the NW.

CONCLUSIONS

The industrial site and surrounding area comprise subsoil karst with a shallow water table and perennially saturated top-of-bedrock (TOB). Although caves are found within the region surrounding the site, no caves are known within the groundwater basin containing the site. Most regional caves reflect fluviokarst genesis, and the site area is far enough from baselevel streams to be free of conduits large enough to allow human access. Site groundwater flow appears to be approximately diffuse, although some local conduit flow is implied by hydraulic characteristics.

Based on site head levels and the directions of elongate TOB forms, fracture traces, joints, and regional cave passages, the inferred flow and transport direction is roughly to the west. However, groundwater impacted by site operations is not expected to reach a water-supply spring located to the west, as independent dye tracing has shown this spring drains a separate basin.

Significant lateral DNAPL migration has occurred by gravitative movement within limestone beds dipping ~1° NW. This migration has occurred at depths beneath ground of 12 to 15 m, over lateral distances to 230 m. This distance of lateral migration is striking and apparently reflects considerable DNAPL volumetric flux along pathway networks perched on tight beds -- the latter inference is supported by pumping-test analyses of this semi-confined aquifer zone of ~3-m thickness. Migration along the TOB appears to be restricted to distances of less than 46 m, and therefore is unlikely to direct significant offsite migration.

ACKNOWLEDGMENTS

Shirley Kormanec and Gregory Smith finalized the manuscript layout and figures. The Western Kentucky Speleological Society kindly provided unpublished information on caves and karst features in the region.

REFERENCES

Chieruzzi, G.O., Duck, J.J., Valesky, J.M., and Markwell, R., 1995, Diffuse flow and DNAPL recovery in the Ste. Genevieve and St. Louis Limestones: in Beck, B.F. (ed.), Karst Geohazards: Engineering and Environmental Problems in Karst Terrane, A.A. Balkema, Rotterdam, p. 213-225.

Dougherty, P.H. (ed.), 1985, Caves and Karst of Kentucky: Kentucky Geological Survey Special Publication 12, Series XI, 196 p.

Kemmerly, P.R., Weber, L.C., and Higgins, Jr., C.S., 1987, Multi-disciplinary analysis and design of Class V drainage wells for a large commercial development: in Beck, B.F., and Wilson, W.L. (eds.), Karst Hydrogeology: Engineering and Environmental Applications, A.A. Balkema, Rotterdam, p. 311-320.

Mylroie, J.E. (ed.), 1981; 1982 and 1983 combined; 1984; Western Kentucky Speleological Society Annual Reports: Murray State University, College of Environmental Sciences, Murray, Kentucky.

Ray, J.A., 1997, Overflow conduit systems in Kentucky: a consequence of limited underflow capacity: in Beck, B.F., and Stephenson, J.B. (eds.), The Engineering Geology and Hydrogeology of Karst Terranes, A.A. Balkema, Rotterdam, p. 69-76.

Environmental characterization of karstic terrains: A case study for the practical application of stable isotope ratios and anion/cation analysis of ground water

JASON M. MASON O'Brien and Gere Engineers, Incorporated, Blue Bell, Pa., USA
WILLIAM J. GABRIEL O'Brien and Gere Engineers, Incorporated, Syracuse, N.Y., USA
D.I. SIEGEL Syracuse University, N.Y., USA

ABSTRACT

This paper presents the results of a case study showing how major solute geochemistry and stable isotope ratio analysis was used in an environmental site investigation to independently test the conclusions based on standard hydrogeologic studies of a karstic aquifer system in Missouri. Ground water beneath the site studied occurs in at least three distinct ground water-bearing zones of the Ozark Aquifer. Detailed field examination and correlation of the rock types across the site identified a fault zone that appeared to influence the ground water flow pathways. Piper diagrams of major solute composition suggested that despite the fault, the major hydrostratigraphic units were largely isolated from each other. Values of $\delta^{13}C$ of dissolved inorganic carbon (DIC) were also largely in agreement with hydraulic isolation. The $\delta^{13}C$ of DIC in the deepest water-bearing unit was about -11 o/oo, compared to -15 o/oo in the uppermost unit. Furthermore, equilibrium calculations showed that the deepest ground water has equilibrated with limestone, in contrast to ground water from the upper two water-bearing units which are generally undersaturated with respect to carbonate minerals. Either ground water moves rapidly in the upper two units compared to that in the lowermost, or the age of the ground water in the lower unit is much older than that found above.

INTRODUCTION

The objectives of environmental site investigations typically include the assessment of ground water quality, extent of impact, and potential migration pathways. To achieve these objectives, environmental professionals generally conduct subsurface investigations that detail the geologic conditions underlying the site and the surrounding area. However, characterization of subsurface conditions in karstic terrains present unique challenges because dissolution features control ground water flow. Several techniques are generally used to assess the subsurface conditions in karstic terrains to overcome these challenges. These techniques include detailing geologic conditions underlying the site, assessing the aquifer material and hydraulic properties, performing dye trace studies, and measuring the chemistry of the ground water. Although conducting a geochemical analysis of ground water is not typically done during an environmental site investigation, such an analysis can prove invaluable to understand karstic hydrogeology. This paper presents a case study demonstrating how the geochemical analysis of major solutes coupled with stable isotopic analysis of dissolved inorganic carbon, can be used to resolve fundamental issues regarding the hydrogeology of karstic terrains.

STUDY AREA AND BACKGROUND

The study area selected is located in a small city in central Missouri (population less than 10,000) where an environmental site investigation was done at a former manufacturing facility located within the city limits. The bedrock consists of Cambrian and Ordovician-age dolomitic limestones overlain by a siliciclastic residuum: (from oldest to youngest) the Potosi Dolomite, Eminence Dolomite, Gasconade Dolomite, and the Roubidoux Residuum (Thompson, 1991). Identification and description of these formations in the area of the site have been documented by Thompson (1991), Robertson (1991), Thompson and Robertson (1993). Ground water in these formations comprise a portion of the Ozark Aquifer, which is the primary source of potable water for the area. Karstic features, including caves, springs, and sinkholes, are common.

The environmental site investigation done at the former manufacturing facility included the installation of forty bedrock ground water monitoring wells to characterize the ground water quality, assess ground water flow conditions, and evaluate the regional and local subsurface geology. The depth of the monitoring wells ranged from 150 ft to 550 ft below grade. Shallow bedrock monitoring wells (approximately 150 ft below grade) were screened at the water table which occurs near the contact between the Roubidoux Residuum and Gasconade Dolomite. Intermediate bedrock monitoring wells (approximately 325 ft below grade) were installed in the lower portions of the Gasconade Dolomite and deep bedrock monitoring wells (approximately 550 ft below grade) were installed in the Eminence Dolomite and Potosi Dolomite. At three locations, proximate bedrock monitoring wells were installed at the 150 ft, 325 ft, and 550 ft depth intervals to evaluate the vertical hydraulic gradients and potential for downward recharge.

Because changes in the physical nature of the rock encountered during drilling are subtle, a detailed petrographic analysis of the drill cuttings was done to better distinguish the formation boundaries (Mason and Gabriel, 1996). The results were used to construct geologic cross sections of the site, which ultimately led to the identification of a previously unknown fault adjacent to the facility. The occurrence of the fault was considered to have a major influence on the potential migration of the impacted ground water at the site.

Ground water quality data obtained from the monitoring wells indicated that the ground water beneath the facility has been impacted by industrial chemicals. The highest concentration of the industrial chemicals in ground water occurred mainly in the upper 150 ft of water-bearing rocks beneath the manufacturing facility, although ground water was impacted to a lesser degree as deep as the 325 ft and 550 ft zones. Because of the karstic nature of the dolomitic limestone and the fault adjacent to the facility, the potential exists for impacted ground water from beneath the facility to migrate to and impact the city's municipal wells, the only current source of potable water.

Dye tracing studies were done by the state regulatory agency to help identify potential migration pathways at the site. In these studies, a quantity of dye was released into a small losing stream near the facility and detectors were monitored at several springs, caves, and municipal water supply wells. The results of the dye trace studies showed a migration pathway between the small losing stream and a number of springs and caves, but dye from these studies was not detected in the municipal water supply wells monitored. The dye trace studies were, however, inconclusive with regards to providing an understanding of the potential deep migration pathways of the impacted ground water beneath the facility.

The geochemistry of ground water can be altered by interactions with host rock during migration (Clark and Fritz, 1997). Therefore, potential migration pathways can be inferred by evaluating the changes in the chemical composition of ground water. This paper reports on the results of such a study done at the site.

METHODS

For this study, water samples were collect from selected ground water monitoring wells, municipal water supply wells, private drinking water wells, springs, and local surface water features. The municipal water supply wells sampled were typically cased to a depth of 350 feet and completed as an open borehole to depths between 850 feet and 1000 ft. Therefore, the ground water collected from the municipal wells was considered to represent a composite of the 325 ft, 550 ft, and deeper ground water zones. Private drinking water wells identified in the area are typically completed from 150 ft to 325 ft deep within the Ozark aquifer and therefore the water samples collected from the selected private wells were considered to reflect a composite of the shallow and intermediate zones.

Several springs and surface water features were identified during the environmental investigation. The springs sampled during this study are located within 4.5 miles of the facility. Surface water bodies in the area include the small losing stream near the facility and a small lake within the city.

A total of 15 ground water monitoring wells, 9 municipal wells, 5 private wells, 5 springs, and 2 surface water bodies were sampled across the site to provide the water chemistry data. Surface water bodies were sampled to assess recharge water chemistry prior to entering the subsurface. The monitoring wells (at differing depths), municipal wells, and private wells were sampled to assess the water chemistry as the water travels through the subsurface; and the springs were sampled to evaluate the water chemistry when the ground water emerges at the surface.

The water samples were analyzed for major anions (chloride, carbonate, bicarbonate, and sulfate) and cations (calcium, sodium, magnesium, and potassium) along with the stable isotope ratio of carbon in dissolved inorganic carbon (DIC) and oxygen and hydrogen in water. Stable isotope ratio analysis and anion-cation data have successfully been used at other sites to evaluate ground water sources, potential ground water migration pathways, and ground water recharge/discharge mechanisms (see Clark and Fritz, 1997; and Mazor, 1991 for numerous examples).

The anion-cation concentration data were converted to milliequivalence units by multiplying the measured concentration by a conversion factor (Hem, 1985). The milliequivalence concentrations were used to generated piper plots to illustrate the results of the anion-cation data. The stable isotope ratios were calculated using standard protocols, and are expressed in parts per thousand relative to the international standards for water and carbon (Clark and Fritz, 1997).

During sampling, measurements of dissolved oxygen, specific conductivity, pH, temperature, and oxidation-reduction potential were also recorded. The analytical results of this study are summarized on Table 1.

INTERPRETATION AND DISCUSSION
Field Parameters and Anion-Cation Analysis

The piper plot for the major solute data (Figure 1) shows that the predominant ground water type in the area is a Ca-Mg-Bicarbonate, and consistent with that expected by dissolving dolomite in the Roubidoux, Gasconade, and Eminence formation. A plot of calcium plus magnesium concentrations (as meq/L) against bicarbonate concentrations generally defines a straight line with a slope of one in accordance with the following equation:

$$CaMg(CO_3)_2 + 2H_2O + 2CO_2 = Ca^{2+} + Mg^{2+} + 4 HCO_3^-$$

A plot of pH versus log HCO_3^- (Figure 2) shows that most of the water samples trend toward equilibrium with dolomite and calcite (Langmuire, 1971). The arrow on Figure 2 defines the general direction that the chemistry follows as CO_2 in recharge water is consumed when it dissolves carbonate minerals. Note that the water in many of the springs fall below the equilibrium shaded area. This indicates that more carbonate mineral can be dissolved by the water discharged from these springs, whereas water in the deep monitoring wells

Table 1. Geochemical Parameters, Anion/Cation Data, and Stable Isotope Ratios

Sampling Location	T (C)	Sp. Cond. (uS/cm)	pH	D.O. (mg/l)	Turb. (NTU)	ORP	Ca (mg/l)	Mg (mg/l)	K (mg/l)	Na (mg/l)	HCO3 (mg/l)	CO3 (mg/l)	Cl (mg/l)	SO4 (mg/l)	TDS (mg/l)	Carbon Isotope Ratio	Oxygen Isotope Ratio	Hydrogen Isotope Ratio
City Well #2	16.2	265	7.62	3.3	1.41	84	37	19	0	4	150	0	10	15	235	-10.9	-7.2	-45
City Well #3	16.5	264	7.44	4.86	3.04	85	30	18	0	3	140	0	3	8	202	-11.2	-7.1	-45
City Well #4	15.9	341	7.81	3.3	0.32	86	41	24	0	3	180	0	5	14	267	-12.2	-7.1	-44
City Well #5	15.9	389	8.83	5.85	3.18	76	50	28	0	3	210	0	3	23	317	-12.3	-7.1	-45
City Well #7	15.8	350	8.11	8.4	0.74	91	44	26	0	3	200	0	3	4	280	-13.5	-7.2	-45
City Well #8	16.2	295	7.87	4.1	0.85	86	42	24	0	3	180	0	4	10	263	-12	-7.1	-45
City Well #9	16.1	276	8.65	4.7	2.81	84	34	20	0	3	160	0	2	10	229	-11.8	-7.3	-44
City Well #10	16.2	321	8.01	6.55	0.53	88	44	22	0	3	180	0	4	9	262	-13	-7.2	-45
City Well #Oak	16.9	273	7.5	6.4	2.3	51	31	19	0	3	150	0	3	9	215	-12.2	-7.2	-44
Spring #1	15	491	7.2	8.3	1.51	69	61	37	0	5	290	0	7	5	405	-14.2	-6.5	-41
Spring #2	15.8	356	7.2	11.21	2.42	30	34	19	0	10	150	0	14	10	237	-13.8	-7	-40
Spring #3	16.9	295	7.5	7.95	5.85	41	37	20	0	4	170	0	6	9	246	-13.8	-6.9	-42
Spring #4	14	397	7.5	8.26	1.2	16	54	29	0	3	220	0	8	6	320	-14.6	-7.2	-44
Spring #5	15	432	7.5	11.75	10.46	57	55	32	0	4	250	0	4	12	357	-14.5	-7.3	-45
City Lake	25	132	7	6.51	7.01	46	9	5	0	10	36	0	20	6	86	-10.6	-2	-18
Creek	19.8	634	7	9.95	90	36	42	18	0	64	150	4	130	12	420	-11.9	-5.2	-32
Private Well #1	18	460	6.86	5.96	1.01	4	63	28	0	4	220	0	12	40	367	-13.7	-6.7	-43
Private Well #2	17.5	329	7.86	11.21	0.42	108	40	21	0	6	160	0	14	8	249	-13.5	-7.3	-45
Private Well #3	16.5	398	7.65	6.2	0.51	56	53	29	0	3	240	0	5	6	336	-13.8	-7.3	-44
Private Well #4	15.5	433	7.81	6.65	1.13	90	59	34	0	4	280	0	6	9	392	-14.3	-7.3	-45
Private Well #5	17.5	461	8.03	7.2	1.42	100	59	34	0	9	270	0	11	13	396	-14.1	-7.3	-46
OBG-1S	16	561	7.2	7.86	18.3	44	55	31	0	19	170	0	46	64	385	-13.1	-6.8	-42
OBG-1D	15	302	7.5	7.8	2.58	42	34	20	0	4	150	0	6	9	223	-12.2	-6.9	-42
OBG-1DD	15	265	7.98	3.71	9.07	134	31	21	0	4	140	0	8	16	220	-10.6	-7.2	-42
OBG-11S	15.5	452	7.31	7.47	6.36	51	38	19	0	17	130	0	38	27	269	-13.3	-7	-42
OBG-11D	14	291	7.29	8.03	6.13	176	30	16	0	4	110	0	14	18	192	-11.7	-7.2	-43
OBG-20DD	15.6	250	7.81	3.95	2.12	62	31	18	0	4	150	0	4	7	214	-10.9	-7.1	-44
OBG-14S	16.9	186	7	4.4	33.9	77	29	5	0	10	72	4	8	25	153	-19.6	-6.5	-39
OBG-14D	16	261	7.5	3.85	7.7	62	30	17	0	4	140	0	3	7	201	-11.9	-6.2	-39
OBG-14DD	13.9	299	7.72	2.85	5.85	-31	35	21	0	4	150	0	5	27	242	-12.3	-6.8	-42
OBG-18S	18	591	7.1	9.2	15.87	81	55	30	0	22	170	0	69	32	378	-19.6	-6.5	-39
OBG-18D	18	241	7.5	4.9	2.58	41	28	16	0	3	130	0	6	5	188	-10.5	-7.2	-43
OBG-18DD	15	241	7.97	1.43	2.14	29	27	18	0	3	140	0	1	12	201	-10.2	-7.1	-43
OBG-22S	14.5	384	7.5	7.4	15.38	41	46	26	0	6	200	0	13	9	300	-13.5	-7.2	-44
MW-101	17.5	412	7.69	7.02	92.1	213	56	32	0	6	240	0	13	10	357	-15.7	-7.1	-44
MW-B201 - Landfill	19.4	51.4	6.44	5.9	200	86	5	2	0	2	6	0	3	23	41	-24	-7	-43

Notes:
T - Temperature
Sp. Cond. - Specific Conductivity
D.O. - Dissolved Oxygen
Turb. - Turbidity
ORP - Oxidation / Reduction Potential
Ca - Calcium
Mg - Magnesium
K - Potassium
Na - Sodium
HCO3 - Bicarbonate
CO3 - Carbonate
Cl - Chloride
SO4 - Sulfide
TDS - Total Dissolved Solids

(completed in the Eminence Dolomite and Potosi Dolomite) and some of the municipal/private wells are at equilibrium (or saturated) with the aquifer minerals at a pH of about 7.8. These results indicate that the length of time that the water is in contact with the dolomite rocks (residence time) is longer in the deep ground water zones than in the shallow and intermediate ground water zones.

The total amount of dissolved solids (TDS) decreases with depth at the site, despite an average downward hydraulic gradient of about 0.19 ft/ft. This trend is unusual since TDS almost universally increases with depth along vertical flow paths and depth (e.g., Freeze and Cherry, 1979). The inverse trend suggests that water recharging shallow ground water zone (through the Roubidoux Residuum) is effectively isolated from the deeper ground water zones (underlying Gasconade Dolomite, Eminence Dolomite, and Potosi Dolomite). The recharge of water to these deeper ground water zones must be derived mainly through fractures. This geochemical data suggests that the shallow ground water zone near the base of the Roubidoux Formation restricts the lower units from diffuse recharge from above and is likely responsible for limiting the migration of the highest concentration of industrial chemicals in the ground water to the 150 ft ground water zone beneath the facility. Based on a review of the detailed petrographic descriptions completed during the environmental investigation, the occurrence of chert layers near the base of the Roubidoux Residuum may account for this isolation.

The piper plot of the solutes (Figure 1) further illustrate that water in shallow wells is geochemically different from that in the intermediate and deep monitoring wells and from the other water samples (municipal, private, springs, and surface water). This result is consistent with the isolation of the shallow ground water zone from the deeper ground water zones as discussed above.

Isotope Samples

A plot of $\delta^{13}C$ of dissolved inorganic carbon versus HCO_3 concentration, shown on Figure 3, agrees with an effective hydraulic separation of the shallow ground water zone (Roubidoux Residuum water) from the underlying units. Note that water from springs, shallow private wells, and shallow monitoring wells fall on a trend defined by nearly constant $\delta^{13}C$ of about -15 to -14 o/oo commensurate with carbonate minerals dissolving to equilibrium and a high TDS. In contrast, the $\delta^{13}C$ of water from deep wells in the Eminence Dolomite and the Potosi Dolomite is about -12 to -11 o/oo. This value would be theoretically expected (+/-1 o/oo) from the equation given above, assuming a typical isotopic content of organic matter in soil of about -25 o/oo and of rock at about zero.

The smaller $\delta^{13}C$ values for dissolved inorganic carbon (DIC) in the water in the Roubidoux Residuum and water discharged from springs suggest that the organic matter producing the recharging CO_2 has $\delta^{13}C$ lighter than that of the CO_2 recharging the municipal/private wells, and the site monitoring wells completed in the Gasconade Dolomite. The smaller $\delta^{13}C$ values also suggest that the water in the Roubidoux Residuum is geochemically isolated from the deeper aquifer zones except where extensive fractures (or faults) penetrate into the lower aquifer zones. It is plausible that the $\delta^{13}C$ value for DIC in the Roubidoux Residuum and spring waters reflect the oxidation of modern forest organic matter, whereas that in water from the deep monitoring wells reflects organic matter in the geologic past when the area may have had vegetation different than that found today.

Water from the municipal wells fall along a trend on the $\delta^{13}C$-HCO_3 diagram (Figure 3) that connects the shallow ground water zone with deeper ground water zones. Water in some of the municipal wells consists of mixtures of Roubidoux Residuum and Gasconade Dolomite waters. The Roubidoux Residuum water may be transported through vertical fractures and/or conduits, induced by the pumping of the municipal wells. The $\delta^{13}C$ value in water from shallow monitoring wells 14S and 201 have isotopic contents almost the same as that of the organic matter from which the dissolved CO_2 was derived (-25 to -20 o/oo). These data and the low TDS and alkalinity of these waters suggest that they did not come into contact with much dolomite, but mostly dissolved silicate minerals present in the Roubidoux Residuum.

A plot of $\delta^{18}O$ versus $\delta^{2}H$ of water shows that most of the data fall approximately on the Global Meteoric Water Line, with most of the data clustering between $\delta^{18}O$ of -7.5 to -6.0 with commensurate values for $\delta^{2}H$. The similar isotopic content of the ground water indicates that the ground water zones are recharged by similar precipitation events, the shallow ground water zones through diffusion and fracture recharge and the intermediate and deep ground water zones mainly through fracture recharge. The isotopic separation of the water samples is insufficient to distinctly track separate recharge events or sources of water for the different aquifer units. However, the isotopic values of $\delta^{18}O$ and $\delta^{2}H$ are consistent with that of average volume-weighted precipitation in Missouri (Clark and Fritz, 1997). The water collected from the city lake, however, is isotopically enriched probably through evaporation. The sample from small losing stream adjacent to the facility has slightly more $\delta^{18}O$ and $\delta^{2}H$ than the ground water, which may reflect the isotopic content of summer rain.

CONCLUSIONS

The results of this study provide lines of evidence that 1) the shallow ground water zone, which occurs in the Roubidoux Residuum, is geochemically isolated from the intermediate and deep ground water zones, and 2) the residence time of the water in the deep (and deeper) ground water zones is longer than the shallower ground water zones. These results suggest that the water present in these aquifer zones may not be mixing under natural conditions, and that the majority of vertical mixing is induced by the pumping of the city's water supply wells which hydraulically connect the different aquifer zones through vertical fractures and faults.

The evidence that the shallow ground water zone is isolated from the deeper ground water zones include differences in major ionic composition, the concentration of total dissolved solids (TDS), carbon stable isotope data, and geologic information from the environmental investigation. The results of the ion and stable isotope analysis are consistent with ground water occurring in the shallow ground water zone (Roubidoux Residuum) being geochemically isolated from the intermediate and deeper ground water zones. The shallow ground water zone likely represents the infiltrated ground water from the surface and is probably younger in age than the deeper ground water zones. Based on geological information, the chert layers occurring near the bottom of the Roubidoux Residuum may be responsible for the permeability differences at the Roubidoux / Gasconade contact.

These permeability differences at the contact, which are supported by the geochemical data presented in this paper, are also considered

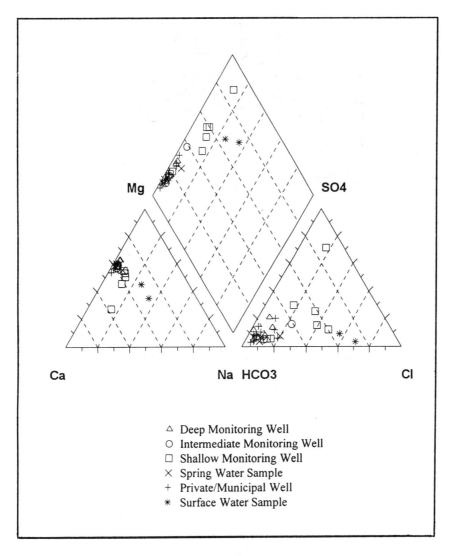

Figure 1. Piper Plot of Anion and Cation Data

to have inhibited the vertical migration of the high concentrations of industrial chemicals in the ground water beneath the site from the shallow zone to the deeper zones. Based on the hydraulic and geochemical separation of the shallow ground water zone from the deeper zones, the presence of the faults and karstic features in the area likely provide the main vertical migration pathway for impacted ground water beneath the facility. Downward vertical migration of the impacted ground water is therefore likely induced by the pumping of the municipal wells.

As indicated by the plot of bicarbonate versus pH, the water in the deep (and deeper) ground water zones are at or near chemical equilibrium with the dolomite host rock. This implies that the water in the deep ground water zones is not actively dissolving the dolomite in the formation (Potosi Dolomite) and suggests that the residence time of the water is longer than the intermediate and shallow ground water zones. The longer residence time in the deep ground water zones is also supported by the stable carbon isotope data. The age difference of the ground water in the shallow, intermediate, and deep aquifer zones was not assessed during this study, however it is anticipated that the water in the deep ground water zone may be significantly older than the intermediate and shallower zones.

ACKNOWLEDGMENTS

The authors wish to thank Jim Moore for collection of the water samples and O'Brien & Gere Laboratories and Geochron Laboratories for performing the analysis of the water samples.

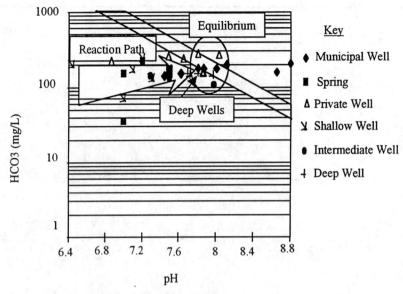

Figure 2. Plot of HCO3 versus pH

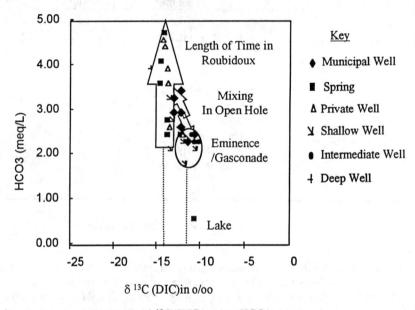

Figure 3. Plot of δ ^{13}C (DIC) versus HCO3

REFERENCES

Clark, I. And Fritz, P., 1997, *Environmental Isotopes in Hydrogeology*, Lewis Press, New York.

Domenico, P.A. and Schwartz, F.W., 1990, Physical and Chemical Hydrogeology, John Wiley and Sons, 824 pages.

Fetter, C.W., 1988, Applied Hydrogeology, 2nd ed., Macmillian Publishing Co., 592 pages.

Hem, J.D., 1985, Study and Interpretation of the Chemical Characteristics of Natural Water, 3rd ed. U.S. Geological Survey Water Supply Paper 2254, 263 pages.

Langmuire, D., 1971, The Geochemistry of Some Carbonate Ground Waters in Central Pennsylvania, Geochim. Cosmochim. Acta., 35, 1023 - 1045.

Mason, J.M. and Gabriel, W.J., 1996, Use of Petrographic Analysis in Environmental Site Investigations: A Case Study, in Sampling Environmental Media, ASTM STP 1282, James H. Morgan, Ed. American Society for Testing and Materials, 143 - 157.

Mazor, E., 1991, Applied Chemical and Isotopic Groundwater Hydrology, Halsted Press, John Wiley & Sons.

Robertson, C.E., 1991, Geologic Map of the Sullivan Quadrangle, Missouri Department of Natural Resources, Division of Geology and Land Survey.

Thompson, T.L., 1991, Paleozoic Succession in Missouri - Part 2 Ordovician System, Missouri Department of Natural Resources, Division of Geology and Land Survey, R.I. No. 70.

Thompson, T.L. and Robertson, C.E., 1993, Guidebook to the Geology Along Interstate Highway 44 (I-44) in Missouri, Missouri Department of Natural Resources, Division of Geology and Land Survey, R.I. No. 71, Guidebook 23.

The protection of a karst water resource from the example of the Larzac karst plateau (south of France). A matter of regulations or a matter of process knowledge?

VALÉRIE PLAGNES   BRGM, Direction de la Recherche, Montpellier, France
MICHEL BAKALOWICZ   CNRS-UM 2, cc057, Montpellier, France

ABSTRACT

Wide karst plateaus extend in limestone and dolostone of the Grands Causses area north of Montpellier. They are surrounded with a hard rock basement that feeds important rivers, such as the Tarn River and its tributaries, crossing the plateaus in deep gorges. The Larzac plateau, the most southern of them, makes the bond with the Mediterranean region, traversed by one of the main national roads, presently doubled with a speedway under construction from Clermont-Ferrand to Montpellier and Spain.

The karst water resource, through several springs, is used for the water supply of the Millau area (30,000 people). It also recharges surface rivers, mainly during the summer and fall low stage, in a region well known for water recreation activities, fish farming and, above all, sheep farming and Roquefort cheese production. The fundamental question was: should the karst and its water resource be fully protected by strict regulations, for example considering the plateau in the same way as a nature reservation, or may we foresee a land management plan of the plateau based upon the knowledge of its hydrogeological functioning and of its karst structure?

Support decisions concerning about a 500-km^2 area were developed from detailed hydrogeological studies, including natural and artificial tracing, landscape analysis, risk inventory and mapping. Recommendations were proposed to local and regional decision-makers in order to define what are the preference shares in terms of water resource protection and water quality restoration.

Following such an approach is not easy for hydrogeologists and for decision-makers. In non-karstic regions protection zones currently extend over small areas of few km^2, which is easily under the control of a small municipality. In the present example, the main water supplies are two springs with recharge areas of 100 and 110-km^2 respectively, concerning several municipalities some of them as a water supply, others for pollution risks and land management projects.

The recent French Water Act, which considers water as a common heritage, allows management of water and the terrain in which it flows in a common way, by all users, in the same area. The regulation tools seem very well suited to karst region management. But time is necessary for educating users and decision-makers, with the help of karst environment scientists, to work together in a frame other than the usual municipal limits.

INTRODUCTION

Karst aquifers, which occur over 35 % of French territory and supply up to 30 % of the total potable water, of which 55 % are groundwater, are especially concerned with water and land use conflicts because of their high vulnerability. But most of time, in karst areas, man activities are not as important as in other areas. Therefore, karst groundwater is generally of good quality, with concentrations in nitrate, pesticides and soluble hydrocarbons much lower than legal standards.

Because exploitation of karst aquifers, reputed as highly vulnerable, has often been avoided until now, in preference of porous and fissure aquifers, the present day development of groundwater resource concerns more and more karst areas. At the same time, karst areas are developing in agricultural and tourism activities that produce more pollutants and consume more water.

New regulations were set up in order to protect the groundwater resource in quantity, for avoiding overexploitation, and in quality against pollution. Knowledge from porous and fissure media directed recommendations on water management regulations.

But are these regulations adapted to karst environment (?), or does karst need specific regulations for managing its water resource? How to manage water use conflicts in karst areas? In order to point out some of the problems of water and land management in karst areas, we present as an example hydrogeological investigations in Millau area, about the northern part of Larzac karst plateau.

THE STUDY AREA: THE LARZAC KARST PLATEAU

Introduction to Grands Causses and Larzac karst area

The Grands Causses area, located in Southern France and surrounded by Paleozoic massifs, is the typical French karst region. The karst landscape develops in limestones and dolostones of Lias and middle Jurassic age, respectively 200 m and 600 m thick, separated by a 200 m thick, impermeable formation, the upper Lias clays. The regional structure is nearly horizontal, cut by major E-W and N-S faults continuously active since the lower Jurassic. The 1000 m uplift which followed the Pyrenean phase (Eocene) developed during Neogene and Plio-Quaternary.

The Grands Causses are such a perfect example of a karstic terrain that the speleologist E.A. Martel (1934) saw them since the beginning of the century as the typical example of limestone phenomena, much more representative, according him, than the Yugoslavian karst itself. Unlike mountainous regions where carbonate rocks are less extensive because of tectonic fragmentation, and other karstic regions in France where the limestone cover rocks are often important in the landscape, the Grands Causses are extremely arid, which accentuates the extent of its plateaus and the deep gorges cut by well fed rivers like the Tarn and its tributaries, Jonte and Dourbie rivers in the north, and the Vis in the south. This aridity, mainly due to deforestation during the Middle Ages, is common to most European karstic regions and gives them the reputation of being unsuitable for any sort of development.

Allogenic rivers, flowing down mainly from the Cevennes Mountains, cross deeply (300 to 500m) the plateau and split it into several hydrogeological units. The wider unit, which is the most southern, is the " causse du Larzac ", the Larzac karst plateau. It is limited to the north by the gorges of Dourbie river, to the west by the Tarn river, a tributary of Garonne river, to the south by the Montagne Noire Paleozoic massif and to the east by the Vis and Lergue rivers, tributaries of the Herault river, one of the most important Mediterranean rivers.

The location of the Larzac karst plateau, perched astride the Atlantic-Mediterranean watershed, places it on a major line of communication and in the hinterland of the Languedoc coast and Montpellier area. For this reason, the area around Millau, comprising the Northern Larzac plateau, is now the target of development projects. Regional economic activity has traditionally been concentrated in the valleys into which are funneled the meat and dairy products of the Causses and into which the water received by the Causses also drains. The groundwater resources required for the maintenance, if not the development, of all economic activity is readily accessible only in the deep valleys where there are high-yield karstic springs.

The regional economy is centered in and on the town of Millau, which has a population of nearly 30,000. The major industries are the food (sheep meat, milk and cheese, the world famous Roquefort cheese) and leather industries, tourism and outdoors activities. Millau and the surrounding Grands Causses are oriented toward the Languedoc-Roussillon Region and the Mediterranean coast, which have seen rapid development in recent years. Numerous development schemes in the planning stage range from building and enlarging highways to developing craft and industrial zones, to which are related the extension of urban centers.

The Northern part of the Larzac plateau, about 500-km^2 in extent, is typical of the whole region: no surface waters, except during heavy rain periods, a few small towns, agriculture and a main highway, from Paris to Montpellier and Perpignan (RN 9) doubled by a speedway (A 75). But this plateau is also the recharge area of several karstic springs pumped for the water supply of the whole region or used for fish farming. These springs also recharge surface rivers especially during low stage, at least 80 % of the low stage flow rate is of karst groundwater origin.

The existence and extent of karstic aquifers in the region have been broadly known for a long time, thanks notably to works of Paloc (1992) and Dubois (1985). But their functioning, their precise boundaries, the size of the resource, their storage capacity and vulnerability conditions were still far from being well known. This information is fundamental: 1) for estimating the part played by karst aquifers in recharging surface streams particularly during low stage, and then their economic part, 2) for knowing the conditions of recharge and pollutant transport from the surface, and 3) for determining their recharge area extent, in order to propose the limits of protection zones and the conditions of protection. Such basic knowledge must be provided before decisions and choices can be made concerning the development, use and management of the Larzac plateau and its natural resources.

From this point of view, the Northern part of the Larzac causse is representative of all of the Grands Causses. In the following the most representative results are stated to show the main characters setting the vulnerability, in order to guide certain development choices that will preserve water quality and the environment of the Larzac karst plateau (Bakalowicz & Ricard, 1996; Ricard & Bakalowicz, 1996; Plagnes, 1997; Bakalowicz & Plagnes, 1997).

Hydrogeology of Northern Larzac

The water infiltrates at the surface of the plateau at an average elevation of 800 m and discharges at the level of the perennial rivers. The plateau has a featureless surface made up of large closed depressions often filled with clays or sands from weathered chert marly limestones and from weathered dolomite. Locally between these depressions there are numerous tower-like limestone and dolomite pinnacles characteristic of this region. The region looks desolate because the soil is scarce and of poor quality except in the depressions such as those on the north and west edges, and because the Causses have long been used for sheep farming. There is now only grass, and in some places even barren rock.

Most of the economic activity on the plateau is located around L'Hospitalet-du-Larzac and La Cavalerie where most of the population is concentrated around a military base, a limestone quarry and a few small companies. Effluents from the ponds of the water treatment plants, which treat the wastewater, flow into karstic depressions. On the plateau there are also dispersed farms raising sheep and, in places, crops. The only other major activity on the plateau is that associated with the main roads and the local airport.

The plateau is made up of a continuous carbonate unit, with a limestone and dolomite series around 500 m thick. This unit overlies the Liassic marl, 200 m thick that forms an impermeable barrier. Geological movements have created a roughly N-S syncline which intersects the Dourbie valley in the middle part of the gorges. These movements abundantly fractured the rocks causing a fracture network. Major faults crossing the plateau have shifted the series or caused weak overthrusts. This fractured carbonate unit has therefore all the qualities of an aquifer overlying an impermeable base.

Springs discharge at the base of the carbonate series above the Tarn River valley (Lavencou, Boundoulaou and L'Homède springs), along the left bank of Dourbie valley (Durzon, Espérelle, Moulin Laumet, Riou Ferrand and small springs of the upper valley that dry up during long low flow stages), along the Cernon valley (Cernon, Mouline and Travers Banc springs) and along the Sorgue valley recharged, by a tenth, from small springs. The Dourbie river is also recharged upstream from creeks and groundwaters of granites in the Cévennes mountain and on its right bank from the Causse Noir karst plateau. Moulin Laumet and Riou Ferrand springs were the only springs that could not be monitored, because of their inaccessibility; but they are minor springs. The hydrological characteristics of the main springs are presented in Table 1. Three major springs (Durzon, Espérelle and Mouline) flow about 70 % of total karst groundwater discharging from the northern Larzac.

Table 1: Hydrological characteristics of the main springs of the Larzac karst plateau (1996 hydrological year).

Spring	Mean annual discharge m^3/s	Extension of recharge area km^2	Storage $10^6\ m^3$
Durzon	3.9	116.8	24.6
Espérelle	3.1	99.4	4.6
L'Homède	0.6	21.3	2.1
Boundoulaou	0.5	17.1	1.8
Lavencou	0.1	3.9	0.5
Cernon	0.5	16.5	5.7
Mouline	0.9	28.9	18.7

The geological criteria were, most of time, insufficient to define the boundaries of the recharge areas of the springs, because the faults and fractures which cross the plateau cause no evident hydraulic discontinuities in this thick carbonate unit. Therefore, tracing experiments using uranine, mass balance of some natural tracers, characteristic of man activities, and water budgets were used in order to define the limits of the karst systems. Because elevations and precipitation are uniform for the whole area and because the pollutant sources were precisely evaluated from systematic inquiries, the budgets are known with a very good precision and consequently the limits of the systems. These limits are not at all in agreement with those of the protection areas defined for the exploited springs of Espérelle, Homède and Boundoulaou during the 70's, only from geological considerations.

All the aquifers drained by these springs have undeniably karstic characteristics. Nevertheless, groundwater flow here seems to be neither very well organized nor concentrated along integrated conduit networks. The karst does not seem as well developed as in other parts of the Grands Causses. Indeed, the tracing experiments have up until now revealed relatively slow transit rates of between 40 and 80-m/h on the average. Furthermore, the volumes of water traced imply the existence of large reserves (several tens to hundreds of thousands of m^3 traced during tests); to be compared to the large dynamic storage (several millions of m^3) calculated from the analysis of low flow stage. Finally, the flow rates, like the chemical quality of water from these springs, vary little compared to springs of aquifers reputed to be well karstified. This again implies the existence of relatively long residence time associated with large storage. These characteristics can be linked to the recent evolution of the karst in the saturated zone and to the absence of surface runoff contributing to the recharge and functioning of aquifers.

Water use and conflicts

In spite of these characteristics, i.e. a mean rate of karst drainage development and a high storage capacity, which would appear to be favorable to a relative immunity of the groundwater resource to pollution, the dependence of surface water to groundwater, the dispersion of regional economic activity on the plateau and along the Dourbie River, and a high demand for water downstream by the town of Millau, creates a *conflict of interest*, at least potentially, concerning the use of this water. Several examples can be cited to illustrate this.

Water from the Durzon spring is pumped to supply drinking water for a large part of the plateau and for fish farming in the Durzon River. Part of this water supply returns to other springs after passing through water treatment plants. This water carries to the aquifer, and therefore to the springs, a polluting bacterial charge. This pollution is eliminated by treatment of the water after pumping, mandatory in the case of karsts. The water of the fish farming which contributes heavily to the recharge of the Dourbie, especially during low flow stages (from a quarter to a third of the total river discharge during low stage) can modify the chemical and bacteriological characteristics of the water as well as those of the treatment plants and fish farms in the valley. The Dourbie River is a popular recreational area during summer and there are many campgrounds along its banks.

Two municipal waste landfills are both located upstream from areas of karstic swallowholes which contribute to the recharge and the pollution of some springs in the Dourbie valley. These two groups of springs are not tapped but play an important role in the quality of Dourbie river water. Their resource is then abandoned.

Farming on the plateau, even if it is not as intensive as in other regions in France, significantly affects some springs, as shown by the increasing nitrate and chloride contents in most of springs. This increase will no doubt continue if no measures are taken to limit this pollution.

The Clermont-Ferrand - Mediterranean highway, at present the RN 9 and the A 75 motorway, which crosses the northern Larzac from the northwest to the southeast, is an important hazard. Indeed, crossing most of the recharge areas of the main springs including those which are pumped, it is heavily traveled and a source of both chronic (hydrocarbons) and seasonal (salting in winter) pollution and a potential source of accidental pollution.

SENSITIVITY OF KARST TO POLLUTION AND MAN ACTIVITIES
Investigation methodology

The first step of the study was to define: 1) the limits of the spring recharge areas, 2) the groundwater flow conditions in the karst systems, the aquifer behavior and the degree of karst drainage development, and 3) the hazards, present and future, which may alter the groundwater quality. From this knowledge, in a second step, the exploitable resource and the storage were evaluated at the same time for the impact of the man activities on groundwater quality. Then in the last step the vulnerability of the karst systems were defined at a regional level. Finally, the different sets of data were organized in a GIS attached to a vulnerability map and to a list of recommendations.

This approach forces to set various and complementary methods. As a matter of fact, the results obtained from a method are not generally conclusive enough by themselves. Only the comparison of data from different methods can give certainty about the limits of each karst system, about their karstic structure and their functioning.

The extension of recharge areas and the location of their limits were determined by different complementary methods:

- Geology and geomorphology, for determining the geometry of the assumed recharge areas.
- Hydrological water balance, at a yearly level, in order to define the water resource available for exploitation and the extent of the recharge area.
- Tracing tests and natural tracers, such as chloride and nitrate, for locating the limits of the recharge area.

Groundwater flow conditions are defined from:

- Geomorphic investigations, providing an inventory of surface and underground karst features characterizing water sinking or rapid groundwater flow (sinkholes, potholes), and their precise location.
- Geological and geomorphic study of surface formations, liable to filter seepage water or, to the contrary, to concentrate surface water in surface streams favoring concentrated recharge.
- Spring hydrographs (sorted discharges, recession curve, correlation and spectrum analysis), providing information about the functioning of the systems in their whole.
- Artificial tracing responses, giving information about local behavior, in particular hydrological conditions.
- Variations of chemical contents, considering inorganic major and minor ions and certain related thermodynamical data, such as calcite and dolomite saturation indexes and calculated CO_2 partial pressure, as natural tracers of groundwater flow (Bakalowicz, 1994).

Thus, knowing the groundwater flow conditions, we are, in particular, able to know how groundwater carries out dissolved solids, e.g. pollutants, from surface to the springs. It then becomes possible to put in place warning systems and protection schemes for groundwater and water supply networks.

Hazards are known as potential sources of groundwater pollution. The present hazards were listed, including towns, with their treatment plants, and lone farms with people and cattle number, abandoned landfills, roads, factories, and cultivated fields. Consequently a state of contamination for each karst system was set up in order to: 1) to qualify the degree of alteration of the resource quality, 2) to define the sensitivity rate of the aquifers, and 3) to predict the time trend in water quality depending on different land managing policy. In that way, the quantity of some markers of pollution easy to measure (chloride and nitrate) could be evaluated in each system.

Main results

Weekly water samples were collected from the seven most important springs from 1994 to December 1996 for five of them and from 1991 to December 1996 for the Espérelle and Homède springs. Major elements were analyzed. Discharge of these springs has been recorded from November 1994 to December 1996.

Chloride and nitrate contents are excellent indicators of pollution because they are related to several sources, common and easy to analyze and then to quantify in a balance. Bacteria or hydrocarbons are important pollutants, but their analysis is not very easy, they are far to conservative and their balance is difficult to calculate at the system scale. Calculation of chloride and nitrate budgets allowed two approaches:

- For the same period: comparison between the seven main springs that drained the Larzac plateau.
- For the same spring: comparison between the different hydrological years.

Cl and NO_3 contents and flow rate of the seven main springs have been studied. The longest common time series was chosen from 16/02/95 to 27/04/96 (437 days). First, chloride and nitrate mean contents and discharged water volume have been calculated for this period. Then, the chemical flux exported from the karstic systems has been calculated for each spring. In order to compare the different springs, the transport flux of each spring was divided by surface of its recharge area. Results are summarized in the Table 2.

Mouline spring appears as the reference spring: chloride and nitrate contents are the lowest. In fact, the recharge area of this spring is completely free of man activity. Cl and NO_3 contents at Mouline spring correspond to the natural contribution from rain and

Table 2: Contents and Cl and NO₃ fluxes at the main springs of the Larzac plateau.

	Recharge area	discharge volume	Cl content	Cl flux	Specific Cl flux	NO$_3$ content	NO$_3$ flux	Specific NO$_3$ flux
	Km²	10⁶ m³	mg/l	10³ kg	10³ kg/ km²	mg/l	10³ kg	10³ kg/ km²
Mouline	28	32.6	3.00	98	**3.5**	2.05	67	**2.39**
Espérelle	100	109.9	4.17	458	**4.58**	7.9	869	**8.69**
Homède	19	21.4	4.31	92	**4.84**	10.34	221	**11.63**
Boundoulaou	16	19.4	4.35	84	**5.27**	9.46	183	**11.46**
Cernon	17	19.5	3.88	76	**4.46**	6	117	**6.9**
Durzon	120	140.5	5.15	724	**6.03**	8.64	1215	**10.12**

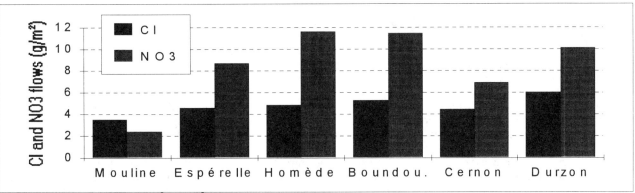

Figure 1: Cl and NO$_3$ fluxes in 10^3 kg/ km²

Table 3: Balance of Cl and NO$_3$ (10^3 kg) discharged by Espérelle spring during 1993 hydrological year.

	Natural contribution	Domestic waste water for concentrated habitation	Domestic and farming waste water and fertilizers	roads
Cl	62	5	7	20
NO$_3$	95	2	153	0

soils. Even if Cl and NO$_3$ contents do not presently reach high levels, the water from the other main tapped springs obviously shows contamination due to man activities on the plateau in comparison with Mouline spring. Cl content is 1.3 to 1.7 times that of Mouline spring, and NO$_3$ content is 3 to 5 times higher than the natural contribution (Fig. 1).

From a detailed inventory of the pollution sources (number of permanent and seasonal people, of cattle, number and location of farms, of sheep fields and barns, waste disposals, roads....) and from the data from the Water Authority concerning the mean production of some pollutants, a balance of Cl and NO$_3$ was calculated at the Espérelle spring (Table 3). The balance of Cl (conservative) and NO$_3$ (non-conservative) as tracers of pollution indicates the importance of the human impacts on water quality (Bakalowicz et al., 1996).

Hydrological conditions are obviously responsible for mass transport. The dissolved solid flux discharged at the karstic spring is related to the water volume flowing through the system. The 1996 hydrological year was a very rainy period, with twice the mean annual rainfall during the winter 1995-96. This high water flow rate increases soil leaching which is responsible for the very high Cl and NO$_3$ daily fluxes for the 1996 year. Therefore time variability is related to hydrologic conditions. This example shows that time variability does not result from seasonal variations as often described in the literature (Boyer and Pasquarell, 1995). In the same way, comparison of nitrate fluxes at all the karst springs of Larzac does not show any seasonal variability.

Evolution of chemical budgets with time

At the same spring, variations of chloride and nitrate fluxes exported from the karstic system are observed for successive hydrological years. Espérelle spring has been chosen because chemical and hydrodynamic data were available for four successive hydrological years (1992-1996). Discharge is recorded, but some gaps exist in the series (15 to 47 days). To compare the years, chloride and nitrate fluxes at the Espérelle spring were divided by the number of recorded days. Results are summarized in the following table (Table 4).

DISCUSSION

Anthropic activity impacts

The example of the Larzac karst plateau, like many others studies in karst terrain, naturally shows that karst groundwater quality is changed because of land uses. In the Appalachian Region (West Virginia) Boyer and Pasquarell (1995) have shown a strong linear

Table 4: Chloride and nitrate fluxes for 4 successive years at the Espérelle spring (Hydrological years: from 01/09/91 to 31/08/96)

	number of days	Water volume 10^6 m^3	Cl content mg/l	yearly Cl flux 10^3 kg	Cl flux per day 10^3 kg/d	NO$_3$ content mg/l	NO$_3$ flux 10^3 kg	NO$_3$ flux per day 10^3 kg/d
1992	292	16.1	5.09	82	**0.28**	7.56	122	**0.42**
1993	301	38.1	4.54	173	**0.58**	8.24	314	**1.04**
1994	322	43.6	4.33	188	**0.59**	5.99	260	**0.81**
1995	365	23.2	4.16	97	**0.27**	7.81	181	**0.49**
1996	264	102.8	4.17	429	**1.63**	8.00	822	**3.11**

relationship between nitrate concentration and percent agricultural land. In the same region, the same authors showed (1996) the influence of cattle farming (beef cow-calf operation) and of a dairy on water quality. In the same way, Zakutin and Paoukchtis (1993) assigned to the development of agriculture in Lithuania the deterioration of karst groundwater quality.

Most of the examples in literature are supported by nitrate increases. Different sources of nitrates are considered. First, autogeneous production of nitrate by nitrifying bacteria may be taking place, but this process is probably unimportant to the total nitrate load in karst springs. The second source of nitrate is domestic waste. The most important sources are the agricultural ones, fertilizers and cattle farming, which always impact groundwater quality. This is the case in the Larzac example where spread habitation (200 persons), sheep farming and fertilizers introduce 70 times more nitrates than a 1000-people town with a wastewater treatment plant. Prins and Wadman (1990) discuss the impact of cattle manure.

Nitrate content is not the only tracer of water pollution. Pesticides are often associated with nitrate increases. In Iowa, most of herbicides commonly used occur in groundwater and persist year round (Hallberg et al., 1985). In the same way, the relationship between pollutant (nitrates and pesticides) concentration maxima and storm events has been shown (Quinlan and Alexander, 1987). Pesticides in the groundwater environment may pose a problem in the future, especially in karst which is a high variability medium.

Bacteriological aspects must be also considered. Pasquarell and Boyer (1995) showed that the relationship between fecal coliform data and percent agricultural land use was less certain than for the nitrate data. Human population density affects bacterial density, especially when the population is concentrated near a major connecting conduit. In addition, landfills in karst terrain influence the quality of groundwater. A bacterial load injected in a karst is concentrated and moved inside the system and that is the reason why the impact of landfills in karstic terrain is very different than in any other geological context.

Pesticide and bacteriological levels were not measured in the present work, except a few analyses showing low fecal bacteria contents during low stage and high ones during rainy time and low pesticide levels. As a matter of fact, as shown by most studies all over the world in the last twenty years, the high variability of all karst groundwater characteristics requires systematic and frequent sampling in order to be able to it take into account. Nitrate and chloride ions are the most common and the easiest and cheapest to analyze. Therefore, change in the quality of karstic groundwater may be easily shown in the nitrate and chloride budgets.

The chloride budget is an original approach to estimating pollution. In addition to the natural contribution (rain and soils), human waste waters and salt for road de-icing contribute to the increase in Cl content in groundwater. This kind of pollution appears clearly in the Larzac groundwater. Except at Mouline spring, with a recharge area free of polluting activities, all the other springs show indices of contamination.

Definition of vulnerable zones: influence of underground flow conditions

He impacts of man activities were then showed at the system scale by Cl and NO$_3$ budgets. The chemical budgets allowed defining the more vulnerable zones of the karstic system. In fact, Cl and NO$_3$ contents are not only related to land occupation, but also to infiltration conditions and location, and especially fast and/or concentrated infiltration (Bakalowicz, 1996). These natural tracers present a very different behavior at the springs. Let us take again the example of Espérelle spring. Cl poorly varies during the year whereas NO$_3$ content is affected by very important variations (Fig. 2).

This different behavior for Cl and NO$_3$ can be explained by their input conditions in the system. Cl input is spread out and rather homogeneous on the whole recharge area, from rain and salt along roads in winter. Moreover, chlorides may be stored in the epikarstic aquifer or flow through the thick infiltration zone either with delayed infiltration water (slow infiltration) or fast infiltration.

To the contrary, nitrate input is concentrated in the cultivated and inhabited zones, from treatment plants, generally developed in karst depression areas. Karst depressions promote fast and concentrated infiltration, directly related to the saturated zone through the karst conduit network. Therefore water leaching cultivated soils and washing treatment plant effluents may reach, very rapidly, the saturated zone and its conduits directly connected to the spring: these conditions are responsible for peaks and variations in nitrate contents at karst springs.

At the spring, chloride content does not vary in time as well as nitrate content. Finally, karst depressions are zones of the karstic aquifer the most sensitive to pollution transfer. Even if they are often coated by soils or detrital sediments, these areas collect and concentrate surface water directly inputting it into the saturated zone.

The role of infiltration conditions has already been discussed in the literature by Boyer and Pasquarell (1996) and by Hallberg et al. (1985). This is precisely what the Larzac study has shown.

CONCLUSIONS

The example of Larzac karst plateau shows that protection and management of the groundwater resource in karst areas requires a thorough knowledge of these types of aquifers and their relations with the human activities at a regional level. In France, a special law

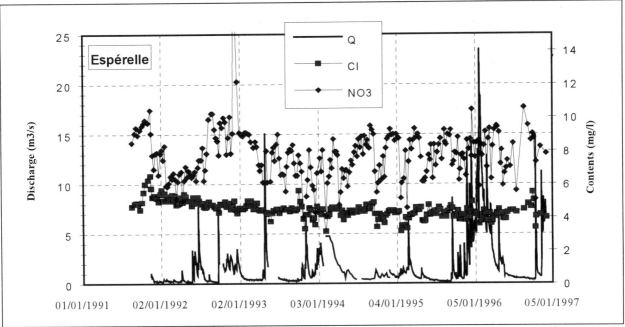

Figure 2: Variations of Cl and NO₃ contents at Espérelle spring during six hydrological years compared to flow rate Q.

about water, the French Water Act (1992), considers the water as "a common heritage", implying a patrimonial approach, i.e. a common management of a common resource (an aquifer or a water system including a surface river catchment and its related aquifers) by all the users and development and management actors. Therefore, karstic groundwater resource should be protected by its direct users. According to the act, the water policy is defined at two different levels.

The national level concerns the main basins. France is divided in six basins corresponding to the main rivers with their tributaries (Rhin - Meuse, Artois - Picardie, Seine - Normandie, Loire - Bretagne, Adour - Garonne and Rhône - Méditerranée - Corse). Each basin is controlled and managed, since the sixties, by a water authority that collects taxes from polluters, i.e. the all water users, according to the rule " *every polluter must pay for the pollution he produces* ". The Water Act imposes to the water authorities to define the general regulations for exploiting and protecting water, i.e. for managing it, after having consulted all the users, water and environment technicians and regional decision makers. Each water authority promulgated in 1996 the final regional water act, named S.D.A.G.E., " *Schéma Directeur d'Aménagement et de Gestion des Eaux* ", or **general scheme for managing water**.

The regional level concerns local catchment areas. Depending on local problems in water use conflicts, users and decision makers may decide to organize together, with the help of the concerned water authority, a *local scheme for managing water*, or S.A.G.E. (" Schéma d'Aménagement et de Gestion des Eaux "). In that aim, a *local commission for water*, or C.L.E. (" Commission Locale de l'Eau ") is created. It lists the water demand, the problems of pollution, of conflicts between users and it groups all information and knowledge about the water system in its whole.

Karst aquifers are especially concerned with water use conflicts, as shown with the Larzac example. It is necessary to point out that the vulnerability of karst regions must not be considered as an irreparable failing, which eggs us on to look for other water resources. First, the vulnerability of karst groundwater is no more constraining than for surface water. Next the ecological effect of man activities on karst waters is restricted because of the quasi absence of a relation with the biosphere, contrarily to surface water. surface water. Last and overall, the vulnerability of karst offers certain interesting properties:

- The fast elimination of accidental pollution.
- The retarding effects (adsorption, dispersion) are, most of time, negligible.
- The fast change, at a hydrological year level, of water quality, following the changes in permanent or seasonal pollution flux.
- The obvious difference of functioning, then of water quality, during low stage and floods.

The patrimonial approach proposed by the French Water Act seems to be particularly well suited to karstic areas. In fact, activities on the recharge areas of karstic springs are generally supplied in drinking water by these same springs, because of the absence of water at the surface. These regulations allow the people using the land and water to determine, all together, the best way to manage and to protect their common water resource. But there does not exist any contingency for creating and organizing protection schemes. Presently, in several karst regions of France, local schemes for managing water, so called S.A.G.E., are in progress. In the near future it should be interesting to analyze the results obtained in water conservation.

ACKNOWLEDGMENTS

Hydrogeological investigations were led in the framework of scientific cooperation on the karst between the French Geological Survey (BRGM) and several research institutes from Universities and CNRS (the National Scientific Research Center), with the support of the Adour-Garonne Water Authority, the municipalities of Millau and of the Northern Larzac plateau and the district of Millau, the Midi-Pyrenees Region and the Grands Causses Regional Park.

REFERENCES

Bakalowicz, M., 1994, Water geochemistry : water quality and dynamics: Groundwater Ecology, J. Stanford, J. Gibert and D. Danielopol, Academic Press, 1, p.97-127.

Bakalowicz, M., V. Plagnes, and Ricard, J., 1996, Land management and sustainable development of karst groundwater. The Larzac plateau (France) as an example: International Conference on Karst-Fractured aquifers-Vulnerability and Sustainability, Ustron, Pologne.

Boyer, D. G. and G. C. Pasquarell, 1995, Nitrate concentrations in karst springs in an extensively grazed area: Water Resources Bulletin, 31, 4, p.729-736.

Boyer, D. G. and G. C. Pasquarell, 1996, Agricultural land use effects on nitrate concentrations in a mature karst aquifer: Water Resources Bulletin, 32, 3, p.565-573.

Doerfliger, N., 1996, Advances in karst groundwater protection strategy using artificial tracer tests analysis and multiattribute vulnerability mapping (EPIK method), PhD thesis, University of Neuchâtel (Switzerland), 308p.

Dubois, P., 1985, Notes karstologiques sur les grands Causses: Bulletin de la Société Languedocienne de Géographie, 19, 3-4, p.197-226.

Hallberg, G. R., Libra, R. D., and Hoyer, B.E., 1985, Nonpoint source contamination of groundwater in karst-carbonate aquifers in Iowa: Perspectives on nonpoint source pollution, EPA 440/5-85-001, p.109-114

Martel, E. A., 1936, Les Causses majeurs: Millau, Artières et Maury.

Paloc, H., 1992, Caractéristiques hydrogéologiques spécifiques de la région karstique des Grands Causses (France méridionale). Hydrogeology of karst terrains, AIH, 13, p.61-87.

Plagnes, V., 1997, Structure et fonctionnement des aquifères karstiques. Caractérisation par la chimie des eaux. Géofluides Bassins Eau. PhD thesis, University of Montpellier II, 372 p.

Quinlan, J. F. and E. C. Alexander, 1987, How often should samples be taken at relevant locations for reliable monitoring of pollutants from an agricultural, waste disposal, or spill site in karst terrane ? A first approximation: 2nd Multidisciplinary Conference on Sinkholes and the Environmental Impacts of Karst, Orlando, Florida, p.277-286.

Ricard, J. and Bakalowicz, M., 1996, Connaissance, aménagement et protection des ressources en eau du Larzac septentrional, Aveyron (France): BRGM, Orléans, report R38953, 94 p.+ annexes.

Zakutin, V. P. and Paoukchtis, B. P., 1993, Effets géochimiques de la pollution agricole des eaux souterraines dans la région karstique de la Lituanie: Hydrogéologie, 1, p.65-70.

Resources and quality of waters in limestone areas of Peddavanka watershed, A.P., India

A.S.SUDHEER & S.SRINIVASA GOWD   Department of Geology, Sri Venkateswara University, Tirupati, A.P., India

ABSTRACT

The area of study falls in the western margin of the crescent shaped Cuddapah basin comprising quartzite, shale, limestone, and basic intrusive (sills) as the major rock formations of the Pre-Cambrian age group. About one third of the total area is occupied by limestone, exhibiting karst topography at some locations.

Based on the geoelectrical interpretations, the water potential zones in the watershed are divided into three categories namely good, moderate, and poor. The entire limestone area is categorized as moderate despite the fact that the limestone is cavernous and forming caves in some places with the formation of stalactites and stalagmites. The cavernous nature of the topography is seen wherever limestone comes in contact with quartzite and gives rise to perennial springs and waterfalls. India's longest cave, namely the Belum cave, is located adjacent to the study area.

Seventy-six water samples were analyzed and rated for their suitability for drinking and irrigation. The groundwater in the limestone area exhibits better quality with lower values of TDS, SAR, Percent Sodium, and RSC.

The solution activity is observed to be very intensive in the geological past, which is very much slackened in recent times.

INTRODUCTION

The Peddavanka watershed comprises an area of about 298 km^2, which forms the western margin of the crescent shaped Cuddapah basin in Andhra Pradesh, India. It is bounded by 14°55'24" and 15°12'50" N latitudes and 77°51'20" and 78°03'26" E longitudes (Fig.1). The climate of the area is generally hot and semi-arid for most of the year with an average annual precipitation of 516 mm, second lowest in the country. The population of the study area is 35,299.

Though the watershed is a chronic drought prone area, its main occupation is agriculture. Systematic hydrogeological study of the watershed is undertaken to identify potential water zones in the limestone and shale areas in order to mitigate the problems in the area. Hydrochemistry of the waters is also studied to assess its quality for drinking and agriculture.

GEOLOGY, STRUCTURE, AND GEOMORPHOLOGY

The stratigraphic succession in the watershed is given below as depicted in Fig. 2.

		Paniam Quartzite	(10-35 m)	
		---------- Local disconformity ----------		
		Auk Shale	(10-30 m)	
Kurnool		Narji Limestone	(100-200 m)	massive limestone & flaggy limestone
Group		Banaganapalle Quartzite (and conglomerate)	(10-57 m)	
		---------- Unconformity ----------------		
Cuddapah Super Group	Chitravati Group	Tadipatri formation	(4600 m)	Shale & basic intrusive: sills

The Tadipatri formation consists mainly of shale of purple color and is interbedded with layers of quartzite and flaggy limestone trending NNE - SSW with about a 10° dip to the east. Basic intrusives are sills of glassy appearance with the presence of vesicles

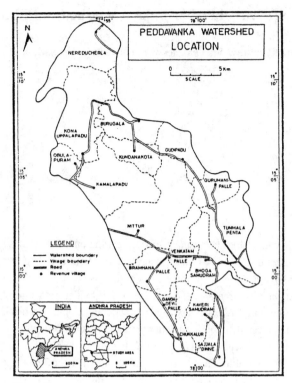

Figure 1   Figure 2

extending up to half-a-kilometer. The Tadipatri formation is overlain by the rocks of the Kurnool group, which includes quartzite, limestone, and shale. A distinct unconformity separates the Cuddapah formations from the overlying Kurnool rocks, the latter occurring as horizontal beds over the tilted Cuddapah rocks.

The landforms identified in the study area include denudational, depositional, and structural landforms namely peneplain, buried peneplain, valley fills, flood plains, alluvial plains, Mesa/Butte, cuestas, and structural hills. The lineament density study reveals a broad idea of the variations in the groundwater potential of the watershed.

EXPLORATION

Ninety-nine vertical electrical soundings were conducted using Schlumberger configuration of the surface geophysical method covering both limestone and shale areas of the watershed. The data was interpreted with the help of master curves and auxiliary point charts (Orellana and Mooney, 1966). The sounding curves A, K, and H type suggest a few three layer geoelectrical sections, and a number of four layer sections of KH, HK, QH, HA, AK, and AA types. Vertical electrical sounding results were verified for their accuracy by conducting probes in close proximity to wells and comparing with their well logs.

Iso-resistivity contour maps at 1.5m, 10m, 50m, and 90m depths as well as their 3D maps were prepared by using SURFER package and they were interpreted in terms of resistivities and thickness of various sub-surface layers. Based on the results, top layer resistivity values ranging from 2 to 630 ohm-m with its thickness ranging from 1.3 to 3.2m in the watershed while in the limestone area it ranges from 6 to 170 ohm-m with a thickness of 1.2 to 3.1 m. The resistivity variation is attributed to variation of moisture content and change in surface conditions.

Depth to basement:

By using the interpreted results of the vertical electrical soundings, depth to basement reveals that the bedrock occurs at a very shallow depth of 6.0 m at the northern portion of the watershed while it is deepest with a depth of 60.7 m near Kundanakota village situated in the Kavulupalle plateau in the middle of the watershed. Depth to basement varies from 9.6 to 60.7 m in limestone area. The watershed area is classified into three zones based on the depth to bedrock. The deepest zone occurs in the southern and central potions of the basin and also as isolated patches in other areas. The rest of the area is classified as moderate zone and shallow zone occurring as isolated patches.

Groundwater Potential Zones:

Based on the depth to bedrock, the thickness of the saturated layer, and the resistivity of the second layer, groundwater potential zones have been identified. Figure 3 depicts different groundwater potential zones that are classified as good, moderate, and poor zones. The study reveals that the weathered and fractured portions in shale and limestone occurring in southernmost and central portions of the watershed area exhibit the productive water bearing zones categorized as good potential zones. Moderate groundwater potential zones occur mainly in the western and central portions and subordinately in northern and southern portions of the basin. The remaining parts of the area are demarcated as very low saturated zones classified as poor potential water zone. The interpretation of the depth to basement is in good agreement with the actual depth observed in the dug wells, dug-cum-bore wells and bore wells. The electrical resistivity data thus gives reasonably accurate results to understand the sub surface layers, basement configuration in groundwater exploration.

SOLUTION ACTIVITY

The potential of limestone as an aquifer can be recognized from surface hydrogeological studies aimed at identification of solution activity that are a result of chemical weathering. The solution features manifested are mainly by enlargement of joints with the intervening ribs, hemi-cylindrical grooves, accumulation of residual deposits - terra rosa, lost rivers, highly disturbed stratification and by formation of caverns, sinkholes, and springs (LaMoreaux et al., 1970; LeGrand and Stringfield, 1973) (Figure 4). The solution phenomena in the study area were activated under humid environment with plenty of carbon dioxide and acidic conditions of circulating waters inflicting on the rocks during some geological past. However almost all the waters in the limestone terrain are found to be alkaline with pH ranging from 8.2 to 8.8 with a mean of 8.21. The solution activity is very much slackened in recent times as evidenced by the analysis of the rainfall data for 94 years. Extensive and intensive solution activity in the limestone of Andhra Pradesh is indicated in the caves of Visakhapatnam and Bethamcherla caves in Kurnool district. In the watershed several caves are noticed in

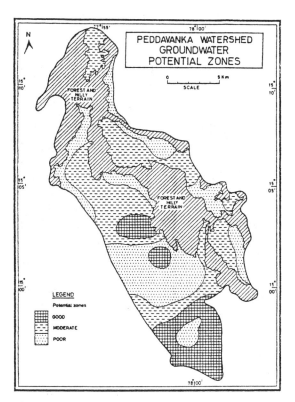

Figure 3

the Narji limestone including the Belum cave which is considered to be the India's longest cave occurring at its boundary. At some places the lost rivers occur - one prominently seen at Kona Rameswaraswami temple near Kona Uppalapadu. Many perennial springs also occur at places where quartzite comes in contact with limestone. At Gudipadu the spring is seen (Figure 5) with its reservoir at the forefront of the picture at its base. Recently, there are reports of sudden appearances of sinkholes, due to subsidence, in some parts of Cuddapah district.

WATER QUALITY

Out of seventy-six water samples collected four are surface water samples while the rest are groundwater samples. The water samples were collected in two seasons during post and pre-monsoon periods (Jan, 1997 and June, 1997) and analyzed for their dissolved constituents by following standard methods for collection, preservation, analysis and interpretations (AWWA, 1971; Hem, 1985). The results of the analysis are given in 1&2.

Based on the concentration of dissolved solids it is found that ninety percent of the waters of the post-monsoon and ninety four percent of the pre monsoon waters are useful for drinking while all waters of both the seasons are suitable for irrigation.

Based on sodium adsorption ratio ninety-seven percent of the water samples of the watershed fall in the excellent category and three percent of the samples fall in the good category in the post-monsoon season. Ninety nine percent of the samples fall in excellent category and one percent falls in the good category in pre-monsoon season. None of the samples is of poor category for irrigation in either season. About eighty percent of the water samples belong to excellent, good and permissible categories in both the seasons for irrigation while none of the samples is unfit for irrigation based on

Figure 4

Figure 5

percent sodium. Based on residual sodium carbonate about twenty five percent of the water samples are unsuitable for irrigation.

The quality of waters in the limestone area both in the pre and post-monsoon seasons is better than that of the waters in other areas either for drinking or for irrigation as revealed from analysis and assessment.

CONCLUSION

Based on the geoelectrical interpretations, the water potential zones in the watershed are divided into three categories namely good, moderate, and poor. The entire limestone area is categorized as moderate despite the fact that limestone is cavernous and forming caves in some places. Where limestone comes in contact with quartzite perennial springs occur with waterfalls. As the present exploration is carried out only up to ninety meters and the solution activity is expected to be very intensive and extensive in the geological past further investigation into deeper levels in the limestone is expected to contain productive groundwater zones.

Table 1. Statistical parameters of groundwater samples (Post Monsoon).

S.No	Constituents	Units	Total Watershed			Limestone Terrain		
			Min.	Max.	Mean	Min	Max	Mean
1	Silica	mg/l	4	49	33.67	4	32	19.33
2	Calcium	mg/l	9	133	35.31	19	133	67.79
3	Magnesium	mg/l	7	215	52.06	7	98	36.92
4	Sodium	mg/l	9	540	119.92	9	219	46.92
5	Potassium	mg/l	1	19	3.11	1	17	3.79
6	Carbonate	mg/l	0	153	38.59	4	39	22.43
7	Bicarbonate	mg/l	64	772	359.28	136	536	284.62
8	Sulphate	mg/l	5	70	41.94	5	70	32.08
9	Chloride	mg/l	11	514	116.93	10	306	91.13
10	Total Dissolved Solids	mg/l	301	1487	637.69	320	911	470.71
11	Hardness as $CaCO_3$	mg/l	80	912	302.29	88	684	320.33
12	Alkalinity as $CaCO_3$	mg/l	85	882	324.88	131	472	233.08
13	N CH	mg/l	0	469	108.97	0	369	101.00
14	SEC	µmhos/cm at 25^0c	510	2520	1074.78	160	1370	780.01
15	pH		6.8	8.8	8.36	8.2	8.8	8.21
16	SAR		0.09	25.97	3.30	0.21	10.26	1.32
17	Percent Sodium		6	94	41.07	6	60	22.21
18	Potential Salinity		0.41	15.23	3.76	0.41	9.36	2.87
19	RS C		0	17.56	3.39	0	8.36	2.01

Based on the total dissolved solids, sodium adsorption ratio, percent sodium, and residual sodium carbonate in the watershed, the groundwater in the limestone area exhibits better quality for drinking and irrigation. As the watershed is a chronic drought prone area

Table 2. Statistical parameters for groundwater samples (Pre Monsoon)

S.No.	Constutuents	Units	Total Watershed			Limestone Terrain		
			Min	Max.	Mean	Min	Max	Mean
1	Silica	mg/l	2	38	21.76	2	26	12.76
2	Calcium	mg/l	8	141	31.24	14	141	58.14
3	Magnesium	mg/l	8	113	43.00	8	81	33.05
4	Sodium	mg/l	9	342	100.48	8	192	43.81
5	Potassium	mg/l	1	8	2.44	1	8	3.00
6	Carbonate	mg/l	4	110	28.52	4	31	17.84
7	Bicarbonate	mg/l	48	684	310.30	116	484	246.67
8	Sulphate	mg/l	5	80	29.24	5	60	25.95
9	Chloride	mg/l	10	418	101.82	10	261	79.61
10	Total Dissolved Solids	mg/l	256	1226	529.58	256	736	420.71
11	Hardness as $CaCO_3$	mg/l	72	524	254.65	72	451	278.33
12	Alkalinity as $CaCO_3$	mg/l	59	581	276.38	112	420	213.95
13	N CH	mg/l	0	310	102.68	0	310	94.95
14	SEC	μmhos/cm at 25°C	365	1830	778.02	365	1020	611.62
15	pH		6.7	8.6	8.45	8.2	8.8	8.38
16	SAR		0.24	10.25	2.88	0.23	9.79	1.39
17	Percent Sodium		7	85	41.38	7	85	23.52
18	Potential Salinity		0.33	12.21	3.18	0.39	7.57	2.60
19	RSC		0	7.86	2.70	0	7.41	2.37

with very low annual rainfall, surface and sub-surface check dams must be constructed at feasible places across the main stream in order to reduce the run off and boost up the water levels. Further the irrigation pattern must be changed from paddy to other commercial crops which require less water by following latest irrigation techniques for the development of the watershed.

REFERENCES

American Public Health Assoc., American Water Works Assoc., and Water Pollution Control Federation, 1975. Standard Methods For Examination of Water and Waste Water: 14th ed., Amer. Publ. Health Assoc., Washington, D.C., 1200 p

Hem, J.D., 1985. Study and Interpretation of the Chemical Characteristics of Natural Water: U.S.Geol. Surv. Water Supply paper - 2254, 264 p.

LaMoreaux, P.E., et al., 1970. Hydrology of Limestone Terranes: Annotated Bibliography of Carbonaceous Rocks, Geological Survey Alabama Bull. 94 (A), 242 p.

LeGrand, L.H., and Stringfield, V.T., 1973. Karst Hydrology - A Review: Jour. Hydrology, V.20, p.97-120.

Orellana, E. and Mooney, H.M., 1966. Master Tables and Curves for Vertical Electrical Sounding Overlayered Structures: Madrid Interciecia, 150 p.

Transport and variability of trace metals in a karst aquifer based on spring chemistry

DOROTHY J. VESPER *The Pennsylvania State University, University Park, Pa., USA*

ABSTRACT

The movement of colloids in a karst ground water system may control the presence of trace, and potentially toxic, metals at karst springs. Over 120 water samples were collected from eight springs at the Fort Campbell Army Base, located on the border of Kentucky and Tennessee. Trace constituents detected included As, Be, Cd, Cr, Co, Cu, Pb, Mo, Ni, Se, Sn and V. The trace metals were present only for a limited duration (often less than two hours) and were generally associated with high concentrations of aluminum and iron. Comparison of unfiltered and filtered samples suggests that the aluminum and iron, along with the trace metals, are present in colloidal form. Colloid-associated metals are affected by different factors than the carbonate-associated metals; therefore, they are discharged from springs at different times, relative to storm pulses, than metals such as calcium and magnesium. The type of flow system, the amount of recharge, and the source area of a spring will determine the extent to which these factors influence overall spring water quality. The Fort Campbell data may have important implications regarding transport mechanisms for inorganic contaminants in karst systems, design of spring water quality monitoring programs, and assessment of human and ecological impacts at springs affected by contaminated ground water.

INTRODUCTION

The water chemistry in some karst springs is known to vary significantly following precipitation events or storm pulses (Shuster and White, 1971). The amount of chemical variability, and the temporal relationship between stage peaks and chemical fluctuations, have been correlated to recharge modes, dominant means of transmission, and aquifer storage (Smart and Hobbs, 1986; Fields, 1993). Understanding the presence, concentrations, duration and timing of contaminant fluxes to a karst spring is essential to being able to predict when contaminants are likely to be discharged. Springs, because they are the discharge locations for ground water, are points of exposure that must be considered when evaluating potential human health and ecological risks in karst systems.

Karst springs have been sampled over storm surges to assess inorganic water quality and to help design a spring monitoring program at the Fort Campbell Army Base on the Kentucky-Tennessee border. Three rounds of spring chemistry data were collected between 1996 and 1998; up to eight springs have been included in the study and over 120 samples have been collected for laboratory analysis. The objectives of the program were to (1) evaluate changes in overall water quality in springs during a storm, (2) determine if contaminants were present at springs, and (3) if contaminants were present, determine the duration of their discharge. This paper reports the findings of that study, relationships between specific metals discharged at the springs, and possible transport mechanisms inferred from the data obtained.

STUDY AREA

Fort Campbell Army Base lies approximately 50 miles northwest of Nashville on the Kentucky-Tennessee border (Figure 1). The base is located on the West Highland Rim of the Nashville Basin and is underlain by limestones of the Mississipian St. Louis and Ste. Genevieve Formations (Klemic, 1966a, 1966b).

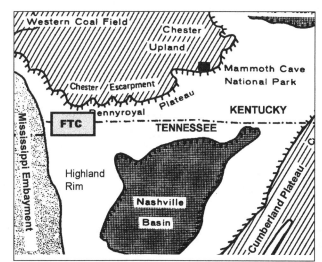

Figure 1: Site Location (FTC=Fort Campbell; modified from George, 1985).

Land use at Fort Campbell is divided into the following areas: the Campbell Army Airfield, the Main Cantonment Area (administration, housing, storage and maintenance), the Old Clarksville Base, the Sabre Army Heliport, and the range. The facility is permitted under the Resource Conservation and Recovery Act (RCRA); over 350 Solid Waste Management Units (SWMUs) have been identified.

The Little West Fork Creek flows across the base from west to east and is the main surface water discharge from the site. The springs included in this study discharge either directly into the Little West Fork or into a tributary thereof. The springs vary considerably in size and morphology; the baseline discharges range from approximately 250 to 5,000 meter^3/day (0.1 to 2 feet^3/day).

METHODS

Eight springs have been sampled up to three times each: in September 1996, May 1997 and May 1998. During the sampling events, water samples for inorganic parameters were collected using automated sampling equipment before, during, and after a storm pulse. Sample frequency ranged from hourly to every 12 hours, depending on the spring.

Field data were collected concurrently with the water samples. Stage, conductivity and temperature were collected continuously using a Campbell Scientific, Inc. digital micrologger with associated probes. Stage was monitored using a Druck pressure transducer that measures changes to 0.03 cm (0.001 foot). Water temperature was monitored using an Omega Engineering, Inc. platinum resistance thermometer that measures to 0.01°C. Specific conductance was measured using a probe designed at Eastern Kentucky University (Idstein, 1992) that measures to 1 [Siemen/m (0.01 [mho/cm). The data were collected every 10 seconds and saved every two minutes or every hour depending upon the amount of change occurring in the parameters. The continuous monitoring data were collected by Ewers Water Consultants, Inc. Turbidity was measured using a HF Scientific DRT-15 turbidity meter; pH was measured a Horiba Water Quality Checker U-10.

Based on stage, conductivity, temperature and turbidity, up to six samples were selected from each spring for laboratory analysis of inorganic parameters. The intent of the program was to determine the range of spring water chemistry, therefore samples were selected based on maximum and minimum values for the field parameters. One baseline sample, collected prior to the storm, was analyzed for each spring during each event.

Generally, the samples were analyzed for total metals. Total, rather than filtered, samples were analyzed so that the data represented total transport, regardless of chemical form. To help evaluate the form of the metals, six samples were selected for analysis of both total and dissolved metals during the 1998 sampling program; they were selected because they had the highest turbidity of the samples collected that sampling round. The dissolved metal samples were field filtered using a 0.45 [m filter. The analyses were conducted by DataChem Laboratories of Salt Lake City, Utah in accordance with protocols specified by the U.S. Army Toxic and Hazardous Materials Agency (1990) and the Army Environmental Center (1993). The analyses were conducted using inductively coupled plasma (ICP) for Al, B, Ba, Be, Ca, Cr, Cu, Fe, K, Mg, Mn, Mo, Na, Ni, Sn, V and Zn; trace ICP for As, Cd, Pb, Se, and Ti; graphic furnace atomic absorption (AA) for Ag and Sb; and cold vapor AA for Hg.

RESULTS AND DISCUSSION

The detection frequency and the range of trace metal concentrations varied by spring and by storm. Most of the trace metals are only present in a limited number of samples (Table 1). The maximum concentrations were detected for only a short duration. For example, the storm pulse sampled in 1997 produced turbid, comparatively metal-rich water at Beaver Spring. Most of the maximum concentrations listed on Table 1 are from sample BEAV2C, collected on May 2, 1997. The same metals, with the exception of lead, were below detection in samples collected one hour before and one after BEAV2C. The aluminum and iron concentrations for the same sample were the highest detected during any of the sampling events.

Sample BEAV2C was collected within six hours of the initial storm response. Monitoring data (Figure 2) show that the sample was collected during a stage peak [20 cm (0.7 feet) above baseline], temperature peak (0.5°C above baseline), and conductivity dip (300 [Siemen/cm below baseline). The sample contained elevated turbidity (200 NTUs above baseline), but it was not the maximum turbidity measured at this spring during this storm.

A storm water channel flows near the spring. However, the water sample is unlikely to be from the storm water channel because of the spring geometry and because the sample was collected from an opening in the limestone from which the spring flows. Chert sand deposits, found near the spring orifice following large storms, indicate that

Table 1: Maximum concentrations in unfiltered water samples (all concentrations in ug/L).

Metal	Number of Detections	Maximum Conc.	Sample Containing Maximum	Corresponding Aluminum Concentration
Arsenic	3	68.4	BEAV2C	111,000
Beryllium	3	4.74	BEAV2C	111,000
Cadmium	8	2.23	DENN1B	539
Chromium	17	227	BEAV2C	111,000
Cobalt	1	60.1	BEAV2C	111,000
Copper	1	102	BEAV2C	111,000
Iron	84	216,000	BEAV2C	111,000
Lead	16	171	BEAV2C	111,000
Manganese	10	5170	BEAV2C	111,000
Mercury	1	0.567	BEAV2C	111,000
Molydendum	4	70.6	N1003A	not detected
Nickel	4	124	BEAV2C	111,000
Selenium	1	7.29	BEAV2C	111,000
Tin	6	114	BEAV2C	111,000
Vanadium	2	419	BEAV2C	111,000
Zinc	31	582	BEAV2C	111,000

significant solid material is transported by Beaver Spring. The sample could represent storm water that entered the system via rapid input from a sinkhole. Tracer tests have confirmed that Beaver Spring is closely linked to at least one sinkhole in the Main Cantonment Area at Fort Campbell.

The water samples for this part of the study were not filtered prior to analysis, therefore the data reflect particulate as well as dissolved concentrations. During the 1998 sampling program several samples were field filtered to assess the distribution between dissolved and particulate metals. In all but one sample, the aluminum and iron concentrations present in the total metal samples decreased to below detection in the filtered metal samples. The one exception, from Eagle Spring, had 84.3 ug/L of iron in the filtered sample; there was 2,800 ug/L of iron in the corresponding unfiltered sample. The maximum aluminum and iron concentrations present in the 1998 total metal samples, respectively 5,240 ug/L and 3,260 ug/L, were well below the maximum concentrations detected in 1997 for the same metals. Three trace metals were detected in the unfiltered samples, but not in the filtered samples (Table 2).

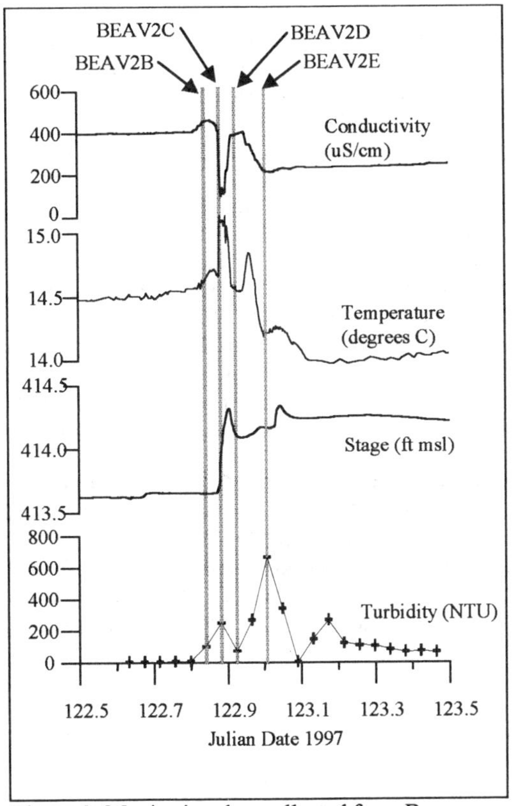

The comparison between total and dissolved concentrations indicates that many of the metals are being transported in a colloidal or particulate form. The abundance, type, and size of the colloids in karst flow systems has been linked to source rock, discharge, flow velocity, and ionic strength of the water (Atteia and Kozek, 1997; McCarthy and Shevenell, 1998). The nature of the colloidal matter is likely to influence the types of contaminants and metals associated with the solids (Buffle et al., 1998).

The correlation between aluminum concentrations and the presence of trace metals was investigated to determine if aluminum could be used as an indicator of the time at which trace metals are transported. Linear regression analyses were conducted between aluminum and various trace metals. The regression was conducted using the least-squares method with a 95 percent confidence limit. The trace metals which were detected three or more times, and had corresponding aluminum concentrations, were included. Other cations, such as calcium and iron, were included for comparison.

The cations calcium and magnesium did not correlate with the aluminum concentrations (Table 3). Regression correlation (R^2) values were at or below 0.02. This can be explained because calcium and magnesium are present in dissolved form and have different transport mechanisms than the colloidal-associated metals.

Stronger correlations were observed between aluminum with arsenic, beryllium, chromium, lead, iron, manganese, and zinc. For these metals, the R^2 values are reported both with and without the maximum concentration included in the analysis. Although the maximum concentration supports the overall trend, regression analysis conducted with the full data set is biased by the data distribution (a single very high maximum) and cannot be considered representative. Therefore, the correlation between aluminum and the other metals can be considered stronger than indicated by the R^2 values generated for the data without the maximum concentrations, but not as strong as indicated by the R^2 values of the entire data set.

In general, the data suggest that the inorganic constituents detected in the springs can be linked to two or more populations depending on their source, form and transport mechanisms. Calcium and magnesium represent dissolution of limestone and transport of dissolved carbonate species. Aluminum, iron and the trace metals appear to be associated with colloidal transport.

The association of the trace metals with the aluminum/iron-rich water pulses may be the result of numerous factors. Possible explanations include (1) the metals may be present in the mineral matrixes, (2) the aluminum and/or iron may be present as colloidal oxides or hydroxides which scavenge trace metals, (3) sorption of metals onto mineral surfaces, or (4) sorption of metals onto bacteria or other organic matter. The transport of similar colloids has been reported by Atteia and Kozel (1997) and McCarthy and Shevenell (1998). Atteia and Kozel analyzed colloids from Noiraigue Spring in Switzerland and identified clays, granular minerals and bacteria. McCarthy and Shevenell (1998) examined karst-aquifer colloids and observed both clay and iron oxides.

The study by Atteia and Kozel (1997) also noted that the number and size of the colloids increased with increasing spring discharge and was greatest during the rising limb of the hydrograph. Interestingly, sample BEAV2C, which contained the highest concentration of trace metals detected during the Fort Campbell study, corresponds to a rise on the hydrograph (Figure 2). If the

Figure 2: Monitoring data collected from Beaver Spring during the 1997 sampling program (vertical gray lines indicate sample times; samples B, C and D taken at 1-hour intervals).

Table 2: Total and dissolved concentrations (all concentrations in ug/L; J values estimated; LT indicates less then the detection limit).

Detected Constituents	Maximum Unfiltered Concentration	Corresponding Filtered Concentration
Aluminum	5,240	LT 112
Iron	3,260	LT 75.5
Arsenic	4.12 J	LT 5.0
Cadmium	0.32 J	LT 2.0
Lead	8.83	LT 3.0

colloids and particulate material are present in the aquifer and are not transported until the threshold velocity is met (Atteia and Kozel, 1997), then is it possible that the maximum concentration of trace metals at a spring may occur after a storm event but prior to any changes in temperature or conductivity. At Beaver Spring, the stage pulse and conductivity/temperature changes are nearly coincident, and so this cannot be evaluated. However, at springs with a lag time between stage pulse and conductivity change, the metal concentrations should be monitored early in the storm and not just concurrently with the discharge of new, recharge water.

At Fort Campbell, the spring with the greatest number of detectable trace metals, and the greatest change in metal concentrations over the storm surges, also had the greatest changes in conductivity and turbidity. Hence, a possible conclusion is that in springs with higher velocities and rapid responses to storms exhibit more colloidal transport and the associated transport of metals. This would also be true for organic contaminants that may be sorbed onto colloidal matter.

Table 3: Results of the linear regression analysis between aluminum and other metals

Relationship	Number of Samples	Regression Coefficients With Maximum	Regression Coefficients Without Maximum
Al-Fe	49	0.997	0.931
Al-As	3	0.993	---
Al-Be	3	0.999	---
Al-Cd	5	0.237	---
Al-Cr	5	0.997	0.654
Al-Pb	15	0.991	0.512
Al-Mn	39	0.993	0.780
Al-Ca	49	0.001	---
Al-Mg	49	0.020	---
Al-Zn	24	0.974	0.530

CONCLUSIONS

Many metals typically monitored at karst springs, such as calcium and magnesium, are present from the dissolution of carbonate rocks. They are transported in a dissolved form, and hence their temporal variation at springs does not match the temporal variation of the colloid-associated trace metals. The carbonate metal concentrations can be predicted using chemical parameters such as conductivity; however, a hydrograph may be more useful for predicting the arrival time of the colloid-associated metals.

Water quality monitoring for inorganics in karst springs must consider transport mechanisms. Not only will the contaminant fluxes change following storm events, but the timing of the pulses may vary depending upon the type of contaminant and the manner in which it is transported. The degree to which this is important depends upon source areas within the ground water basin, recharge types, ground water flow velocities, and if the spring is fed by conduit-dominated or diffuse-dominated flow.

Data collected from seven springs in Kentucky indicate that colloidal transport may control the movement of trace metals and inorganic contaminants. The entrainment and movement of colloids is a function of flow velocity, therefore the degree of metal transport varies with storm events and hydrologic regimes. Springs with a rapid response to storms and significant change in chemistry are more likely to discharge colloids and the associated potential contaminants. Although this study focused on trace metals, organic compounds sorbed to colloids may also have similar transport patterns.

ACKNOWLEDGEMENTS

The author greatly appreciates Fort Campbell Army Base for support of this project and the Hunt family for access to Quarles Spring.

REFERENCES

Atteia, O. and Kozel, R. (1997) Particle size distribution in waters from a karstic aquifer: from particles to colloids. *Journal of Hydrology* 201, 102-119.

Buffle, J., Wilkinson, J. J., Stoll, S., Filella, M., and Zhang, J. (1998) A generalized description of aquatic colloid interactions: The three-colloidal component approach. *Environmental Science and Technology* 32, 2887-2899.

George, A. I. (1985) Caves of Kentucky. In *Caves and Karst of Kentucky*, Dougherty, P. H. editor. Kentucky Geology Survey Special Publication 12, Series XI.

Field, M. S. (1993) Karst hydrology and chemical contamination. *Journal of Environmental Systems* 22, 1-26.

Idstein, P. J. (1992) *Investigation, using florescent dyes and continuous groundwater monitoring, of the sources and transfer mechanisms that contribute to the Cathedral Hall Passage cave stream: Unthanks Cave, Lee County, Virginia*. M.S. Thesis, Department of Geology, Eastern Kentucky University.

Klemic, H. (1966a) *Geologic Map of the Hammacksville Quadrangle, Kentucky-Tennessee*. US Geological Survey Map GQ-540, Washington, DC.

Klemic, H. (1966b) *Geologic Map of the Oak Grove Quadrangle, Kentucky-Tennessee*. US Geological Survey Map GQ-565, Washington, DC.

Shuster, E. T. and White, W. B. (1971) Seasonal fluctuations in the chemistry of limestone springs: A possible means for characterizing carbonate aquifers. *Journal of Hydrology* 14, 93-128.

Smart, P. L and Hobbs, H. O. (1986) Characterization of carbonate aquifers: A conceptual base. In *Proceedings of the Environmental Problems in Karst Terranes and Their Solutions Conference, October 28-30, 1986, Boiling Green, KY*, 1-14.

U.S. Army Environmental Center (1993) *Guidelines for Implementation of ER-1110-1-263 for USAEC Projects.*

U.S. Army Toxic and Hazardous Materials Agency (USATHAMA) (1990). *Quality Assurance Program*, USATHAM PAM 11-41.

Use of GORE-SORBER® modules to screen for organic contaminants in karst springs

DOROTHY J. VESPER   The Pennsylvania State University, University Park, Pa., USA
JAMES E. RICE   Arthur D. Little, Incorporated, Cambridge, Mass., USA
RAY F. FENSTERMACHER   W. L. Gore and Associates, Incorporated, Elkton, Md., USA

ABSTRACT

The application of the GORE-SORBER® Screening Survey was evaluated as an alternative method for screening of organic compounds in karst springs at the Fort Campbell Army Base, Kentucky. This technology uses GORE-SORBER® Modules (modules), which are passive, time-integrating, sorbent sampling devices patented and manufactured by W. L. Gore & Associates, Inc. The devices are designed to sample volatile and semi-volatile organic compounds (VOCs, SVOCs) in soil gas, but may have applications as screening devices for spring water quality in karst springs. Our study found that the modules can detect the presence of VOCs and SVOCs in karst springs. Discharge volume and sample duration do not appear to have significant impacts on the data, but the compounds detected at the highest masses appear to have the most consistent frequency of detection. The suites of the detected compounds varied by spring and may be used to segregate ground water basins and to link possible source areas with specific springs. Based on the data from Fort Campbell, there are limitations for this technology, but it is an effective method for providing qualitative, cost-effective data.

INTRODUCTION

Karst springs, because they are discharge points for ground water, can be impacted by the presence of contamination in their ground water basins. Representative spring water quality data in springs can be difficult to obtain because of the amount of variability in water chemistry during storm pulses. The amount of chemical variability, and the temporal relationship between stage peaks and chemical fluctuations, have been correlated to recharge modes, dominant mode of transmission, and aquifer storage (Smart and Hobbs, 1986; Fields, 1993). Understanding the presence and temporal variability of contaminants in karst springs is an essential component of understanding spring water quality.

Sampling springs through storm pulses for volatile organic compounds (VOCs) can be expensive and a logistical problem. GORE-SORBER Modules are passive, time-integrating sampling devices that may be useful to screen for the presence of VOCs in karst springs. Use of the modules may eliminate some of the cost of sampling as well as the logistical problems associated with conventional sampling. Additionally, the typical two or more days of module exposure in spring waters may eliminate the potential of "missing" a short-duration contaminant pulse. This paper reports a series of studies that were conducted to determine the feasibility of using modules in karst springs, identify the limitations of the method, and present recommendations for future use and assessment.

STUDY AREA

Fort Campbell Army Base (FTC) lies approximately 50 miles northwest of Nashville, TN on the Kentucky-Tennessee border. The base is located on the West Highland Rim of the Nashville Basin and is underlain by carbonate rocks of the Mississipian St. Louis and Ste. Genevieve Formations (Klemic, 1966a and 1966b). The primary land uses include the Campbell Army Airfield, the Main Cantonment Area (administration, housing, storage and maintenance), the Old Clarksville Base, the Sabre Army Heliport, and range areas. The facility operates under a RCRA Part B Permit from Kentucky; over 350 potentially contaminated Solid Waste Management Units (SWMUs) have been identified.

The Little West Fork Creek flows across the base from west to east and is the main surface water discharge from the site. Eight springs have been included in the chemical spring monitoring program to date. The springs vary considerably in size and morphology; baseline discharges range from approximately 250 to 5,000 meter^3/day (0.1 to 2 feet^3/second).

METHODS

The GORE-SORBER Modules were placed in the springs with the sorbers approximately four inches above the sediment-water interface. They were held down with stainless-steel weights but kept above the sediment by attachment to a floatation device. Unless specified otherwise, they were placed in the main current of the spring and were left in place for two days. Grab water samples were collected as near to the spring orifice and the module placement as possible.

Each module consists of several separate passive collection devices (sorbers). Each sorber contains a patented, granular, absorbent material. The sorbers are contained within a 4-foot long, vapor-permeable cord constructed of inert, hydrophobic, microporous expanded polytetrafluorethylene membrane. After exposure, the sorbers are thermally desorbed and analyzed using a gas chromatograph coupled with mass spectroscopy (GC/MS). The analytical methodology follows a modified EPA method 8260a and 8270b for VOCs and SVOCs, respectively. Method detection limits have been established for each compound on a variety of target compound lists. Quantification of the mass from each sorber is based on a calibration curve, and results are reported for each compound as mass desorbed from each sorber in micrograms. Quantitation of any mass which is less than the MDL is reported as "bdl", and when no mass is present the result is reported as "nd".

The water analyses were conducted by DataChem Laboratories of Salt Lake City, Utah. The analyses were conducted using GC/MS in accordance with protocols specified by the U.S. Army Toxic and Hazardous Materials Agency (1990) and the Army Environmental Center (1993).

RESULTS AND DISCUSSION

Several different studies were conducted to assess the usefulness of the GORE-SORBER modules for VOC screening in springs. A preliminary test was conducted in 1997 to evaluate if the modules could detect VOCs in spring waters. Twenty-two VOCs were detected on the modules; however, only four compounds were detected in the concurrent aqueous samples. The initial data were sufficient to warrant subsequent studies.

The studies conducted in 1998 were designed to address specific concerns identified during the preliminary study. Parameters that were varied in the 1998 studies included the location of the modules, the sampling duration, and the timing relative to storm pulses (baseline versus storm). Additionally, grab water samples were collected concurrently with the GORE-SORBER samples to assess the relationship between aqueous concentrations and sorbed masses.

Evaluation of Spring Flow Velocity

The first study addressed the influence of flowing water on the response from the GORE-SORBER modules. At two springs, the modules were placed in still water and in the main discharge. The study was conducted during baseline conditions and the modules were left in place for two days.

Table 1: Comparison of Flowing and Still Water Data (All values in ug of mass sorbed; MDL=Method Detection Limit; nd=not detected; bdl=below detection limit).

Detected Compounds		Beaver Spring			Quarles Spring		
		Flowing Water	Still Water (N)	Still Water (S)	Flowing Water	Still Water (NE)	Still Water (SW)
	MDL						
Toluene	0.03	nd	nd	nd	nd	0.04	0.07
Undecane	0.06	bdl	bdl	bdl	bdl	bdl	0.11
cis-1,2-dichloroethene	0.02	nd	nd	nd	0.03	0.04	bdl
Chloroform	0.04	0.32	0.37	0.30	0.04	bdl	bdl
Trichloroethene	0.02	0.42	0.45	0.56	0.07	0.04	bdl
Tetrachloroethene	0.04	0.06	0.10	0.09	0.28	0.10	0.05

At Quarles Spring, the compounds were detected with less consistency and with greater variability in the detected masses than were the compounds at Beaver Spring (Table 1). At Quarles Spring, the frequency of detection ranged from 1-in-3 to 3-in-3. In contrast, the same compounds were detected at Beaver Spring in all modules.

With the exception of tetrachloroethene at Quarles Spring, the sorbed masses were generally consistent between flowing and still water in both springs; for each three-sample set, the data vary by less than an order-of-magnitude and typically vary by less than two times the MDL. Based on the geometry of the springs, Quarles Spring is more likely to have stagnant areas than is Beaver Spring. This may account for the difference in the results and suggests that the data from Quarles Spring are more representative of differences between flowing and still water than the data from Beaver Spring.

The above data also suggests that GORE-SORBER module data can be used to observe what suites of contaminants are present at different springs. For example, both petroleum-related and chlorinated compounds were detected at Quarles Spring while only chlorinated compounds were detected at Beaver Spring. The suite of compounds detected may provide valuable information linking spring chemistry to source areas. Although water samples can provide this data more quantitatively, the modules are more cost effective than collecting storm-related water samples if the presence or absence of the compound is the objective of the sampling program. The modules are also easier logistically than storm-related water samples, so a larger number of springs can be included in the screening study.

Evaluation of Sample Duration

A second study was conducted to evaluate the impact of sampling duration. Modules were placed within six-inches of each other in the main spring discharge at Beaver Spring. One set of modules was exchanged daily, one on a 2-day basis, and the third was left in place throughout the 4-day period. The first two days of the study were conducted during baseline conditions; the third and fourth days of the study were conducted following a rainstorm.

In general, the modules left in place for longer durations sorbed greater masses of VOCs (Table 2). The 4-day masses were calculated for the 1-day and 2-day samples and compared against the single 4-day sample. There is no distinct trend in the total masses. Three compounds, trichloroethene (TCE), tetrachloroethene (PCE), and chloroform were detected in each set of modules. The chloroform mass is greatest when measured using a series of 1-day samples; the TCE and PCE concentrations were very consistent between the total 4-day masses. This comparison suggests that the combined masses from short duration samples can only be quantitatively compared against masses from longer duration samples for some compounds.

Table 2: Comparison of Sample Duration Times (all values in ug of mass sorbed; MDL=Method Detection Limit; nd=not detected; bdl=below detection limit).

Detected Compounds	MDL	1st 1-day	2nd 1-day	3rd 1-day	4th 1-day	1-day Total	1st 2-day	2nd 2-day	2-day Total	One 4-day
Toluene	0.03	nd	nd	nd	0.11	0.11	nd	nd	---	bdl
Total undecane, tridecane, and pentadecane	0.05	bdl	bdl	bdl	0.05	0.05	bdl	bdl	---	0.07
Total cis and trans-1,2-dichloroethene	0.02	nd	nd	nd	nd	---	nd	nd	---	0.02
Chloroform	0.04	0.22	0.26	0.25	0.24	0.96	0.32	0.33	0.65	0.50
Carbon tetrachloride	0.04	nd	nd	nd	nd	---	nd	nd	---	0.07
Trichloroethene	0.02	0.30	0.24	0.31	0.35	1.20	0.42	0.77	1.19	1.19
Tetrachloroethene	0.04	bdl	0.06	0.05	0.11	0.22	0.06	0.16	0.22	0.30

The compounds detected most frequently were measured at masses substantially greater than the MDL, suggesting that the total concentration may affect reliability of the method. Carbon tetrachloride and 1,2-dichoroethene were only detected in the 4-day sample, suggesting that either longer sampling durations may be needed for compounds present at low concentrations or that the module may preferentially sorb some compounds over others.

The duration study indicates that consistent sampling intervals are essential if comparable data are to be generated between springs or through time. Compounds present at higher concentrations are more likely to be detected with a consistent frequency and with less variability than compounds present at low concentrations. Longer sampling times may be necessary to detect low-concentration compounds. It should also be noted that only detected compounds are presented in Table 2. Nineteen other compounds were included in the analysis, but were not detected in any of the nine modules.

Comparison of Baseline and Storm Conditions

A third study was conducted to assess if the GORE-SORBER modules can reflect changes in spring chemistry before and during a storm pulse. Five springs were included in this study. The modules were placed as close as possible to the spring orifice and were left in place for two days.

A comparison of the data suggests that some springs have greater changes in VOC masses relative to storms than do other springs (Table 3). In Beaver and Gate One Springs, there is a general increase in contaminant mass between baseline and storm conditions. Contaminant masses do not have a consistent trend at Eagle Spring and are generally consistent or decrease at Gordon and Quarles Springs. The nature and timing of the spring response is likely to vary with numerous factors, however, the modules can be used to determine if there are significant changes between baseline and storm conditions. This application may be limited by the effects of dilution.

Comparison of GORE-SORBER Module Masses and Water Sample Concentrations

The usefulness of GORE-SORBER modules for spring monitoring would be improved if the mass sorbed onto the modules could be quantitatively correlated to aqueous water concentrations. To evaluate this relationship, we conducted a study in which we took aqueous samples concurrently with a GORE-SORBER module sampling period. The goal of this study was to establish the module mass – aqueous concentration relationship. A strong correlation could be used to predict water concentrations using the passive samplers. This study was conducted during baseline conditions to eliminate the impacts of storm surges and so that the water chemistry is reasonably consistent throughout the sampling interval.

In our study, no compound was detected in both the GORE-SORBER modules and aqueous samples in all four springs (Table 4). Two compounds, 1,1,1-trichloroethane and PCE, were detected in the modules but not in the water samples; this suggests that the modules can be used to monitor the presence of contaminants below the aqueous detection limits. There were insufficient data to quantify the relationship between mass sorbed and water concentration. The data qualitatively indicate that the mass sorbed increases with increasing aqueous concentration. Further evaluation of the relationship would be needed if the modules are to be used for a monitoring program beyond the screening level. Establishing the relationship will require more data points than included in this study, a higher frequency of detection in both the modules and water, and a wider range of contaminant concentrations. These data should initially be collected during baseline conditions to avoid the complexity caused by the effects of dilution and storm surges. Additionally, the mass to concentration relationships are site specific and vary with numerous geochemical parameters. Therefore, the mass to concentration relationship should be evaluated on a site-by-site basis.

CONCLUSIONS AND LIMITATIONS

The data reported herein support that GORE-SORBER modules can be used for screening-level spring investigations. The studies have shown that the modules are capable of detecting VOCs in karst spring water and can provide information regarding what compounds are present. These data may be useful for linking specific compounds to source areas or focusing a more rigorous spring water sampling program. Use of a passive, time-integrating sampling device limits the potential that a contaminant pulse is missed, a possibility when collecting grab water samples. The ease of use, relative to conducting a storm-based sampling program, decreases field program workloads and overall costs. In order to alleviate or quantify some of the apparent limitations of this technology, the following items should be considered for future investigations.

1. The amount of de-sorption and volatilization that occurs has not been quantified. The estimate will be important as part of future studies to investigate the effects that desorption creates in these data. For instance, if a strong, short pulse of VOCs is discharged at a spring, it is unclear if the sorber will continue to retain that mass, or if the mass will desorb with time after the pulse.
2. The sorbers have different sensitivities to different compounds. It is important to understand the principles behind this technology in order to understand the limitations. For instance, strongly polar compounds (such as alcohols and ketones) are not good candidates for this technology. They will not partition out of a dissolved state as readily as other groups of compounds, and therefore will remain unavailable for sorption. Sorption of VOCs onto particulate material in the water may also affect the degree to which they are available to the module.
3. It is unknown what aqueous concentrations correspond to the mass sorbed. It is likely that the modules can detect VOCs at concentrations below aqueous detection limits. This relationship requires further quantification.
4. The data cannot be used for compliance monitoring, risk assessment, or comparison with conventional regulatory standards. This application could help to identify potential migration pathways as part of a risk assessment, but not for quantifying the concentrations. The technology cannot identify maximum concentrations over the sampling period, and as such, is limited in this application. It is critical to develop a site-specific relationship between mass sorbed and water concentrations. It is equally

Table 3: Comparison of Baseline and Storm Data (all values in ug of mass sorbed; MDL=Method Detection Limit; base=baseline conditions (pre-storm); storm=collected during rain storm response; bdl=below detection limit; nd=not detected).

Detected Compounds	MDL	Beaver Sp. Base	Beaver Sp. Storm	Eagle Sp. Base	Eagle Sp. Storm	Gate One Sp. Base	Gate One Sp. Storm	Gordon Sp. Base	Gordon Sp. Storm	Quarles Sp. Base	Quarles Sp. Storm
Total BTEX	0.02	nd	nd	0.03	0.17	nd	0.04	0.46	0.19	0.04	0.06
Total diesel range alkanes	0.05	bdl	bdl	0.06	0.14	nd	0.06	bdl	0.05	0.08	0.06
Total trimethylbenzenes	0.03	nd	nd	bdl	0.04	nd	bdl	0.03	bdl	bdl	bdl
Total 1,2-dichloroethene	0.02	nd	nd	0.09	0.07	nd	nd	0.08	0.16	0.03	0.09
Naphthalene and 2-methyl naphthalene	0.06	nd	nd	nd	bdl	nd	nd	0.21	nd	nd	nd
Chloroform	0.04	0.32	0.33	0.98	0.87	0.41	0.47	0.41	0.42	0.04	0.06
1,1,1-Trichloroethane	0.04	nd	nd	0.08	0.07	0.15	0.26	bdl	nd	nd	nd
Carbon tetrachloride	0.04	nd	nd	0.33	0.29	nd	nd	0.06	nd	nd	nd
Trichloroethene	0.02	0.42	0.77	0.12	0.12	nd	nd	1.30	0.75	0.07	0.05
Octane	0.29	nd	nd	nd	0.32	nd	nd	bdl	bdl	nd	nd
Tetrachloroethene	0.04	0.06	0.16	0.38	0.34	bdl	0.04	2.33	2.05	0.28	0.21

Table 4: Comparison of GORE-SORBER Masses and Aqueous Concentrations (LT=Less Than; data for aqueous samples is mean concentration $\pm$ standard deviation for three samples).

Detected Compounds	Beaver Spring Gore (ug)	Beaver Spring Aqueous (ug/L)	Eagle Spring Gore (ug)	Eagle Spring Aqueous (ug/L)	Gate One Spring Gore (ug)	Gate One Spring Aqueous (ug/L)	Quarles Spring Gore (ug)	Quarles Spring Aqueous (ug/L)
1,1,1-Trichloroethane	LT 0.04	LT 1.00	LT 0.04	LT 1.00	0.10	LT 1.00	LT 0.04	LT 1.00
Trichloroethene	0.44	0.97 $\pm$ 0.02	0.27	0.45 $\pm$ 0.02	LT 0.02	LT 1.00	LT 0.02	LT 1.00
Tetrachloroethene	0.10	LT 1.00	0.09	LT 1.00	LT 0.04	LT 1.00	0.12	LT 1.00
Chloroform	0.39	1.3 $\pm$ 0.06	0.10	0.50 $\pm$ 0.01	0.37	1.4 $\pm$ 0.1	LT 0.02	LT 1.00

important to recognize that this technology can help to identify presence/absence of specific compounds, or gross (order of magnitude) changes in water quality, but it is not recommended for compliance monitoring unless the regulatory agency has direct input into the interpretation and understands these limitations.

ACKNOWLEDGEMENTS

The authors greatly appreciate the Fort Campbell Army Base for support of this project and the Hunt family for access to Quarles Spring.

GORE-SORBER Screening Survey is a registered service mark of W.L. Gore & Associates, Inc.
GORE-SORBER and GORE-TEX are registered trademarks of W.L. Gore & Associates, Inc.

REFERENCES

Field, M. S. (1993) Karst hydrology and chemical contamination. *Journal of Environmental Systems* **22**, 1-26.

Klemic, H. (1996a) *Geologic Map of the Hammacksville Quadrangle, Kentucky-Tennessee.* US Geological Survey Map GQ-540, Washington, DC.

Klemic, H. (1996b) *Geologic Map of the Oak Grove Quadrangle, Kentucky-Tennessee.* US Geological Survey Map GQ-565, Washington, DC.

Smart, P. L and Hobbs, H. O. (1986) Characterization of carbonate aquifers: A conceptual base. In *Proceedings of the Environmental Problems in Karst Terranes and Their Solutions Conference, October 28-30, 1986, Boiling Green, KY*, 1-14.

U.S. Army Environmental Center (1993) *Guidelines for Implementation of ER-1110-1-263 for USAEC Projects.*

U.S. Army Toxic and Hazardous Materials Agency (USATHAMA) (1990). *Quality Assurance Program*, USATHAM PAM 11-41.

W. L. Gore & Associates, Inc. (1997) *Description of Services.*

9 Waste disposal and storage in karst terrane

Design of geotechnical fabrics for septic systems in karst

JOSEPH A. FISCHER, JOSEPH J. FISCHER & RICHARD S. OTTOSON Geoscience Services, Bernardsville, N.J., USA

ABSTRACT

Rural areas of New Jersey currently under development pressure often do not have public wastewater collection and treatment systems. Individual septic systems are frequently used in new construction. These same areas often rely on ground water as the primary potable water source. The typical residential septic system constructed in New Jersey uses a tank to collect solids and a leach field for ground water recharge. The leach field usually entails a lateral distribution system above a "treatment" zone of natural soils or select fill.

At the present time, many residential developments are underlain by solution-prone carbonate rocks as new construction heads for the bucolic valleys of northwestern New Jersey and Eastern Pennsylvania. At one such development over solutioned dolomites, the following procedures were agreed upon as a result of private, municipal, county, and state concerns for the provision of reliable septic systems at this site. It was decided that a series of excavation inspections would be undertaken to determine whether a geotechnical fabric system warrants installation to provide "early warning" of potential sinkhole formation below the system.

In this particular instance, the purpose of the geofabric system was two fold; to provide strength to bridge possible sinkholes long enough to give some warning of potential septic system failure, and to prevent the migration of the filter zone treatment materials into possible subsurface voids.

In addition, geofabric design and selection was intended to not inhibit flow from the treatment zone fills while providing the specified "structural" support. Thus, a combination of strength, permeability and pore size governs the selection of the geofabrics. This paper presents the details of the rationale as well as a useful calculational process for selecting the appropriate geofabric(s).

INTRODUCTION

The influx of development into the scenic valleys of the New Jersey Highlands and Ridge and Valley provinces has lead to a series of problems in water supply contamination and structural reliability. Public water or sewer does not serve many of these rural areas. A water supply well and septic system within the same lot is quite common. State and local communities with some Federal assistance have addressed these issues in a variety of manners, ranging from no action, but with legitimate concern, to pro-active legislation on both a state and municipal level. The major concern for septic system failure in karstic areas is the possibility of untreated effluent reaching potable ground water through passages and openings in the soil and rock. Sinkhole occurrence within the system is an obvious end product of running water through soil voids into rock cavities.

A new development being constructed atop solutioned dolomites faced an aware municipal review (that had no real "limestone" ordinance at that time), county health department review (state delegation of authority) and Federal regulations on detention basin construction (United States Department of Agriculture Natural Resources Conservation Service). Septic system design and construction are being handled with an awareness of the underlying subsurface conditions and perhaps, an over-developed concern for possible future malfunctions.

SUBSURFACE CONDITIONS
Tectonic framework

The site lies within the New Jersey Highlands Physiographic Province, a folded and faulted belt of rocks which stretch from the Phillipsburg area, northeasterly to the New Jersey/New York border. Physiographically, the Highlands are bordered by the Valley and Ridge Province to the north and the Piedmont Province to the south. The Highlands are part of the Reading Prong, a northeast/southwest trending band of deformed Proterozoic-aged rocks with the intermontane valleys infilled with Cambrian to early Ordovician-aged sediments. The tectonics of the region are well described in Herman and Monteverde, 1989. Both the Highland rocks and the similar Valley and Ridge rocks to the northwest overly a master décollement dipping to the southeast and passing below the site locale. The core of the Highlands Province are ancient Proterozoic-aged (more than 500 million years old) metamorphic and igneous rocks.

The site itself is located between two thrust faults and is underlain by two carbonate rock formations. An overturned anticline, the

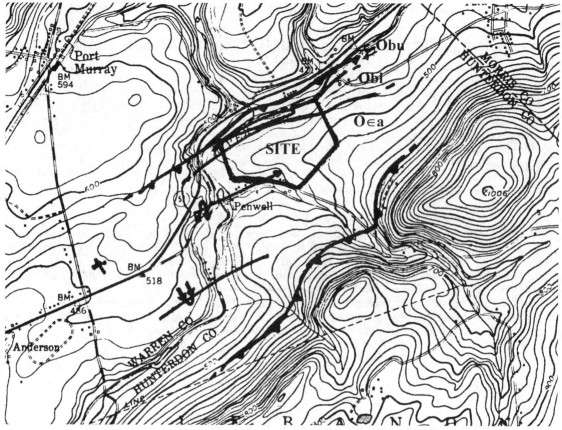

Figure 1: Geology Map

consequence of previous tectonic events, and the effects of weathering on the tilted beds of the area have resulted in the rolling topography of the site. The site geology is shown on Figure 1.

Geologic conditions

As construction is to be confined to the area underlain by the Allentown Dolomites (O∈a), the site investigation was intended to only confirm the mapping in that area of the site. The youngest rocks in the site locale are found to the north, along the Musconetcong River. The rocks in the northeasterly corner of the site are of the Beekmantown Group. The upper member of the Beekmantown is often termed the Epler Formation (Obu) in the literature. The lower member, found in a small portion of the site not presently intended for development, is also known as the Rickenbach Formation (Obl). The Rickenbach is a laminated to medium-bedded dolomite and minor interbedded limestone. Silty dolomite laminars and reticulate mottling are common. Fine- to very coarse-grained dolomite occurs in the very thick beds at the base of the formation. The Rickenbach is known to be pinnacled and cavernous with sinkhole occurrence considered common. The Allentown is composed of very thin- to thick-bedded dolomite with minor clastics. This formation is known to be pinnacled with sinkholes and caverns commonly occurring. The Allentown continues to the Musconetcong Fault to the south of the site.

Chemical weathering of the rock surface has produced a variable thickness of residual soils. These residual soils are found immediately below the topsoil in the more southerly areas of the site. In the northerly areas of the site, the residual soils are found below varying thicknesses of what appears to be glacially derived(?) soils (tills). As the tills (or perhaps colluvium) often contain limestone fragments, distinguishing between the two in the field is difficult.

The rock surface at the site is expected to be quite variable in depth and to some extent, in quality. The rocks have been folded and faulted and bedding would be expected to be at varying dips, often resulting in abrupt changes in the physical properties of the rocks as one encounters different beds across strike (i.e., from a northwesterly to southeasterly direction). Evidence of this typical "karst" condition is provided at a number of test pit locations where rock was encountered in one test pit of a group and not the other. As another example, one test pit encountered rock at 5½ feet at one end of the pit and was at a depth greater than the backhoe reach (12½ feet) in the other end.

In general, rock was found near the surface in the areas of steeper slopes. In the higher sections of the site, it appears that a significantly thicker soil section exists. The true ground water table is relatively deep at this site and probably flows toward the Musconetcong River through the underlying fractured Allentown rocks. It is presumed that the lower lying areas near the Musconetcong River would have a water table essentially at the same elevation or slightly higher than the water level in the river. Thus, ground water would not be expected to be a construction concern.

SEPTIC SYSTEM DESIGN RECOMMENDATIONS

The geotechnical report for the project stated:

> "Septic system design and construction should consider the karstic nature of the subsurface. The main concern with any septic system installed in carbonate terrane is the possible, rapid, undiluted movement of untreated effluent through subsurface cavities. To reduce this hazard, it is necessary to assure the proper operation of the system during its' lifetime. A well-designed and operated system will eliminate viruses in the effluent and reduce nitrate and phosphate levels. Thus, the concern is to keep a well-designed system operating in the manner the designer intended.
>
> This can be accomplished by providing suitable support for the septic system, allowing it to function as planned. As no exploration program can be expected to define all possible subsurface conditions at a carbonate rock site, experienced construction inspection personnel should observe the excavations to see if weathered zones, soil voids or shallow, solutioned rock exists. Should anomalous subsurface conditions be encountered, they should be reviewed by the developer's geotechnical consultant. Any alterations to the planned design caused by unexpected subsurface conditions can be corrected on a case-by-case basis."

The Township Planning Board inspection protocol was finalized as follows:

> "During the excavations for septic systems and for house foundations, an inspector hired by the township, to be paid from the escrow account of the applicant, shall be present to observe the soil conditions encountered.
>
> If any suspicious areas, voids or sinkholes are encountered for the septic system excavation, the applicant developer is required to fill with suitable fill as may be determined.
>
> If any suspicious areas, voids or sinkholes are encountered for (*sic*) the house foundation, the developer shall be required to do the appropriate grouting as determined by the developer's geology engineer (*sic*).
>
> The inspector upon making observation of suspicious areas, voids or sinkhole conditions for the septic or foundation, must give notice immediately to the Township Construction Code Official, Township Engineer, the developer and developer's geology engineer (*sic*). No further excavation will be conducted until an appropriate written geological protocol is designed for the particular situation as aforementioned and approved by the Township Engineer. The construction code official shall note on the C.O. for any house whether a written protocol was designed and a copy of the protocol attached".

In addition, the Township Geotechnical Consultant (TGC) required a geotextile fabric support system be designed to withstand the formation of a sinkhole below the septic system area and to provide some warning before the failure of the treatment process. The TGC recommended that the septic system designers;

> "...investigate a means of supporting the 4 feet of soil infiltration media below the effluent distribution pipes of the disposal field from being washed away or developing piping paths in the event of a soil loss below this level. This may be accomplished by the incorporation of soil reinforcement grids and geotextile filter media below the 4-feet of soil materials below the field distribution pipes. In this way, the wastewater will pass through at least 4 feet of intact soil matrix for aerobic biological treatment prior to potential rapid transportation to the groundwater via voids or piping paths which may develop below the bottom of the disposal field and top of rock surface."

This portion of the work fell to the developer's geotechnical consultant.

GEOTECHNICAL FABRIC DESIGN

In their review of the suitability of the septic system design for specific lots, the county health department (CHD) inquired as to the nature of the geotextile(s) recommended and questioned the inspection protocols planned for the site. Also, it was the CHD's opinion that the addition of a geotextile to a conventional septic system would have to be cleared with the New Jersey Department of Environmental Protection (NJDEP) as it represented a deviation from normal practice.

Upon these bases, and as a result of a meeting of the interested parties at the CHD, the following course of action was established:

1. The septic system for the first two lots will be submitted and reviewed as conventional septic systems, i.e., without the use of a geotechnical fabric.
2. Representatives of the Township Engineer will inspect these areas and initiate the Planning Board protocols where necessary. If no other areas are noted, the septic systems will be installed as designed.
3. The applicant will prepare a design for geotextile reinforcement (remediation alternate) of septic systems to be submitted to the concerned parties, including the NJDEP.

In this instance, geofabrics will be used to:

1. Provide strength to bridge possible sinkhole formation for at least enough time to give some warning of potential septic system failure.

2. Prevent the migration of the "filter sands" into a possible subsurface void as well as reduce the possibility of sudden failure of the system.

In addition, the geofabrics will be designed so as not to inhibit flow from the treatment zone fills while providing the specified support.

Essentially, what is required is a material (or materials) that will not impede the flow of water through the septic system by reducing the infiltration rate into the subsurface, while providing sufficient strength to span a potential sinkhole. In addition, the pore size of at least one of these fabrics should be small enough to preclude the select fill soils from passing through the fabric into the postulated underlying void. Thus, a combination of strength, permeability and pore size governs the selection of the geotechnical fabrics for the purpose intended at this site.

For the initial two lots investigated, the in-place permeabilities were 1.5 to 5 centimeters per hour (cm/hr). Permeability estimates for the select fills (23 cm/hr) to be used in septic construction are based upon tests of the same materials in another installation. As may be seen from these data, any hydraulic conductivity (permittivity) greater than 5 cm/hr for the geotextile would not inhibit the movement of water into the subgrade. Any material with a permittivity value higher than 23 cm/hr would, in addition, have greater permeability capability than the replacement fill selected for the septic system. To prevent the loss of the select fill, the geotextile "filter" must prevent the downward migration of soils in the event of sinkhole occurrence. For a further explanation of the procedures used in selecting geotextiles, see the Calculations Appendix.

As noted in this Appendix, many geotextiles have the required permeability and retentive ability (without clogging) that would allow the treated effluent through the filter at a rate greater than the absorption of the subgrade soils. However, the specifications manual used (GFR, 1996) had only a few materials in which the manufacturer provided tensile strength values as well as addressing permeabilities for a "geotextile". Thus, the several geotextile fabrics noted in the Appendix represent only a small portion of the materials that provide the permeability characteristics needed. As strength data is available from manufacturers, a variety of other geotextiles could be evaluated on the basis of economics. Therefore, the question of permeability and non-interference with the intent of the septic system can be relatively easily addressed by the use of readily obtainable geotextiles. The strength issue is not as simple and more complex analyses that considers void dimensions and shape, tensile strength of the geofabric, as well as the physical properties of the soils above and below the geofabric, are necessary.

For purposes of this study, where strength is of importance in bridging the postulated future sinkhole, either:

1. Two separate geofabric materials could be used (one for permeability and retention, and one for strength);
2. One material could function for both purposes; or
3. The selected filter (geotextile) could be reinforced with a geogrid(s), or similar.

For analysis purposes, the sinkhole was assumed to be circular in shape and about 5 feet in diameter. The sinkhole will also be presumed to occur anywhere within the septic system.

As shown in the Appendix, for the size of the sinkhole assumed in design and the presumed thickness (weight) of material above the geotechnical fabric layer(s), the use of a geotextile alone does not seem feasible. A geogrid layer or layers could be used to reinforce the geotextile placed below the filter sands. Although typical geogrid reinforcement materials are presented in the Calculation Appendix, the final selection for any location where the procedure is deemed necessary can be based upon economics.

SUMMARY AND CONCLUSIONS

There are a number of geofabrics that can be used to meet the requirements listed in the TGC report without altering the permeability and treatment characteristics of a conventional septic system. For the presumed select fill parameters and existing subgrade permeabilities, in relation to the postulated sinkhole size, several suitable geofabrics are available. On the bases of the sample calculations presented herein, a typical system would include a layer of Mirafi 401 or 402 geotextile underlain by two layers of Tensar Geogrid UX1400SB spaced about 6 inches apart and laid perpendicular to each other. A possible septic system cross-section is provided on Figure 2.

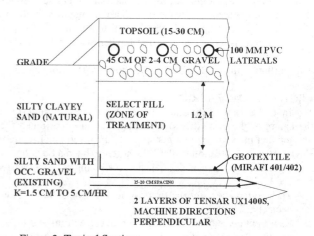

Figure 2: Typical Section

CALCULATIONS APPENDIX
Drainage design

A. Use permittivity (K) of fabric ≥ 10 times hydraulic conductivity (K) of soil (Nordquist, 1986). Therefore, K of fabric ≥ 10 x 5 cm/hr (2 inches/hour [in/hr]) ≥ 51 cm/hr (20 in/hr, highest measured K of fill, see Table 1).

B. Retention Ability
   EOS (Equivalent Opening Size) < 2 (Nordquist, 1986; Koerner, 1990) D85 for woven & thin non-woven where D85 relates to

filter sands (select fill), assume D85 = 2 mm (average of two tests) and material is not gap-graded. Therefore, EOS < 4 mm (to be checked when filter sands are selected).

Appropriate geotechnical fabrics
Bradley Industries Phoenix 70 or 70/20
Mirafi 401 or 402
TNS Advanced Technologies M404 or M706

Strength design
A. Use procedures of Giroud, et al, 1990.
B. Use a 1.5-meter (5-foot) diameter sinkhole for strength design.
C. Resistance will be provided by soil/geofabric interface (i.e., pullout resistance) & bridging effects by tensile resistance of the geofabric.
D. Lot 7.14 - LDC TP-8 - Soils at 1.2 meters (4 feet) below grade are firm sandy clayey loam. GS B-1 - Soils at 1.2 meters (4 feet) below grade are silty sands, Standard Penetration Test blows (SPT) > 13 per 0.3 meters (13/foot), $\sigma = 33°$. **Lot 7.07** - LDC TP-46 & TP-47 - Soils at 1.2 meters (4 feet) below grade are sandy clayey loam with gravel. GS TP-15 - Soils at 1.2 meters (4 feet) below grade are sandy soils. Assuming test pit data is more appropriate, $\sigma = 35°$. For safety factor use $\sigma = 30°$ in subsequent analyses.
E. Allow 10% strain to see movement at top of septic beds to indicate potential failure. This amounts to about a 0.3 meter (1-foot) drop in the fill in the center of a 1.5-meter (5-foot) diameter sinkhole.
F. Compute vertical unit loads
   1. Assume δsoils = 1.67 kilograms per cubic meter (kg/m³) or 130 pcf saturated unit weight.
   2. $p = \delta h = 33$ kg/m² (780 psf)
   3. Use 4.22 kg/m² (100 psf) surcharge to account for use of light equipment on top of system.
G. Use a spread sheet (Giroud, et al, 1990)
   1. Pressure on geofabric = 18.71 kg/m² (443 psf) - includes the 4.22 kg/m² (100 psf) surcharge.
   2. Required tensile strength of geofabric = 140 kilograms per meter (kg/m) or 1000 pounds per foot (p/f), including a safety factor of 1.25.
   3. Could use Tensar UX1400SB - creep strength on the order of 195 kg/m (1400 p/f) with a safety factor of 1.1, placed against sandy soils, Tensar recommends a long-term allowable strength of 168 kg/m (1,212 p/f). Satisfactory (see Figure 2, below, for a typical section).
H. Compute pullout resistance using Equation 1 in Bonaparte and Berg (1987) using a spreadsheet. Use a tensile loading on the geogrid of 140 kg/m (1000 p/f) with a safety factor of 1.25 and an H of 1.2 meters (4 feet) of buried geogrid extending outside the bed proper. In consideration of possible unknowns, add an additional safety factor of 1.5 into the calculations.

For an assumed 1.2 meter (4-foot) depth (see Figure 2) the fabric would have to extend some 1.2 meters (4 feet) outside the edge of the excavation to have enough "pullout" resistance should a sinkhole form at the corner. Extending the geogrid up the sides of a sloped excavation and providing additional resistance by burying a small roll of fabric in a shallow trench, however, could reduce this additional excavation. This type of resistance could be evaluated, as appropriate, during the actual design of the system in relation to the excavation procedures proposed and economics.

REFERENCES

Bonaparte, R. and R.R.Berg, 1987. The Use of Geosynthetics to Support Roadways Over Sinkhole Prone Areas, Karst Hydrogeology: Engineering and Environmental Applications, A.A. Balkema, Boston, MA.

Falyse, E., A.L. Rollin, J.M. Rigo, J.P. Gourc, 1985. Study of the Different Techniques Used to Determine the Filtration Opening of Geotextiles, Proc. of 2nd Canadian Symp. on Geotextiles and Geomembranes, Edmonton, Alberta, Canada.

Fischer, J.A., J.J. Fischer and R.J. Canace, 1996. Designing Safe Ground-Water Recharge Basins for Stormwater Detention Using Sinkholes in Karst, Program and Abstracts, 41st Midwest Groundwater Conference, Lexington, KY.

Geoscience Services, 1996. Report, Engineering Geologic Evaluation, Rolling Hills Estates, for Midstate Mortgage Investors Group, Matawan, NJ.

Geoservices, Inc., 1985. Exxon Geotextile Design Manual for Paved and Unpaved Roads, Exxon Corp.

Geotechnical Fabric Reports, 1996. 1997 Specifier's Guide, v. 14, no. 9, December.

Giroud, J.P., R. Bonaparte, J.F. Beech, and B.A. Gross, 1990. Design of Soil Layer-Geosynthetic Systems Overlying Voids, Geotextiles and Geomembranes, 0266-1144/90, Elsevier Science Publ., Ltd., England.

Gregory, G.H. and S. Bang, 1994. Effect of Geogrid Reinforcement on Seepage Through Earth-Fill Dam Cores, Hydrogeology, Waste Disposal and Politics, Idaho State Univ., Pocatello, ID.

Herman, G.C., & D.H. Monteverde, 1989. Tectonic Framework of Paleozoic Rocks of Northwestern New Jersey; Bedrock Structure and Balanced Cross Sections, Valley and Ridge Province, Southwest Highlands Area, in Paleozoic Geology of the Kittatinny Valley and Southwest Highlands Area, NJ, Proc. Of 6th Annual Mtg. Of Geol. Assoc. of NJ, Lafayette College, Easton, PA.

Koerner, R.M., 1990. Designing with Geosynthetics, 2nd ed., Prentice Hall, Englewood Cliffs, NJ

Mirafi, Inc., 1984. Design Guidelines for Subsurface Drainage Structures, Mirafi, Inc.

Mlynarek, J., 1996. Geotextile Filters: Which Criteria is Best?, Geotechnical Fabrics Report, v. 14, no. 6, August.

Nordquist, J.E., 1986. Geotextile Filtering in Gap Graded Soils, Proc. of 22nd Symp. on Engineering Geology and Soils Engineering, Boise, ID

Richards, E.A. and J.D. Scott, 1985. Soil Geotextile Frictional Properties, Proc. of 2nd Canadian Symp. on Geotextiles and Geomembranes, Edmonton, Alberta, Canada.

Rowe, R.K., S.K. Ho and D.G. Fisher, 1985. Determination of Soil-Geotextile Interface Strength Properties, Proc. of 2nd Canadian Symp. on Geotextiles and Geomembranes, Edmonton, Alberta, Canada.

Sarsby, R.W., 1985. The Influence of Aperture Size/Particle Size on the Efficiency of Grid Reinforcement, in Proc. of 2nd Canadian Symp. on Geotextiles and Geomembranes, Edmonton, Alberta, Canada.

The environmental hazards of locating wastewater impoundments in karst terrain

BASHIR A. MEMON, M. MUMTAZ AZMEH & MARY WALLACE PITTS P.E. LaMoreaux & Associates, Inc., Tuscaloosa, Ala., USA

ABSTRACT

A wastewater storage lagoon failed due to development of a sinkhole at a site in the Lehigh River valley in Allentown, Pennsylvania. The polluted wastewater from the lagoon entered into the underlying aquifer and moved within a narrow pathway controlled by cracks, fissures, and solution channels within the karstified Allentown Formation, of the Cambrian Period. The Allentown Formation serves as the principle aquifer for the public water supply of the area.

To develop appropriate remedial measures a thorough understanding of the geologic setting was required.. Therefore a geologic and hydrogeologic characterization of the area was completed; aerial photography and satellite imagery interpretations were performed; stratigraphic core holes were drilled, geophysically logged and data correlated to define structural control and movement of ground water and pollutants. A number of wells were drilled and constructed, and water levels were monitored on a continuous basis to correlate with climatic changes and determine the direction of flow. Water samples were collected periodically and analyzed to delineate the vertical and lateral extent of migration of pollutants.

Five saturated zones were identified within the bedrock based on the analysis of cores and interpretation of geophysical logs. Ground water from the lower zones is polluted; the concentration of pollution increases with depth.

Monitoring stations were established in the creek, south of the site, to measure flow rate, several times during different seasons, and at different reaches, to determine the losing and gaining sections of the creek. Pumping tests were conducted to determine hydraulic characteristics of the aquifer. Based on the hydrogeologic model of the karstified aquifer, flow regime and structural control, a plan of action was defined and initiated to remediate the aquifer.

The ground water is being remediated using a pump and treat methodology. The clean up effort is continuous and the pollutant level is fluctuating with an overall declining trend. The application of this technology has also created a pressure trough thereby controlling off site migration of pollutants.

INTRODUCTION

Ground water is a valuable water supply source and serves as an important resource in all climatic zones throughout the world. Its use for agricultural, industrial, municipal and domestic purposes continues to grow at increasing rates because of its good quality, convenient availability near the point of use, and relatively low cost of development. Karst aquifers are an important source of water supply for private and municipal uses.

Incidents of pollution of karst aquifers due to leaking of leachate from abandoned dump sites, waste management facilities, improperly constructed and/or located waste lagoons and impoundments; uncontrolled disposal of used solvents, and chemicals; accidental spills of hydrocarbons, and chemical or waste materials; may reduce the availability of good quality ground-water. Regional growth may be impaired due to the lack of a good quality water supply. Remediation of polluted Karst aquifers requires a three-step approach:
    a. developing a hydrologic model to thoroughly understand the geologic and hydrologic conditions, and so
     to define flow paths within the aquifer.
    b. management of the plume to prevent the movement of pollutant into the usable aquifer zones.
    b. remediation and restoration of water quality to its baseline quality by removal of the source of pollution
     and remediating the polluted ground water. The ground water may then be returned to its beneficial uses
     within a reasonable timeframe using cost-effective available technologies.

POLLUTION OF THE KARST AQUIFER

A process waste-water pond was constructed on the site of an industrial manufacturing Plant in Allentown, Pennsylvania (PA). The pond was constructed with a clay liner to prevent, or at least minimize, vertical infiltration of waste-water into the underlying karst aquifer. In 1979 a long dry weather spell was followed by heavy rains which caused a sinkhole to develop under the waste-water pond. The pond failed and waste-water moved quickly through the sinkhole into the underlying aquifer in all directions because of mounding conditions. After the incident, groundwater samples were collected from the wells in the immediate area and analysed to determine if any wells were polluted. No pollution was identified in groundwater from any of the wells in the area. Several years after the incident, two public water supply wells, and a test well, were constructed to the north east of the Plant site, and following pumping from these wells, pollutant was detected in the water. Pumping was suspended. A hydrogeological study was conducted, to characterize the site, and determine the extent of contamination.

SITE CHARACTERIZATION
Physiography and Climate
   The site is an industrial manufacturing plant in the east-central section of Pennsylvania, and is in the Lehigh River valley. Twelve miles to the north is Blue Mountain, a ridge with an elevation of 1,000 to 1,800 feet above mean sea level (amsl). South Mountain, with an elevation of 500 to 1,000 feet AMSL, fringes upon the southern edge of Allentown, PA. Otherwise the topography of the area is characterized in general by rolling hills with numerous small streams.
   The site is in the low-lying portion of the Great Valley section of the Valley and Ridge Physiographic Province in Lehigh County, which is underlain primarily by carbonate rocks (Wood et al., 1972). Surface drainage in this broad area of gently rolling hills, bordered to the north by Blue Mountain and to the south by South Mountain, is controlled by a few irregularly spaced and shallow entrenched streams that have gentle valley slopes and transect the area.
   The site is bordered to the north and south by tributaries of the Lehigh River. Considerable structural deformation of the carbonate rocks has facilitated the process of weathering and solution of carbonate rocks, and therefore the area has typical karst-terrain features including underground caverns and sinkholes.
   The climate in the area is characterized by a humid continental-type where annual precipitation significantly exceeds natural losses by evapotranspiration. Rainfall records indicate a mean normal annual precipitation of 44.31 inches for the period of 1951-1980. The wettest month of the year is August, which has a normal rainfall of 4.44 inches. The driest month of the year is October, which has a normal rainfall of 3.05 inches. The average monthly temperature is 51.0°F. Temperatures as high as 105°F and as low as -12°F have been recorded (Wood et al., 1972).

Geology
   The site is in the Great Valley section of the Appalachian Valley and Ridge Province of Pennsylvania. Data from drilling and construction of 11 wells indicate that the site is underlain by the Allentown Formation of the Cambrian Period. Residuum is formed at the top of bedrock from weathering of the Allentown Formation.
   The residuum is typically composed of two distinct units. The upper 10 feet of the residuum is predominantly composed of fine-grained sandy-clay to silty-clays which are strong brown (10YR4/3) to yellowish-brown (10YR4/6). Colors at depths below 10 feet are yellow (10Y7/8), yellowish-red (5YR5/6), and reddish-yellow (7.5YR6/8). The thickness of the residuum varies significantly over relatively short distances, and ranges from 10 feet to 97 feet at the site.
   The Allentown Formation exhibits little or no intercrystalline porosity. Ground-water movement in the karst aquifer occurs along solutionally enlarged fractures and bedding planes. During drilling numerous voids or cavities were penetrated, some of which were clay-filled. Width of the cavities ranges from a few inches to 5 feet. Caliper logs indicate that many smaller cavities may exist which were not identified during drilling.
   In the vicinity of the site the Epler Formation has been thrust over the Allentown Formation along the Portland Thrust Fault (Berg and Dodge 1981). Subsequently, the Epler, Allentown, and the intervening Portland Thrust Fault were folded into a series of east-west trending anticlines and synclines (Figure 1). Erosion has locally removed the Epler Formation, creating windows, in which the underlying Allentown Formation is exposed. The irregular shape of the windows and the trace of the Portland Thrust Fault are a result of the effects of erosion and the underlying geologic structure.
   In general, the axes of the folds are nearly horizontal, and the northern limbs of the anticlines dip more steeply than do the southern limbs. The angle of dip varies from 15 to 80 degrees, and locally beds are overturned.

Well construction
   Two public supply wells, and a test well, were constructed northeast of the failed waste-water pond on the site. These wells were drilled to depths of 350 ft, 400 ft, and 360 ft respectively. As pollution was identified in these wells, monitoring wells on the site were sampled in an effort to determine the extent of contamination. All wells on the site were found to be free of pollution. Other wells were located along identified lineaments, and constructed between the failed pond and the polluted water supply wells. These wells were also found to be free of pollution. Then a corehole was drilled in the vicinity of the water supply wells to a depth of 400 ft bls. Geophysical logs and description of cores were used to identify five saturated zones within the Allentown Formation. Approximate depths of these zones are as follows: zone 1 is from 175 to 190 ft below land surface (bls), zone 2 is from 205 to 220 ft bls, zone 3 is from 231 to 258 ft bls, zone 4 is from 294 to 320 ft bls, and zone 5 is from 330 to 400 ft bls,
   To identify the polluted zone(s), the corehole was reamed and constructed into a monitoring well, cased down to a depth of 325 ft bls, thus tapping zones 4 & 5, which were polluted. A shallow well was constructed to a depth of 187.4 ft bls with 138 ft casing, tapping zone 1. This shallow well was found to be clean. A third intermediate well was constructed to a depth of 265 ft bls with 223 ft of casing, tapping zones 2 and 3. This well was found to be moderately polluted.
   The residuum was cased off in the water supply wells and the test wells above the first saturated bedrock zone. As pollutant was found only in the lower zones of the aquifer in the vicinity of the public supply wells, it was concluded that the pollutant had migrated to and pooled in that area at the time of the pond failure.

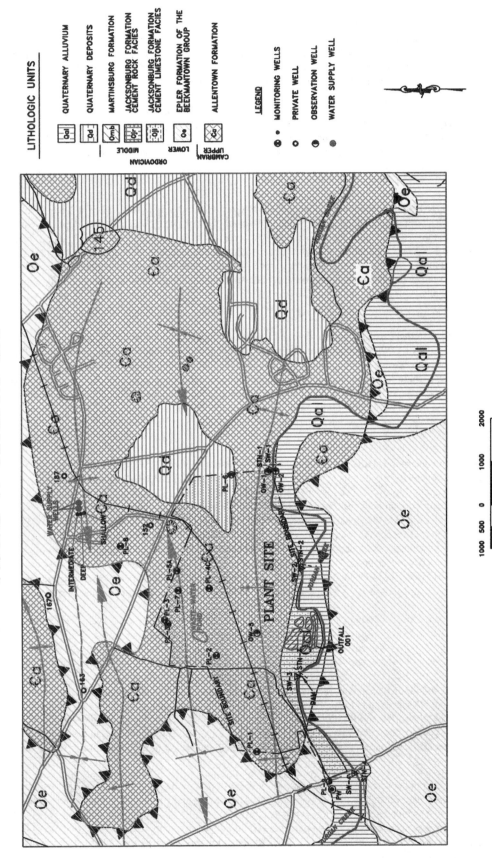

FIGURE 1. GEOLOGIC MAP

Hydrogeology

The site is in the low-lying portion of the Great Valley section of the Valley and Ridge Physiographic Province in Lehigh County. A topographic ridge extends from east to west of the site. The area to the north of ridge drains into tributaries of Coplay Creek. There are large sinkholes north of ridge that may cause hydraulic connection between surface water and ground water in this area. The area south of this ridge drains into Jordan Creek. Both creeks eventually discharge into the Lehigh River. Considerable structural deformation of the carbonate rocks has facilitated the process of weathering, and the area has karstic geomorphic features, as well as underground cavities and well developed fracture systems.

The carbonate rocks of the Allentown Formation act as a hydrogeologic unit under artesian conditions. These rocks contain and transmit ground water through fracture systems and cavities. Most of which are developed within 150 to 480 feet bls.

Precipitation is the main source of recharge to the ground water. The overburden (residuum) acts as sponge and recharges ground water where it is in hydraulic connection with openings in the underlying bedrock. Ground water is recharged by runoff which percolates downward through the joints, fissures, fractures, and cavities in the carbonate rocks.

Water level monitoring

All wells at the site were equipped with Stevens recorders to continuously record changes in water level. The water level in these wells along with four water supply wells, four private wells, and three public water supply wells were measured. The depths to water level from measuring points were converted into water level elevations using surveyed elevations of the measuring points. Using water level data collected, a ground-water surface map was prepared (Figure 2). The map indicated the presence of a groundwater divide at the site. Ground-water flow north of the divide moves to the northeast, whereas ground water south of the divide flows to the south. The divide which lies north of the failed pond controls the groundwater movement under normal flow conditions and permits migration of the wastewater toward the public supply wells.

Stream gaging

Four stations were established in Jordan Creek to measure flow in the creek. The measurements of flow at each station in the creek were taken periodically from May 4 through 13, 1993, using current Price AA meter and Pygmy meter model number 625-F(Buchanan and Somers 1980, and 1982). Flow measurements were also taken using a flowmeter (FLO-MATE Model 2000).

The discharge of Jordan Creek is controlled by the complex interaction of channel characteristics, including cross-sectional area, shape, slope, and roughness. Flow measurements in the creek are effected by creek bed irregularities, flow velocities, variable channel storage, and changing stages. Therefore, the measured flow of the creek, determined by conventional stream flow techniques, which also have inaccuracies inherent to the measurement procedures, has a margin of error of 5 percent or more.

Aquifer tests

Aquifer tests were performed to determine hydraulic characteristics (transmissivity, hydraulic conductivity, and storativity) of the carbonate aquifer and evaluate the hydrogeologic conditions at the site. Test wells, observation wells, stilling wells, and pumping wells were equipped with Stevens recorders and/or electronic data loggers to collect water level data on a continual basis prior to initiation of pumping tests, during, and after termination of pumping. The flow in the creek was measured at four stations during the pumping tests to determine if a hydraulic communication existed between surface water and the underlying carbonate aquifer. Evaluation of flow data collected during the pumping tests at selected sections in the creek demonstrates that reaches within the site limits do not lose significant amounts of water. Although there may be a small loss of stream flow within some reaches, water would not have the opportunity to travel very far from the channel, as that water is flowing as underflow, and resurfaces into the channel downstream.

The evaluation of ground-water level data collected during pumping test at the monitoring wells, and surface-water hydrographs and flow data for Jordan Creek, indicated that there was no hydraulic communication between the creek and the aquifer during the pumping tests.

The drawdown and recovery data collected during the tests were plotted and analysed, using Theis and Walton methods (Lohman 1979, Walton 1970, Driscoll 1987). The value of transmissivity, computed from time-drawdown and time-recovery field data graphs of the aquifer, ranges from 14,500 to 47,750 gpd/ft. The coefficient of storage, ranges from 1.1×10^{-2} to 9.8×10^{-3}.

Water samples for chemical analysis

Water samples were collected from monitoring wells located within the site, public supply wells, monitoring wells and test well close to public supply wells, and from the Creek. Water samples were analyzed for the following parameters: Total Dissolved Solids (TDS), Carbon Oxygen Demand (COD), Total Organic Carbon (TOC), Sulfate (SO_4), and Pentaerythritol(PE). The concentrations of these parameters, except PE, are within the permissible level. The pollutant identified to be originating from the failed waste-water pond is Pentaerythritol(PE) which has been continuously monitored at selected wells since 1993.

REMEDIATION

To remediate groundwater at the public water supply wells (# 1 & 2), these wells and the test well were plugged with bentonite from the bottom of each well up to a point approximately 275 feet bls (Figure 4). These plugs prevented entry of groundwater into the well from the deepest two zones (zones 4 and 5) which have the highest concentration of pollutants. A pump was installed in each of the two public water supply wells. A third pump was installed in a deep monitoring well, in the vicinity, tapping the lowest two zones. Pumping was initiated simultaneously from all three wells. The rate of pumping from the deep monitoring well is equivalent to the combined rates from the two water supply wells. The recovered water is treated and then discharged into the creek. The remediation effort is continuous

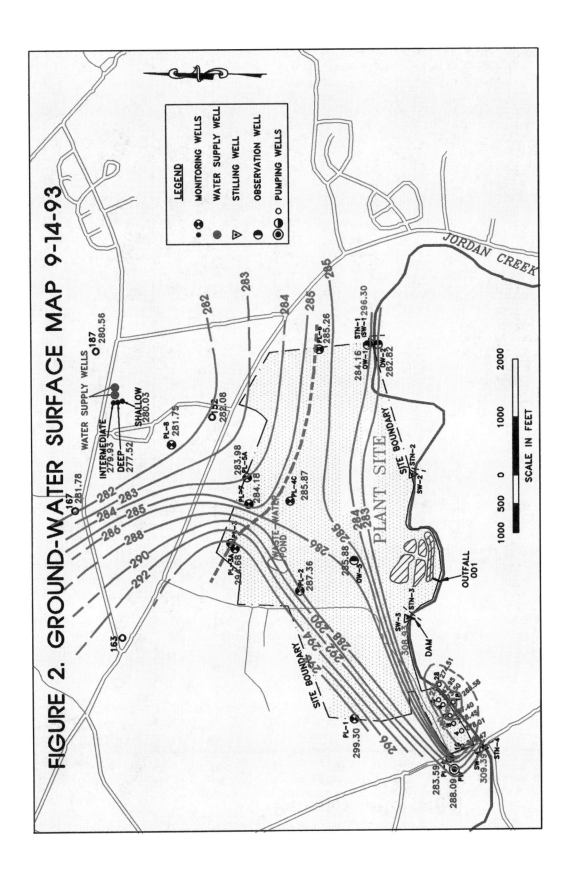

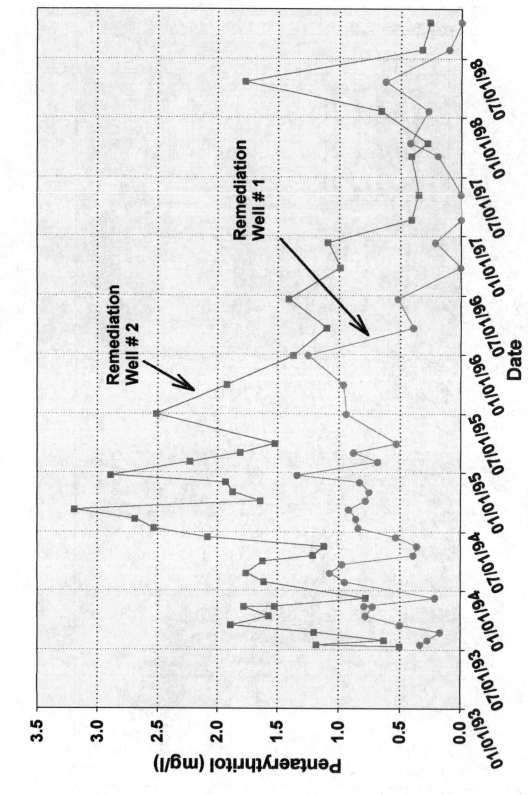

Figure 3: Pentaerythritol (mg/l) in Two Remediation Wells August 1993 - October 1998

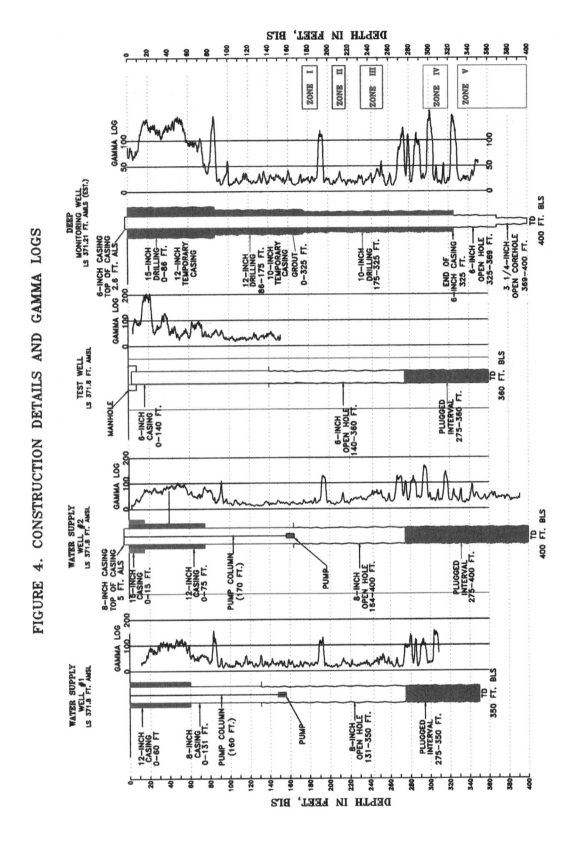

FIGURE 4. CONSTRUCTION DETAILS AND GAMMA LOGS

and the pollutant level is fluctuating with an overall declining trend. Figure 3 shows the analytical results for PE in the two pumping wells, over time. The application of this technology has also created a pressure trough thereby controlling off site migration of pollutants.

CONCLUSIONS
- The ground-water flow regime in the area is primarily controlled by secondary porosity and permeability due to karstification.
- The failure of the waste-water pond was due to development of a sinkhole beneath the pond.
- Hydrogeological studies indicated the presence of a ground-water divide. However, failure of the pond and the resultant mounding made the divide temporarily ineffective in controlling flow paths.
- The complexity of the karst system required a thorough understanding of the geologic and hydrogeologic setting, and geologic structural control to design and implement effective remediation measures.
- To facilitate future extraction for public supply from the shallow and intermediate zones, these zones were targeted for remediation. The deeper impacted zones were plugged to control migration of pollutants.
- A pump and treat technology for remediation was implemented in the water supply wells tapping the shallow and intermediate zones. In order to prevent upconing, pumping was also initiated from a nearby well completed in the deeper zones. Pumping from the deep well is maintained at a rate equivalent to the combined discharge from the supply wells.
- Concentrations of Pentaerythritol in the remediation wells have declined over time.

BIBLIOGRAPHY

Berg, T.M., and Dodge, C.M., 1981, Atlas of Preliminary Geologic Quadrangle Maps of Pennsylvania: Pennsylvania Geologic Survey Map 61.

Buchanan, T.J., and Somers, W.P., 1980, Discharge Measurements at Gaging Stations (Chapter A8): U.S. Geological Survey, Techniques of Water-Resources Investigations of the United States Geological Survey, Book 3, Applications of Hydraulics, 65 p.

Buchanan, T.J., and Somers, W.P., 1982, Stage Measurement at Gaging Stations (Chapter A7): U.S. Geological Survey, Techniques of Water-Resources Investigations of the United States Geological Survey, Book 3, Applications of Hydraulics, 28 p.

Driscoll, F.G., principal author and editor, 1987, Groundwater and Wells (2d ed.): Johnson Division, St. Paul, MN, 1,089 p.

Lohman, S.W., 1979, Ground-Water Hydraulics: U.S. Geological Survey Professional Paper 708, 70 p.

Walton, C.W., 1970, Groundwater Resource Evaluation: McGraw-Hill Book Company, New York.

Wood, C., Flippo, H.N., Jr., Lescinsky, J.B., and Baker, J.L., 1972, Water Resources of Lehigh County: U.S. Geological Survey Water Resources Report 31.

Stability evaluation for the siting of municipal landfills in karst

MICHAEL ZIQIANG YANG & ERIC C. DRUMM *Department of Civil and Environmental Engineering, University of Tennessee, Knoxville, Tenn., USA*

ABSTRACT

During the siting, construction, and maintenance of facilities in karst terrain, the stability of the residual soils that overlie the cavitose limestone is often a concern. The development of arching in the residual soils and the associated distribution of stress is important to the stability. The economics of landfill construction are usually improved by excavating a portion of the residual soil to increase the capacity of the landfill. However, it is generally accepted that for a given diameter of potential limestone cavity, the residual clay soil becomes less stable as the thickness the soil cover decreases. Geotechnical engineers are often called upon to evaluate the stability of the residual soils as a part of the permitting process or prior to construction of critical structures such as municipal landfills, yet simple, rational methods are not available. Simplified methods are needed to estimate the thickness of soil required to provide stability against soil collapse and the subsequent formation of sinkholes or distress to surface structures.

A simplified method is proposed, which consists of the evaluation of stability with respect to two potential modes of failure within the residual soil. Stability with respect to the first mode depends upon the development of arching in the residual soil, while the second failure mode corresponds to the yielding and plastic flow of the soils into the soil dome and/or rock void. The finite element method was used to conduct parametric studies of both stability modes for a typical residual clay soil over a range of cavity diameters. The results are presented in the form of a design chart. The application of the chart is demonstrated by an example. If a candidate site with a proposed cover soil depth appears to be stable for the anticipated range of cavity diameters, then a more comprehensive subsurface investigation and additional analysis can be conducted to evaluate the stability with respect to the proposed waste loading.

INTRODUCTION

The stability of the residual soils that overlie the solution cavities in limestone is often a concern during the siting, construction, and maintenance of facilities in karst terrain. The development of arching in the residual soils and the associated distribution of stress is important to the stability. In the absence of construction activities, which change the surface hydrology and reduce the thickness of the residual soil cover, these residual soils may be very stable and in fact capable of supporting significant additional surface loading. However, the economics of landfill construction is usually improved by excavating a portion of the residual soil to increase the capacity of the landfill. It is generally accepted that for a given diameter of potential limestone cavity, the stability of the residual clay soil decreases as the thickness of the soil cover decreases. Geotechnical engineers are often called upon to evaluate the stability of these soils as a part of the permitting process, or prior to construction of critical structures such as municipal landfills. Simplified methods are needed to estimate the thickness of soil required to provide stability against soil collapse and the subsequent formation of sinkholes or distress to surface structures.

Because of the limited strength data available early in the site selection process, and due to uncertainty with respect to the dimensions of the underlying limestone cavities, simplified design charts would prove useful in the evaluation of prospective landfill sites in karst terrain. Once it has been determined that the residual soils are stable under the proposed soil cover depth and anticipated range of cavity diameters, a more comprehensive analysis can be performed to evaluate the loading due to the municipal waste.

SUBSURFACE PROFILE IN KARST

The subsurface investigation of residual soils in karst terrain typically suggests that the soil above the bedrock is very heterogeneous, and the interface between the rock and soil is irregular. In contrast with most residual soils, the strength and stiffness of the residual soil in karst areas generally decreases with depth (Sowers, 1996). According to Sowers (1996), a general subsurface profile consists of five layers, Figure 1. Layer 1 nearest the surface is a remolded structureless layer with highly variable stiffness, which is the product of the near surface environment. Layer 2 and Layer 3 are residual soil, with the overconsolidation ratio and strength in Layer 2 typically greater than that of in Layer 3. The thickness of Layer 2 may vary from 1 to 50 m, and the thickness of Layer 3 varies from 1.5 to 5 m (Sowers, 1996). Layer 4 usually consists of randomly distributed rock pinnacles with rock blocks and soft soils between them. The soils accumulated near the slots often exist at moisture contents exceeding their liquid limit, but in some cases they may experience strength increase due to consolidation from the weight of the residual soil above (Sowers, 1996). The slots may be very deep, which is often reflected by greatly varying refusal depth during subsurface exploration. The standard penetration test (SPT) is widely used in the evaluation of soil strength and stiffness in residual soils above the soluble limestone bedrock (Siegel and Belgeri, 1995; Sowers, 1996; Smith, 1997). In some areas, the cone penetration test (CPT) has been used to evaluate subsurface conditions (Chang and Basnett, 1997). Often, undisturbed samples are obtained and tested in the laboratory to determine the effective shear strength parameters φ and c.

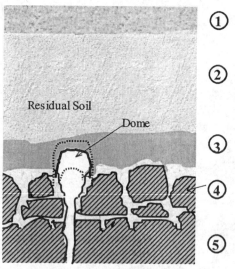

(1): Near surface layer (4): Partially weathered zone
(2): Firm residual soil with very soft and wet soil
(3): Soft residual soil (5): Competent bedrock

Figure 1. Idealized subsurface profile in Karst (After Sowers, 1996)

Field investigations in the U.S. (Newton and Hyde, 1971; Newton, 1976; Williams and Vineyard, 1976, Newton, 1984) indicate most sinkhole problems are caused by the failure of voids developed in the residual soil. Usually the soil void begins at a slot between the rock blocks as shown in Figure 1. Details of the sinkhole evolution process have been described by others (Donaldson, 1963; Jennings et al., 1965; Newton, 1984; and Sowers, 1996). In a uniform soil layer, the typical soil void or dome resembles an arch, the sides at the bottom generally coincide with pinnacles or irregularities in the rock, and the dome walls are usually vertical (Newton, 1984). Groundwater fluctuations may enlarge the void over time. If the thickness of the stronger residual soil above the dome is insufficient for arching to develop, the soil above the dome will fail, resulting in a sinkhole. Both physical models (Sorochan et al., 1989; Goodings and Abdulla, 1996) and numerical models (Drumm et al. 1987; Ketelle et al., 1987) have been used to investigate the stability of this type of karst system.

PROPOSED SOIL DOME FAILURE MECHANISMS

If stability is achieved by the development of arching, creep or plastic flow of the soft soils into the limestone cavities may result in surface subsidence, or the formation of a doline. Both of these failure mechanisms should be addressed during the siting of municipal landfills in karst areas.

Sinkhole investigations (Newton and Tanner, 1986; Sowers, 1996) have suggested sinkholes may form above relatively small rock fractures. Rock cavities as small as 150 mm (6 in) in diameter can generate domes more that 15 m (50 ft) in diameter, depending on the surrounding soil strength and groundwater activities in the rock cavern (Sowers, 1996). It is hypothesed here that the residual soil domes in karst may fail by one of two failure modes. The first mode is related to excessive shear or tensile stress as a result of the inability of arching to develop. The second mode corresponds to plastic flow around the dome and into the rock opening. Each of these mechanisms is discussed below.

Stability Mode I - Evaluation of Potential Tensile Failure

Since the solution cavities in most karst systems are hydrologically connected, seasonal variations in the groundwater likely result in softening and erosion of the adjacent residual soil, with a gradual formation and enlargement of the soil dome, Figure 2 (a). With low overburden depths, arching cannot take place and tensile stresses develop in the residual soil, which may result in surface cracks, Figure 2 (b). When the tensile stress in the soil exceeds its tensile capacity, the top of the soil dome may suddenly collapse, Figure 2 (c). This process can be accelerated by a heavy precipitation event after a long drought. The surface runoff water could more easily infiltrate into fissures in the near surface soil, increasing the unit weight of dry soil and inducing seepage forces. In addition, the increase in saturation reduces any suction in the unsaturated soil voids, which decreases the effective stress and the strength of the soil above the dome. According to an investigation in East Tennessee (Newton and Tanner, 1986), rainfall triggered 85% of the sinkhole problems surveyed. From an engineering point of view, Stability Mode I must be evaluated for the minimum depth of soil cover proposed prior to construction of the landfill. Since the stability with respect to Mode I increases with the thickness of soil cover, this mode is not likely to govern the long term stability for landfill construction and operation. Potential sinkhole problems associated with this mode can be remedied during construction if necessary. Following dome collapse, additional failure in the sinkhole may take place, Figure 2 (d), extending the ground surface opening. Although not of engineering interest because the first failure will generate remedial action, methods have been suggested to evaluate the stability of these side wall failures (Drumm et al., 1990).

Stability Mode II – Evaluation of Potential Plastic Flow

Soil dome systems which are stable with respect to Mode I (arching is well developed, Figure 3 (a) and (b)) may still exhibit distress due to excessive surface deformation. High shear stress in the residual soil may lead to yielding or plastic flow, Figure 3 (c). Small increments of additional load could then cause the plastic zone to continue to grow. Yielding or plastic flow occurs in the areas of highest shear stress, which are located in the lower part of the dome above the bedrock. As shown in Figure 3 (c) and (d), the initial plastic zones may enlarge with an increase of the external load and/or long term creep. The soil at the bottom of a dome may be squeezed into the rock opening as a result of the progressively enlarged plastic area. As illustrated in Figure 3 (d), if additional loads are applied to the surface, the plastic zone may spread throughout the circumference of the dome, and the size of a dome will reduce even to zero. This could lead to surface settlement or formation of a doline.

Both Stability Mode I and Mode II should be evaluated during the initial site selection process, which can be done on the basis of limited subsurface information as typically available early in the site evaluation process. These stability modes are evaluated on the basis of the existing overburden soil thickness as determined from the subsurface investigation, and estimation of the anticipated soil void or dome diameter. If insufficient overburden thickness exits, evaluation of these stability modes may facilitate the estimation of the required thickness of engineered fill to assure stability with respect to these two failure modes. If the prospective site appears to be stable with respect to Mode I and Mode II stability, then additional testing, sampling, and analysis may be warranted to assure stability with respect to additional loads. The prediction of surface deformation and distress to landfill liner systems requires additional site information such as overburden and liner material properties, as well as liner serviceability criteria and the design waste loads. This is a more comprehensive analysis than is typically possible during the initial site selection phase of a typical landfill project, and is not within the scope of this paper.

SOIL DOME ANALYSIS: GENERATION OF DESIGN CHART

The investigation of these two failure modes requires some practical judgement and simplifying assumptions. The stability with respect to these modes was investigated by the finite element method for a range of overburden heights and dome diameters. In the case of Stability Mode I (soil dome collapse under self weight), instability was defined as the presence of tensile failure at the ground surface or at the top of the dome. Due to the limited tensile strength of most soils, the following criterion was assumed: when the mean principal stress $(\sigma_1+\sigma_2+\sigma_3)/3$ and the horizontal stress at the ground surface or top of the dome are tensile, the dome can then be judged to have failed. If overburden soil is of sufficient depth for arching to develop, then the tensile stress will not exist.

In the case of Stability Mode II, the plastic flow stability criterion is based on the inelastic or plastic strains generated. The yielding zone in the finite element analyses can be represented by the plastic strain magnitude, ε_p (Chen, 1994), where:

$$\varepsilon_p = \sqrt{\frac{2}{3}\varepsilon^{pl}\varepsilon^{pl}} \text{ where } \varepsilon^{pl} \text{ is the plastic strain}$$

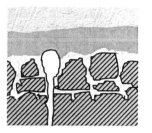
(a) Initial stage
Dome formation near limestone fracture

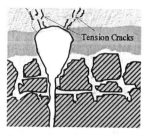
(b) Enlargement of soil dome until critical stage of tension failure
- Shear stresses develop between the dome and surface

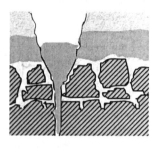
(c): Dome collapse

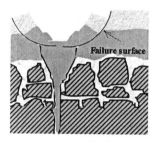
(d): Additional slope failure
(After Drumm et al., 1990)

Figure 2 Schematic of sinkhole development - Stability Mode I

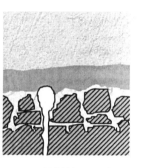
(a) Initial stage
Dome formation near fracture

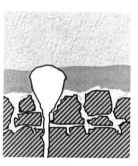
(b) Dome enlarges, but arching provides stability
(Stable with respect to Stability Mode I)

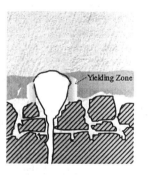
(c) Plastic yielding in residual soil, conditions suggesting unstable with respect to Stability Mode II

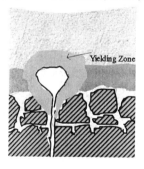
(d) Significant surface deformation and dome failure caused by additional loading

Figure 3 Schematic of Soil Dome Failure due to Plastic Flow - Stability Mode II

If the geometry of the soil and dome system results in stresses that lie below the specified shear strength envelope only elastic strains will result. In areas where the shear stress is at the yield stress, plastic strains will be produced. For the case of Stability Mode II, it was assumed that the area with plastic strains must be less than two times the area of the dome. An additional limit of 10% plastic strain was also imposed.

Numerical Modeling

To simplify the finite element model, the original 3-D problem was simplified as an axisymmetric 2-D problem, with a half sphere void above rigid bedrock. Due to the axial symmetry of the problem, one quarter of the sphere was modeled as shown in Figure 4.

The following assumptions were made in the finite element analysis:

1. The water table is assumed to be below the bedrock surface. Soil strength and hydrologic conditions associated with the water table fluctuation are neglected. Failure of the soil dome is assumed to be due solely to a stress state that exceeds the assumed shear strength.

2. The plane of the limestone-soil contact is horizontal, and the dome is the result of a slow geologic process of enlargement, which starts at the soil-rock interface with very a small void under the constant overburden soil thickness. This assumption leads to the imposition of a linear distribution of initial stress in the residual soil.

3. The stiffness of the limestone is much greater than that of the residual soil above. Therefore, the rock can be considered as a rigid body, with no deformations during the loading process. This implies that failure in the bedrock is not considered in the analysis.

4. At the soil-rock interface, the horizontal constraints on the soil mass are removed. This assumption is based on the presence of a zone of wet and very soft soil at the soil rock contact.

5. The problem in Figure 1 can be idealized as shown in Figure 4. The 8-node quadratic elements were used in the analysis. In order to avoid the effects of the model boundary constraints, the length of the soil domain, L, should be much greater than the radius, R. A ratio of L/R greater than 10 was used in the analysis, which is consistent with that used by others (Rowe and Davis, 1982; De Borst and Vermeer, 1984; Koutsabeloulis and Griffiths, 1989).

6. The extension of the linear Mohr-Coulomb yield envelope into the tension zone may overestimate the soil tensile strength, and thus overestimate the sinkhole stability. To reflect the limited tensile strength of the soil, a hyperbolic tension cut off yield criterion was used

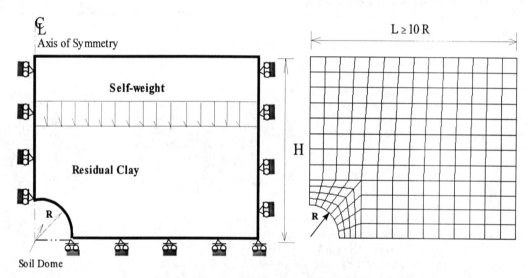

Figure 4 Idealization of Problem and Typical Finite Element Mesh

in the analysis. For stress states below yield, the soil was represented by a linear elastic model.

Analysis of Soil Dome - Stability Mode I

The minimum overburden height over bedrock, which is required to create the arching effect, can be obtained for given material properties and dome diameter. A lower bound of overburden height required for dome stability (Mode I) can therefore be obtained by finite element analysis. The effect of friction angle was investigated, and the soil materials properties used in the analysis are listed in Table 1.

Table 1 Assumed Material Properties for Residual Soils

Cohesion c kPa(psf)	Friction Angle φ (degree)	Elastic Modulus E MPa (ksi)	Poisson's Ratio μ	Unit Weight γ kN/m³ (pcf)
25 (550)	0,10,20,30	11 (76)	0.3	18 (115)

The stable relationships between the overburden height and the dome diameter for the different values of internal friction angle φ with a fixed cohesion of 25 kPa (550 psf) are shown in Figure 5. For a given dome diameter, the data points indicate the lowest overburden height such that arching was developed and tensile stress did not occur. Analyses with different values of cohesion indicated that a minimum soil cohesion was necessary to divert the stress away from the soil dome and to maintain stability. Within a wide range of soil dome diameters, the Mode I relation was not sensitive to the value of the internal friction angle. This suggests that the contribution of the internal friction angle to the lower bound of dome stability under soil self-weight is not significant. Therefore, it is possible to develop a relationship between the overburden height and dome diameter for a given soil cohesion that is independent of the internal friction angle. Similar curves can be developed for other values of c. Although Mode I stability also depends upon the unit weight, the value of $\gamma = 18$ kN/m^3 (115 pcf) is representative of many residual clays.

A power function was fit to the data in Figure 5, yielding:

$$H = 1.45 D^{1.37} \tag{1}$$

where H and D are expressed in meters. For a given soil with $c = 25$ kPa, the stable zone for Mode (I) is bounded by the line defined by equation (1).

The numerical analysis results for dome stability under Stability Mode (I) are consistent with the observations made by Broms and Bennermark (1967). They conducted experiments in which they extruded clay under pressure through vertical circular openings. They considered field observations both where failure had occurred and where stability had been maintained, and suggested a stability number for a vertical opening in clay as only a function of soil undrained strength c_u.

Analysis of Soil Dome -- Stability Mode II

Stability Mode II was also investigated by the finite element method to determine the upper bound of overburden height for dome stability. The results are shown in Figure 6, where each symbol represents an analysis which met the Mode II stability criterion. For a given c, the maximum overburden height increases significantly with the internal friction angle of the overburden soil. A higher internal friction angle in the residual soil allows a larger soil dome to develop without loss of stability. It is also clear that the overburden height for plastic flow is not dependent upon the dome diameter, such that the upper limit for Stability Mode II for each friction angle could be represented by a horizontal line in the Figure 6 stability chart.

Design Chart for Mode I and Mode II Stability

Combining results from the Mode I and Mode II analyses, a dome stability chart can be developed. As shown in Figure 7, three lines bound a zone of stability: the vertical axis corresponding to a dome diameter of zero, the power curve at bottom and right (Mode I), and a horizontal line above (Mode II), dependent upon φ. This chart can be

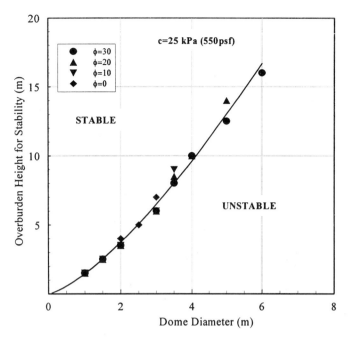

Figure 5 Lower Bound of Overburden Height for Dome Stability (Mode I) $c = 25$ kPa (550 psf)

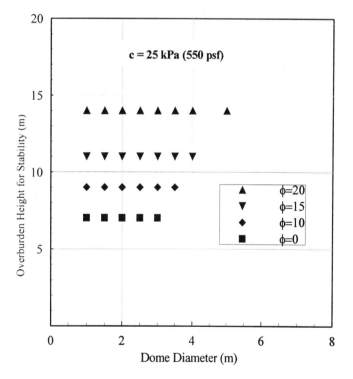

Figure 6 Results from Numerical Analysis Indicating Upper Bound of Overburden Height for Dome Stability (Mode II) $c=25$ kPa (550 psf)

used to evaluate the stability of a candidate site for a range of anticipated dome diameters. Without further external load, a dome configuration for which the anticipated H and D fall inside the bounded zone is stable. The chart also suggests that the upper bound overburden height increases rapidly with an increase in friction angle.

Stability charts for the different soil cohesion values can be developed in a similar manner. It is suggested that the soil cohesion is critical for the Mode (I) stability whereas Mode (II) stability is governed by the internal friction angle and unit weight of the overburden residual soil.

It is suggested that sinkhole problems are more likely to be identified in the field as a Mode I pattern, which is a shallow failure collapse mode. According to a sinkhole investigation in East Tennessee (Newton and Tanner, 1986), about 70% of the sinkholes surveyed had overburden thickness less than 10 m. With greater overburden thickness, instability with respect to Mode II is deformation related and not as obvious.

EXAMPLE: EVALUATION OF CANDIDATE SITE FOR MUNICIPAL WASTE LANDFILL

The application of the proposed design chart can be demonstrated by an example case history. The leading candidate site for a proposed municipal waste landfill was underlain by soluble carbonate bedrock. The

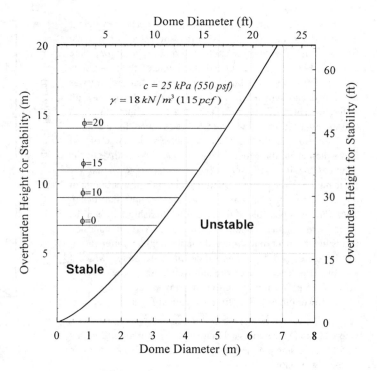

Figure 7 Karst Dome Stability Chart for c=25 kPa (550 psf)

geotechnical report excluded the potential for rock cave failure, but suggested that the anticipated maximum soil dome diameter might range from 1.2 to 3.4 m (4 to 12 ft). It was suggested that cavities larger than this range would have been detected by geophysical methods. The overburden residual soil deposit consisted mostly of silty sand and silty clay. Laboratory testing on representative samples of the overburden soil indicated strength values of c =25 kPa (550 psf) and φ = 20 degrees. The typical thickness for the residual soil ranged from 7.5 to 12.2 m (25 to 40 ft), but to increase the capacity of the landfill it was proposed to excavate 2 m of the residual soil, leaving from 5.5 to 10.2 meters of overburden. Upon completion of site preparation work, a compacted clay and geomembrane liner system would be constructed. Prior to an evaluation of the stability under the proposed waste loading, the residual soils should be evaluated with respect to Stability Modes I and II under a self weight loading. Potential instabilities identified at this stage can be corrected during construction, or changes in the base elevation of the landfill could be modified to assure stability.

The range of anticipated dome diameters (1.2 to 3.4 m) and excavated overburden thickness (5.5 to 10.2 m) for the candidate site are shown on the design chart in Figure 8. Although the zone representing the anticipated site conditions plots on both sides of the proposed Mode I stability boundary, the majority of the region is in the stable zone. This suggests that the probability of stability Mode I is high. In addition, the large diameter soil domes under low overburden heights (the most critical combination) are most likely to be discovered during geophysical exploration or construction/excavation. If it is assumed that a Mode I instability could be repaired during construction, it might be concluded that this combination was acceptable. Alternatively, the economics of the landfill could be evaluated assuming a smaller depth of the residual soil would be excavated. For the measured friction angle of φ = 20 degrees, Figure 8 suggests that for the maximum excavated overburden thickness of 10 m, the candidate site would be stable with respect to Mode II. Figure 8 also implies that the candidate site could be assumed to be stable with respect to Mode II provided the thickness does not exceed about 14 m, and the additional overburden material has the same strength and density as the residual soil.

The stability with respect to Mode II should not be confused with the stability of the landfill under the proposed waste loading. The strength and deformation properties of the waste materials, as well as those of the engineered compacted clay and geomembrane liner system, will contribute to the stability. The waste materials have highly variable density and strength, but friction angles as high as 70 degrees have reported (Daniel, 1996). Since the candidate site was stable with respect to Mode I, sudden sinkhole collapse under the waste loading is unlikely. The long term deformation and distress imposed on the liner system must be evaluated with the proposed height of the landfill. This analysis requires additional site information such as overburden and liner material properties, as well as liner serviceability criteria and the design waste loads, and is not within the scope of this analysis. Because this example site was judged to be stable with respect to both Mode I and Mode II, the evaluation of the linear system under the proposed load could be justified.

CONCLUSIONS

A simplified method was presented to evaluate the stability of residual soils in karst areas for which development is proposed. The method is intended to be used during the site selection process, when only limited subsurface information is available. The evaluation is performed on the basis of the expected final thickness of overburden soil and the anticipated range of soil void or dome diameters.

It was suggested that sinkhole stability should be evaluated with respect to two different conditions or stability modes. Stability with respect to Mode I suggests that for the expected thickness of soil cover, the anticipated soil void or dome diameter is small enough such that a sudden collapse will not occur. For a given value of soil cohesion, the lower bound of soil dome diameter and overburden thickness was described by a power function that is independent of soil friction angle. Stability mode I suggests that arching will be developed in the residual soil, diverting the stresses around the void.

Stability Mode II describes conditions such that yielding and plastic flow of the residual soil will be limited. Systems that are stable with respect to Mode I also have an upper bound of overburden thickness such that excessive yielding and ground deformation does not occur. For a given value of soil cohesion, this limit depends upon the soil friction angle and unit weight.

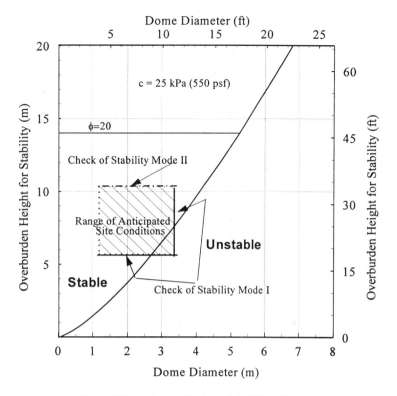

Figure 8 Example Application of Stability Chart

The two stability modes were evaluated for $c = 25$ kPa (550 psf) $\gamma = 18$ kN/m^3 (115 pcf) and presented in a design stability chart. The stability chart can be used to evaluate the stability of a potential site for a waste landfill or other engineered facilities in karst. Using this chart may facilitate the concentration of dome detection by geographical methods.

If insufficient overburden thickness exits, evaluation of these stability modes may facilitate the estimation of the required thickness of engineered fill to assure stability with respect to these two failure modes. If the prospective site appears to be stable with respect to Mode I and Mode II stability, then additional testing, sampling, and analysis may be warranted to evaluate the conditions under the imposed loadings. The prediction of surface deformation and distress to liner systems requires additional site information. Information such as overburden and liner material properties, the liner serviceability criteria, and the design waste loads are required. This is a more comprehensive analysis than is typically possible during the initial site selection phase of a typical landfill project, and is not within the scope of the proposed stability evaluation.

REFERENCES:

Broms, B. B. and Bennermark, C. M., 1967, Stability of Clay in Vertical Openings, J. of Soil Mech. & Founda. Div. ASCE, Vol. 193, SM1, pp.71-94.

Chang, K. R. and Basnett, C., 1997, Delineation of Sinkhole Boundary Using Dutch Cone Soundings, Engr. Geo. and Hydrogeo. of Karst Terrain: Proc. 6th Multidisc. Conf. on Sinkholes and the Engr. & Environ. Impact of Karst, Springfield, MI., USA, pp 305-311.

Chen, W. F., 1994, Constitutive Equations for Engineering Materials, Vol. 2: Plasticity and Modelling, Elsevier, New York. pp 740-743.

De Borst, R. and Vermeer, P. A., 1984, Possibilities and Limitations of Finite Elements for Limit Analysis, Geotechnique, Vol. 34, No. 2, pp.199-210.

Daniel, D. E. 1996, Geotechnical Practice for Waste Disposal, Chapman & Hall, London, UK. pp. 252-254.

Donaldson, G. W., 1963, Sinkholes and Subsidence Caused by Subsurface Erosion, Proc. of 3rd Regional. Conf. for Africa on Soil Mech.

and Founda. Engr. Salisbury, South Rhodesia, pp123-125.

Drumm, E.C., Kane, W.F., and Yoon, C.J., 1990, Application of Limit Plasticity to the Stability of Sinkholes, Engr. Geo., Vol.29, pp. 213-225.

Drumm, E.C., Ketelle, R.H., Manrod, W.E., and Ben-Hassine, J., 1987, Analysis of Plastic Soil in Contact with Cavitose Bedrock, Proc. Spec. Conf. on Geotech. Prac. for Waste Disposal, ASCE Geotech. Spec. Pub. No. 13, Ann Arbor, MI, June, pp. 418- 431.

Goodings, D. J. and Abdulla, W. A., 1997, Predicting the Development of Sinkholes in Weakly Cemented Sand. Geotech. News, pp. 27-31.

Jennings, J. E., Brink, A. A., Louw, A., and Gowan, G. D., 1965, Sinkholes and Subsidence in the Transvaal Dolomites of South Africa, Proc. of 6th Inter. Con. on Soil Mech. and Found. Engr. pp51-54.

Ketelle, R.H., Drumm, E.C., Ben-Hassine, J., and Manrod, W. E., 1987, Soil Mechanics and Analysis of Plastic Soil Deformation over a Bedrock Cavity, Karst Hydrogeo.- Engr. & Environ. App.: Proc. 2nd Multidisciplinary Conf. on Sinkholes & Environ. Impacts of Karst, Orlando, FL, Feb., pp. 383-387.

Koutsabeloulis, N. C. and Griffiths, D. V., 1989, Numerical Modeling of the Trap Door Problem, Geotechnique, Vol. 39 No. 1, 77-89.

Newton, J. G., 1976, Induced Sinkholes – A Continuing Problem along Alabama Highways, Proc. Inter. Assoc. of Hydro. Sci. No. 21, pp 453-463.

Newton, J. G., 1984, Natural and Induced Sinkhole Development in the Eastern United States, Proc. of the Third Interna. Sym. on Land Subsidence, Venice, Italy, pp. 549-564.

Newton, J. G., and Hyde, L. W., 1971, Sinkhole Problem In and Near Roberts Industrial Subdivision, Birmingham, Alabama – A Reconnaissance, Alabama Geo. Circular, No. 68.

Newton, J. G. and Tanner, J. M., 1986, Regional Inventory of Karst Activity in the Valley and Ridge Province, Eastern Tennessee, Phase I, ORNL/Sub/11-78911/1, Oak Ridge National Laboratory, Oak Ridge, Tenn.

Rowe, R. K. and Davis, E. H., 1982, The Behavior of Anchor Plates in Clay, Geotechnique, Vol. 32, No. 1, pp. 9-23.

Siegel, T. C., and Belgerie, J. J., 1995, The Importance of a Model in Foundation Design over Deeply Weathered Pinnacled Carbonate Rock, Karst Geohazards: Proc. 5th Multidisci. Conf. on Sinkholes & Engr. & Environ. Impacts of Karst, Gatlinburg, TN. pp 375-382.

Smith, T. J., 1997, Sinkhole Damage Investigations for the Insurance Industry in West-Central Florida, The Engineering Geology and Hydrogeology of Karst Terrain: Proc. 6th Multidisci. Conf. on Sinkholes & Engr. & Environ. Impact of Karst, Springfield, Missouri, USA, pp 299-304.

Sorochan, E. A., Khomenko, V. P., Tolmachyov, V. V., Troitzky, G. M., 1989, Karst Failures: Model Testing and Conceptual Models, Proc. 12th Inter. Conf. on Soil Mech. & Found. Engr., Rio de Janeiro, Brasil, Vol. 2, pp 977-981.

Sowers, G. F., 1996, Building on Sinkholes: Design and Construction of Foundations in Karst Terrain, ASCE Press, New York.

Williams, J. H., and Vineyard, J. D., 1976, Geological Indicators of Catastrophic Collapse in Karst Terrane in Missouri, National Acad. of Sci. Trans. Res. Record 612, pp31-37.

10 Stormwater management and flood hazards in karst terrane

Simulating time-varying cave flow and water levels using the Storm Water Management Model (SWMM)

C.WARREN CAMPBELL   Vista Technologies Incorporated, Huntsville, Ala., USA

SEAN M.SULLIVAN   Mevatec Corporation, Huntsville, Ala., USA

ABSTRACT

The Storm Water Management Model (SWMM) is an Environmental Protection Agency code used to estimate runoff through storm water drainage systems that include channels, pipes, and manholes with storage. SWMM was applied to simulate flow and water level changes with time for a part of Stephens Gap Cave in Jackson County, Alabama. The goal of the simulation was to estimate losses from a surface stream to the cave. The cave has three entrances that can remove water from the surface stream. These entrances connect through several passages to an 8 m (27 ft) high waterfall in a dome room. After a storm, the walls of this dome room had leaves on the wall as high as 4.6 m (15 ft) above the floor. The model showed that the height of the leaves did not represent a water level that could have occurred following any recent storm.

Campbell, et al. (1997) developed the CLG model to estimate losses from karst surface streams. This model treats losses as pipe flow from a reservoir and gives the loss flow rate as $\sim h^{0.5}$ where h is the depth of flow in the surface stream. Losses to Stephens Gap Cave calculated with SWMM varied as $h^{1.8}$. This depth dependence is more characteristic of flow over a weir than of pipe flow.

The SWMM-calculated losses to Stephens Gap Cave showed no hysteresis, that is, the rising and falling limbs of the stage-discharge plot followed the same curve. Loss curves with significant hysteresis are difficult to simulate with simple models such as CLG or a weir flow model. However, a SWMM model of a simple hypothetical cave demonstrated that storage in Stephens Gap Cave is far below that required to cause hysteresis. Losses from many karst surface streams can probably be adequately estimated with a calibrated weir flow model.

BACKGROUND

Purpose

The purpose of this study was two-fold: 1) to demonstrate the usefulness of simulating cave flow and water levels using the Storm Water Management Model (SWMM), and 2) to evaluate the much simpler CLG model used to calculate water losses from karst surface streams (Campbell, et al., 1997). CLG treats losses from karst surface streams as full-conduit flow from a reservoir and gives losses that are proportional to $h^{0.5}$, where h is stream depth. CLG predicts a single-valued increase and decrease in losses (no hysteresis) with flow depth. Cave streams can alternate between full-conduit flow (siphons) and open channel flow with significant storage. SWMM can predict flow with hysteresis. Physically, hysteresis occurs when the stage-discharge plot follows different paths for the rising and falling limbs.

SWMM was used to determine the importance of hysteresis for a typical cave-surface stream interaction. If hysteresis were important, then the CLG model or any other simple model could not be used to adequately predict surface stream losses.

SWMM is a very sophisticated model that is too complex to conveniently use with large river forecast models used for flood prediction. One goal of this study was to determine if a simpler model such as CLG could be developed from the SWMM simulation. A simpler model could adequately predict surface stream losses only for flows without significant hysteresis.

Modeling

The Storm Water Management Model (SWMM) is a code available from the Environmental Protection Agency Center for Exposure Assessment Modeling (CEAM). It consists of several modules or blocks, and the Extended Transport (EXTRAN) block was selected for this study. EXTRAN uses an explicit finite difference solution of the Saint Venant equations to calculate time-varying flow in storm systems that include surface channels, pipes, and manholes (Roesner, et al., 1992).

EXTRAN requires the input of at least one hydrograph provided by either another block of SWMM or by external calculations or measurements. The Soil Conservation Service (SCS) method is a widely used and flexible method (see, for example, McCuen, 1998) that can estimate runoff from medium sized watersheds. This method was used to provide the input hydrograph for this study.

A cave is nature's storm drainage system. Application of SWMM to cave flow seems logical, since the features modeled by EXTRAN correspond to cave drainage features (pipes - passage, storage - dome rooms, etc.).

Stephens Gap Cave, Alabama Number 585

Stephens Gap Cave in Jackson County, Alabama has seven known entrances (see Figure 1). Three of these entrances can withdraw water from a surface stream. The water taken from the stream runs through a system of passages to an 8 m (27 ft) high

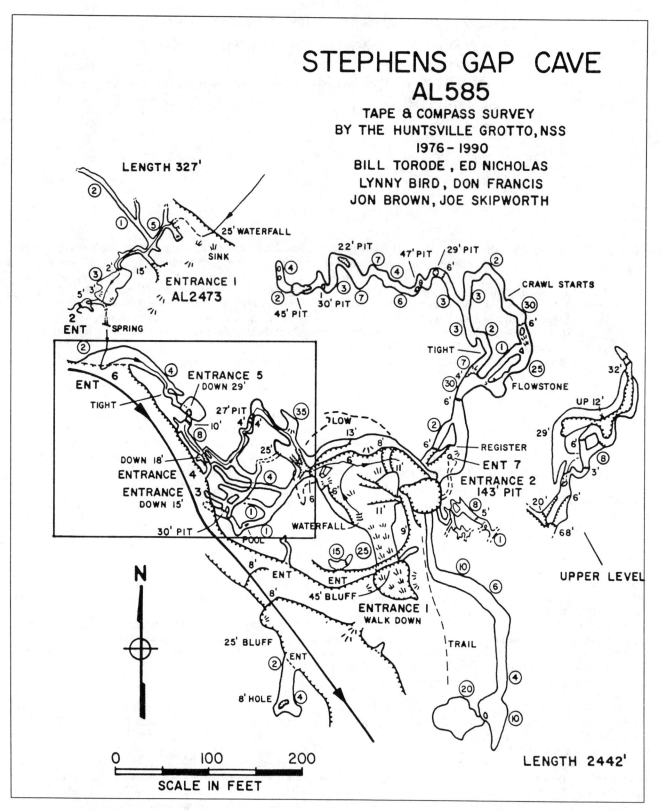

Figure 1: Map of Stephens Gap Cave (Alabama Number 585) showing the section modeled (inset)

waterfall. Water leaves this dome room through canyon passage and falls over a low waterfall into the main room of the cave. During June 1997, leaves were observed 4.6 m (15 ft) high on the wall of the dome room. These leaves appeared to indicate high water levels in this room. The corresponding flow would have had to be at least 8.5 cubic meters per second (cms) (300 ft^3/sec).

Stephens Gap Cave provides a good test for SWMM since it has three entrances removing water from the surface stream, and it has significant storage in several pits and domes in passages leading to the waterfall dome.

APPROACH
Watershed Runoff Modeling with the SCS Method

The surface stream watershed at Stephens Gap Cave consists of approximately 1 km^2 (0.4 mi^2) of wooded, rural land. Most of this land lies on the level top of Nat Mountain which is part of the Cumberland Plateau. The surface stream is incised deeply in the flank of the mountain. The stream is steep (slope > 0.1) with large boulders and debris in the channel.

A Tennessee Valley Authority rain gage was located within 3 km (2 mi) from the site. Since leaves in the Stephens Gap Cave waterfall dome were observed during June 1997, rainfall from the first six months of that year was used with the SCS method to estimate runoff from the watershed. The SCS model is usually run with normal antecedent moisture conditions. However, simulations using normal soil moisture did not give enough runoff to account for the high water level in the waterfall dome. Subsequently, the SCS runoff calculations were repeated under saturated soil conditions, and the amount of runoff was still not high enough to cause a 4.6 m high water level in the dome. Finally, the two-year 24-hour flood (100 mm = 3.87 in) for north Alabama was used. This was a larger storm than any that occurred from January to June 1997. This flood was used for all subsequent analyses.

SWMM Modeling

The first step in applying SWMM to cave flow modeling is to develop an abstraction of the physical cave for input to the code. This abstraction consists of conduits, channels, and nodes. SWMM can accept conduits of circular, rectangular, or other shapes. Channels include rectangular, trapezoidal, or user defined cross-sections. Nodes used in this study were of three types: 1) junction, 2) storage, and 3) output. Junctions occur where passages join with no significant storage. Storage nodes are similar to junctions, but with significant storage. Storage nodes were used to represent domes and pits in Stephens Gap Cave. Both junction and storage nodes can have multiple inputs and outputs. In contrast, output nodes have only one input. These nodes are used to apply the downstream boundary conditions. For the Stephens Gap model, the downstream boundary conditions were all elevated flow.

The naming convention for this model is as follows: 1) the first letter of junction node names is "J" and is followed by a number, 2) the first letter in a storage node name is "S" again followed by a number, 3) the first letter of an output node is "O" followed by a number, and 4) the first letter in a channel or conduit name is "C" followed by the name of the upstream junction and then the name of the downstream node. Figure 2 demonstrates this naming convention as it was used in the Stephens Gap Cave model.

Input data includes the elevations of the top (crown) and bottom (invert) of junction, storage, and output nodes. For storage nodes, the cross-sectional area of the node is also an input. Storage nodes can also be user defined. For these, the user provides pond area for different depths (stages). In engineering terminology, these are called stage-storage relationships.

Input parameters for channels and conduits include length, cross-sectional geometric characteristics, elevations for entering and exiting conduits and channels, and channel roughness. Conduit calculations are based on Manning's equation.

$$Q = \frac{\phi}{n} A R_h^{2/3} S^{1/2} \tag{1}$$

where

Q = flow rate (cms)
ϕ = a proportionality constant = 1 m$^{1/3}$/sec = 1.486 ft$^{1/3}$/sec
n = Manning's roughness coefficient
A = cross sectional area of the conduit (m^2)
R_h = hydraulic radius = A/P (m)
P = wetted perimeter (m)
S = slope of the pipe

Manning's equation represents uniform flow in a pipe or conduit. It does not accurately estimate flow rates for pipes with bends, contractions, expansions, or other structures that accelerate the flow. These features are called "minor" losses, but in a cave they may predominate. Minor losses are usually represented by an equation of the following form.

$$h = K \frac{Q^2}{2gA^2} \tag{2}$$

where

h = head loss (m)
K = dimensionless loss coefficient (obtained from handbooks)
g = acceleration of gravity = 9.81 m/sec^2 = 32.2 ft/sec^2

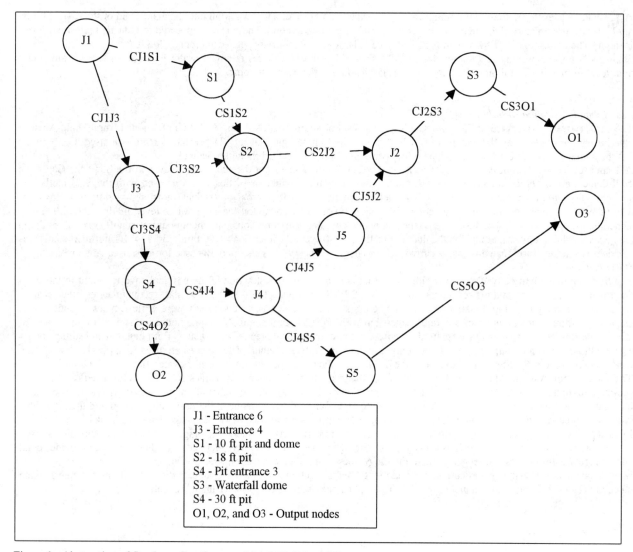

Figure 2: Abstraction of Stephens Gap Cave used for SWMM modeling

The head loss has units of length and is the energy lost per unit weight of the fluid. The loss coefficient depends on the physical structure ($K \sim 0.4$ for right angle bends, ~ 0.5 for pipe entrances, ~ 1.0 for pipe exits, and so on).

Since SWMM does not account for minor losses directly, they must be incorporated into Manning's roughness coefficient. A single cave passage may leave a dome room (pipe entrance - $K = 0.5$). It may have several bends ($K = 0.4$ for each), and it may discharge into another dome room (exit loss - $K = 1.0$). Each of the losses can be incorporated into an effective roughness coefficient as follows.

$$n_{eff} = \left[n^2 + \frac{\phi^2 R_h^{4/3}}{2gL} \sum K_i \right]^{1/2} \qquad (3)$$

where

n_{eff} = the effective roughness coefficient
L = conduit or channel length (m)
$\sum K_i$ = the sum of all the loss coefficients for the conduit or channel.

The roughness coefficient n adds in quadrature (like the length of the hypotenuse of a right triangle) with the square root of the second term in the bracket in Equation (3).

This completes the description of the modeling approach. The next section describes the results of the analysis.

RESULTS

Losses from the surface stream to Stephens Gap Cave

The SCS model predicted a peak flow for the 2-year, 24-hour flood of 5.6 cms (200 cfs). This is the flow rate to the surface stream. At peak flow, the losses to the cave were 2 cms, or approximately 36 percent of the total flow in the surface stream. This is the flow to all three entrances of the cave.

Figure 3 shows the predicted losses plotted against depth at junction J1. The curve in the figure shows almost no hysteresis. The simulated data were fit with the CLG model and the figure also shows this very poor fit ($r^2 = 0.57$ where r is the correlation coefficient between the simulated losses and the fit). CLG assumes pipe flow losses and predicts loss variation with $h^{0.5}$ (h = depth of flow at J1).

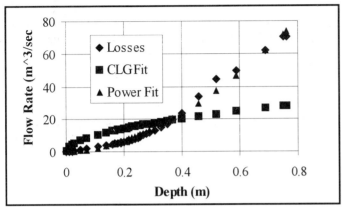

Figure 3: Simulated losses with CLG and power fits

The third curve in Figure 3 is a power function fit with an $h^{1.8}$ dependence which does match the curve very well ($r^2 = 0.99$).

Obviously, a pipe-flow model of losses is inadequate. The losses more closely match weir flow. A rectangular weir has flow dependence on $h^{1.5}$. Flow through a parabolic weir varies as h^2.

Flow and depths in the waterfall dome

Of the 2 cms (70 cfs) peak flow entering the cave, a peak of 0.53 cms (19 cfs) flowed from the waterfall dome. This gave a maximum depth of only 0.23 m (0.76 ft). Since the flow levels were so low, the leaves 4.6 m up on the walls could not have been caused by water at that depth. The leaves may have come into the room with the waterfall flow and been carried by strong air currents to this high level.

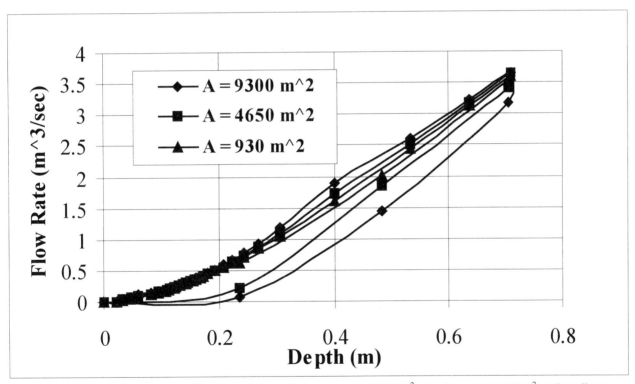

Figure 4: Effect of storage on stage-discharge relationship for large room (9300 m²), medium room (4650 m²) and small room (930 m²) storage

Hysteresis

The flow-depth curve for the waterfall dome showed very little hysteresis, that is, the plot of the rising limb fell on top of the plot of the falling limb. Additional analysis was done to determine the amount of storage needed to create hysteresis in this curve. A very simple cave with flow into and out of a large room was analyzed. Figure 4 shows hysteresis for a cave room with areas 930, 4650, and 9300 m^2. The largest area has the greatest hysteresis and the greatest loop amplitude. The section of Stephens Gap Cave modeled has much less storage than any of these three. This implies that hysteresis will be unimportant for many caves that receive water from surface streams. Consequently, a weir model will probably adequately describe these losses.

SUMMARY

The Storm Water Management Model (SWMM) was used to simulate flow from a surface stream into Stephens Gap Cave in Jackson County Alabama. Water levels and flow rates in the cave passages were simulated as a function of time. The analyses showed that leaves observed high on the wall of the waterfall dome do not indicate that water levels reached that height. Losses from the surface stream through three entrances can be modeled as weir flow diversion from the stream. The CLG model (Campbell, et al., 1997) did not match loss data for this cave. A single valued stage discharge relation was obtained. The effect of storage on hysteresis in the stage-discharge curve was demonstrated with a simple hypothetical cave model. The storage in Stephens Gap Cave was well below that needed to cause hysteresis. Curves without hysteresis can be well represented by simple models such as the weir flow model.

The utility of SWMM for analyzing cave flows was established. SWMM produced stable solutions with very low continuity errors for this cave.

ACKNOWLEDGEMENTS

The authors thank Bill Torode, Librarian of the National Speleological Society for the map of Stephens Gap Cave. We also thank Dr. Qi Mao of the Tennesse Valley Authority, and Drs. Richard McNider and Bob Clymer of the Global Hydrology and Climate Center who provided rainfall data for this study. Finally, we thank Dr. Alauddin Khan and Dr. Shabbir Ahmed of SAIC in Oak Ridge, Tennesse for the opportunity to use SWMM.

REFERENCES (EXAMPLES)

Campbell, C. Warren, Livingston, Leslie R., and Garza, Reggina, 1997, Modeling Karst Stream Losses, Final Report for University Center for Atmospheric Research Grant No. 896-71865, 29p.

McCuen, Richard H., 1989, Hydrologic Analysis and Design, 2nd Edition, Prentice-Hall, Upper Saddle River, New Jersey, 814p.

Roesner, Larry A., Aldrich, John A., and Dickinson, Robert E., 1992, Storm Water Management Model User's Manual Version 4: EXTRAN Addendum, Environmental Research Laboratory, Office of Research and Development, U.S. Environmental Protection Agency, Athens, Georgia, 188p.

Stormwater management design in karst terrane adjusting hydrology models and using karstic features

JOHN C. LAUGHLAND Jefferson County, Charles Town, W.Va., USA

ABSTRACT

Characteristics of karst terrane effect the extent to which rainfall either seeps into the ground or runs off to the surface watercourses. In stormwater management design this means that the commonly used hydrology models, such as SCS TR-55, which provide satisfactory estimates of runoff in most geologic regions, can be inaccurate on the high side in karstic regions. Most of Jefferson County, West Virginia, lies in a karst region. County officials have adopted methods for adjusting surface hydrology computations used in stormwater management design to reflect the impact of karst on amounts of predevelopment runoff flows. The County also has established standards and developed methods for the protection and utilization of karstic features such as rock breaks, sink holes and depressions.

Interest in the impact of karst resulted from review of the hydrology study for an industrial park where a drainage area of 2.3 square miles drained to a single 18 inch diameter culvert which exhibited very low flows even during intense storms. In seeking information about this phenomena it was learned that although a few researchers and practitioners have studied the relationships between karst and hydrology this information has not reached most practicing designers. This paper cites the findings of a review of these previous studies including descriptions of various ways of adjusting model results.

INTRODUCTION

Webster's Dictionary (Webster's,1972) defines karst as consisting of "irregular limestone with sinks, underground streams and caverns." The American Society For Testing and Materials (ASTM) (ASTM, 1995) defines karst terrane in more detail as:

" a landscape and its subsurface characterized by flow through dissolutionally modified bedrock and characterized by a variable suite of surface landforms and subsurface features, not all of which may be present or obvious. These include: sinkholes, springs, caves, sinking streams, dissolutionally enlarged joints or bedding planes, or both, and other dissolution features. Most karsts develop in limestone or dolomite, or both, but they may also develop in gypsum, salt, carbonate-cemented sandstones, and other soluble rocks."

The characteristics of karst itemized in these definitions affect the extent to which rainfall either seeps into the ground or runs off to the surface watercourses. For the stormwater management designer this means that the commonly used hydrology models, such as TR-55, which provide satisfactory estimates of runoff in most geologic regions can be in error as much as an order of magnitude in karstic regions. On the other hand, these same characteristics-- such as rock breaks, sink holes and depressions--when recognized by site planners can be used as part of a stormwater management plan.

ADJUSTING HYDROLOGY MODEL FOR KARST EFFECTS

Background

In 1989 an industrial park was planned for Jefferson County, West Virginia, a jurisdiction which geologically is predominantly karst terrane. The hydrology study for the industrial park showed that a drainage area of 2.3 square miles drained to a single 18 inch diameter culvert. Observations of the culvert during intense storms indicated that very little water was reaching the culvert. But the hydrology computations indicated that a peak flow from a 2-year storm over a 24 hour period would be several hundred cubic feet per second, a value far exceeding the capacity of the existing 18-inch-diameter culvert. This raised several questions about the application of the commonly-used hydrology models.

Design consultants for the industrial park asked the State Hydrologist at the Natural Resources Conservation Service (NRCS) about this apparent inconsistency. The hydrologist indicated that similar observations had been made elsewhere in karst terrane and that NRCS had conducted flood hazard area studies of streams in karst terrane using equations reflecting a relationship between karst terrane and runoff levels less than those from the models. These equations had been derived from stream gage and rainfall from carbonate (karst) areas of Pennsylvania.

Since then the Jefferson County Office of the County Engineer has sought and obtained information from experts in karst terrane hydrology, made appropriate adjustments to it's hydrology computations and suggested to stormwater management designers and developers that they protect and utilize the karst features on their development projects as part of their site designs.

Previous Studies of Relationships Between Karst and Hydrologic Characteristics

The Pennsylvania study referred to previously was conducted by Herbert N. Flippo of the Geological Service, U.S. Department of the Interior (Flippo,1977). He studied eight different hydrologic areas and developed regression models for each one using rainfall measurements and stream gauge records. One of the models applied to primarily undeveloped drainage basins predominantly underlain by limestone or dolomite bedrock. The equation for this model is shown in Table 1.

Sonia Jacobsen of the Natural Resources Conservation Service studied the terrane of southeastern Minnesota (Jacobsen,1990) where the karst strata are horizontal. Her study compared actual stream gage readings against the results of computations from the TR-20 hydrology model using various peak rate factors. The premise was that by adjusting the peak rate factor, closer fits between the computed and actual flows could be obtained. She found that a unit hydrograph with a peak rate factor of 259 seemed closer to actual gage readings than did the default 484 value with Type II rainfall. Prior to this, the Minnesota NRCS had allowed use of the Type I storm to make simple adjustments to improve the fit.

In a study of two karst watersheds in Pennsylvania, David H. Bailey and Edward T. Van Blargan (Bailey,no date) strove to account for the impacts of karst factors such as natural underground reservoirs, closed depressions and rates of infiltration and evaporation from the depressions. They spelled out certain basic characteristics of the karst drainage areas such as:

1. karst areas tend to absorb initial rainfall,
2. runoff from karst areas becomes more like runoff from non-karst areas as storm magnitude increases, and;
3. land development can negate karst effects, thus making the difference between pre-development and postdevelopment hydrograph peaks greater in karst areas than in non-karst areas.

TABLE 1 - REGRESSION EQUATION FOR KARST AREA FROM FLIPPO STUDY

$$Q = cA^x$$

where: Q = Runoff in cubic feet per second
T = Storm frequency year
c = Coefficient variable by year
x = Exponent variable by year
A = Drainage area in square miles

T	c	x	Standard Error
2	23.5*	0.880*	N.A.
10	39.8	0.933	26
100	64.4	0.979	33

* Values extrapolated.

Bailey and Van Blargan used "global" increase of the Ia factor, the Initial Abstraction, which is related inversely to the RCN. They concluded that precise modelling would require measurements not available normally and that "no clear cut model approach exists." They opined that their method was useful in watershed analysis, but not useful "for design of facilities for control of accelerated runoff."

In 1972 R.H. Tice conducted a study of flood frequencies of seven major river basins and their subbasins extending from the Hudson River to the Potomac River (Tice,1972). The relevant finding from this study was that the "limestone regions of Pennsylvania, Maryland and Virginia (hydrologic area 9) with sinkholes and subterranean channel storage have lower floods than surrounding areas." The results of the Tice study were incorporated in a computer program by David Mikotz during graduate studies at Pennsylvania State University. The program, "USGS-IND" is available from Pennsylvania State University, which also has another computer model, "PSU-IV" , that specifically takes into consideration the percentage of a drainage area which has a carbonate (karst) substrata.

Making Adjustments To Hydrology Computations

Upon recognition of the relationship between karst and hydrology, Jefferson County asked design consultants to take this into consideration in computing the predevelopment runoffs for their stormwater management designs. In the case of the industrial park referred to above, Gene R. Weakley, Jr., P.E. of Kelley, Gidley, Blair and Wolfe (KGBW) (Kelley,1989) made the following adjustments to hydrology computations for the industrial park:

1. Identified all sink holes and respective contributory drainage areas and separated them from the main drainage area.
2. Used a flatter (less peaked) hydrograph by using peak hour factor 205 instead of 484, the default in TR-20.
3. Identified Runoff Curve Numbers (RCN's) for each rainfall event which would yield results similar to those obtained using the equation prepared by Flippo.

Weakley also looked at adjusting times of concentration, travel times and rainfall intensities. However, no matter which value was adjusted, a trial and error method was needed to develop the adjustments. Another consultant, Appalachian Surveys, Inc. (ASI), used both the Flippo equation and the Pennsylvania State University program, "PSU-IV" to make adjustments to the predevelopment RCN's on another land development project. The Flippo equation was used for the two year storm and PSU-IV was used for the 10 and 100 year storms after conferring with professors at Pennsylvania State University. It was opined that the Flippo equation tends to underestimate peaks for the less frequent storms. The results of the pre-development RCN adjustments by KGBW and ASI varied from 3 to 7 points; the greatest difference occurring with the 100 year event and the least with the 2 year event.

Finally, the Office of the County Engineer, using the PSU-IV model and a generic drainage area, developed a table of factors to be multiplied times the predevelopment peak discharge values resulting from normal use of hydrology models that do not consider karst. However, design consultants are permitted to develop their own adjustment values with adequate documentation. The County table is shown in Table 2.

TABLE 2 -- MULTIPLIERS FOR ADJUSTING PRE-DEVELOPMENT RUNOFF QUANTITIES FOR KARST IMPACT

% of Karst	Storm Year 2	10	100
100	0.33	0.43	0.50
80	0.38	0.51	0.62
60	0.55	0.66	0.74
40	0.73	0.80	0.85
20	0.91	0.92	0.93
0	1.00	1.00	1.00

These efforts at adjusting conventional hydrologic models have been made by journeymen engineers and designers using research studies as bases, but without the opportunity to conduct structured research of their own. Therefore, these efforts need to be viewed as first efforts to be used until these practitioners either change them based on field observations or results of further formal research conducted by universities and research arms of major governmental agencies.

DESIGN OPPORTUNITIES

Karstic characteristics, ones that make the difference in hydrologic computations, if preserved and protected in the development process can be used as part of designs for stormwater management The characteristics most likely to be encountered are described below:

1. The most obvious situation is the natural depression or sink. These are natural basins which, except for overflow during high water conditions, have no surface outlet. And in many cases not even then is there any overflow. Water collects in the sink to either form a pond, infiltrate or evaporate.
2. Sinkholes, the result of dissolution of underlying material and subsequent collapse, are surface openings in sinks which usually convey surface runoff waters from the sinkhole watershed into the underlying aquifer. The rate of conveyance can vary widely and in some cases is no greater than surrounding soils due to natural or man-caused plugging of the sinkhole.
3. Near-surface solution channels and vertically inclined bedding planes are sometimes uncovered during construction and when exposed receive and convey surface waters to the underlying aquifer.
4. Finally, there are the ridges of exposed rocks usually referred to as "rock breaks" which can function either as dams or conduits to the aquifer.

None of these characteristics can be ignored when developing a parcel of land. Their presence can limit the scope, or add to the cost, of a project. On the other hand they can be used as part of the stormwater management design, if the ground water can be protected from contamination. Our approach has been to allow recharge through the karst after runoff has passed through the water quality component of the stormwater management facility.

In Jefferson County, the following steps have been taken in the land development process toward these ends:

1. l. Developers are required to identify any sinkholes on a piece of land intended for development. The local office of the Natural Resource Conservation Service (NRCS), formerly the Soil Conservation Service (SCS), maintains a sinkhole inventory. However, sinks are considered potential sinkholes whether or not they are on the inventory and are identified and discussed early in the planning phase.
2. Plats for development projects must show an easement or reserve area around all identified sinkholes on the project.
3. Stormwater management and erosion and sediment control plans must be designed for protection of identified sinkhole areas. Protection may be accomplished by either (a) preventing water from entering a sinkhole or (b) filtering water before it reaches the bottom of the sinkhole opening.

The County has several generic designs for sinkhole protection. They include a concrete plug for small sinkholes, a stone/earth filter cap for small to moderate sinkholes, a sinkhole perimeter berm and filter for medium to large sinkholes and sinkhole standpipe and filter system for situations where the opening on the solid rock is exposed.

The filter cap detail was obtained from the local NRCS staff which is doing research on the effectiveness of sinkhole biofilter caps. This study is still in the study phase so there are no data available at this time.

The only sinkhole protection device used so far by plan has been a perimeter berm/filter with a stone filter in the middle of the sink. The size of the area draining to the device is 12 acres containing subdivision roads and large lot residential development. Over the last four years, but before the current drought, the facility has not been observed to be either dry or overflowing. It had been expected to detain runoff for relatively short periods, but to be dry at most other times.

Rock breaks are very common in Jefferson County. They generally need blasting if they are in the way of needed site grading. But they generally stick up a foot or so from the surrounding grade, generally follow a level contour and probably are a major reason for the low runoff characteristic of karst terrane. So far, rock breaks have only been used in Jefferson County in a passive way in lieu of zero-grade berm/swales (a stormwater management device used to retain or detain sheet flow with a minimum of flow concentration). In these cases site grading has been far enough away from the breaks to allow a grass filter strip to be left to provide for trapping of sediment in the grasses and for infiltration (the County also uses filter strips to protect wetlands).

More active uses of rock breaks have only been discussed although there are several proposed projects on which the only choices are to provide no control or use the rock break. These active uses would involve channeling waters to the break, and possibly some grading of the break to ensure capture. This approach would probably require more than a grass strip filter, most likely elements such as geotextile, sand or peat filters.

SUMMARY AND CONCLUSIONS

1. Actual peak flows of runoff from undeveloped karst terrane tend to be much lower than values derived using standard hydrology models such as SCS-TR-55.

2. The "karst" effect cited above varies with event frequency, the frequent storms being affected most.

3. Predevelopment runoff values for karst areas need to be adjusted to ensure postdevelopment release rates that achieve stormwater management objectives.

4. Several methods of adjusting predevelopment runoff values for karst areas have been proposed or applied. These methods breakdown into two categories: adjustments to hydrology model variables and adjustments to the end results of hydrology models. In either case empirical studies such as Flippo are needed as a baseline for making the adjustments.

5. Where adjustment of model variables has been used, only a single variable (not always the same one) has been adjusted (Runoff Curve Number, Initial Abstraction or the Unit Hydrograph). None of these methods have been subjected to comparative study.

6. Where a sink occurs in a larger drainage area the drainage area of the sink needs to be subtracted from the drainage area to the design point in the predevelopment analysis. If the runoff to the sink is released to the larger drainage area in the postdevelopment design it needs to be added to the postdevelopment design point runoff and analysis.

7. Consideration of the impact of karst characteristics on surface runoff leads logically to consideration of ways to take advantage of these characteristics in responsible ways as part of the stormwater design. However, there is a need for research to help practitioners make informed design decisions.

REFERENCES

ASTM Committee D-18 on Soil and Rock, "Standard Guide for Design of Ground-Water Monitoring Systems in Karst and Fractured-Rock Aquifers, American Society for Testing and Materials, West Conshohocken, PA, 1995.

Bailey, David H. and Edward T. Van Blargan, "Watershed Modeling in Karst Terrain", Hartman and Associates, Inc., Camp Hill, Pennsylvania.

Castelle, A.J., A.W. Johnson and C. Conolly. "Wetland and Stream Buffer Size Requirements - A Review", Journal of Environmental Quality, Volume 23, September-October, 1994.

Eastern Panhandle Soil Conservation District, "Flowing Springs Run and Evitt's Run Flood Hazard Area, Jefferson County Flood Plain Management Study", U.S. Department of Agriculture, Morgantown, West Virginia, April, 1987.

Flippo, Herbert N., Jr., "Floods in Pennsylvania", United States Department of the Interior, Geological Survey, 1977.

Jacobsen, Sonia M.M., letter to John C. Laughland, dated September 17, 1990.

Kelley, Gidley, Blair and Wolfe, Inc., "Stormwater Management Plan, Burr Industrial Park, Jefferson County Development Authority", submitted with a preliminary plat, July, 1989.

Sowers, George F., Building On Sinkholes: Design and Construction of Foundations in Karst Terrain, ASCE Press, New York, New York, 1996.

Tice, R.H., "Magnitude and Frequency of Floods, Part 1-B", United States Geological Survey, 1972.

Webster's New World Dictionary of the American Language : College Edition, World Publishing, New York.

A review of stormwater best management practices for karst areas

MOLLY S. MCCANN & JAMES L. SMOOT The University of Tennessee, Knoxville, Tenn., USA

ABSTRACT

Stormwater runoff can pollute receiving waters and cause local flooding. Such runoff commonly contains heavy metals, animal and human wastes, grease, oil, sediment, and other potential pollutants. In karst areas, these impacts to groundwater can be magnified because of the nearly direct input of these contaminants to groundwater when runoff flows into sinkholes or through sinking streams. In addition, sinkholes can flood, creating a nuisance and/or damage to surrounding areas. Stormwater management practices that are applicable to controlling water quantity and quality are not necessarily effective in karst areas. Because karst drainage is not always recognized or understood, prevention of flooding and pollution in such areas through application of best management practices (BMPs) and regulatory controls has not been fully realized.

Stormwater BMPs which may work well for controlling water quantity in karst areas are dry detention ponds, runoff spreaders, porous pavement, and increased vegetation density. For controlling water quality, the use of skimmers, wet retention ponds, wetlands, and filtration systems may be applicable. These BMPs would be effective if installed near the sinkhole area or throughout the watershed. The discussion of two case studies in Knoxville, Tennessee, shows that some advances have been made in applying stormwater BMPs in karst areas and classifying sinkholes as Class V Injection Wells.

INTRODUCTION AND BACKGROUND

In order to best manage karst stormwater quantity and quality in the future, it would be useful to examine the range of BMPs that have been permitted or are applicable. Advances have been made in the areas of controlling flooding in karst areas, classifying and managing sinkholes as groundwater "injection wells," and identifying and managing drainage wells which provide for sub-surface drainage. Evaluation of the success of these BMPs could assist in determining what actions are most effective for controlling stormwater quality and quantity in karst areas.

A BMP can be defined as an optimal, structural or non-structural stormwater management practice which considers technical function and economic and social interaction. BMPs discussed in this paper can be divided into two function areas: controlling water quantity and improving water quality. A stormwater quantity BMP would reduce runoff volume and/or attenuate the peak flow rate, while a stormwater quality BMP would reduce concentrations and/or loads of target pollutants. BMPs for typical urban stormwater situations are numerous and widely used; however, all BMPs applicable to a non-karst setting are not necessarily desirable or effective in a karst setting. Typically, non-karst settings provide for some infiltration and natural removal of pollutants through the soil mantle, where many contaminants are adsorbed or further removed by microbes and vegetation. In contrast, karst settings usually provide little or no natural pollutant attenuation and can quickly transport contaminants through solution features to other locations. In addition, encouraging infiltration in karst areas may lead to sinkhole collapses.

WATER QUANTITY PROBLEMS IN KARST AREAS

Typical causes of water quantity (flooding) problems that occur in karst areas include the following:

1. Plugging of intake or throat of sinkhole by debris and/or sediment

2. Interruption of water conveyance through karst conduits by collapses, plugging, and intrusion by inflows from other sinkholes or sinking streams

3. Control of conduit capacity by elevated water levels in receiving waters

4. Exceedance of sinkhole storage capacity and rate of flow in conduits by increased runoff volume and peak flows due to watershed development

5. Filling or building in sinkholes

6. Reverse flows

WATER QUANTITY BEST MANAGEMENT PRACTICES

In order to prevent flooding in karst areas, the most effective practice usually entails the control of runoff before it reaches a sinkhole. Therefore, the three major functions that water quantity BMPs should serve are to increase storage, enhance hydrologic "losses" (e.g. interception, depression storage, infiltration, evaporation) and increase the travel time of runoff to the sinkhole (time of concentration). Practices within each function category may overlap.

BMPs to increase storage

The design criteria for flood control are typically based upon the reduction of runoff peaks for rare rainfall events. A detention pond is often effective in attenuating the peak flow resulting from such a design storm. If the treatment of the runoff is not a concern, a dry detention pond may be the most efficient technique to prevent flooding at the downstream sinkhole (Yu & Nawang 1993). Dry ponds are not very effective in removing pollutants due to their short detention times, but they can intercept debris that might plug the sinkhole throat (Wanielista & Yousef 1993). Detention ponds are most effective if installed on the upstream side of the sinkhole. It should be noted, however, that space limitations may rule out the possibility of constructing a detention pond, and in addition, ponding runoff on a pervious surface may trigger the development of, or expansion of, a sinkhole collapse.

BMPs to enhance hydrologic losses

Practices that could reduce stormwater runoff volumes from the watershed include increasing vegetation density, terracing slopes, using runoff spreaders, and using porous pavement (Yu et al. 1993). By increasing vegetation density around the sinkhole area, one can simulate a non-karst situation by encouraging more water to infiltrate and be treated by the soil column. Slope terracing around a sinkhole lowers the hydrograph peak, increases length of overland flow, and prevents erosion (Wanielista & Yousef 1993). Runoff spreaders consist of a ditch or swale that intercepts runoff and then distributes the overflow evenly to a wide grass strip (Yu et al. 1993). Furthermore, when sinkholes are present in urban areas, porous pavement in the form of modular blocks with large perforations could be used in nearby streets, sidewalks, and parking lots to allow infiltration of runoff before reaching the sinkhole.

BMPs to increase time of concentration

Practices that enhance hydrologic losses may also increase the time of concentration. In general, longer and shallower drainage paths and a higher surface roughness serve such a purpose. Grass swales, rip rap barriers, terraced slopes, sand filtration systems, and runoff spreaders are all common techniques for increasing the time of concentration.

WATER QUALITY PROBLEMS IN KARST AREAS

Karst features are found in both rural and urban settings and are therefore subjected to a wide range of pollutants. Typical pollutants that characterize stormwater runoff include suspended solids, biochemical oxygen demand, chemical oxygen demand, total phosphorus, soluble phosphorus, total Kjeldahl nitrogen, nitrate, nitrite, copper, lead, and zinc (Urbonas & Stahre 1993). These pollutants can enter stormwater through atmospheric deposition, flushing from paved and unpaved surfaces, and overflow of sewers (Ibid.). It is desirable that contaminated runoff to sinkholes be treated before it is injected directly into the groundwater system. This issue is of particular importance to areas that depend on an aquifer system for their drinking water supply.

The BMPs mentioned for water quality control are mainly structural facilities, though the value of controlling these pollutants at their source through non-structural, "good-housekeeping" techniques cannot be overestimated.

WATER QUALITY BEST MANAGEMENT PRACTICES

Physical Methods

Physical BMPs are useful for removing pollutants that either settle or float on the water surface (see Table 1). The most efficient technique for promoting settlement before entering a sinkhole would be the construction of wet retention or settling ponds. Wet ponds maintain a permanent pool, therefore allowing for particulate and dissolved pollutant removal through enhanced particle settling, decay processes, and biological uptake (Yu & Nawang 1993). Pollutant removal efficiency is much higher for wet ponds than for dry ponds or extended dry ponds, and therefore wet ponds may be a good choice when stormwater pollution is an issue (see Table 2) (Wanielista & Yousef 1993). If runoff contains floating pollutants such as oils, litter, and yard wastes, the use of skimmers in the retention pond can contribute to the overall pollutant-removal efficiency. As an alternative to a retention pond, sand filtration systems that run perpendicular to the stormwater drainage path are extremely useful for removing heavy sediments, organics, and debris (Shaver & Baldwin 1995).

Physiochemical Methods

Many dissolved pollutants can be removed through physiochemical processes (see Table 1). The use of peat moss and/or activated carbon in filtration systems around the sinkhole can promote adsorption of heavy metals, anthropogenic organics, and other contaminants. In addition, constructed wetlands have been proven very effective in particle settling, adsorption of ammonium ions, phosphate, metals, and viruses, chemical precipitation of metals, volatilization of oils, filtration of organic matter and phosphorus, and for nutrient uptake (Urbonas and Stahre 1993). Though the use of wetlands to treat urban stormwater runoff is a recent issue in general, their applicability to karst areas seems promising. These wetlands could be constructed directly around the sinkhole, in conjunction with terraced slopes, or within the permanent pond area of a wet retention pond.

CLASS V INJECTION WELL PERMIT REQUIREMENTS

Before any stormwater BMP can be implemented in a karst area, a Class V Injection Well Permit must be obtained from the State Department of Environment and Conservation. Any alteration to the drainage pattern and/or the composition of the runoff to a

Table 1: Fate of pollutants by applicable stormwater BMPs (Debo & Reese 1995)

Pollutant	Vegetative Controls	Detention Basin	Infiltration System	Wetlands
Heavy Metals	Filtering	Adsorption Settling	Adsorption Filtration	Adsorption Settling
Toxic Organics	Adsorption	Adsorption Settling Biodegradation Volatilization	Adsorption Biodegradation	Adsorption Settling Biodegradation Volatilization
Nutrients	Bioassimilation	Bioassimilation	Adsorption	Bioassimilation
Solids	Filtering	Settling	Adsorption	Adsorption Settling
Oil and Grease	Adsorption	Adsorption Settling	Adsorption	Adsorption Settling
Pathogens	Not applicable	Settling	Filtration	Not applicable
BOD	Biodegradation	Biodegradation Settling	Biodegradation	Biodegradation

Table 2: Typical water quality BMPs and pollutant removal efficiency (Debo & Reese 1995), (Urbonas & Stahre 1993)

BMP	TSS	Heavy Metals	Organics	BOD	Bacteria	Nutrients
Porous Pavement	H	H	H	H	n/a	M to H
Dry Detention Ponds	L to H	L to H	L to M	L	n/a	L to M
Wet Retention Ponds	H	H	H	L to M	n/a	L to H
Extended Detention Ponds	M to H	M to H	M to H	—	M to H	L
Wetlands	H	H	M to H	L	n/a	M to H
Grass Swales	H	H	H	H	H	M
Sand Filtration System	M to H	M	M	M to H	n/a	L to M

KEY: L = low removal efficiency, M = medium removal efficiency, H = high removal efficiency

sinkhole requires this permit. The permitting process takes approximately fifteen days, and it includes a thorough review of the possible impacts of the design on the sinkhole and receiving groundwater (personal communication, Thomas Sorrels 1998).

CASE STUDIES

I-40/I-640 Sinkhole Filtration System, Knoxville, TN

This sinkhole is located directly adjacent to a major highway interchange that hosts approximately 76,000 vehicles per day, and therefore is subjected to a large stormwater contaminant load. Stormwater runoff from 60 acres originally passed through a series of culverts, ditches, and a small detention basin and gabion before it flowed into a small sinkhole. Recently, however, a larger sinkhole collapsed within the basin and began intercepting the runoff. In order to intercept and treat the "first flush" of runoff before flowing into the karst system, a sorption/filtration bed system was designed by P.E. LaMoreaux and Associates and the University of Tennessee under contract with the Federal Highway Administration (Stephenson et al. 1997). The preliminary design showed a filtration system installed directly in the sinkhole, but this design was later changed due to feasibility. In the new design, runoff out of a pipe flows through several rip-rap barriers, through spaces between concrete barriers, into a shallow detention area, and then into a peat and gravel filtration system. The outflow channel from the filter to the sinkhole is lined with an impermeable membrane. This site was permitted successfully as a Class V Injection Well and though recent observations show that the system is working as planned, long-term monitoring is needed to determine true effectiveness.

All Saints Catholic Church Injection Well, Cedar Bluff Area, Knoxville, TN

This area is particularly prone to sinkholes, and a clogged sinkhole at this particular site was causing flooding problems. As a result, a drainage well was drilled directly in the throat of the sinkhole and an adjacent sand filtration system was installed. The project was designed for a 100-year, 24-hour rainfall in order to keep flooding off of a nearby major road, Cedar Bluff Road. Initial considerations for design included pumping the water or acquiring a drainage easement to the nearest stream; however, these solutions were deemed infeasible because the nearest receiving stream was more than a half mile away. The sand filter was placed in a subsurface trench, 100 feet long at 1% slope and 2.5 feet deep. A perforated pipe at the base of the trench captures the filtered runoff and carries it to the injection well. The well is operated by a valve that is opened only when a major storm event occurs and flooding seems imminent (personal communication, Mike Eiffe 1998). This site was also permitted successfully as a Class V Injection Well, and to date, the project appears highly successful in removing sediments and debris.

CONCLUSIONS

There are numerous stormwater quality and quantity BMPs that can be applied to traditional urban and rural settings, and many of these are applicable and effective in karst areas. Every site in which stormwater runoff to sinkholes is encountered is unique; methods that have been applied elsewhere to similar situations must often be adapted to conform to the new site's characteristics. The most effective approach in applying stormwater BMPs in karst areas, however, is to manage the entire watershed by applying BMPs throughout, not solely at the sinkhole site. In this manner, many water quantity and quality problems can be diminished before they reach a fragile karst environment.

REFERENCES

Debo, T.N., and Reese, A.J., 1995, Municipal storm water management: CRC Press, 756 p.

Eiffe, Mike, 1 December 1998, Personal communication: Water Resources Engineer, HydroGeologic, Knoxville, TN.

Shaver, E., and Baldwin, R., 1995, Sand filter design for water quality treatment: Stormwater Runoff and Receiving Systems, Impact, Monitoring, and Assessment, Herricks, E.E., ed.: CRC Press, p. 379-390.

Sorrels, Thomas, 30 November 1998, Personal communication: Division of Groundwater, Tennessee Department of Environment and Conservation, Nashville, TN.

Stephenson, J.B., Zhou, W.F., Beck, B.F., and Green, T.S., 1997, Highway stormwater runoff in karst areas - Preliminary results of baseline monitoring and design of a treatment system for a sinkhole in Knoxville, Tennessee: in Proceedings of the Sixth Multidisciplinary Conference on Sinkholes and the Engineering and Environmental Impacts of Karst, Springfield, Missouri, April 6-9, p. 173-181.

Urbonas, B. and Stahre, P., 1993, Stormwater Best Management Practices and Detention for Water Quality, Drainage, and CSO Management: Prentice Hall, 449 p.

Wanielista, M.P., and Yousef, Y.A., 1993, Stormwater Management: John Wiley & Sons, 579 p.

Yu, S.L., Kasnick, M.A., and Byrne, M.R., 1993, A level spreader/vegetative buffer strip system for urban stormwater management: Integrated Stormwater Management, Field, R., O'Shea, M.L., and Chin, K.K., eds.: Lewis Pub., p. 93-103.

Yu, S.L., and Nawang, W.M., 1993, Best management practices for urban stormwater control: Integrated Stormwater Management, Field, R., O'Shea, M.L., and Chin, K.K., eds.: Lewis Pub., p. 191-205.

11 Special session on highways in karst – Design, construction and repair

Highway engineering geology of karst collapse features in the Sacramento Mountains, Otero and Lincoln Counties, New Mexico

ROBERT M. COLPITTS Sun Valley, Nev., USA
W. RICHARD HAHMAN Tucson, Ariz., USA

ABSTRACT

Karst features are present throughout the crest area of the Sacramento Mountains of Central New Mexico. Three major east-west highways cross this region: U.S. Highways 380, 70 and 82. These routes are impacted directly by the presence of karst along and beneath their rights-of-way. The karst occurs in the Permian San Andres and Yeso Formations and is at least Pliocene to Pleistocene in age. Other geologic evidence suggests that these features may be as old as Late Permian age (250 – 260 Ma). Karst impacts on road construction include loss of nearby suitable grade and roadbed material for highway construction, decreased useful life of highway following paving and roadbed integrity problems, which develop during and after construction. In addition, road cuts must be benched to stabilize slopes and prevent landslides. Many of these problems are easily solved with simple changes to pre-construction geologic investigations. Coring to supplement air rotary test drilling, and geologic mapping and analysis before construction or reconstruction of the road in question can aid in construction planning and development.

INTRODUCTION

Karst features are widespread throughout the Sacramento Mountains of south central New Mexico. Their importance to and impact on road construction causes continuing problems for construction contractors and the State Highway and Transportation Department, who must maintain these highways. The purpose of this paper is to document some of these features, their proximity to rapidly developing population and recreation centers and their impact on highway construction that will be required in the future. We will also provide a case history for the purpose of learning what to look for while working in this region.

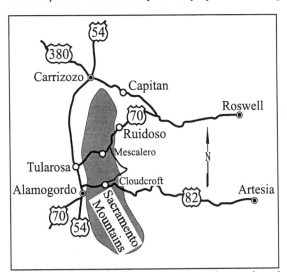

Figure 1: Location of Sacramento Mountains, roads and towns in south central New Mexico.

Location

The Sacramento Mountains are located in south central New Mexico in Lincoln and Otero Counties. Three major highways cross these mountains (Figure 1): U.S. 380 skirts the northern end, U.S. 70 crosses the center and U.S. 82 crosses the southern part; no major highways cross the southern-most part of the range. In addition, the area is crisscrossed by numerous State and secondary roads, which connect communities and provide access to more remote areas.

The Sacramento Mountains are surrounded by the communities of Alamogordo, Tularosa and Carrizozo in the west, Capitan on the north and Ruidoso on the northeast. In addition, the villages of Mescalero and Cloudcroft are situated in the mountains proper.

GEOLOGIC SETTING

Karst development in the Sacramento Mountains is related to dissolution of both carbonate and evaporitic rocks in the higher elevations of the range. Of primary interest is the karst developed in the carbonate sequences; evaporite karst is volumetrically minor in comparison.

More than 14 stratigraphic units crop out in the Sacramento Mountains proper (Figure 2). Of these, the Permian Yeso, San Andres and Grayburg Formations, Dakota Group, Mancos Shale, and Mesaverde Group are specifically involved in either karst development or as debris or large blocks in the sinkholes. These units are intruded by Tertiary-age mafic to intermediate dikes and sills. Kelley (1971), Moore et al. (1988), and Sloan and Garber (1971) describe specific details of the regional geology and hydrology.

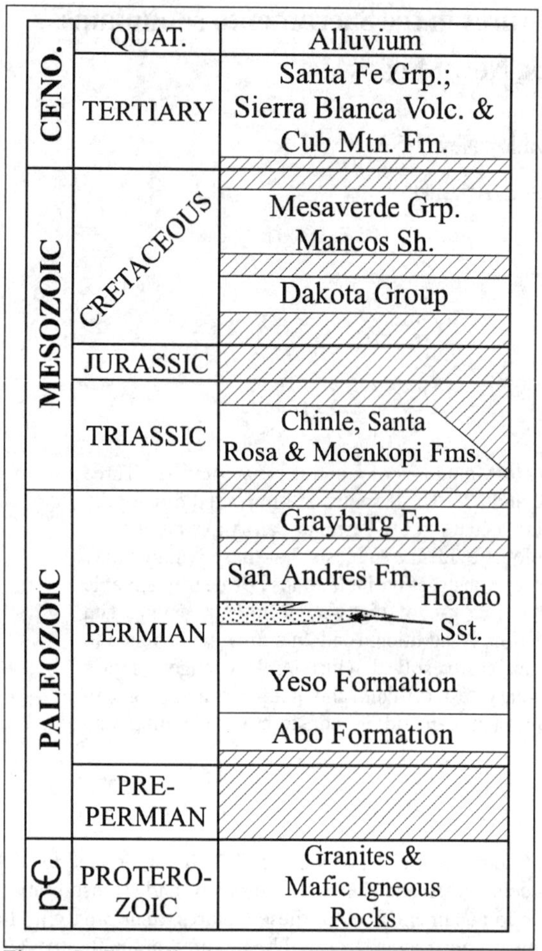

Yeso Formation

The Yeso Formation consists of pale yellowish gray to pale reddish brown very fine-grained sandstone and coarse siltstone interbedded with brownish gray dolomitic limestone and very light gray to white gypsum. The Yeso Formation is about 457 m (1500 ft) thick.

San Andres Formation

The San Andres Formation consists of 182 m (597 ft) of dark-gray limestone and dolostone. One or two quartzarenite beds 15 to 30 m (50 to 100 feet) above the base of the unit belong to the Hondo sandstone tongue of Lang (1937). The San Andres Formation is divided (in ascending order) into the Rio Bonito, Bonney Canyon and Fourmile Draw Members (Kelley, 1971). The Fourmile Draw Member is not present in the Sacramento Mountains south of Capitan and U.S. Highway 380 owing to erosion during Late Permian time. It reappears farther east near Roswell.

The Rio Bonito Member consists of gray, dark-gray and dark gray-brown limestone and dolostone (Kelley, 1971). Bed thickness ranges from 0.61 m to over 1.5 m (2 ft to over 4.9 ft). The Bonney Canyon Member consists of dark-gray and gray-brown dolostone and limestone. Bed thickness ranges from less than 2.5 cm to a maximum of 0.61 m. The top of the unit may be missing in this area due to erosion during Permian time.

The Hondo sandstone tongue is an important stratigraphic marker in the Sacramento Mountains and is used to determine stratigraphic position in the lower part of the San Andres Formation (Rio Bonito Member). It is a grayish orange to dark yellowish orange, fine-grained, very well sorted calcite-cemented quartzarenite and is 2.0 to 2.5 m (6.6 to 8.2 ft) thick.

Grayburg Formation

The Grayburg Formation consists of interbedded dark yellowish orange to grayish red siltstone and very fine-grained sandstone, and medium light gray limestone and dolostone. It disconformably overlies the Bonney Canyon Member of the San Andres Formation, is about 34.5 m (113.2 ft) thick and pinches out just north of Cloudcroft.

Figure 2: Generalized stratigraphic chart for Sacramento Mountains (after Austin et al., 1991). Stratigraphic units for the west flank of the range are not included.

Dakota Group

The Dakota Groups consists of a series of interbedded light gray, medium- to fine-grained, well-sorted quartzarenite, light gray siltstone and light gray silty shales. The Dakota Group is about 27 m (88.6 ft) thick.

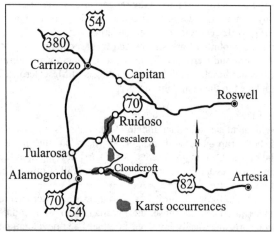

Figure 3: Karst occurrences in the Sacramento Mountains, New Mexico.

Mancos Shale and Mesaverde Group

The Mancos Shale consists of medium dark gray, dark gray and brownish gray shales and mudstones interbedded with medium dark gray fossiliferous limestone. It both intertongues with and is overlain by units of the Mesaverde Group. The Mesaverde Group consists of interbedded light to moderate brown, medium to very fine-grained, moderately to well-sorted sandstone, brownish gray mudstone and coal. The aggregate thickness of the Mancos Shale and Mesaverde Group is about 380 to 390 m (1247 to 1280 ft).

Structural Setting

Faults, joint systems and karst features deform the rocks in the region surrounding the quarry site and control the position of surface stream drainage; folds play no important role here. The few major faults in the region around the quarry display normal stratigraphic separation with generally down-to-the-west displacement. They strike N40°W, N-S and N30°—40°E. Major joint sets have the same general orientations as the fault. Additional joint trends include N50°E, N65°E,

N20°E and N50°W. Most drainages follow these trends, making them easy to map. Some faults and major joints are intruded by Tertiary age rhyolite, andesite/latite or diabase dikes (Kelley, 1971; Moore et al., 1988).

KARST

Karst features are scattered along the crest of the Sacramento Mountains. Although poorly documented because of heavy forest cover, highway construction and development has exposed many outcrops (Figure 3) so cavern structure and morphology and cave deposit geology are becoming better known. Karst development is usually associated with joint systems or faults. This correlation is well documented by Chronic (1987), Hallinger (1964), Harbour (1970), Melton (1934), and Mott (1959).

Karst impacts not only highway construction but also water resource development. Water supplies for the Village of Cloudcroft are obtained from a large collapse structure 4 kilometers east of the town. Several springs are used in addition to wells as the source. Field examination of the collapse shows that other springs were once present at other locations in the valley but are now no longer flowing because they have been occluded by travertine deposition.

Karst development is primarily confined to the Yeso and San Andres Formations and produces broken limestone and dolostone clasts that range from several tens of meters to greater than 2 millimeters grain size. Clasts are loose or cemented by banded travertine fault and joint traces, and at the intersections of these structures. Many of these major joint system intersections are marked by confluences of several streams and internally drained depressions. Some workers have mistaken extensive solution and collapse of carbonate lithologies accompanied by juxtaposition and/or removal of strata for evidence of faulting or reef development

Age to the karst is probably variable. Direct evidence suggests it is at least Pliocene to Pleistocene in age. However, the presence of an occasional block of

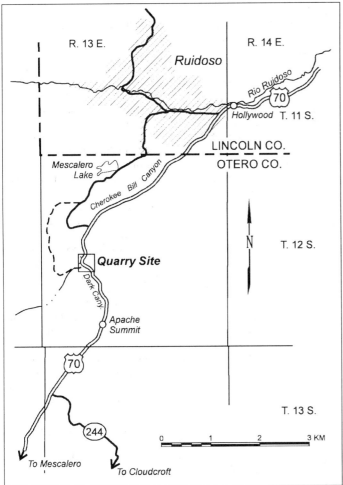

Figure 4: Location of road cut quarry site along U.S. 70, Otero County, New Mexico.

gypsum, possibly derived from the now eroded Fourmile Draw Member of the San Andres Formation suggests some of the higher levels of karst may be as old as Late Permian (Colpitts et al., 1991).

Karst impacts on roads and highways

Karst impacts road construction in a variety of ways. First, many sinkholes are filled with loose or soft, friable material. This material is not properly compacted and creates an unstable surface for road building purposes. The vibration for heavy trucks, for instance, may loosen or cause pronounced settling of the substrate beneath the roadbed. If the cave fill material is saturated, constant vibration from heavy traffic may cause liquefaction of the ground below the road and cause the road to settle, shift or heave. If the cavern below the road is not filled completely and cave debris forms a free bridge over the cavern, traffic vibration can loosen a single rock and bring about cavern collapse, resulting is loss of the part or all of a section of highway. All of these potential hazards should be considered and accounted for when building roads in this or any karst area.

CASE STUDY – U.S. 70 ROADCUT QUARRY

An excellent example of some of the problems involved in highway construction through a karst terrain occurs along U.S. 70 between Mescalero and Ruidoso, New Mexico. We have chosen this site as a case study because it has been thoroughly examined and because the problems are well documented.

Background

In 1981, the New Mexico State Highway and Transportation Department (NMSHTD) started a project to widen U.S. 70 across the Mescalero Apache Reservation. Reconstruction extended from just west of Mescalero to 8 km south west of Ruidoso (Figure 4.)

One part of the project involved straightening a blind curve in Sec. 20, T. 12 S., R. 13 E. (NMPM). The curve was located at the confluence of Dark and Cherokee Bill Canyons and ranges in elevation from 2195 and 2286 m (7200 to 7500 ft). Hereafter, we will refer to this location as "the quarry site".

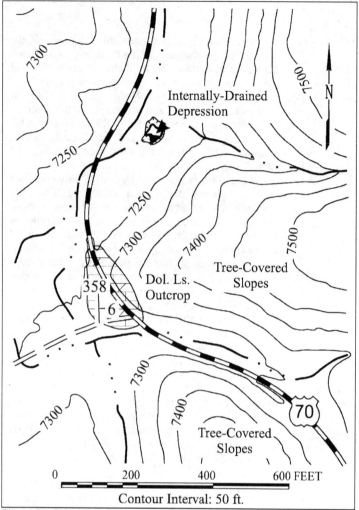

Figure 5: Original Geology of the quarry site prior to reconstruction of the highway. Refer to Figure 4 for location of site.

Construction history

The contractor building the road was required to use all of the material excavated from the quarry site for fill and roadbed construction material. The initial examination of the site indicated that there was a prominent cliff of dolomitic limestone of the San Andres Formation exposed in the old roadcut (Figure 5). On the assumption there was adequate material for construction, no other sources were located and identified. San Andres carbonates are tough, compact, non-porous rocks used extensively in this region as crushed stone for aggregate and base course in highway construction.

Several randomly sited test holes were drilled in the hill above the dolomitic limestone cliff with air rotary tools and a down-hole hammer bit. The holes ranged in depth from 12 to 78 meters. The cuttings from the air rotary holes were collected along with several short cores and narrative logs of the cuttings were prepared in the field by the site geologist; the cores were not described. The cuttings and were subsequently discarded and the cores were destroyed during testing. The narrative logs indicate that the drill encountered reddish-colored silt and dust; little carbonate rock was encountered. Rust Tractor also ran a demonstration seismic velocity survey line across the proposed quarry site and recorded velocities of 1676 m/s (5500 ft/s) suggesting unconsolidated material. No further velocity surveys were carried out. Also, after quarrying began, no geologic sketch maps were made of the quarry floor to show changes in rock types as the work progressed or for correlation with the drill-hole data.

Quarrying operations started and immediately experienced trouble. Large boulders of quarried rock were pushed toward a jaw crusher from the excavation site with bulldozers. The boulders gradually shrank in size during transport until little was left to crush. The quarry highwall also developed stability problems, which forced the contractor to develop a benched face for the road cut. Faced with the coming winter when construction activities would not be possible, the contractor stockpiled enough large, crushable material to resume construction when the weather warmed again in the spring.

Upon resumption of quarrying operations the following spring, the construction workers soon discovered that the stockpiled material had decrepitated and disintegrated so much that most of it was of little value for crushing or construction purposes. Once solid material had turned to dust and boulders had turned to dust and gravel. After resuming quarrying operations, the contractor discovered that material removed was neither the carbonate rock characteristic of the San Andres Formation as mapped by Kelley (1971) nor did it meet minimum specifications established for the project. In addition, much of the material was not suitable for fill or for mixing with oil and asphalt; the oil and asphalt would not coat the crushed rock, rather it soaked into each of the grains and failed to set up a good bond. The contractor experienced so much difficulty maintaining material quality and completing construction that he eventually had to seek other quarries with suitable materials to complete the project.

POST CONSTRUCTION SITE INVESTIGATION

In 1988, we were asked to examine the site and answer three questions. First in what stratigraphic unit was the quarry excavated? Second, were the problems encountered at the quarry related to faulting or something else? Third, why did quarry material degrade so much during excavation that it became unsuitable for construction purposes? The contractor was told that the material at the excavation was suitable for road construction yet the material turned out to be substandard.

Methods of investigation

First, Hahman prepared a detailed geologic map of the benched roadcut at 1 inch = 10 feet. Next, we both examined strata in the surrounding region to establish a stratigraphic framework for the quarry site. Finally, we preformed insoluble residue analyses of the material from the quarry site to see if the quarry materials matched the insoluble residues of any of the surrounding stratigraphic units. We constructed strip logs from the field notes made by the geologist during drilling. These logs demonstrated considerable clay present at the quarry site.

Results of investigation

Our work showed that rocks exposed in the quarry are not characteristic of any of the stratigraphic units in the surrounding region. The regional stratigraphic investigation showed us that the quarry site was located in the upper part of the Rio Bonito Member of San Andres Formation. Earlier suppositions that the quarry was developed in the underlying Yeso Formations proved to be unfounded.

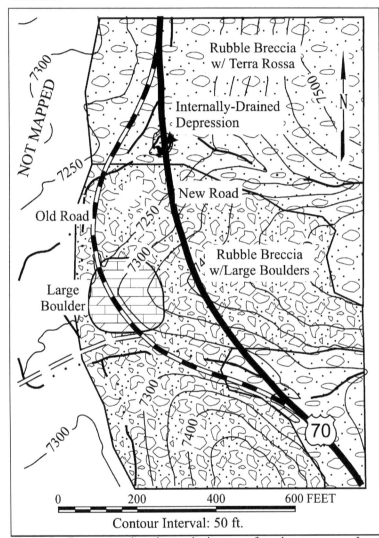

Figure 6: Post excavation site geologic map of road cut quarry. Large boulders are also present southwest of roadway but are not mapped in.

At the quarry site we determined that the quarry had been developed in a very large, debris-filled sinkhole. We identified five lithologic types unique to the quarry site. These include 1) rubble and collapse breccia, 2) travertine-cemented breccia, 3) bedded terra-rossa clay, 4) terra rossa clay with scattered breccia fragments, and 5) recent sand, gravel and clay deposited in underground stream channels. Embedded in the breccias and clays are large blocks of limestone and dolomitic limestone, some as large as a small frame house (Figure 6). The dolomitic limestone outcrop that the contractor counted on to tell him what he had available for construction material proved to be a very large, house-sized block surrounded by cave-fill debris.

Detailed mapping of the quarry site (Figure 6) revealed that the central part of the quarry is rubble and collapse breccia, overlying a well-weathered clay "trash" zone. Clay-matrix rubble breccias with isolated blocks flank this rubble and clay trash. Travertine-cemented breccias, bedded terra-rossa clay, silt, sand and gravel channel-fill deposits occur throughout the system. The top of the hill to the south has terra-rossa clay and rubble breccia. The drainages on both sides of the highway contain travertine flow, rubble, and collapse breccia. All of the material is loose and quite friable. The unconsolidated nature of the deposit is supported by Rust Tractor's seismic velocity survey; normal dolomite and/or limestone should have velocities of about 6100 m/s (20,000 ft/s).

We were not able to correlate the strip logs with each other despite their close proximity to one another. The field notes indicated considerable clay and fine-grained sand at the quarry site to total depth. There was very little solid rock in any of the cuttings

Quarry mapping and regional reconnaissance combined with the low velocities obtained by Rust Tractor indicates the quarry was developed in a large debris-filled sinkhole. This alone explains why the contractor had trouble maintaining quality and integrity of material obtained from the quarry; collapse breccia fragments are quite porous because of percolation of weakly acidic ground waters through the system. Increased dissolution produced friable stone. This type of rock tended to disintegrate during quarrying, crushing, screening and transport. Mechanical weathering processes (freeze-thaw cycle) during the winter when operations were shut down destroyed stockpiled crushed and sized rock. Degradation of the stocks was not apparent until work on the highway resumed in the spring. In addition to poor material integrity, aggregate that could be used would not take a coating of asphalt for paving; asphalt was absorbed rather than adsorbed. The contractor finally ran out of usable material at the quarry site and had to locate other sources for aggregate to complete the job.

CONCLUSIONS AND RECOMMENDATIONS

Our study revealed several things. First, the roadcut quarry was excavated in a debris-filled sinkhole in the Rio Bonito Member of the San Andres Formation. Second, the breccias exposed in the roadcut resulted from karst processes rather than faulting. And third, leaching of breccia fragments produced rock that was too porous and friable for road grading and paving.

In addition to the above-stated problems, others are also apparent. Compare Figures 5 and 6 and note the position of the new roadbed in relation to the internally drained depression north of the quarry site. Because the road was constructed across an active sinkhole, ground and roadbed instability has produced undulations in the pavement. Ultimately, the road may require reconstruction or reinforcement to restore stability.

We recommend that if more highway construction is carried out in this region, that at least two core holes be drilled to supplement data from the air rotary drill holes. The resulting RQD (Rock Quality Designation) data produced will help the contractor determine if the material is suitable for construction purposes. Also, the clearer picture of the types of rock he or she will encounter during quarrying should help in the planning of drilling and blasting operations and preparation for possible slope stability problems both during construction and after. Problems arose for three reasons. First, inappropriate drilling methods (air rotary versus coring) yielded poor to unusable data for the operation. Second, no one recognized the unconsolidated nature of the material in the proposed quarry from Rust Tractor's seismic velocity survey. Third, geologic maps of the pit site and vicinity prior to quarry operations were not prepared in advance of work. Had the maps been prepared, it is likely that the presence of karst at the quarry site would have been recognized and appropriate measures could have been taken to prepare for its occurrence.

REFERENCES

Austin, G.S., Barker, J.M., Bauer, P.W., Bowsher, A.L., Colpitts, R.M., Jr., Lucas, S.G., Cather, S.M., Hawley, J.W., Kottlowski, F.E., Kues, B.S., McLemore, V.T., Smith, C.T., and Toomey, D.F., 1991, Stratigraphic nomenclature chart: New Mexico Geological Society, Guidebook 42, end paper.

Bowsher, A.L., and Fly, S.H., III, 1986, Roadlog from Alamogordo to Tularosa, in, Ahlen, J.L. and Hanson, M.E. eds. Southwest Section of AAPG, Transactions and Guidebook of 1986 Convention, Ruidoso, New Mexico: New Mexico Bureau of Mines and Mineral Resources, p. 5 – 7.

Chronic, H., 1987, Roadlog for U.S. 380, Roswell to Carrizozo, in David Alt and Donald Hyndman, eds.: Roadside Geology of New Mexico, Mountain Press Publishing Co., p. 208-213.

Colpitts, R.M., Austin, G., Barker, J.M., Bauer, P., Hahman, W.R., Hawley, J.W., and Lozinski, R.P., 1991, First-day Roadlog from Inn of the Mountain Gods to Bent Dome, Tularosa, Alamogordo, Cloudcroft and return to the Inn of the Mountain Gods: New Mexico Geological Society, Guidebook 42, p.1- 25.

Hahman, W.R., and Colpitts, R.M., 1991, Engineering geology of karst features in a highway roadcut quarry, Otero County, New Mexico, in, Colpitts and others, First-day Roadlog from Inn of the Mountain Gods to Bent Dome, Tularosa, Alamogordo, Cloudcroft and return to the Inn of the Mountain Gods: New Mexico Geological Society, Guidebook 42, p. 24 - 25.

Hallinger, D.E., 1964, Caves of the Fort Stanton Area, New Mexico: New Mexico Geological Society Guidebook 15, p. 181-184.

Harbour, R.L., 1970, The Hondo sandstone member of the San Andres Limestone of south-central New Mexico: United States Geological Survey Professional Paper 700-C, p. C175 - C182.

Kelley, V. C., 1971a, Geology of the Pecos country, southeastern New Mexico: New Mexico Bureau of Mines and Mineral Resources, Memoir 24, 75 p.

Melton, F.A., 1934, Linear and dendritic sinkhole patterns in southeastern New Mexico: Science, new series, v. 80, n. 2066, p. 123 - 124.

Moore, S.L., Foord, E.E., and Meyer, G.A., 1988, Geologic and aeromagnetic map of a part of the Mescalero Apache Indian Reservation, Otero County, New Mexico: United States Geological Survey, Miscellaneous Investigations Series Map I-1775, Scale 1:50,000.

Mott, W.S., 1959, Geomorphology of the east side of the Sacramento Mountains, New Mexico, in Guidebook for joint field conference in the Sacramento Mountains of Otero County, New Mexico: Roswell Geological Society and Permian Basin Section of Society of Economic Paleontologists and Mineralogists, p. 223 - 233.

Sloan, C.E., and Garber, M.S., 1971, Ground-water hydrology of the Mescalero Apache Indian Reservation, south-central New Mexico: United States Geological Survey Hydrologic Investigations Atlas HA-349, 1 sheet.

Road and bridge construction across gypsum karst in England

ANTHONY H.COOPER   British Geological Survey, Keyworth, Nottingham, UK
JONATHAN M.SAUNDERS   Joynes Pike and Associates Limited, Doncaster, UK

ABSTRACT

Gypsum karst problems in the Permian and Triassic sequences of England have caused difficult conditions for bridge and road construction. In Northern England, the Ripon Bypass crosses Permian strata affected by active gypsum karst and severe subsidence problems. Here, the initial borehole site investigation for the road was supplemented by resistivity tomography studies. The roadway was reinforced with two layers of tensile membrane material within the earth embankment. This will prevent dangerous catastrophic collapse, but will allow sagging to show where problems exist. The River Ure Bridge was constructed across an area of subsidence pipes filled with alluvial deposits. It was built with extra strength, larger than normal foundations. If one pier fails the bridge is designed for adjacent arches to span the gap without collapse. The bridge piers are also fitted with electronic load monitoring to warn of failure. In the Midlands area of England, road construction over Triassic gypsum has required a phase of ground improvement on the Derby Southern Bypass. Here, the gypsum caps a hill where it was formerly mined; it dips through a karstic dissolution zone into an area of complete dissolution and collapse. The road and an associated flyover were built across these ground conditions. A major grouting program before the earthworks began treated the cavities in the mine workings and the cavernous margin of the gypsum mass. Within the karstic dissolution zone gypsum blocks and cavities along the route were identified by conductivity and resistivity geophysical surveys, excavated and backfilled. In the areas of complete dissolution and collapse the road foundation was strengthened with vibrated stone columns and a reinforced concrete road deck was used.

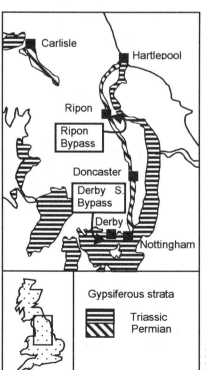

Figure 1. (Left) The distribution of the Permian and Triassic gypsiferous sequences in the United Kingdom and locations of the Ripon Bypass and the Derby Southern Bypass.

INTRODUCTION

Gypsum ($CaSO_4 \cdot 2H_2O$) is readily soluble in water and develops karst features much more quickly than does limestone. Road construction over gypsiferous terranes has to deal with karst problems including progressive dissolution, sinkhole formation and poor ground conditions caused by collapsed strata. Furthermore, the complete dissolution of gypsiferous beds can leave a residue of weak and brecciated strata which also produces difficult conditions for construction. Where gypsum is present in the bedrock, either as massive beds or as veins, it can be associated with sulphate-rich groundwater that can be harmful to concrete, and precautions to prevent damage should be considered (Forster et al., 1995).

The Permian and Triassic strata in England both contain thick gypsum sequences, but individually their associated rocks and respective engineering problems are different. The most problematical gypsum karst conditions are found in the Permian rocks, notably in the Ripon area of North Yorkshire. Gypsum karst and difficult ground conditions have been encountered in the Triassic rocks of the English Midlands in the area to the south-west of Derby. These are described below with details of the difficulties encountered during road construction and their engineering solutions.

RIPON BYPASS
Geology and gypsum karst problems

The Ripon Bypass was constructed to the east of Ripon. It crosses the Permian sequence which includes approximately 35m of gypsum in the Edlington Formation (formerly the Middle Marl) and 10m of gypsum in the higher Roxby Formation (formerly the Upper Marl). These two gypsum sequences rest on two limestone aquifers, the Cadeby Formation (formerly the Lower Magnesian Limestone) and the Brotherton Formation (formerly the Upper Magnesian Limestone) respectively. The limestone dip slopes act as catchment areas and the underground water flows down-dip into the gypsiferous sequences, before escaping into a major buried valley along the line of the Rive Ure (Cooper, 1986, 1995, 1998). Complex cave systems have developed in the gypsum, and artesian, sulphate-rich springs are locally present. Because of the thickness of gypsum the caves are large and surface collapses up to 30m across and 20m deep have been recorded. The subsidence is not random, but occurs in a reticulate pattern related to the jointing in the underlying strata (Cooper, 1986). However, it is impossible to predict where the next subsidence event will occur. Around Ripon, a significant subsidence occurs approximately every year (Cooper, 1995). The dates of the subsidence events show that some areas are more active than others, especially areas bounding the Ure valley where cave water escapes into the buried valley gravels. The new Ripon Bypass crosses the subsidence belt and the new Ure Bridge is situated in a very active area. The new road and bridge pass close to several subsidence hollows.

Engineering solutions

In addition to the standard site investigation that was undertaken for the Ripon Bypass a desk study of the subsidence features was carried out. This study included an assessment of the likely magnitude and frequency of the subsidence events along the route of the road. Resistivity tomography was also undertaken on the bridge site, but the results were inconclusive and no additional subsidence features were pinpointed. Even if a costly investigation of closely spaced boreholes had been undertaken, it is unlikely that any cavities or breccia pipes which were identified could be stabilised. Grouting was impractical due to the large size of the cavities and the fact that filling them was likely to cause accelerated dissolution in the adjacent ground. Because the line of the bridge could not be changed a decision was made to design and construct the bridge with inbuilt protection against future ground subsidence.

A conventional bridge in a non-hazardous area would comprise piers with individual sections of road deck (Figure 2A). The new Ripon Bridge has a strengthened heavy-duty steel girder construction designed with sacrificial supporting piers. The structure will withstand the loss of any one support pier without collapsing (Figure 2B), (Thomson et al., 1996). The individual piers have larger than normal foundation pads, in order to span a small subsidence event. On the alluvial deposits of the river flood plain the ground was improved by the use of stone columns. Piling was also undertaken for one of the piers. In addition to the bridge strengthening, the piers are equipped with electronic monitoring devices to detect separation from the deck. The philosophy of the approach has been to maintain public safety, while not over-engineering the structure to an excessive degree with the resultant heavy financial costs. In addition to these measures, an added degree of security could have been obtained by extending the foundations of each pier laterally to an extent that could span the normal-sized collapses (Figure 2C).

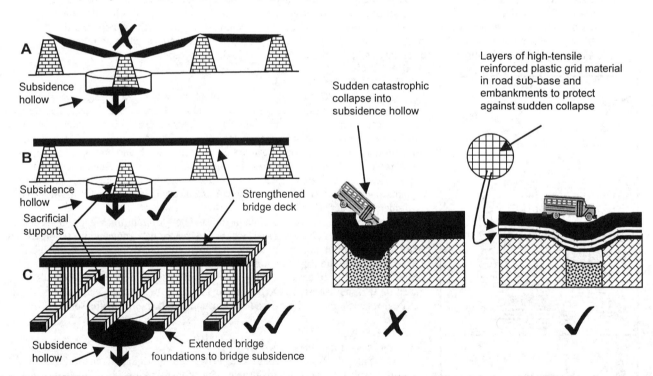

Figure 2. Sacrificial supports and extended foundations in bridge construction to protect against sudden subsidence failure.

Figure 3. The use of tension membrane material or "Geogrid" in road construction to protect against subsidence failure and sudden collapse.

Regarding the construction of the road, the priority was again the maintenance of public safety without the unacceptable cost of investigating possible subsidence features that would be difficult or impossible to remediate. The course of the road could not be changed as it follows a disused railway line. The solution was to use tension membranes with a design brief that the integrity of the road should be supported for at least a 24-hour period road (Kempton, et al., 1996). This would allow time for any problem to be identified and isolated from the public. The tension membrane material, or geogrid, used was ParaLink 700S and 325S incorporated in the road embankment (Kempton, et al., 1996: Thomson et al., 1996). The embankment/membrane combination was modelled using a FLAC (Fast Lagrangian Analysis of Continua) computer programme to simulate the occurrence of a typically sized collapse void beneath the road (Kempton, et al., 1996). The incorporation of these membranes in the road will allow the road to sag into any subsidence void beneath it, thus showing there is a problem, but catastrophic collapse should not occur (Figure 3). The use of geotextile materials in this way is a proven method of protecting public safety in car parks and public spaces.

DERBY SOUTHERN BYPASS
Geology and gypsum karst problems

The Derby Southern Bypass is a 24km long strategic link road between the M1, near Derby, to the M6 at Stafford. It includes 72 major engineered structures, and the design and supervision of the construction works was carried out by Scott Wilson on behalf of the Highways Agency. The road crosses the Triassic Mercia Mudstone Group of rocks (formerly called the Keuper Marl), that are mainly sandstones and calcareous or dolomitic mudstones, with two main sequences of gypsum in the upper part of the group. These are the Tutbury and Newark gypsum beds of the Cropwell Bishop Formation (Elliot 1961; Charsley et al. 1990).

The Mercia Mudstone Group has a reputation for having complex weathering and requiring careful site investigation (Chandler, 1969; Bacciarelli, 1993). In some places, the rock has a low bearing capacity and a complex weathering profile with less weathered rock commonly overlying severely weathered material. In some parts of the sequence this can be attributed to differently cemented strata. Elsewhere, this weathering may be attributed to the dissolution of gypsum in the near-surface strata, possibly to a depth of 30m (Elliot, 1961; Reeves, et al., 1993). In these areas, caution has to be exercised as engineering works, such as piling, may open channels for water ingress into the gypsiferous strata. If this happens, and there is a throughput of water, gypsum dissolution may occur similar to that recorded at Ratcliffe-on-Soar Power Station, near Nottingham (Seedhouse and Sanders, 1993).

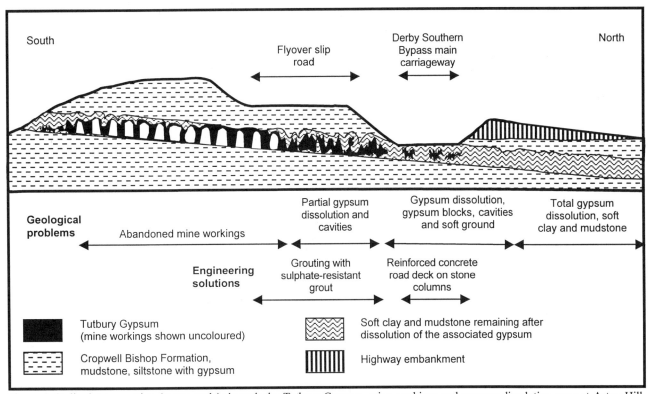

Figure 4. Stylised cross-section (not to scale) through the Tutbury Gypsum, mine workings and gypsum dissolution zone at Aston Hill south-west of Derby on the Derby Southern Bypass.

The Tutbury gypsum is best developed in the area south and southwest of Derby where it reaches about 10m in thickness. The overlying Newark gypsum here comprises mainly thin beds and veins. The thick Tutbury gypsum bed has a long history of mining and quarrying dating back to the Middle Ages. Initially it was worked for monumental alabaster and in 1367 Edward III had blocks of

Chellaston alabaster taken to the Chapel of the Garter at Windsor Castle. From the 19th Century onwards it was increasingly exploited for plaster manufacture (Smith, 1918; Sherlock & Hollingworth, 1938). There are still several working gypsum mines extracting the Tutbury seam in the region, but not locally. The Tutbury gypsum is not well documented, but the descriptions of Smith (1918) record much of it as massive and nodular. Features that we now consider to be of karst origin are also described including dissolution features and areas of collapse. In the gypsum mines at Fauld, about 20km to the south-west of the construction site, Wynne (1906) described caves and collapse features similar to those seen in the Permian gypsum of the Vale of Eden (Ryder and Cooper, 1993).

The construction of the Derby Southern Bypass, near Aston-on-Trent, took the main east-west carriageway along the base of Aston Hill with an interchange linking it to the A6 Derby Spur road. The associated slip roads for the spur passed through a deep cutting in the northern flank of the hill. Available published geological maps (British Geological Survey, 1976: now resurveyed) and unpublished documents were consulted during the initial desk study and subsequent design stages. They indicated that the workings of the Aston Holme Gypsum Mine beneath the hill would underlie the slip roads. They also suggested that the gypsum dipped about 3 degrees to the north and would crop out on the side of the hill in the middle of the cutting. The detailed ground investigations showed the mainline of the road to be underlain by up to 8m of soft to very soft clays and silts. These belonged to the Mercia Mudstone Group, had low bearing capacities and a long-term settlement potential. The ground investigation showed the gypsum workings beneath Aston Hill to be above the groundwater table. For safety reasons, it was decided that the mine workings should be treated by a phase of drilling and grouting prior to the excavation of the cutting. This was because of the potential for significant voids to be encountered at various levels during the earthworks, a factor that would pose a risk to personnel and heavy construction traffic.

During the drilling and grouting programme, it became apparent that the Tutbury seam was in fact present at a greater depth than indicated by the published geological records. However, the final rockhead cover was insufficient to negate any long-term risk of void migration. In addition, during the initial earthworks for the main carriageways three minor collapses were encountered. These three areas were investigated by a grid of 12 drill-holes to depths of between 15m and 30m. No gypsum was encountered and grout takes were minimal. It was concluded that the gypsum seam was present at a depth greater than originally envisaged and that it passed beneath the mainline of the road (Figure 4). It was concluded that the gypsum mine workings were not restricted by the seam outcrop, on the side of Aston Hill, but by a zone of dissolution at the edge of the mine. Down dip from the mine, the gypsum passes into a zone of partial dissolution and collapsed strata. This comprises a mixture of large, partially dissolved gypsum blocks up to 4m in diameter, collapsed areas and some natural cavities. Further down dip, beneath the River Trent floodplain the gypsum has been totally dissolved resulting in a sequence of insoluble residue and collapsed confining strata. The sequence formerly associated with the gypsum is present as a mixture of soft to very soft clay with brecciated mudstone clasts. Deep boreholes showed that farther down dip, where there was little groundwater circulation, the gypsum had not been dissolved and was again present. Some of the subsidence hollows encountered during the excavations contained glacial and periglacial deposits. It is probable that the karst mechanisms that dissolved the gypsum may be related to past fluctuations in the groundwater levels. These fluctuations might relate to decreasing water table levels associated with post-glacial river incision and a change in base level. This is suggested by the presence of extensive stepped terraces along the valley here.

Engineering solutions

The area of mine workings and partial dissolution beneath the slip road area was treated by a phase of drilling and grouting. At the start, a perimeter grout curtain was formed by grouting holes initially at 3m centres, reduced to 1.5m centres in areas of significant grout take. The enclosed area was then grouted on an initial 6m grid with secondary grouting at 3m centres in areas of significant grout take. Tertiary grouting was undertaken at 1.5m centres for areas of significant grout take during the secondary phase. The boreholes were drilled by open-hole percussive techniques and air flush. Grouting was achieved by the installation into the boreholes of a tube, nominally 50mm in diameter, with an inflatable packer. Grout was injected at a maximum pressure of 965kPa (140 psi). Approximately 2900 holes were drilled to a maximum depth of 32m and an average depth of 20m. The maximum void height encountered was 7.5m with several migrating voids encountered at depths as shallow as 2m below original ground level.

The mainline carriageway and approach roads to the A6 Derby Spur Bridge were in an area of complete gypsum dissolution and soft strata. Here, the foundation strata were improved by the installation of 450mm diameter stone columns on a grid to a typical depth of 10m. In addition, sections of the mainline carriageway were constructed using a continuously reinforced concrete carriageway to span any minor localised shallow subsidence.

Because the Tutbury gypsum seam was found beneath all the slip roads and the mainline of the road, it was deemed prudent to carry out a two-phase geophysical survey to look for possible cavities and soft ground. The geophysics comprised ground conductivity and ground resisitivity techniques undertaken over approximately 900m of highway route. The survey identified two areas where anomalies were encountered at depths of between 6m to 8m below finished road level. These were correlated to possible voids and water-filled ground, or zones of very soft clay. The anomalies affected a total length of 75m of highway and these areas were treated by the installation of stone columns.

DISCUSSION

The construction of bridges and viaducts over actively dissolving gypsum karst is difficult. Unlike the relatively slow formation of limestone karst, gypsum karst can develop on a human/engineering time-scale. Furthermore, the site investigation of cavernous ground is difficult, especially where thick Quaternary deposits overlie gypsum that is dissolving to depths of 80m or more. Many of the engineering solutions used in limestone karst areas are not appropriate. Grouting is commonly used to fill cavities and stabilise the ground in limestone karst, but in gypsum karst it can only be used in areas like the Derby Southern Bypass where there is very little groundwater movement. Where there is high groundwater movement, such as at Ripon, the high dissolution rate of gypsum, the aggressive nature of sulphate-rich water, the cavity sizes and the difficulties of actually sealing them make grouting impractical. Incomplete grouting may actually aggravate the situation by causing enhanced dissolution in the adjacent ground around the treated area. Because grouting is largely ineffective it can be argued that in many areas exhaustive site investigation to find the cavities is not cost-effective, unless the road can be re-routed to avoid them. The alternative approach is to look for the obvious problems, such as known sinkholes, avoid them and build strength and safety measures into the structures; this is the approach that was taken by the road engineers at Ripon. A similar approach has been taken by French engineers for road construction around Paris. Some of the gypsum karst has been grouted (Cadilhac et al., 1997; Poupelloz & Toulemont, 1981), but additional safety measures have also been used. Viaducts around Paris have been constructed with foundations that will span the likely size of collapse (Arnould, 1970). In addition, benchmarks to monitor movement have been installed and these are surveyed regularly. A network of inclinometers and extensometers are automatically monitored and linked to a warning system that is activated if the settlement exceeds 60mm. During the viaduct construction all known cavities were filled, but the structures were also designed with grout-holes for future possible use to enable new cavities to be grouted (Arnould, 1970). The similar use of extended foundation designs and strengthened structures has similarly been used in Germany and Russia (Reuter and Tolmačëv, 1990; Sorochan, et al., 1985).

The complete dissolution of gypsum also leads to problems where construction has to deal with the residual deposits caused by gypsum removal and collapse. In the UK, where this process has occurred in the Mercia Mudstone Group it has commonly produced ground with a low bearing capacity. The distinctive weathered character of the Mercia Mudstone has been known for some time (Chandler, 1969; Bacciarelli, 1993). However, very little attention has been paid to the presence of gypsum in the unweathered rock, its removal by dissolution and the resultant deposits. The excavations on the Derby Southern Bypass show the effects of dissolution removing thick gypsum deposits and the resultant soft, collapsed ground. Gypsum is not uniformly distributed throughout the Mercia Mudstone Group and it is likely that there is some correlation between the areas with the lowest bearing capacity, the stratigraphy and the former presence of gypsum.

ACKNOWLEDGEMENTS

The Highways Agency and Scott Wilson are thanked for permission to publish construction details of the Derby Southern Bypass. Tim Charsley, Alan Forster and Dave Entwhistle are thanked for reviewing the manuscript. AHC publishes with permission of the Director, British Geological Survey (N.E.R.C.).

REFERENCES

Arnould, M. 1970. Problems associated with underground cavities in the Paris region. 1-25 in Geological and geographical problems of areas of high population density. Proceedings of the symposium, Association of Engineering Geologist, Sacramento, California.

Bacciarelli, R. 1993. A revised weathering classification for Mercia Mudstone (Keuper Marl) 169-174 in Cripps, J.C., Coulthard, J.C., Culshaw, M.G., Forster, A., Hencher, S.R. and Moon, C. (Editors). The Engineering Geology of Weak Rock. Proceedings of the 26th annual conference of the Engineering Group of the Geological Society, Leeds, September, 1990. A.A.Balkema, Rotterdam.

British Geological Survey, 1976. Loughborough, Sheet 141, Solid and Drift. 1:50 000 (Southampton: Ordnance Survey for the British Geological Survey).

Cadilhac, M., Poupelloz, B. & Toulemont, M. 1977. Fondations d'ouvrage d'art en site karstique. Le viaduc de l'autoroute A15. Bulletin de Liaison Labo. P. et Ch. 87. janv.-févr. Réf. 1915.

Chandler, R.J. 1969. The effects of weathering on the shear strength properties of Keuper Marl. Geotechnique, Vol. 19, 321-334.

Charsley, T.J, Rathbone, P.A. and Lowe, D.J. 1990. Nottingham: A geological background for planning and development. British Geological Survey Technical Report WA/90/1.

Cooper, A H. 1986. Foundered strata and subsidence resulting from the dissolution of Permian gypsum in the Ripon and Bedale areas, North Yorkshire. 127-139 in Harwood, G M and Smith, D B (eds). The English Zechstein and related topics. Geological Society of London, Special Publication. No. 22.

Cooper, A.H. 1995. Subsidence hazards due to the dissolution of Permian gypsum in England: investigation and remediation. 23-29 in Beck, F.B. (ed.) Karst Geohazards: engineering and environmental problems in karst terrane. Proceedings of the fifth multidisciplinary conference on sinkholes and the engineering and environmental impacts of karst Gatlinburg/Tennessee/2-5 April 1995. 581pp. A.A.Balkema, Rotterdam.

Cooper, A.H. 1996. Gypsum: geology, quarrying, mining and geological hazards in the Chellaston area of South Derbyshire. British Geological Survey Technical Report WA/96/30

Cooper, A.H. 1998. Subsidence hazards caused by the dissolution of Permian gypsum in England: geology, investigation and remediation. In Maund, J.G. & Eddleston, M (eds.) Geohazards in Engineering Geology. Geological Society, London, Engineering Special Publications, 15, 265-275.

Elliot, R.E. 1961. The stratigraphy of the Keuper Series in Southern Nottinghamshire. Proceedings of the Yorkshire Geological Society. Vol. 33. 197-234.

Forster, A., Culshaw, M.G. and Bell, F.G. 1995. Regional distribution of sulphate rocks and soils of Britain. 95-104 in Eddleston, M., Walthall, S., Cripps, J.C. and Culshaw, M.G. (eds) Engineering Geology of Construction. Geological Society Engineering Geology Special Publication No10.

Kempton, G.T., Lawson, C.R. Jones. C.J.F.P. & Demerdash, M. 1996. The use of geosynthetics to prevent the structural collapse of fills over areas prone to subsidence. In De Groot, Den Hoedt and Termaat (eds) Geosynthetics: Applications, Design and Construction. Balkema, Rotterdam.

Poupelloz, B. & ToulemonT, M. 1981. Stabilisation des terrains karstiques par injection. Le cas du Lutetian gypseux de la region de Paris. Bulletin of the International Association of Engineering Geology. No. 24. 111-123.

Reeves, G.M., Hilary, J. and Screaton, D. 1993. Site investigation techniques for piled foundations in Mercia Mudstones, Teesside, Cleveland County. 457-463 in Cripps, J.C., Coulthard, J.C., Culshaw, M.G., Forster, A., Hencher, S.R. and Moon, C. (eds). The Engineering Geology of Weak Rock. Proceedings of the 26th annual conference of the Engineering Group of the Geological Society, Leeds, September, 1990. A.A..Balkema, Rotterdam.

Reuter, F. and Tolmačëv, V.V. 1990. Bauen und Bergbau in Senkungs und Erdfallgebieten, Eine Ingenieurgeologie des Karstes. Schriftenreihe für Geologische Wissenschaften, Vol 28. Academie-Verlag, Berlin.

Ryder, P F, & Cooper, A H. 1993. A cave system in Permian gypsum at Houtsay Quarry, Newbiggin, Cumbria, England. Cave Science, Vol. 20, No. 1, 23-28.

Seedhouse, R.L. and Sanders, R.L. 1993. Investigations for cooling tower foundations in Mercia Mudstone at Ratcliffe-on-Soar, Nottinghamshire. 465-471 in Cripps, J.C., Coulthard, J.C., Culshaw, M.G., Forster, A., Hencher, S.R. and Moon, C. (eds). The Engineering Geology of Weak Rock. Proceedings of the 26th annual conference of the Engineering Group of the Geological Society, Leeds, September, 1990. A.A..Balkema, Rotterdam.

Sherlock, R.L. and Hollingworth, S.E. 1938. Gypsum and anhydrite. Special reports on the mineral resources of Great Britain. Vol. 3., 3rd edition. Memoirs of the Geological Survey. 98pp.

Smith, B. 1918. The Chellaston gypsum breccia and its relation to the gypsum-anhydrite deposits of Britain. Quarterly Journal of the Geological Society, London. Vol. 77. 174-203.

Sorochan, E.A., Troitzky, G.M., Tolmachyov, V.V., Khomenko, V.P., Klepikov,S.N., Metelyuk, N.S. and Grigoruk, P.D. 1985. Antikarst protection for buildings and structures. 2457-2460 in Proceedings of the eleventh international conference on soil mechanics and foundation engineering. San Francisco, 12th-16th August 1995. A.A.Balkema, Rotterdam.

Thomson, A., Hine, P.D., Greig, J.R. and Peach, D.W. 1996. Assessment of subsidence arising from gypsum dissolution: Technical Report for the Department of the Environment. 228pp. Symonds Group Ltd, East Grinstead.

Wynne, T.T. 1906. Gypsum, and its occurrence in the Dove Valley. Transactions of the Institute of Mining Engineers, Vol. 32, 171-192.

Remediation of sinkholes along Virginia's highways

DAVID A. HUBBARD JR Virginia Division of Mineral Resources, Charlottesville, Va., USA

ABSTRACT

Highway construction in karst alters local hydrology and may trigger sinkhole development. Highway surfaces and associated landscaping increase runoff, which is artificially channeled along new flow paths. The altered distribution and increase in volume of runoff is commonly allowed to percolate through soils, especially along unlined ditches, altering the local hydrology. Percolating waters are eventually channeled along the soil/bedrock interface to solutionally enlarged fractures in the folded and faulted carbonate rocks of the Commonwealth of Virginia. Soil piping and winnowing may enlarge voids sufficiently to allow progressive soil arch failures to propagate to the surface and form sinkholes. A wide range of remedial solutions has been employed to mitigate sinkholes.

Many sinkhole fixes are merely creative fills that ignore the natural processes active in karstic sinkhole development and the human induced hydrological alterations that trigger these features. Of critical concern is the movement of water at the soil/bedrock interface and through solutionally enlarged fractures and partings into the cave systems that under-drain the terrain. Fixes utilizing materials that can erode into under-draining conduits may undergo further subsidence. Grout plugs that are not on bedrock may undergo settling as soil underlying plugs is eroded into under-draining conduits. Effective grout sealing of solutionally enlarged fractures restricts groundwater movement and may result in ponding and saturation of adjacent soils as well as the formation of new sinkholes proximal to the site as groundwater moving along the soil/bedrock interface flows into adjacent enlarged fractures. In rare circumstances, grout sealing may extend into the under-draining cave system resulting in back flooding of up-system sinkholes drained by the now occluded conduit.

The use of sinkholes as discharge points for highway runoff exacerbates the groundwater contamination hazard inherent in karst and maximizes the pulse effect, possibly to the point of exceeding the capacity of the under-draining conduit and the back-flooding of other system sinkholes normally drained by the same conduit. Such a fix can trigger additional subsidence and new sinkholes.

The most process oriented fix seems to be a composite or graded filter constructed once the sinkhole has been excavated to the bedrock interface. The filter is constructed using rock clasts large enough to bridge the solutional conduit, successive lifts employ smaller clast sizes that are coarser than the interstitial spaces of each lower lift. Such a filter mitigates subsidence and allows groundwater recharge from water moving along the soil/bedrock interface.

INTRODUCTION

In Virginia, the most extensive karstlands are in the Valley and Ridge Physiographic Province and are developed on folded and faulted Paleozoic limestone and dolostone (carbonate rocks) ranging in age from Cambrian (570 million years) to Mississippian (310 million years). The fracturing in these deformed rocks commonly is solutionally enlarged in carbonate rock, but also provides pathways for groundwater through clastic beds typically serving as aquitards or in partitioning aquifers in similar, less deformed, horizontally-bedded rocks. The natural plumbing in deformed rock karst may involve leaky conduits and deep phreatic loops between a specific recharge point and a traced resurgence. A resurgence may not be the sole point of discharge under all flow regimes.

Sinkholes are one of the more recognizable features in karst, a terrain formed by the dissolution of soluble bedrock and characterized by subsurface drainage. The three major hazards associated with covered karst, where much of the bedrock is mantled by soil, are subsidence, sinkhole flooding, and groundwater contamination. Traditionally the most feared of these hazards, subsidence is typically linked to changes in the local hydrology. Such changes may occur naturally from weather variation, but human induced hydrological changes triggering sinkhole formation range from groundwater fluctuations induced by water-well pumping to landscape altered surficial drainage patterns. Although sinkhole flooding does occur in Virginia, it is not a problem of the magnitude as seen in Bowling Green, Kentucky (Crawford, 1984), Springfield, Missouri (Aley, 1981), or some other U.S. karstlands (Currens and

Graham, 1993; Ford and others, 1997). The use of sinkholes as outfalls for altered terrain generated runoff enhances flooding and groundwater quality problems. Groundwater contamination is emerging as the hazard of greatest concern to karstland residents. The practice of disposing trash and other waste in sinkholes and caves has diminished as residents have learned that sinkholes are recharge points and how highly susceptible their groundwater is to contamination.

HIGHWAYS IN KARST

The most obvious karst hazard to highways is subsidence. In the historic past, roadways traversed around most sinkholes and other subsidence features. Nowadays, roads and highways commonly are built across some subsidence features. Where existing sinkholes are a concern, construction engineered with a focus on sinkhole processes may mitigate subsidence; however, construction that treats sinkholes only as geomorphic features may prove costly in terms of safety and maintenance. From a cursory view, highway construction in karst is deceptively simple. Engineer fills in those subsidence features within the highway footprint, direct highway runoff to the adjacent natural drains where necessary, and employ mandatory siltation controls to prevent the sealing of those natural drains or the degradation of the groundwater during the construction phase. Complications developing after road construction include: accidents to vehicles transporting potential groundwater contaminants and increased storm runoff associated with encroaching landscape modifications proximal to the transportation corridor.

Highway construction and sinkhole development

In Virginia, many of the new sinkhole collapses along highway corridors do not appear to be related to the sinkholes present prior to road construction. During highway construction, cuts and fills alter the local surficial drainage and infiltration patterns. Infiltrating water is channeled through the epikarst, along the soil/bedrock interface to solutionally modified bedrock fractures, and into under-draining conduits. Soil piping at the soil/bedrock interface creates voids, which may continue to enlarge or stope upward depending on the soil properties, size and configuration of modified fractures, the amount of infiltration, and the elevation and variability of the local "water table." The soil arch roofs of these voids may progressive fail to the surface creating a sinkhole during construction or later.

Sinkholes induced by highway construction have been extensively reported at past Sinkhole Conferences, particularly at the First Multidisciplinary Conference on Sinkholes. Newton (1984) includes such sinkholes in his category of "Construction" induced, although he favors the concept that "Grading results, in cuts, in the thinning of unconsolidated deposits. Emplacement of weight on thinned roofs of existing cavities in residual clay or on those of shallow bedrock cavities can cause their failure." Myers and Perlow (1984) note that "Excavation ... exposes an underlying material that is a more granular and consequently more permeable and with less shear strength than the overlying materials. Surface waters, which can be introduced naturally, or during construction processes, can rapidly erode these silty, granular materials and, in areas where the underlying bedrock is highly pinnacled and riddled with solution enhanced openings, sinkholes will readily develop." They note the occurrence of 184 sinkholes that developed along a 7 km section of a Pennsylvania highway. Moore (1984) documents the formation of collapse features along unpaved ditches in excavated cuts. He reports "The prevention of karst related subsidence and collapse of highways is centered around controlling the drainage...." Of even more importance is his discussion of subsidence, flooding, and runoff contaminants and their interrelationships in highway construction and designed function.

The variability in the concepts of karst processes and how they relate to sinkhole formation are apparent in the above four papers. Of these, the Moore (1984) paper demonstrates a wider grasp of the importance of water in karst and the processes by which some cover collapse sinkholes form.

Subsidence remediation

Sinkholes forming during the construction phase of many state and federal highway projects are likely to be remediated by site specific engineered solutions. Sinkholes forming post construction or along privately constructed roads may undergo remediation without the benefit of process sensitive engineering support and adequate budgets (Figure 1). Unfortunately, different VDOT districts treat karst subsidence remediation with highly variable methodologies as part of their independent highway maintenance programs. T The most apparent process oriented remediation of cover collapse sinkholes entails the excavation of the feature to the soil/bedrock interface and construction of an inverse graded filter. The base course of the filter is selected of a size to bridge the solutional features (Figure 2). Successive beds are sized of finer clasts that are coarse enough not to pass through the interstitial spaces of the previous bed. Topped off with the appropriate geotextile filter fabric and final surface to exclude the infiltration of highway runoff, a properly constructed and

Figure 1: Asphalt patching of a recently patched sinkhole.

compacted filter mitigates the subsidence problem and allows epikarstic waters access to underlying conduits and continued aquifer recharge.

Aggregate fills placed in unexcavated or partially excavated sinkholes may undergo subsequent subsidence as subsurface water movement erodes sediments between the fill and bedrock. If these aggregates are fine enough to pass through the under-draining modified fractures subsidence may be extensive. These fill solutions ignore the sinkhole forming processes most likely active along highways. An even riskier type of fill involves the use of grout to bind aggregate fills not constructed on bedrock (Figure 3). The winnowing of underlying sediment into the epikarst may create a void large enough to allow catastrophic settlement of the fill block.

The use of grout in attempts to seal solutionally enlarged fractures in mitigating sinkhole subsidence may be cause for concern in most highway corridors. With the exception of base level streams and rivers, most water in karst occurs in the subsurface. After water percolates through the soil cover it is channeled along the upper portion of the epikarstic zone to modified fractures, which eventually drain into conduits. The effective sealing of epikarstic fractures may result in pooling and saturation of soil peripheral to the fix and increased subsurface drainage to adjacent fractures resulting in void formation and development of one or more new sinkholes. Mellet and Maccarillo (1995) note that sinkholes that developed along a western New Jersey Interstate Highway were "associated with failures or ruptures at joints in the subsurface drainpipes." Remediation involving pressure grouting was successful, in that these sites remained stable; however, "new sinks have continued to develop along the right-of-way, so that remediation has become an ongoing process." They conclude that "grouting some areas will only divert underground water flow elsewhere, and new sinks will appear at other locations along the highway." A further hazard of using grout to seal sinkhole drains is the potential to actually restrict or occlude the conduit that under-drains the sinkhole. The restriction or occlusion of a system conduit could lead to back-flooding of other up-system sinkholes, which previously might only flood during exceptionally wet weather.

Figure 2: Inverse graded filter under construction in a sinkhole that was excavated to bedrock. Sinkhole developed along an unlined ditch on a Virginia Interstate Highway. Engineered fill was completed with a concrete lined open culvert.

Hydrologic remediation along highways

Moore (1984) recognized that drainage control was a key to preventing subsidence and flooding. His recommendations included "The use of paved ditching for all ditches and channels within a karst area...to prevent or greatly reduce the build-up of groundwater seepage pressures and subsurface erosion which can lead to collapse problems" and "The use of asphalt curbs ...for better control of runoff through karst...." Despite his recognition that highways can alter "groundwater conditions by induced siltation or runoff contaminants entering the subsurface via sinkholes, caves..." he appears to not only endorse the use of swallets, but also their improvement to better handle runoff. He further points out that "Often the treatment of one kind of karst problem can led directly to the cause of another kind of karst problem (i.e. treating sinkhole flooding problems can lead to induced collapse problems)."

The use of groundwater recharge features for drainage outfalls for highway runoff represents a serious potential hazard to groundwater resources. Storm water runoff transports contaminants such as heavy metals, road salt, nutrients, bacteria, and hydrocarbons from road surfaces (Stephenson and Beck, 1995). An even greater potential hazard is the hazardous materials and other potential groundwater contaminants that are transported along these highways. U.S. Department of Transportation (DOT) and Environmental Protection Agency (EPA) data indicate that highway transportation of hazardous materials is a relatively high-risk industry (Padgett, 1993). Designs that commit highway drainage directly into adjacent sinkholes and caves, thereby potentially endangering karst water resources, may lead to increased litigation as environmentally unsound yet still accepted construction practices are challenged by karstland residents.

Figure 3: Excavation of a sinkhole that opened under a lane of Virginia Interstate Highway. Excavation was not completed to bedrock (greater than 12 m deep) before an aggregate and concrete fill was used to treat feature.

CONCLUSIONS

The three major hazards associated with covered karst, a terrain formed by the dissolution of soluble bedrock and characterized by subsurface drainage, may be induced by highway construction activities. Drainage is the most important element with respect to subsidence, sinkhole flooding, and groundwater pollution hazards associated with highways. Subsidence has been the traditional hazard of concern to highway personnel and karstland residents. The most process oriented fix for subsidence utilizes an inverse graded filter constructed on bedrock, with a base course sized to bridge or chock the under-draining solutionally enlarged fracture(s). The use of paved ditches and other drainage structures are important in isolating highway landscape generated runoff from the karst of the highway corridor and mitigating subsidence, flooding, and groundwater contamination hazards. The use of sinkholes and other groundwater recharge features as drainage outfalls are design flaws that place karst groundwater at risk to leaks and spills of hazardous materials and other contaminants transported along highways.

ACKNOWLEDGEMENTS

Frankie Gilmer has been extremely helpful in discussing sites and problems along Virginia highways. Clay Stowers and Sam Graybill provided information about other Virginia sinkholes. Additional thanks to other VDOT folks who have provided information about sinkhole sites over the years.

REFERENCES

Aley, T., and Thomas, K.C., 1981, Hydrogeologic mapping of unincorporated Green County, Missouri, to identify areas where sinkhole flooding and serious groundwater contamination could result from land development: Project summary prepared for the Green County Sewer District, 11 p., 5 plates.

Crawford, N.C., 1984, Sinkhole flooding associated with urban development upon karst terrain: Bowling Green, Kentucky: in Beck, B.F., ed., Sinkholes: Their Geology, Engineering & Environmental Impact, Proceedings of the First Multidisciplinary Conference on Sinkholes, Orlando, FL. A.A. Balkema, Rotterdam, p. 283-292.

Currens, J.C., and Graham, D.R., 1993, Flooding of the Sinking Creek karst area in Jessamine and Woodford Counties, Kentucky: Kentucky Geological Survey, Rept. of Invest. 7, 33 p.

Ford, C.G., Ogden, A.E., Ogden, L.R., Ellis, S., and Scarborough, J.A., 1997, Ground water basin delineation for sinkhole flood prevention, Johnson City, Tennessee: in Beck, B.F., and Stephenson, J.B., eds., The Engineering Geology and Hydrogeology of Karst Terranes, Proceedings of the Sixth Multidisciplinary Conference on Sinkholes and the Engineering and Environmental Impacts of karst, Springfield, MO. A.A. Balkema, Rotterdam, p. 259-263.

Mellett, J.S., and Maccarillo, B.J., 1995, A model for sinkhole formation on interstate and limited access highways, with suggestions on remediation: in Beck, B.F., ed., Karst GeoHazards, Proceedings of the Fifth Multidisciplinary Conference on Sinkholes and the Engineering and Environmental Impacts of karst, Gatlinburg, TN. A.A. Balkema, Rotterdam, p. 335-339.

Moore, H.L., 1984, Geotechnical considerations in the location, design, and construction of highways in karst terrain – 'The Pellissippi Parkway extension', Knox-Blount Counties, Tennessee: in Beck, B.F., ed., Sinkholes: Their Geology, Engineering & Environmental Impact, Proceedings of the First Multidisciplinary Conference on Sinkholes, Orlando, FL. A.A. Balkema, Rotterdam, p. 385-389.

Myers, P.B., Jr., and Perlow, M., Jr., 1984, Development, occurrence, and triggering mechanisms of sinkholes in the carbonate rocks of the Lehigh Valley, eastern Pennsylvania: in Beck, B.F., ed., Sinkholes: Their Geology, Engineering & Environmental Impact, Proceedings of the First Multidisciplinary Conference on Sinkholes, Orlando, FL. A.A. Balkema, Rotterdam, p. 111-115.

Newton, J.G., 1984, Review of induced sinkhole development: *in* Beck, B.F., ed., Sinkholes: Their Geology, Engineering & Environmental Impact, Proceedings of the First Multidisciplinary Conference on Sinkholes, Orlando, FL. A.A. Balkema, Rotterdam, p. 3-9.

Padgett, D.A., 1993, Remote sensing application for identifying areas of vulnerable hydrogeology and potential sinkhole collapse within highway transportation corridors: *in* Beck, B.F., ed., Applied Karst Geology, Proceedings of the Fourth Multidisciplinary Conference on Sinkholes and the Engineering and Environmental Impacts of karst, Panama City, FL. A.A. Balkema, Rotterdam, p. 285-290.

Stephenson, J.B., and Beck, B.F., 1995, Management of the discharge quality of highway runoff in karst areas to control impacts to ground water – A review of relevant literature: *in* Beck, B.F., ed., Karst GeoHazards, Proceedings of the Fifth Multidisciplinary Conference on Sinkholes and the Engineering and Environmental Impacts of karst, Gatlinburg, TN. A.A. Balkema, Rotterdam, p. 297-321.

The vulnerability map of karst along highways in Slovenia

STANKA ŠEBELA, ANDREJ MIHEVC & TADEJ SLABE   Karst Research Institute ZRC SAZU, Postojna, Slovenia

ABSTRACT

More than 25 years of experienc studying surface and underground karst features across Slovenia has shown good mechanical stability of the highways. We do not have any examples of a cave or doline collapse occuring directly under the highway pavement. All collapses occured near or between highway lanes. The prediction of karst features which can show up after highway construction is good only for dolines and denuded caves. The prediction of karst caves is not very good. On highways which are still under construction, collapses occur even during the last consolidation of the gravel roadway by vibration rollers. In a 3-year project, a vulnerability map of karst along highways was prepared. It includes identification of all karst geomorphological features, karst springs and regulation of discharge of waste waters from highways.

INTRODUCTION

The Karst Research Institute ZRC SAZU (Postojna, Slovenia) has more than 25 years of experience with karst investigations in the planning and construction of highways across Slovenia. From 1996-1999 the Slovene Ministry of Sciences and Technology was one of the sponsors for the applied project: The Vulnerability Map of Karst Along Highways in Slovenia.

In that project karst research took place along highways which have been in existance for more than 25 years and along highways which are still under construction. Our principal investigations are related to determining all surface (dolines, collapse dolines, denuded caves, karrens) and underground (cave) karst features. Some karst features (especially dolines and sometimes denuded caves) can be described before construction of the highway begins, that means on the original karst morphology. Most of the karst caves are found during construction of the highway, especially if the highway needs to have many road cuts where they remove or blast the upper layers. On a 30-km highway constructed in the last few years, more than 200 new karst caves were found.

Before construction of the highway begins, geoelectrical and geophysical investigations are undertaken. In areas where bridges, viaducts, subways or causeways are planned, geomechanical investigations of basic rock stability are performed. In recent years karst investigations of highways, before construction started, were performed with the special purpose of predicting karst features, especially those which can show up after highway construction. Our studies show that prediction is good only for dolines and denuded caves. After the highway pavement is made, georadar measurements are done. When new cavities are discovered, drilling is undertaken to determine the collapse hazards.

On our highways, we have just a few examples of cave or doline collapse in the period when the highway has been opened for many years. Most are caused due to mistakes in construction or due to the wash out of sediments along the highway. There are no examples of a cave or doline collapse occurring directly under the highway pavement; all collapses occurred near or between highway lanes. But in highways which are still under construction, we have many examples of cave collapse. Collapses occur even during the last consolidation of the gravel roadway by vibration rollers. The Karst Research Institute ZRC SAZU makes detailed cave maps with special regard to thickness of the cave roof and the stability of the newly opened cave. So far the good stability of Slovene highways is probably due to the fact that our karst is not covered with thick sediments or soils.

Southern and especially southwest Slovenia (highways: Vrhnika-Postojna, Čebulovica-Dane, Dane-Fernetiči, Divča-Kozina) represents a karst region with numerous caves and surface karst features. It is built mostly of Cretaceous and Paleocene carbonate rocks. In the Cretaceous the region was part of the Dinaric carbonate platform. With disintegration of the platform, major tectonic events took place. Karst regions in southeast Slovenia are covered with thicker sediments; in that area, because the principal construction of highways just started and will not be finished for some years, we do not have sufficient data about possible collapses.

EXPERIENCES FROM THE 26 YEAR OLD HIGHWAY VRHNIKA-POSTOJNA

During construction of the Vrhnika-Postojna highway (29 km long) in 1972, 22 new karst caves were discovered, representing 7 caves per 1 km^2 (Kranjc 1983). Later 2 places in the roadway collapsed. At Postojna at the contact of limestone and non-carbonate rocks, after blasting the debris which partly filled up the karst cave was not removed. Twenty years after the road was finished, a cavity of about 10 m³ occurred at Verd on the grassy zone between 2 road lanes. The cave entrance opened due to falling of the lower layer of gravel into the unfilled cave below.

Directly under the highway some very deep shafts were included in highway construction, with a depth of up to 38 m (Jama Medvednica). Many of the shafts are now not accessible because the highway runs over them (Brezno II pod železniško postajo Planina (13 m deep), Avanzova jama (28 m deep), Škantlovo brezno (10 m deep). The thickness of the ceiling between the cave and the highway is, for example in Jama Medvednica, 5 m. The position of all caves as well as dolines which are situated under or near the highway is marked on topographical maps 1:5,000. All cave maps are taken from Cave Cadastre of Karst Research Institute ZRC SAZU.

A cave passage of Avanzova jama is situated directly under the highway (Figure 1). The cave is 47 m long and 28 m deep. Roof thickness between the cave passage and the highway is from 10 to 20 m. The original entrance shaft was filled with material and covered with a concrete plug during the construction. After 26 years of highway operation, no collapse occurred directly under the highway due to the underground cave passage.

Just under the highway and railroad near Unec, there is a 2,285 m long and 83 m deep karst cave Logarček (Figure 2). One of its collapse chambers, called Podorna dvorana, is directly under the edge of the highway (Figure 3). The roof thickness between cave and surface is 13 m. A passage of the cave, called Severni rokav, is situated directly under the highway. The ground plan of the cave is from 1963 (Gams), and with new teodolitic measurements of cave passages, we can suspect some corrections. The thickness between the cave passage of Severni rov and the highway is around 50 m. The cave Logarček is often influenced by high water. A part of the cave can at times be flooded and not accessible. The Podorna dvorana collapse chamber is never included in high floods. Cavers have said that the sounds of traffic on the highway and railway can be heard in the cave. We do not have good evidence that some collapse blocks in Podorna dvorana fell due to traffic on the highway.

Between kilometres 15 and 15.5, between Logatec and Postojna, the highway crosses unknown passages of the Javorniki underground water flow, which is part of Planinska jama cave. This is an important underground water passage, and waters at Malni springs are used as a water supply.

Highway Vrhnika-Postojna runs across 470 dolines. The rocks in the area are composed of limestone, dolomite and dolomitic breccia. Most of the dolines have a diameter of 20-80 m and a depth of 5-15 m. The average number of dolines is 250-300 per km². If we consider the average depth of a doline to be 5 m and the diameter 50 m, the suggested average volume of a doline is 3,000 m³. Regarding the elevation of the highway, most dolines (74 %) today are situated under a causeway. In 17% the elevation of the highway is the same as the elevation of the doline bottom, and 9% of the dolines lie in a road cut. Of dolines in the Vrhnika-Postojna highway, 36% are bowl dolines with remaines of clay sediments with collapse bottom; 34.6% are funnel dolines with not much clay sediment, and 29.4% are accumulation dolines with clay sediments on the bottom.

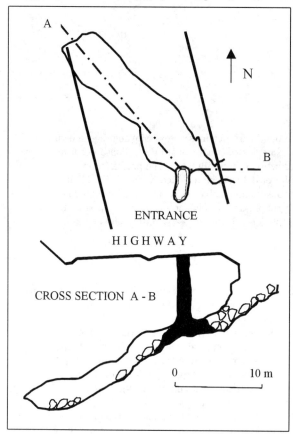

Figure 1: Avanzova jama cave is situated under a highway.

HIGHWAY ČEBULOVICA-DANE

Near Divača there are 11 dolines per 1 km² of road section; between Divača and Sežana there are 5 per 1 km². The greatest depth of the doline established by means of drilling, was 27.5 m (Habič 1974).

Altogether 76 caves were discovered on 14 km of a new road. These were mostly smaller old caves, and only 6 of them had passages 5 m or more across. Two caves were known before highway construction. Between newly discovered caves, 57 were old caves, and 19 were potholes. Between old caves 24 were empty, and 33 were filled with sediments (Slabe 1996).

HIGHWAY DANE-FERNETIČI

Two tunnels (240 and 260 m long) were constructed in a 4.5-km long Dane-Fernetiči highway. Just at the exit of a tunnel, due to blasting, a 5x10 m cave opened. The cave was filled with block material and protected by a concrete plate. Between both tunnels a 30-m-deep vertical shaft was discovered. It was formed along a well-expressed fault. Near the highway a 110-m-deep shaft was discovered. The entrance to the cave was in a doline which is used as a catchment area for waste waters from the highway. Special arrangements for protection against leakage into the cave was carried out.

PRELIMINARY STUDIES OF KARST FEATURES BEFORE CONSTRUCTION OF HIGHWAY DIVAČA-KOZINA

A horizontal karst depression filled with sandstone pebles and cave sediments was opened during the construction of the Divača-Dane highway (Knez and Šebela 1994; Šebela and Mihevc 1995). This was actually an old horizontal cave passage, filled with sediments and without a roof. Flowstone deposits on the original cave passage walls and over the cave sediments proved that this was an old karst cave and not a surface stream valley.

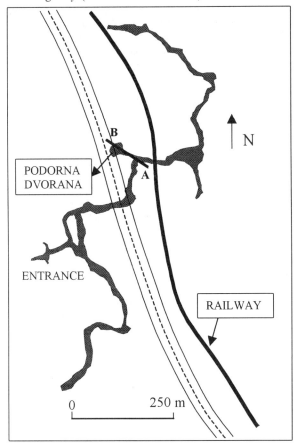

Figure 2: Logarček cave.

Denuded or roofless caves can be determined from the morphological shape of depressions and dolines. Preliminary field studies on a planned 6.7-km-long highway, 4 denuded caves were discovered (Šebela 1996). During highway construction, 2 more denuded caves were opened by construction works.

Beside denuded caves, all regular smaller or bigger caves were also detected. On a 6.7-km-long highway, beside 6 already known caves, 9 new caves were found. This means that preliminary karst studies done before highway construction determined that there were 2.238 caves per 1 km of road (Šebela 1996).

Later during construction of the Divača-Kozina highway on a 7.5-km-long section of the highway, 50 old caves were discovered.

The entrance of Jama nad Škrinjarico is situated 35 m west of the Divačca-Kozina highway. The length of the cave is 270 m, and the depth is 130 m. The thickness of cave roof under the highway is 85 m. The cave is listed as Natural Monument Number 882 in the List of the Natural Heritage of Slovenia. The highway over Jama nad Škrinjarico has been open since 1997.

Škocjanske jame caves (5.8 km long) have been included in UNESCO's World Natural Heritage List since 1986. Škocjanske jame Regional Park extends over an area of 413 hectares. The boundary of the Park runs along the Divača-Kozina highway on the west. Between Škocjanske jame caves and Kačna jama, there is an unknown passage of the river Reka which runs about 200 m beneath the highway. This means that the thickness of carbonate rocks between the cave and the surface is about 150 m. Martel's chamber in Škocjanske jame which, along a straight line, is 450 m from the highway is the biggest known underground chamber in Slovenia and has a volume of 2,000,000 m^3 (Mihevc 1995) with a 140-m-high passage.

ROAD SELO-VIPAVA

Construction of the new Selo-Vipava road is organized through flysch of Vipava Valley. Inside flysch we can find an approximately 10-m-thick layer of calcarenite which is well karstified. In the 15 km long road more outcrops with karst caves can be expected. Several smaller caves were found during the construction. The biggest one was an approximately 9-m-deep shaft. One of the bearing pillars for a road bridge is planned to be in close vicinity to that shaft. Karst drainage can also be organized in limestone breccia inside flysch.

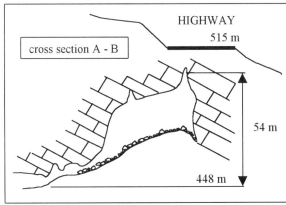

Figure 3: Cross section A-B of Podorna dvorana in Logarček cave.

HIGHWAY ACROSS SOUTHEASTERN SLOVENIA

A highway is also being constructed in southeastern Slovenia between Ljubljana and the border with Croatia. In many areas, the region is composed of carbonate rocks overlain by Pliocene and Quaternary clays and alluvium. Karst caves are present in the region, but many times are not very well determined due to a thick cover of sediments. During high water, karst depressions can be filled with water. Because construction of the highway is just beginning, no collapses have occurred yet.

IMPORTANCE OF APPLIED KARST STUDIES DURING HIGHWAY CONSTRUCTION FOR THEORETICAL KARST STUDIES

For the project "The Vulnerability Map of Karst along Highways in Slovenia" about 70 km of highways across karst terraines were studied. The project included old highways which have been in use for more than 25 years and new highways, some of which are still under construction. During the studies of karst surface and underground morphological features along highways, special attention was paid to protection of karst springs which are captured for water supplies and to the drainage of waste waters from highways.

Besides providing solutions for applied karst problems in highway construction, a big step in the theoretical approach in karst geomorphology was developed. We discovered many denuded caves which are old horizontal cave passages filled with cave sediments, but due to the processes of tectonic uplifting, erosion, and corrosion, they do not have a roof anymore. Before 10 years ago, such karst features were described as dolines or karst depressions filled with the remains of old surface stream sediments. Now we have proven that many dolines are, in fact, the remains of old denuded caves (Mihevc and Slabe and Šebela 1998). U/Th dating and paleomagnetism show that cave sediments filled old caves before 350,000 years ago and can be at least 700,000 years old.

REFERENCES:

Cave Cadastre of Karst Research Institute ZRC SAZU, Postojna.

Gams, I., 1963, Logarček: Acta Carsologica SAZU, 3, Ljubljana, 5-84 p.

Habič, P., 1974, Poročilo o kraških pojavih na AC Senožeče-Divača-Sežana: Elaborat IZRK ZRC SAZU, Postojna, 20 p.

Knez, M., and Šebela, S., 1994, Novo odkriti kraški pojavi na trasi avtomobilske ceste pri Divači: Naše jame, 36, Ljubljana, 102 p.

Kranjc, A., 1983, Speljava hitrih cest čez kras: Obzornik 4, Ljubljana, 316-319 p.

Mihevc, A., 1995, New surveys in the Martel Hall, Škocjanske jame: Naše jame, 37, Ljubljana, 39-44 p.

Mihevc, A., and Slabe, T., and Šebela, S., 1998, Denuded caves – An Inherited element in the Karst Morphology; The case from Kras: Acta Carsologica XXVII/1, Paper presented at Fourth International Conference on Geomorphology "M3-Classical Karst", Lipica, 24.8.-28.8.1997, Ljubljana, 165-174 p.

Slabe, T., 1996, Karst features in the motorway section between Čebulovica and Dane: Acta Carsologica SAZU, 25, Ljubljana, 221-240 p.

Šebela, S., and Mihevc, A., 1995, The problems of constructions on karst – The examples from Slovenia: Karst Geohazards, Proceeding of the Fifth Multidisciplinary conference on Sinkholes and the Engineering and Environmental Impacts of Karst, Gatlinburg/Tennessee/2-5 April 1995, A.A. BALKEMA, Rotterdam, 475-479 p.

Šebela, S, 1996, Predhodne krasoslovne raziskave trase avtoceste Divača-Kozina: Annales, 9, Annals for Istrian and Mediterranean Studies, series historia naturalis 3, Koper, 103-106 p.

The system of antikarst protection on railways of Russia

VLADIMIR V.TOLMACHEV, SERGEY E.PIDYASHENKO & TATIANA A.BALASHOVA   State Venture 'Antikarst Protection', Dzherzhinsk, Russia

ABSTRACT

The problem of train traffic safety on railways in karst terrain is an actuality for many railway lines in Russia. It is mostly found on a 25-kilometer section on the Moscow-Nizhny Novgorod express line in the vicinity of the city of Dzherzhinsk. It is the collapse hazard that makes this section of railway line a difficulty in the management of high-speed railway traffic. In the last 55 years, karst collapses have caused two collisions and long disruptions of train service. At this time, the speed of the trains is reduced to 40-60 km/h. A system of protection against karst collapses (antikarst protection) has been instituted which includes:

1. Special operating measures which include linear karst monitoring and a special system of control of the track conditions, among other measures.
2. Engineering antikarst measures including grouting of karst cavities and strengthening of the railway tracks, among other measures.
3. Operating measures including antikarst alarms and standby reserves, among other measures.

One or another system of antikarst measures is set based upon zoning of the track by the degree of perceived collapse danger and the characteristics of the railway track—the curve of the track, the availability of bridges, etc. Operating measures were demonstrated to be priority in antikarst measures.

Twenty percent (20%) of the European area of Russia is karst terrain (Tolmachev, this volume). Therefore, providing train traffic safety on the railway lines located in karst terrain is a large problem. It becomes even more of a problem on lines carrying high-speed trains. Thus, when instituting a comprehensive train-traffic-safety program, the following problems of a railway built in karst terrain (a karst railway geotechnical system) must be taken into account:

1. In the past most of the railways built in Russia in dangerous segments of karst terrain were built without proper consideration of karst processes. This is not only true for the track on land but also on railroad bridges over large rivers.
2. The karst process is dynamic; in principle it is impossible to stop it in any narrow section along the railway. Therefore, so-called "active antikarst protection", i.e., measures aimed at intervention in the nature of the process, is extremely difficult and cannot be insured with absolute reliability.
3. The karst process is a stochastic process (Tolmachev, 1980; Sorochan, 1985). In this regard, tactical provisions of train traffic safety should be made. Specifically, one can make an estimate of the probability of collapse in any given stretch and then find the probability of construction damage and compare such to the minimal allowable probabilities. Based on this estimate, a decision can be made on the relationship between engineering and operating measures, the frequency of site observation, the necessity of karst monitoring, and the reduction of speed of the trains, etc.

The probability of a site being damaged by karst collapse in any time, t, can be expressed with the formula:

$$P_t = \left(\frac{A_n}{A} + \frac{\bar{d}}{d_{max}} \cdot \frac{A_0}{A} \right)(1-P)$$

Where
$$P = \exp(-\lambda A t)$$

$$A = A_0 + A_n$$

λ - is the frequency of karst collapses in a year per 1 km^2
d and d_{max} – are respectively median and maximum diameters of karst sinkholes;
A_n – is an area of railway bed in the route
A_0 – is an area of the site along the railway delineated at the distance $d_{max}/2$ from the toe of the embankment (edge of cutting).

4. Most of the hazardous karst segments of railways are in terrains where soluble rocks are buried at great depths, from 15 to 70 meters, under arenaceous, loamy and argillaceous soils (so-called "covered karst"). In these cases, the detection of karst cavities under linear objects such as railways for the purpose of prediction of exact location and time of collapse is economically unjustified and a radically intractable problem. In practicality, the most needed and significant actions are the delineation of the most hazardous karst sites which cover tens to hundreds of meters and the estimation of the probability of karst collapse within any given time.

In the process of making conduits in karst, subsidences are not always connected to the most accessible cavities in soluble rocks (Khomenko, 1986). Obvious cavities are not always dangerous, i.e., they do not always result in collapse of the railway bed in any given time.

Thus, the strategy for investigation should not be developed by attempting to detect karst cavities in the soluble rock and eliminate them. In most cases, the efficiency of engineering and geological investigations aimed basically at detecting cavities in the soluble thickness is extremely poor for linear objects such as railways.

5. The man-made influences of railways upon the geological environment stimulates karst processes to a certain extent. Primary among these influences are:
   - increased local infiltration of water into soils where there is drainage, snow retention, and culvert construction;
   - Extra static load;
   - Vibrational dynamics of a rollingstock.

The nature and degree of these listed factors are not unique. They are primarily defined by the particular engineering/geological situation.

6. As with time, the degree of karst danger varies with space. Thus, to provide traffic safety for trains, it is necessary to monitor karst development in the soil thickness, on the soil surface, and also on the constructed features (normal settlement of track, tilts of the masts of the electrical system, etc.)

7. There are other conditions which influence train-traffic safety besides karst terrain: bridges, trough bridges, tunnels, culverts, deep cuttings, high embankments, pipeline crossings, restricted-sight locations, etc. These conditions must be considered and prioritized in coordination with control of possible karst subsidence and special antikarst measures.

The spatial-temporal prediction of karst danger is the main problem of train traffic safety in karst regions. Presently, the problem of spatial prediction of karst collapse is being solved with some degree of success in karst engineering (Tolmachev, 1986). Long-term prediction of karst danger is connected with determining the probability of karst collapse within a sufficiently large given time (several decades). However, methods of short-term prediction of karst collapse have not been developed.

The State Venture of Antikarst and Shore Protection has planned two interconnected ways of solving this problem:

1. Spatial/temporal prediction based upon the physical simulation of collapse development and the detection, by various geophysical methods, of the development of underground karst features at various depths.
2. Prediction of collapse in or near the foundation of a railway bed based upon visible conditions and observation by regular monitoring conducted by karst railway engineering specialists.

All antikarst measures on railways may be classified into three groups as shown in Table 1.

The problem of train traffic safety is most acute for the Gorky railway, particularly for the site 395-420 km in the direction of Moscow – Nizhny Novgorod, which will soon be handling high-speed traffic. Here, the karst manifests itself as collapses of various intensity over the whole length of the track. It also occurs as alow subsidences at three sites with lengths of 600 – 1,700 m with rates of subsidence from 5 to 25 mm/year. In addition, the vertical component of subsidence displacements are also typical for these sites (Reuter, 1990). It is important to keep these facts in mind concerning continuous-welded rails, especially in the summer when the probability of horizontal buckling increases. This is why these sites of subsidence need increased budgets for track maintenance.

However, karst sinkholes are most dangerous for railway operations by virtue of the following special characteristics:
- In most cases, sinkholes form practically instantaneously.
- The probability of the occurrence of a collapse immediately under a railway increases due to the vibration dynamics of train traffic.
- Visual indications of a probable collapse, until only several hours before occurrence, are most often absent on the land surface and the railway bed. However, they can be used to take steps to ensure the safety of the trains, but in most cases such indications appear only minutes before collapse.
- The diameters of karst collapses vary over wide limits and extend to 50 meters. The average mean collapse diameter is 12 meters. Diameters increase especially fast in sandy soils during the initial period after the occurrence of collapse; therefore, a collapse which forms close to a railway bed, even if not directly underneath it, is also dangerous to train safety.

On the railway right-of-way, about 350 sinkholes have been noted includein 36 new sinkholes formed during the last 60 years. In 1943 and 1961, karst collapses occurred beneath trains and caused the trains to wreck. In 1996, a collapse occurred which stopped train traffic for 12 hours. Train speed is reduced to 40-60 km/h over several kilometers of the track arount this site.

Table 1. Classification of main antikarst measures on operabling railways under conditions of covered karst.

Group	Type	Applicable conditions
Non-recurrent capital	Grouting a karst cavity in soluble rocks	The cavity is in the zone of compressible rock mass of a railway bed or pier. Soluble rocks are buried at great depths, but the real danger of the detected cavity is proven by karst activity in adjacent sites related to hydrogeological conditions.
	Grouting soil cavities in overlaying soils	In all cases where work should be done on a short-term basis.
	Removing cavities by causing dynamic action (provocation of a collapse)	Soluble rocks are buried at a shallow depth (up to 10m) and karst conduits are absent.
	Erection of end-bearing piles driven to solid rock	Soluble rocks are buried at shallow depths.
	Erection of subgrade strengthening constructions	Predicted sinkhole size is not more than 10m. Very large karst subsidences are absent.
	Erection of berms	Predicted collapses have great diameters.
	Erection of drainage or grout curtains	Chloride karst, soluble rocks are buried at shallow depths.
Constant operating	Regular technical studies with line workers of all levels and instruction concerning personnel actions in the case of karst subsidence.	Under any conditions.
	Visual observation of track condition and adjacent areas	Under any conditions.
	Arrangement of temporal or constant observational posts in especially hazardous karst sites	Where poor sight, large bridges and tunnels, etc., exist.
	Special visits to hazardous karst sites	Under any conditions.
	Periodic geodetic observations of railway bed collapses	In subsidence sites.
	Reduction of train speed in hazardous karst sites	Before capital antikarst protection.
	Installation of "KARST" warning signs in the most hazardous karst sites	Under any conditions.
	Repairing detected local underground and surface karst formations	Under any conditions.
	Karst/railway monitoring in the most hazardous sites	Under any conditions.
	Particularly thorough maintenance of track ditches	Under any conditions.
	Particularly thorough maintenance of continuous-welded tracks	On collapse sites.
Episodical capital/ operating	Arrangement of linear antikarst alarm signaling and regular control	The most hazardous sites.
	Arrangement of spot alarm signaling and regular control	Piers and detected hazardous karst cavities.
	Installation of temporal rail packets	Probable sinkholes of a diameter of not more than 3 to 6 meters before capital antikarst protection.
	Standby reserves and their use at karst subsidences	Near the most hazardous karst sites.

In the site under discussion, the karst qualifies as sulphate/carbonate-covered karst. Early Permian gypsums (P_1), limestones, and dolomites of the Late Permian (P_2), (Fig. 1) are the soluble rocks. The surface of these rocks is very irregular and marked by more than 50 cavities 1-7 m in height as located by boreholes penetrating the rocks. The limestones and dolomites are very fractured and broken into rock debris and carbonate sand. In some places, they are entirely absent due to the formation of karst conduits. The buried depth of the soluble rocks along the railways varies from 30-60 m. Soluble rocks are covered by Late Permian clay (P_2) which is a comparative aquifuge. The thickness of the clay varies from 1-18 m. In the western part of the site (395-403 km) argillaceous rocks are entirely absent. Quaternary alluvial sands (Q) 25-50 m thick cover the Permian rocks. The arenaceous interval for the most part is water saturated. Therefore, karst sinkholes in this area may occur not only because of the collapse of a karst cavity roof, but also because of the undermining of arenaceous soils by erosion into cavities and fractured zones which are located beyond the base of the railway bed.

In 1994-1995, the State Venture Antikarst and Shore Protection wrote a report on the evaluation of the karst danger at this site of the railway, which included:
- A special analysis of previously conducted works and air photography
- A karst survey along the railway covering a 100-meter width

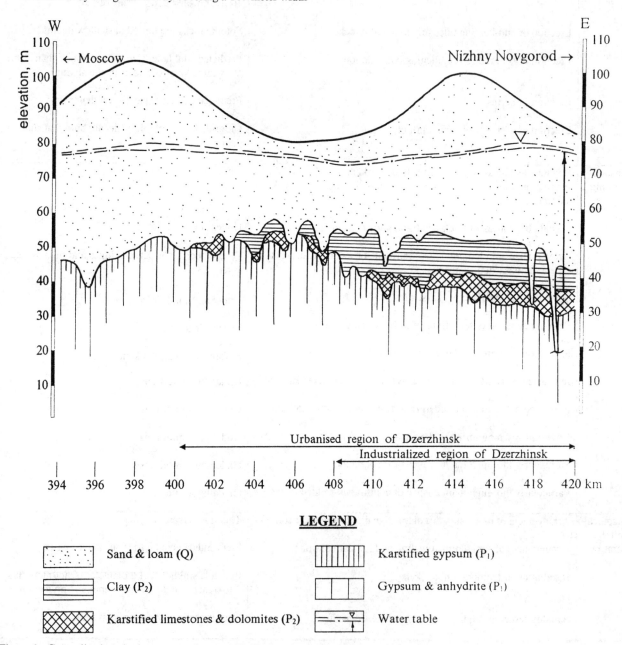

Figure 1. Generalized geologic cross-section on 395-420 km site on the Moscow-Nizhny Novgorod railway.

- A geophysical survey (electrical resistivity, gravity survey, and seismic survey)
- A cone penetration test (CPT)
- Boreholes
- A simulation of collapse formation under laboratory conditions

The evaluation of the karst danger for the railway was made to determine the following:
1. The classification of the karst danger of the railway lines per Russian codes in force at this time.
2. A statistical probability of collapse, and the diameter of that collapse, and the determination of the danger zones along the lines.
3. The relative danger of karst cavities detected in the soluble rocks, in order to define the priority of repair.
4. A temporal prediction as to the danger from underground karst anomalies detected during the investigations.

The railway most discussed is the railway in Dzherzhinsk, which is a city that is important in the chemical industry. This fact predetermined that it was necessary to take into account the following when predicting karst danger and assigning antikarst protection to the Dzherzhinsk railway:
1. The karst terrain of Dzherzhinsk is well investigated. There are sufficiently detailed maps divided into districts by the degree of karst danger, as dictated by building regulations now in force. Monitoring of the karst process is being conducted. The characteristics of karst development and its visual appearance underground and on the surface are being investigated. This makes it very important to draw engineering/geological analogies between the conditions of karst development along sites of the railway and corresponding sites in this well-investigated karst terrain.
2. To a certain degree, karst formations along the railway are exposed to the hydrological regime of the Oka River.
3. The karst development along the railways is exposed to various man-made influences, especially those from industry: hydrogeological changes due to water-well pumping, leaking water pipes, and groundwater contamination by urban wastes.
4. Pedestrian and automobile bridges cross the railway, along with a great many pipe lines carrying ecologically dangerous materials. These sites are the most important sites along the railway and need special controls against karst subsidences.
5. The proximity of ecologically dangerous industry to the railway limits the application of such antikarst measures as grouting karst cavities, in that the use of these measures may lead to activation of karst processes in other industrial sites (Sorochan, 1982, Recommendations, 1985; Recommendations, 1987).

As a consequence of the above-mentioned engineering/geological, geomorphological and geophysical investigations, it has been possible to delineate areas of various categories of karst danger in the well-investigated karst areas in Dzherzhinsk. These areas have been delineated based upon the frequency of karst collapses λ corresponding to building regulations (Fig. 2). In this the values of the index of karst collapse frequency were determined by the use of probability/statistical methods (Savarensky, 1983; Tolmachev, 1986; Reuter, 1990). Any given site of the railway was assigned to one of five categories of karst danger (Table 2).

The width of a dangerous site along the railway (B) was defined with regard to the measured maximum diameters of sinkholes (d_{max}) and the width of a railway bed base (b):

$$B = b + d_{max}$$

The meaning of d_{max} is calculated statistically with a confidence level of 0.998 with regard to the engineering/geological analogies (Tolmachev, 1986).

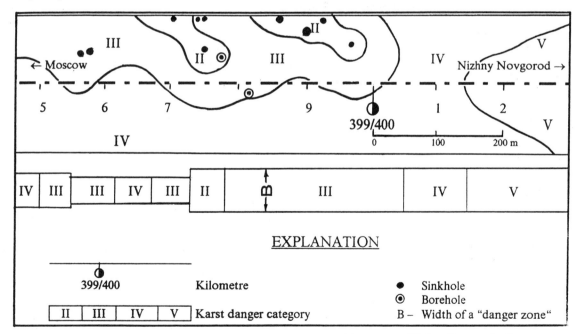

Figure 2. Karst danger zoning on the 399-400 km site of the Moscow-Nizhny Novgorod railway.

Table 2. Karst danger categories in the frequency of karst collapses on the 395-420 km site of the Moscow-Nizhny Novgorod railway.

Karst Danger Category	Index of Karst Collapse Frequency $\lambda \left(\dfrac{\text{collapses}}{\text{km}2 \cdot \text{year}} \right)$	Length p/km
I	>1.0	1.29
II	0.1 – 1.0	4.42
III	0.05 – 0.1	11.51
IV	0.1 – 0.05	6.77
V	<0.01	1.00

In depth, the danger zone includes the area where karst piping is active. In the site under discussion, the danger zone varies in depth from 37-72 m. In this zone, detected karst anomalies located at various depths vary in the degree of danger with relation to their probable impact upon the railway over time. The danger zone was differentiated into three subzones according to depth. A brief description is shown in Table 3.

A more exact prediction of the time of karst subsidences at the base of the railway bed or on the soil surface is determined by following the dynamics of the development of detected anomalies and the laboratory simulation of collapses (Fenk, 1981; Sorochan, 1989). In the site of discussion, railway capital and operating measures were recommended based upon the results of the evaluation of the karst danger (Tables 4 and 5).

Table 3. Hazardous karst subzones according to the depth of detected karst anomalies on the 395-420 km site of the Moscow to Nizhny Novgorod railway

Subzones	Depth (meters)	Soils	Detected karst anomalies	Evaluation time of anomalies presenting as collapses and localized subsidences at the base of the railway bed
A	0-15	Sands in the zone of aeration	Zones of decompaction in overlaying soils; voids	Several days to several months (according to a depth and size in the plan
B	10-60	Sands, clays, and loams below the groundwater table	Flooded voids; voids	One year to several years (depending upon the depth and size in the plan and in a vertical line
C	25-72	Limestones, dolomites and gypsums	Voids	Several years to 100 years (depending upon a depth, a void roof, its size and a volume)

Table 4. Recommended antikarst capital measures in the 395-420 km site of the Moscow – Nizhny Novgorod railway

Categories of karst danger	Antikarst protection				
	Arrangement of subgrade strengthening constructions	Linear alarm signalling	Spot alarm signalling	Grouting detected voids	Filling in sinkholes with clays in the rail right-of-way
I including two detected karst voids	+	+	+ + (before grouting)	+ +	+
II including four pipe lines		+	+ +	+	+
III including: detected karst voids two trough bridges deck bridges			+ + (nearest to the railway piers) +	+ +	+
IV including four trough bridges			+ + (nearest to the railway piers)		+

In 1997, special instructions were published which regulate, in detail, track maintenance in hazardous karst terrains (Track Maintenance Instructions, 1997).

Table 5. Operating antikarst measures in the 395-420 km site of the Moscow – Nizhny Novgorod railway

Categories of karst danger	Reducing the speed of traffic	Permanent observation	Day-to-day observation	Day-to-day observation during specific climatic conditions (snow, thawing, high and low temperatures)	Installation of warning signs "KARST"	Monitoring with the use of:		
	Before capital antikarst protection					The analysis of constant deformations of the railway	Geodosy measurements	Geophysical measurements
I	+	+ (In the sites where voids are detected in subzones "a", "b")	+		+	+	+	+
II			+		+	+	+	+
III				+	+	+	+	
IV				+ (In the sites of detected anomalies)		+	+ (in the sites of subsidence)	
V						+	+	

REFERENCES

Fenk, J. (1981) Eine theorie zue Entstehung von Tagesbrüchen uber Hohlräumen in Lockergebirge.: Freiberger Forschungshefte, A 639, p. 97

Khomenko, V.P. (1986) Karst piping processes and their prediction. Moscow, "Nauka" p. 96

_____, (1985) Recommendations on grouting karst soils at the base of civil and industrial objects. Moscow NIIOSP, 28 p.

_____, (1987) Recommendations on the use of engineering geological information in deciding on antikarst methods. Moscow, PNIIIS, Stroyizdat, p. 81

Savarensky, I.A. (1983) The use of data on surface for karst formations for evaluation of karst terrains stability.: In the book "Building on karst terrains", "Theses of the report of the meeting in Podolsk", Moscow, NIIOSP, pp 60-62

Sorochan, E.A., Troitsky, G.M., Tolmachev, V.V. (1982) Combined measures in building in karst terrains.: Bases, foundations, soil mechanics, No. 4, pp. 16-19

Sorochan, E.A., Troitsky, G.M., et al (1985) Antikarst protection for buildings and structures.: IX International Society for soil mechanics and foundation engineering. San Francisco, USA, pp. 2457-2460

Tolmachev, V.V. (1980) Probable approach in evaluating karst terrain stability and planning of antikarst measures.: Engineering Geology, No. 3, pp. 98-107

Tolmachev, V.V., Troitsky, G.M., Khomenko, V.P. (1986) Engineering/building development of karst terrains. Moscow, Stroyizdat, p. 117

Track Maintenance Instructions for Hazardous Karst Terrains (1997) Moscow, MPS RF, p. 31

Geotechnical engineering and geology for a highway through cone karst in Puerto Rico

LUIZ VÁZQUEZ CASTILLO & CARLOS RODRIGUEZ MOLINA   Vázquez Castillo-Rodriguez Molina, Suelos Incorporated, Hato Rey, P.R., USA

ABSTRACT

A 4.5 kilometer long sector of highway PR-10, a mayor 4 lane highway connecting the north and south coasts of Puerto Rico, was designed to cut through Puerto Rico's North Coast Karst Belt. The highway would cut through well developed cockpit or cone karst within a highly scenic rainforest. The selected route was the final alternative studied, since for the others alternatives evaluated either sinkholes and landslide terrain would be crossed and even more sinks intercepted. It will be mentioned that the Karst Terrain and the rain forest produce a well-developed recharge system to the P.R. north-central groundwater aquifer.

A method to design and construct the roadway through approximately 12 sinkholes was devised, the main object being: constructing the roadway without blocking and thus, permitting runoff water flow into the sinkhole system. Twelve sinkholes were treated, some were provided with geomembranes and filters, in most cases enhancing the sinkhole's drainage capacity. Others, although filled with granular and permeable material, were capped by less permeable material designed to reduce and/or divert infiltration into adjoining sinkholes.

The exploration leading to the design consisted of a multi-disciplinary approach commencing with early conceptual project traverses through well developed cockpit karst for an eyewitness view of the magnitude of the sinkholes the right of way would cut through. Traversing a right of way through cockpit karst proved to be an enormous if not impossible task. Relief between the top and bottom of any one sinkhole chosen varies from 100 to 200 meters. The construction of a pilot road was commissioned by the Puerto Rico Highways and Transportation Authority (PRHTA) after which, although the pilot road constructed was extremely winding and an up and down roller coaster view, the right of way and the sinkholes were better assessed. It became evident that most sinkholes were active and presumably interconnected to an intricate underground drainage regime of caverns and conduits. From the onset it was clear that any attempt to build a road through the rain forest meant designing a road that would not plug the natural drainage of runoff. The latter meant numerous bridge structures followed by short road cut sectors, bridge structures followed by tunnels, or conventional cut and fill construction.

A cut and fill approach was selected, but rather than using a conventional cut and fill design, (which would have eventually proved to be disastrous due to settlement and embankment collapse problems caused by sinkhole plugs), a cut and "filter" fill treatment approach were chosen. Each sinkhole was explored by geophysical and geotechnical methods such as: surface seismic refraction, cross hole seismic refraction, vertical seismic profiling, electrical conductivity and resistivity, geotechnical boreholes, color dye tracing, water well pump tests, and hydrology. The object was first to model each sinkhole subsurface condition, to estimate its drainage capacity, determine storage requirements (when needed), and provide a system of inverted filters designed to permit runoff drainage, including the one generated by the roadway itself, into them.

Natural soft sediments deposited at the sinkhole basins were to be removed as part of the treatment. Deeper cavities and soft zones below the basin were to be treated by promoting early collapse of the cavities or settlement of soft zones with the use of surcharge loads.

The roadway was successfully constructed, and has been in use since 1994.

INTRODUCTION

Much of the industrial, residential, rural, an uninhabited countryside in Puerto Rico is in the karstic limestone of the north coast. The highway cuts across the northern Puerto Rico middle tertiary limestone belt which extends approximately 125 kilometers (80 Miles) from Loiza to Aguadilla. Figure A depicts an island map showing two prominent limestone belts, the north coast limestone belt and the south coast belt.

The purpose of the exploration was to establish the vertical and horizontal extent of the subsurface soil and limestone characteristics for approximately 12 sinkholes that were to be crossed by the highway. Also, subsurface permeability characteristics were to be evaluated to estimate each sinkhole's internal drainage limitations in direct response to the construction of earth embankments needed for the highway. In addition, embankment subsidence potential related to sinkhole activity was assessed.

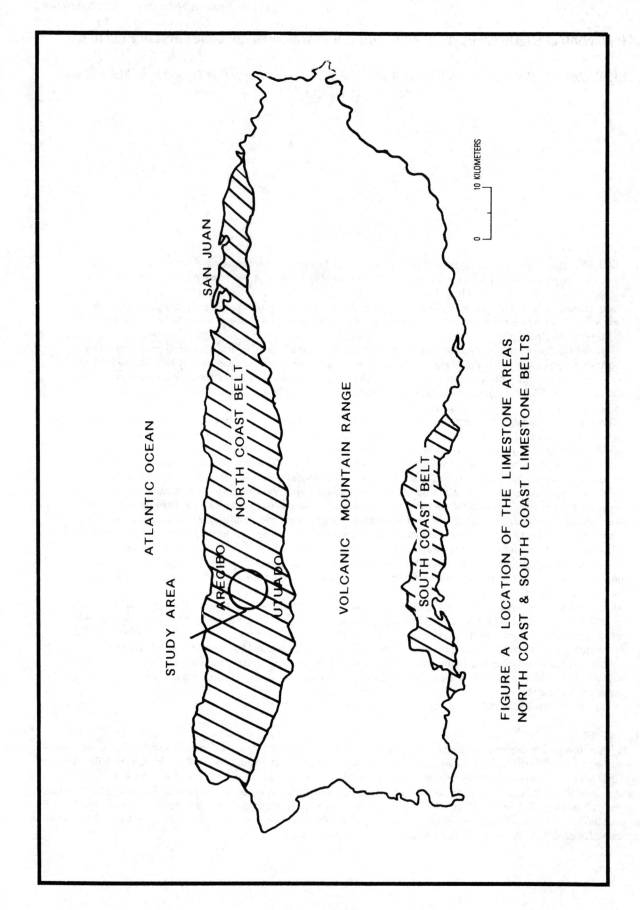

FIGURE A LOCATION OF THE LIMESTONE AREAS
NORTH COAST & SOUTH COAST LIMESTONE BELTS

Figure B shows a portion of the USGS 7.5 series Utuado Topographic Map (Utuado being the town closest to the highway within the southern border of the limestone belt). The figure shows the approximate highway alignment and location of all the sinkholes subject of the exploration. The exploration commenced early in 1984 and it was completed late in 1994 with the actual construction of the highway.

METHODS

The exploration leading to the design and construction of the highway consisted of a multi-disciplinary approach including the following methods:

a. early conceptual traverses through cockpit terrain
b. geotechnical engineering including boreholes
c. geophysical techniques
d: dye tracing through sinkholes to nearby natural water spring
e: hydrology

Because of the voluminous nature of the data generated for the 12 sinkholes this paper will use sample interpretation data from some of the sinkholes explored for illustration.

Early Conceptual Traverses Through Cockpit Terrain

The early stages of the exploration consisted of many field trips where traverses were made on foot, through densely vegetated, well-developed cockpit karst for an eyewitness view of the magnitude of the sinkholes the right of way would cut through. Traversing a right of way through cockpit karst proved to be an enormous, if not impossible, task. Relief between the top and bottom of any one sinkhole chosen varies from 100 to 200 meters. The construction of a pilot road was commissioned by the Puerto Rico Highways and Transportation Authority (PRHTA) after which, although the pilot road constructed was an extreme winding and up and down roller coaster view, the right of way and the sinkholes were better assessed.

Geotechnical Engineering & Geophysical Techniques, Dye Traces, & Hydrology

Each sinkhole was explored by geophysical and geotechnical methods including: surface seismic refraction, cross hole seismic refraction, vertical seismic profiling, electrical conductivity and resistivity, geotechnical boreholes, color dye tracing, water well pump tests, and hydrology. The object was first to model each sinkhole subsurface condition, to estimate its actual drainage capacity, determine storage requirements (when needed), and provide a system of inverted filters designed to permit runoff drainage but at the same time capable of resisting the embankments loads.

All 12 sinkholes are shown on Figure B. This was the only base map available during the early stages of the exploration. The preparation of a site specific topographic map was commissioned to a surveyor. Each sinkhole was accessed on foot and the topographic map prepared. Since relief between the top and bottom of any one sinkhole varies between 100 to 200 meters, the preparation of the entire sinkhole topography was not possible. Thus the topography was prepared only for the bottom basin of the sinkhole and perhaps 30 to 40 meters of relief along the walls of the sinkhole. Figure C shows the topographic map for sinkhole No. 7.

The geophysical methods described above were then employed, preceding and eventually concurrent with the drilling of deep geotechnical boreholes.

Geophysical methods usually were commenced with electrical methods. Figure D shows typical results electromagnetic surveys performed along the basin of the sinkhole. Moist, and thus conductive clay zones indicated areas of possible concentration or infiltration of water (these are initial indicators of active drainage zones within the sinkholes). Thus, electromagnetic anomalies were sought as an initial indicator of target zones to be probed with a surface and eventually borehole refraction surveys. The surface seismic refraction arrays and location of boreholes on sinkhole 7 is shown on Figure C.

A Generalized interpretation of the subsurface conditions at sinkholes 5, and 7 can be observed on Figures E, F, and G. The conditions generally consisted of a soft and wet soil matrix with limestone inclusions, over solution riddled, usually voided limestone. It was possible to determine cavities within the rock in some of the sinkholes. However, it was also possible to determine soft or void (collapse soils) conditions within the soil covering the top of limestone (refer to Figures E, F, and G). The determination of the soft and/or voided soil covering the limestone was of the utmost importance for the embankment subsidence assessment that was part of the exploration.

In-situ water well pump tests were then performed on the sinkholes to determine general permeability characteristics of the solution riddled limestone. It was soon determined, at no surprise to the authors, that the permeability of the solution riddled limestone was conduits controlled, rather than controlled by the permeability of a homogeneous rock medium.

A hydrologic assessment was performed for each of the sinkholes and the general watershed within the right of way contributing runoff to the sinkholes. Data on expected rainfall over the watershed of the various sinkholes for obtaining probable rainfall intensities for selected frequency and estimated runoff duration were developed from published Generalized Estimates of Probable Maximum Precipitation and Rainfall-Frequency data for Puerto Rico and the Virgin Islands. The watershed areas for each of the sinkholes under study were plotted and area figures in Acres as shown in Table 1.

Both the existing groundwater flows and the conditions remaining after construction of the highway embankments, cross slopes, and drainage improvements, were investigated and the corresponding areas and runoff flows were determined. These discharge quantities, measured in cubic feet per second (cfs) are listed in Table No.1 and Figure H. A design frequency interval of 100 years was used for the runoff design, as recommended to the PRHTA.

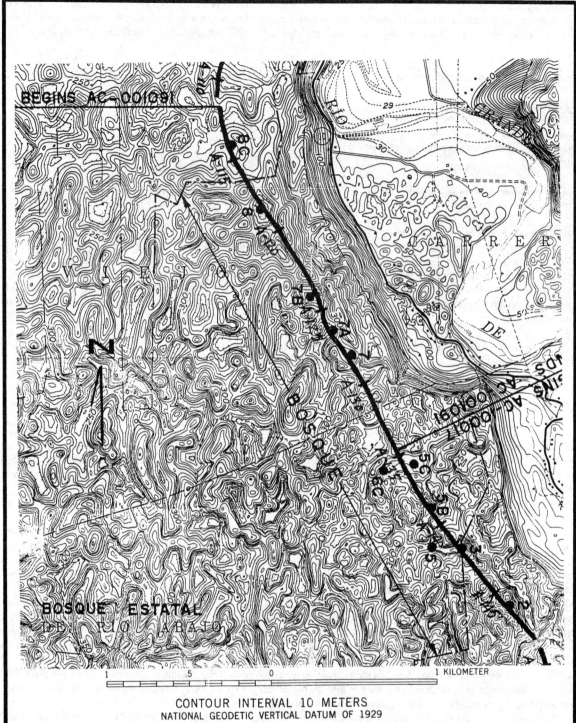

FIGURE B: UTUADO 7.5 SERIES TOPOGRAPHIC QUADRANGLE SHOWING THE ROUTE ALIGNMENT & THE LOCATION OF SINKHOLES

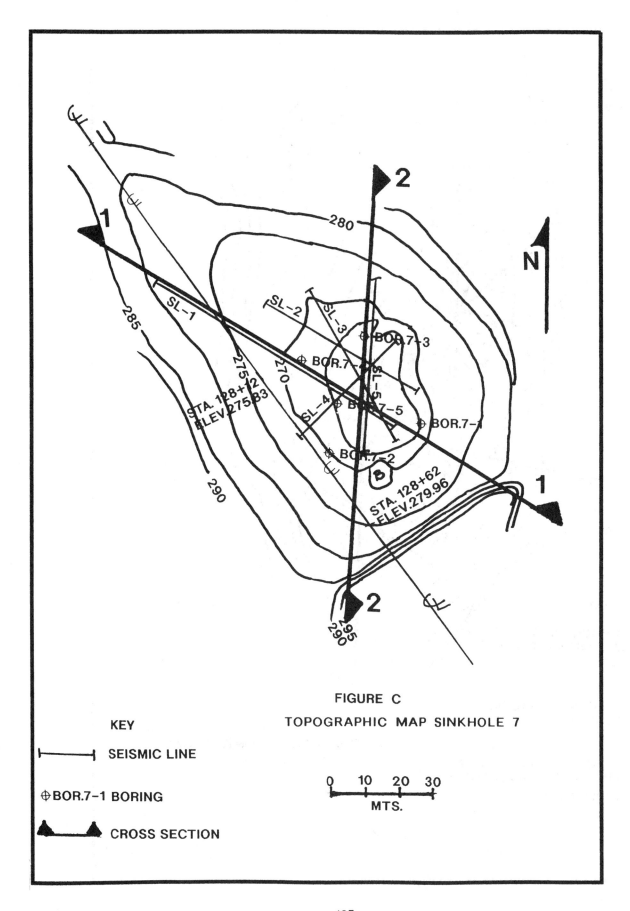

FIGURE C
TOPOGRAPHIC MAP SINKHOLE 7

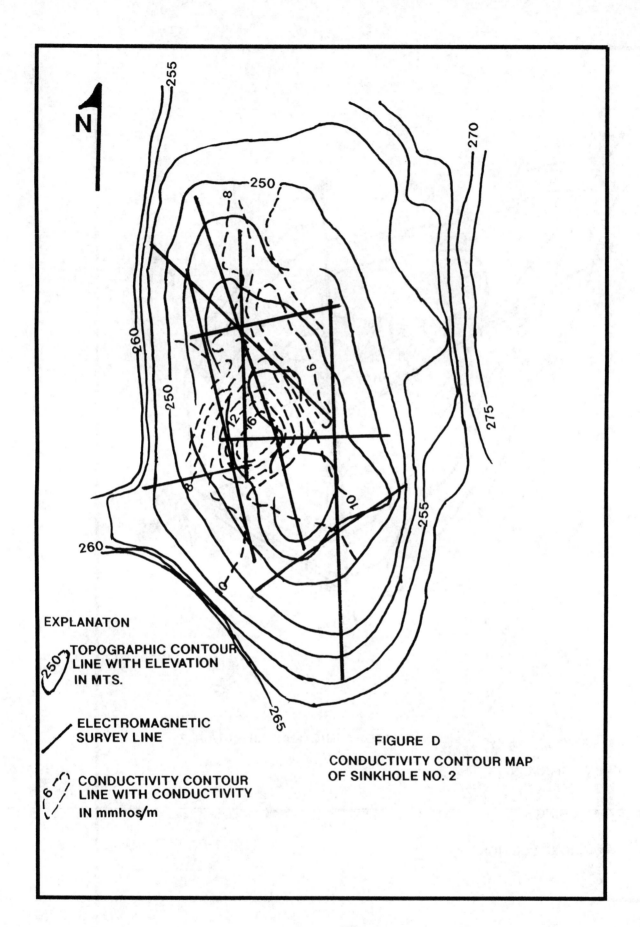

FIGURE D
CONDUCTIVITY CONTOUR MAP
OF SINKHOLE NO. 2

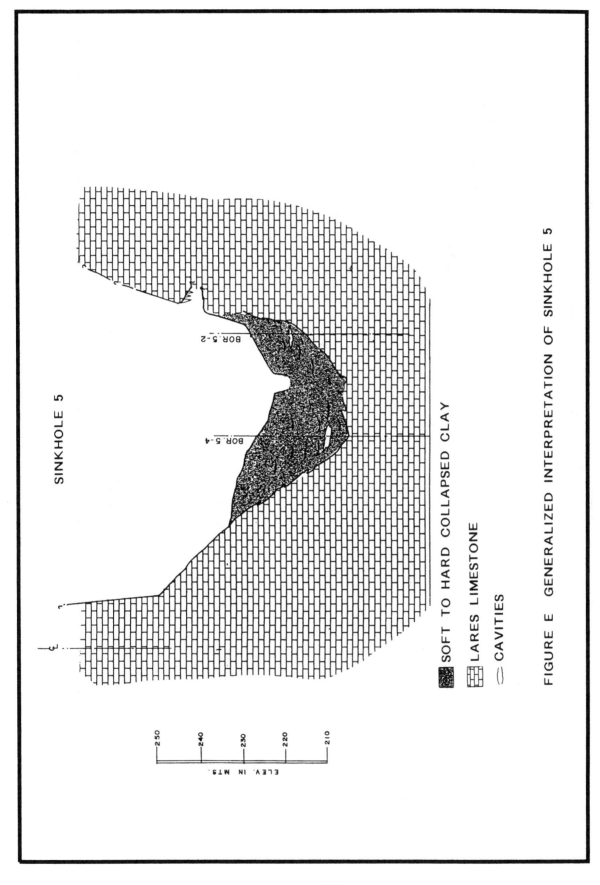

FIGURE E GENERALIZED INTERPRETATION OF SINKHOLE 5

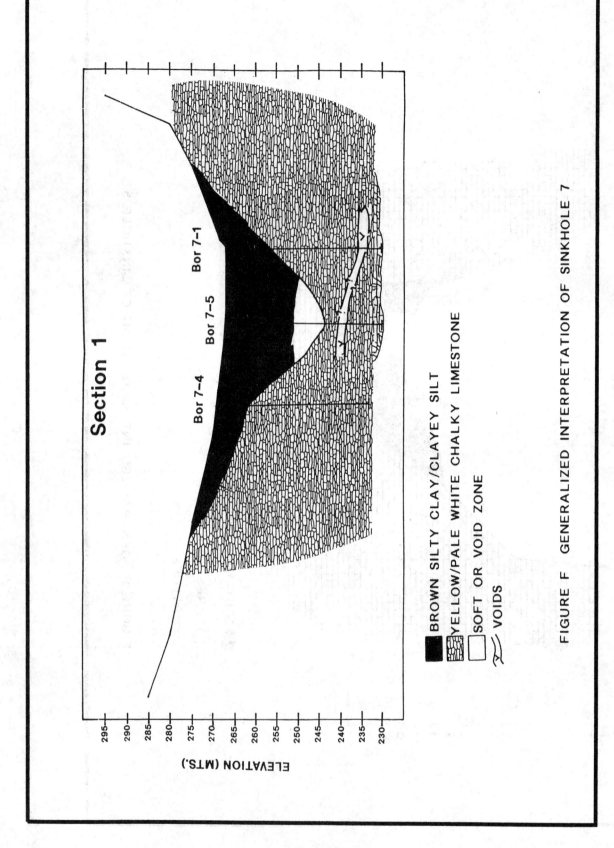

FIGURE F GENERALIZED INTERPRETATION OF SINKHOLE 7

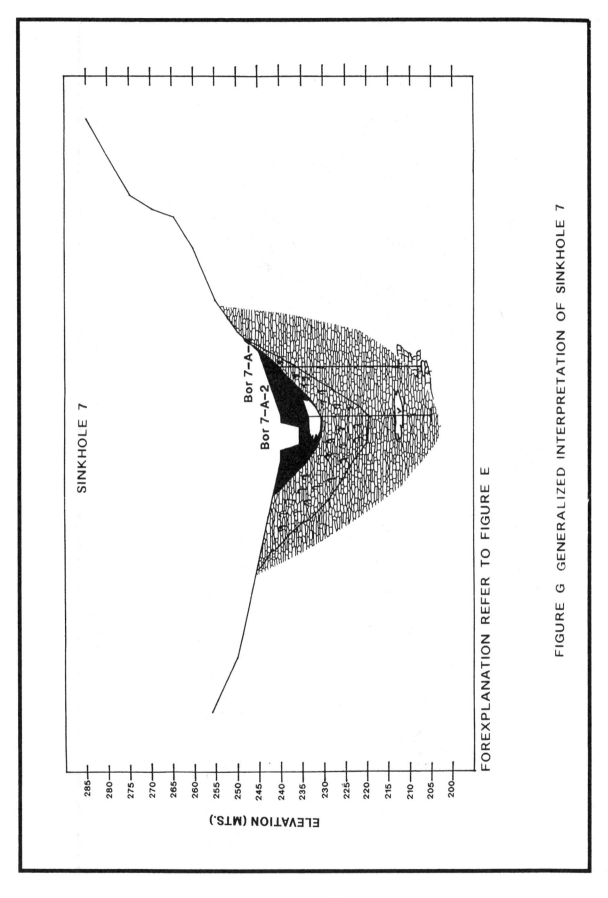

FIGURE G GENERALIZED INTERPRETATION OF SINKHOLE 7

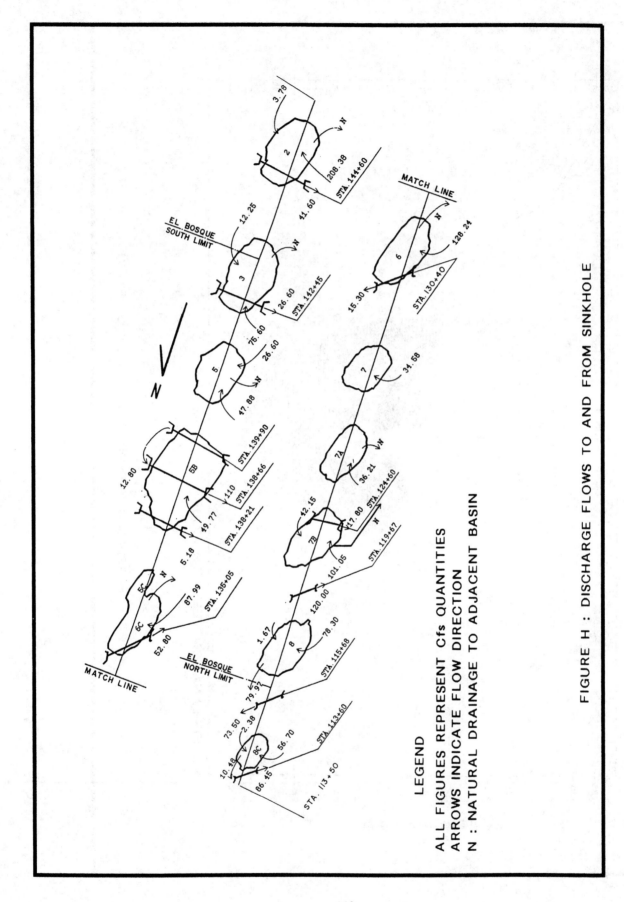

FIGURE H : DISCHARGE FLOWS TO AND FROM SINKHOLE

LEGEND

ALL FIGURES REPRESENT Cfs QUANTITIES
ARROWS INDICATE FLOW DIRECTION
N : NATURAL DRAINAGE TO ADJACENT BASIN

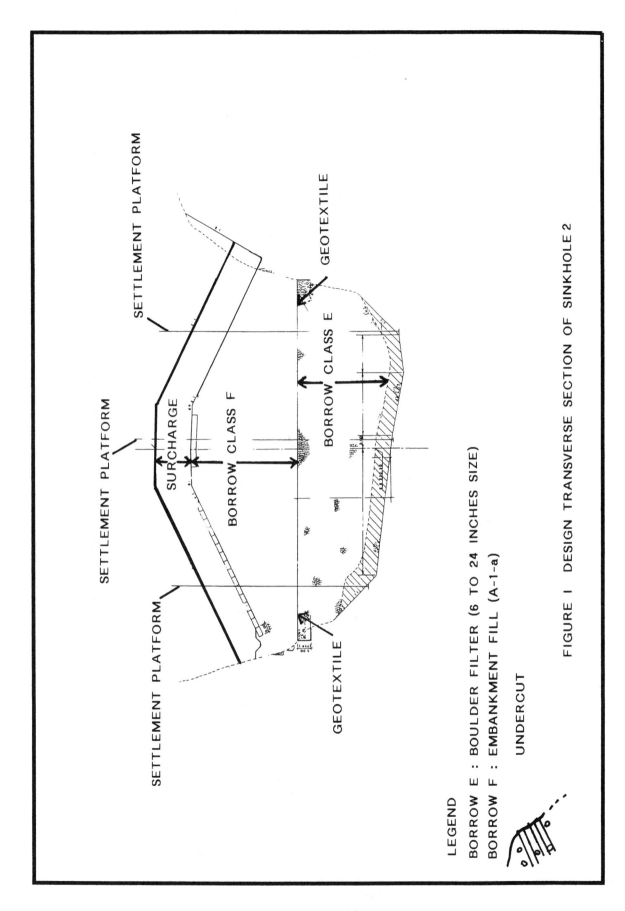

FIGURE 1 DESIGN TRANSVERSE SECTION OF SINKHOLE 2

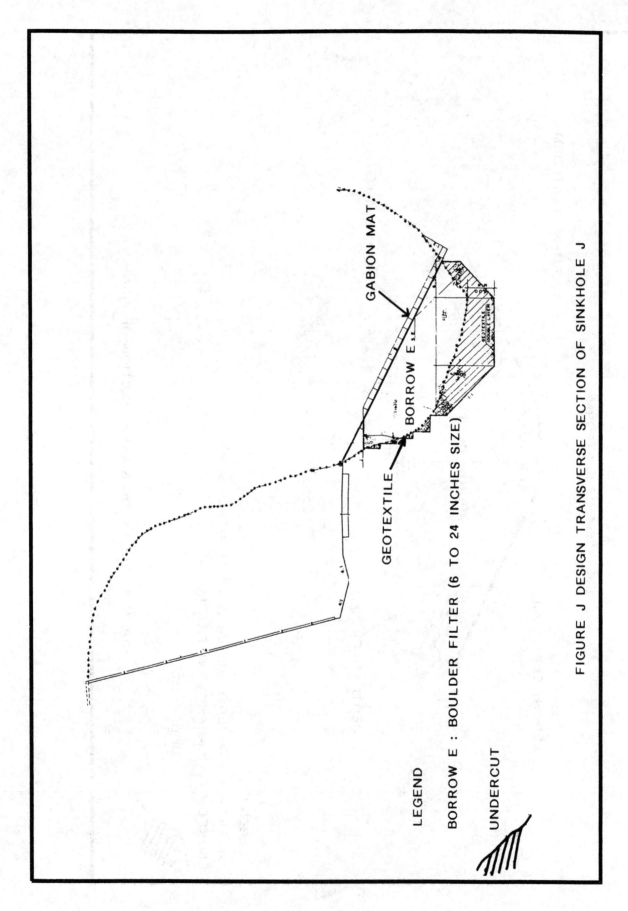

FIGURE J DESIGN TRANSVERSE SECTION OF SINKHOLE J

LEGEND

BORROW E : BOULDER FILTER (6 TO 24 INCHES SIZE)

UNDERCUT

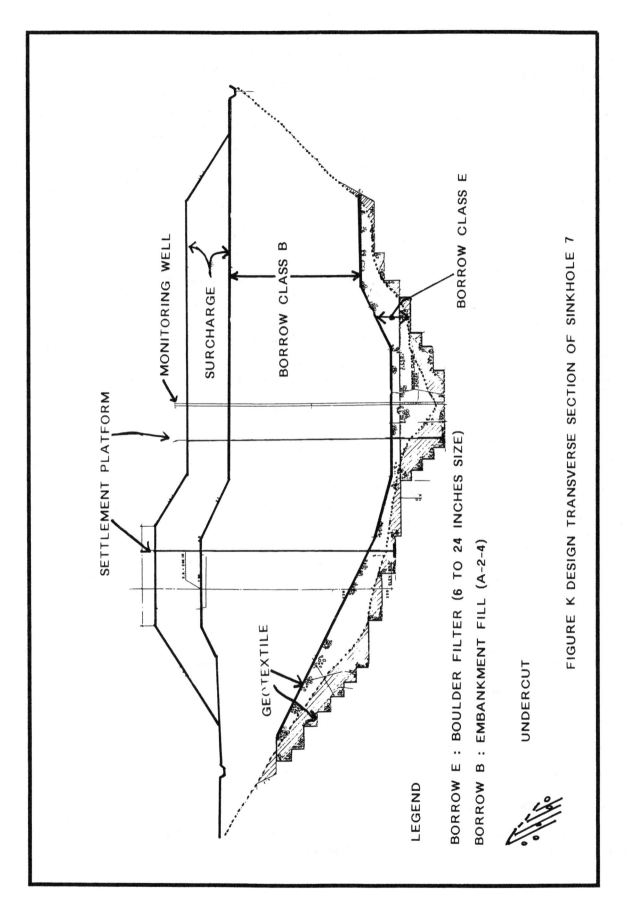

FIGURE K DESIGN TRANSVERSE SECTION OF SINKHOLE 7

TABLE NO.1

Sinkholes	Watershed	Areas/Runoff
Sinkhole No.	Area (Acres)	Original Runoff (Cfs)
8C	8.10	56.70
8	13.16	78.30
7B	16.05	101.05
7A	5.17	36.21
7	4.94	34.58
6	18.32	128.24
6C AND 5C	12.58	87.99
5B	7.90	79.77
5	8.56	47.88
3	10.86	75.60
2	37.31	208.38

Color dye was injected on several sinkholes and traces were measured on selected springs of water and on natural cascades hundreds of meters north and east and away of the highway. It became evident that all the sinkholes were active and interconnected to an intricate underground drainage regime of caverns and conduits. The connection was established to be a south-to-north connection. That is, dye injected on the southern sinkholes were detected on the northern sinkholes. The opposite was not established, except for a natural waterfall east of the highway where dye was detected.

From the onset it was clear that any attempt to build a road through the rain forest meant designing a road that would not plug the natural drainage of runoff. The latter meant numerous bridge structures followed by short road cut sectors, bridge structures followed by tunnels, or conventional cut and fill construction.

A cut and fill approach was selected mainly due to cost consideration, but rather than using a conventional cut and fill design, (which would have eventually proved to be disastrous due to settlement and embankment collapse problems caused by sinkhole plugs), a cut and "filter" fill treatment approach was chosen. Natural soft sediments deposited at the sinkhole basins were to be removed as part of the treatment. Deeper cavities and soft zones below the basin were to be treated by promoting early collapse of the cavities or settlement of soft zones with the use of surcharge loads. The sinkholes were treated, some were provided with geomembranes and filters, in most cases enhancing the sinkhole's drainage capacity. Some were fitted with deep injection wells. Others, although filled with granular and permeable material, were capped by less permeable material designed to reduce and/or divert infiltration into adjoining sinkholes, but still being able to at least handle the original (or before highway) drainage capacity.

The filter criteria mostly used was that the filter size needed to be at least D/4 of any open mouth (where D is the diameter of the boulders in filter) or sink found within the main depression. In addition to the size of the boulders to be used as filters, an abrasion criterion was also required. In-situ void ratio measurements were made to assure filter permeability and internal storage capacity. Some of the sinks where more run-off disposition was to take place were instrumented during designing phase with rain gauges, flow meters and water head pressure meters, from which the theoretical run-off capacity of the sinks were re-evaluated from that assessed only from mathematical computations. Based on said field determined capacity to permit water to flow into the sinkholes, the filter size were re-evaluated so as to try to maintain percolation capacities. This meant that sometimes the filter size needed was determined by its storage capacity (internal voids) since the D/4 criteria proved to be less dominant.

Figure I, J and K, are design cross transverse sections of sinkholes 2, 3 and 7. Note that Figures I and K have a surcharge load included. The surcharge was included in the design to promote early collapse of near surface cavities in rock that were not detected by the multi-disciplinary exploration methods. Early collapse of previously undetected, possibly near surface, cavities were evidenced during construction by settlement platforms set at the top of limestone before placing the embankment, and during subsequent monitoring of these as the embankments were raised.

CONCLUSIONS

The application and results of the multidisciplinary approach methods were presented on this paper, rather than the details of the types of equipment and the selected field parameters. However, with respect to typical geotechnical surveys for highway design and construction, based on more than 25 years on experience in Puerto Rico and the Virgin Islands, the multi-disciplinary methods presented in this paper are minimal requirements for the successful interpretation of this highly complex karst terrain.

The roadway was successfully constructed and has been in use since 1994. Monitoring is still being implemented and assessed as recommended.

REFERENCES

Giusti Enio, 1978. Hydrogeology of te Karst of Puerto Rico U.S.G.S. Professional Paper 1012

Monroe, Watson H., 1976. The Karst Land Forms of Puerto Rico. U.S.G.S. Professional Paper 899.

Vázquez Castillo, Vázquez Castillo & Associates, 1987 Caribbean Soil Testing Laboratories, San Juan, PR., Geotechnical & Geophysical Exploration & Grouting Operations for Abbott Laboratories, Barceloneta, Puerto Rico

Vázquez Castillo, Vázquez Castillo et. al. 1989 Geotechnical & Geophysical Exploration for Industrial Development at Sterling Pharmaceuticals, Barceloneta, P.R.

Vázquez Castillo-Vázquez Castillo 1987. Caribbean Soil Testing, San Juan, PR. Geotechnical & Geophysical Exploration Proposed PR-10, Utuado, P.R. (Sinkholes 1 through 5)

Vázquez Castillo-Vázquez Castillo, 1988. Caribbean Soil Testing, San Juan, PR. Geotechnical & Geophysical Exploration Proposed PR-10, Utuado, P.R. (Sinkholes 6 through 8)

Vázquez Agrait-Vázquez Castillo, 1984. Caribbean Soil Testing, San Juan, Report B, Geotechnical Recommendations for Sinkhole C at Proposed P.R. No.10, Utuado, P.R.

Rodríguez Molina C., Holt Richard, Vázquez Castillo L., Corchado Vargas F, 1990. Engineering and Exploration of Karstic Foundations in Puerto Rico, Proceedings Volume 1, International Symposium on Unique Underground Structures, US Bureau of Reclamation, Denver, Colorado

Fieldtrip guidebook

Pre-conference field trip: Karst geology and hydrogeology of the Ridge and Valley and Piedmont Provinces of south-central Pennsylvania

WILLIAM E.KOCHANOV & HELEN L.DELANO   Pennsylvania Geological Survey, Harrisburg, Pa., USA
CAROL DEWET   Department of Geosciences, Franklin and Marshall College, Lancaster, Pa., USA
STEPHANIE B.GASWIRTH   Department of Geological Sciences, Rutgers University, Piscataway, N.J., USA
DAVID A.HOPKINS   Baker Refractories, York, Pa., USA
JEFFREY L.LIEBERFINGER, HEATHER L.RECCELLI-SNYDER & RICHARD A.HOOVER   Science Applications International Corporation, Middletown Pa., USA
ALBERT E.BECHER   Harrisburg, Pa., USA

INTRODUCTION

Thick sequences of structurally deformed carbonates comprise the surface bedrock of a sizeable area in south-central and southeastern Pennsylvania (Figure 1). Folds in the carbonate bedrock, often overturned and attenuated, and numerous faults have resulted in extensively fractured bedrock. Weathering of the deformed carbonate units for millions of years under temperate climatic conditions has produced a generally subdued, but deeply developed, karst (Wilshusen and Kochanov, in press).

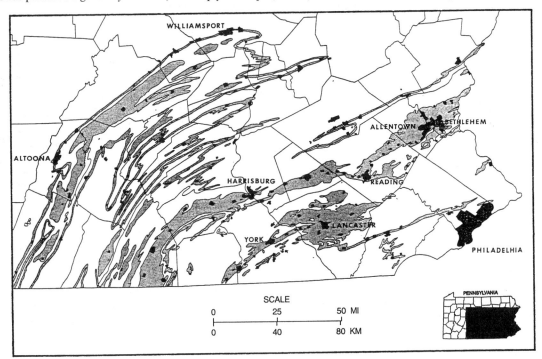

Figure 1. Map of south-central and southeastern Pennsylvania showing the distribution of carbonate bedrock (gray) and the location of major population centers (black).

The contact between residual soil and well-indurated carbonate bedrock is sharp but extremely irregular. Surface water and groundwater movement have transported soil, filling some voids and solution channels, but leaving others partially filled or entirely empty. The depth to carbonate bedrock is variable. Data from water well records in the Great Valley carbonate section give depths to bedrock ranging from five to over two hundred feet (Pennsylvania Geological Survey, Water Well Data System). In karst areas adjacent to mountain ridges, a greater thickness of colluvium has been added to the carbonate residuum such as along the north flank of South Mountain in Cumberland and Franklin Counties. Colluvial material beneath Pond Sink in Franklin County may be as much as 335 feet thick (Pierce, 1965). Deposits of glacial outwash have also contributed sediment to the carbonate residuum such as in the Lehigh Valley of eastern Pennsylvania (Wilshusen and Kochanov, in press).

Certain carbonate units, such as the Cambrian Allentown, Ledger, Rockdale Run, Tomstown, and Zullinger Formations, the Ordovician Annville, Conestoga, Epler, Ontelaunee Formations and the St. Paul Group, are examples of carbonate units more susceptible to sinkhole development within the Great Valley Section and Piedmont Province of Pennsylvania (Figure 2). In a study of sinkholes in the Great Valley Section (Kochanov, 1993), the Allentown and Epler Formations, and their stratigraphic equivalents, account for 85 percent of recorded sinkholes (Wilshusen and Kochanov, in press).

The karst geology of Pennsylvania is unique in that it has developed to different degrees depending on where one is located with respect to physiographic province. Lithologic characteristics and structural attitude are the key factors in karst development. In central and southeastern Pennsylvania, areas of karst are found in northeast southwest trending valleys. Coincidentally, these relatively "flat-bottomed" valeys are more desirable than the adjacent ridges as sites for homes, farms, industry, and transportation routes. The residual soil is excellent for agriculture, and in many places the carbonate rock is a valuable mineral resource and a host rock for some metallic ore deposits (Wilshusen and Kochanov, in press).

System	Series		CUMBERLAND VALLEY		LEBANON VALLEY		LEHIGH VALLEY		NOTHERN LANCASTER AND CONESTOGA VALLEYS
ORDOVICIAN	Trentonian	CHAMBERSBURG FM.	Martinsburg Fm.*		"Martinsburg Fm."*	JACKSONBURG FM.	Martinsburg Fm.		Cocalico Fm.
			"Oranda" Fm.		Hershey Fm.*		"cement rock"		? ? ? ? ?
			Mercersburg Fm.		Myerstown Fm.*		"cement ls."		Myerstown Fm.
	Black River		Shippensburg Fm.						
	Chazyan		St. Paul Gp.*		Annville Ls.*				Annville Ls.
	Canadian	BEEKMANTOWN GP.	Pinesburg Station Dol.*	BEEKMANTOWN GP.	Ontelaunee Fm.*	BEEKMANTOWN GP.	Ontelaunee	BEEKMANTOWN GP.	Ontelaunee Fm.
			Rockdale Run Fm.		Epler Fm.*		Epler Fm.		Epler Fm.
			Stonehenge Ls.		Rickenbach Dol.		Rickenbach Dol.		Stonehenge Ls.
					Stonehenge Ls.*				
CAMBRIAN	Upper		Conococheague Fm.	CONOCOCHEAGUE GP.	Richland Fm.		Allentown Fm. undivided	CONOCOCHEAGUE GP.	Richland Fm.
					Millbach Fm.				Millbach Fm.
					Schaefferstown Fm.				
			Big Springs Station Mbr.		Snitz Creek Fm.				Snitz Creek Fm.
	Middle		Elbrook Fm.		Buffalo Springs Fm.*				Buffalo Springs Fm. or Elbrook Fm.
					concealed				? ? ? ? ?
	Lower		Waynesboro Fm.				Leithsville Fm.		Ledger Dol.
			Tomstown Fm.		Leithsville Fm.				Kinzers Fm.
									Vintage Fm.
			Antietam Ss.		Hardyston Qtzite		Hardyston Qtzite		Antietam Ss.
			Harpers Fm.						Harpers Fm.
			Weverton Fm.						Chickies Fm.
			Loudon Fm.						
Basement			Volcanics		Gneisses		Gneisses		Gneisses

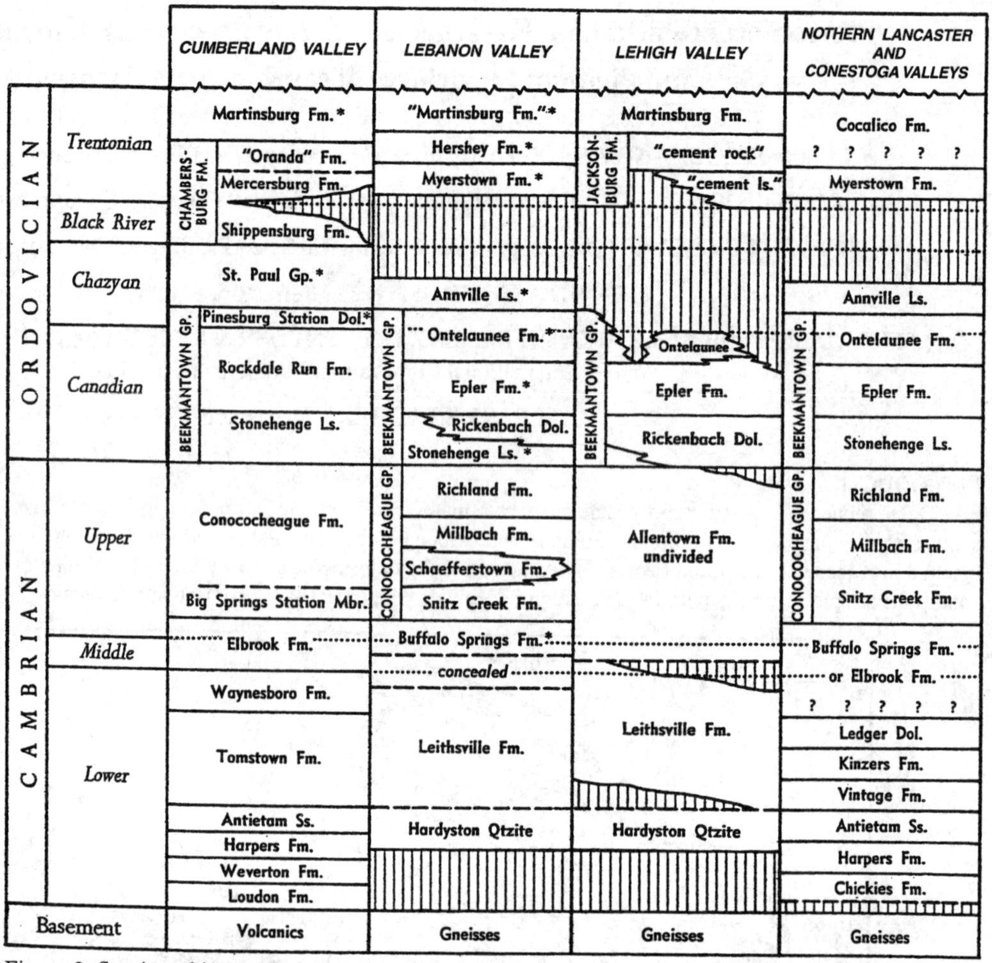

Figure 2. Stratigraphic correlation chart of units within the field trip area (modified from MacLachlan, 1967).

As population increases in these regions, rural areas are targeted for urban expansion. Figure 1 shows the location of population centers within or in close proximity to, the major carbonate regions of Pennsylvania. As these areas become increasingly developed, the potential for karst-related problems increases. Intensive land use has led to many incidents of subsidence (Knight, 1971; Myers and Perlow 1984, White and others, 1984; Wilshusen, 1979; Wilshusen and Kochanov, in press).

Subsidence in carbonate terranes is a natural process that is often accelerated by the activities of man. A number of man-induced factors can account for the majority of sinkhole occurrences. Examples vary from the failure of water-bearing utility lines (Berry, 1986) to groundwater withdrawal from quarries (Foose, 1953). As urban areas expand, stormwater runoff is redirected and concentrated. Sinkholes have been observed in stormwater retention basins and storm drains as a direct result of heavy precipitation and increased stormwater runoff. In some instances, sinkholes have been used as stormwater drains (Wilshusen and Kochanov, in press). Additionally, sinkhole development can be significantly increased through abnormal precipitation events such as hurricane Agnes in 1972, which dumped approximately 43 cm (17 in) of rain in the Susquehanna River drainage basin over a three-day period.

Areas that have already undergone development have special problems in redesign and reconstruction. The after-the-fact methods of subsidence repair are often expensive and offer no guarantee from sinkhole reoccurrence. Sinkhole repair for the Vera Cruz Road in Lehigh County cost nearly $800,000 (US) and had a new sinkhole open, just outside of the repair area, within six months (Wilshusen and Kochanov, in press).

A primary cause for subsidence problems is the failure to be cognizant of karst processes and their impact, prior to land development. The municipal government determines most guidelines for construction. In most cases, the local zoning laws are ineffective or nonexistent in regulating land development in potential subsidence areas. The lack of continuity from one municipality to the next, with regard to zoning laws in carbonate regions, hinders efforts for comprehensive and effective land-use planning. There are special regulations, however, for construction of certain facilities, such as sanitary landfills, that have been established at a state level (Wilshusen and Kochanov, in press).

Field trip prologue. A detailed discussion of the varied nature of the karst landscape in Pennsylvania would exceed the scope of this Proceedings entry. It is the intent of this field trip, however, to examine a portion of two important sections within the Ridge and

Valley and Piedmont provinces to provide the individual with a small sampling of Pennsylvania's varied karst landscape (Figure 3).

The reference list at the end of the field trip log should provide a good reference base. Additional sources of information on karst in Pennsylvania can be obtained through the Pennsylvania Geological Survey and other state agencies such as the Department of Transportation and the Department of Environmental Protection. Additionally, information can also be obtained through the state's colleges and universities.

Data on caves is available through a series of bulletins published by the Mid-Appalachian Region of the National Speleological Society and through a computer database.

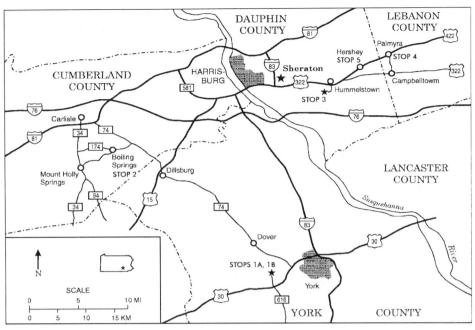

Figure 3. Map showing the location of stops of the field trip.

The database, maintained by The Pennsylvania Cave Survey, contains approximately 1500 caves with over 50 items of data per cave. Site specific information from the database can be made available for use as an aid to site investigation for land use and development.

Field Trip Log
Set Odometer to 0.

0   Leave Sheraton parking lot, turn left on East Park Drive. Get in left lane.
0.3   Turn left at traffic light onto Union Deposit Road
0.7   Turn left at traffic light onto entrance ramp to I-83 south.
1.9   I-83 and Rt. 322 (Derry Street) split, stay on I-83 (left lanes). Immediately after split, right lane towards Harrisburg.
4.9   Bridge over the Susquehanna River. View of downtown Harrisburg to right, smoke stacks of Three Mile Island, site of infamous nuclear power plant accident, visible downstream on left.
5.6   Exit Right - Follow I-83 to York as it splits from Rt. 581
6.6   Capital City Airport is 1-2 miles east (left) of here. The airport has had a long history of sinkhole problems. During 1983, the site underwent an extensive engineering evaluation. The selected treatment for the airport was to perform limited remedial work and overall pavement rehabilitation and couple this with a monitoring program using acoustic emission/microseismic techniques (Belesky and others, 1987).
8.0   Cross over Pennsylvania Turnpike
8.5   Exit 18, Contact with the Triassic Gettysburg Formation. Stay on I-83
8.9   Jurassic age diabase outcrop on left (east)
9.6   Reesers Summit. Green dodecahedral garnets occur at and near the contact of the diabase dike and Triassic quartz pebble conglomerate. See Lapham and Geyer (1969) for more information.
9.8   Fanglomerate of the Triassic Gettysburg Formation outcrop on both sides of the road
16.6   Red beds of the Triassic Gettysburg Formation on both sides of road
17.4   Cross Conewago Creek
18.9   Approximately 2 miles southwest of here lies the hamlet of Zion's View and the "Stahle Bone Bed." Numerous disarticulated skeletons of the Triassic phytosaur *Rutiodon* and the amphibian *Buettneria* were collected here and are now housed in the Pennsylvania State Museum (Kochanov and Sullivan, 1994).
24.0   World famous York Barbell Company on left.
24.1   Contact of Cambrian carbonates, note flat topography.
25.2   Leave I-83 at Exit 10 get in center lane.
25.5   Turn right at traffic light onto N. George Street, follow signs "to Rt. 30"
26.0   Turn right onto Rt. 30 West
26.1   Traffic light at Susquehanna Trail, stay on Rt. 30
27.9   Manchester Mall on right, site of the Mummy Sinkhole (Kochanov, 1997)

28.2 Note patched pavement on Rte. 30. This is from repairs to fix a sinkhole in December of 1993. The 6 m (20 ft) void was discovered when conducting routine road maintenance. At that time, there was at least 1.5 m (5 ft) of earthen material between the void and the subsurface of the highway. Sinkholes and small caves are present along the south side of the highway west of the Olive Garden Restaurant.

31.1 Bear right; stay on Rt. 30 toward Gettysburg.
31.8 Turn right at traffic light onto Baker Road.
32.0 Rail road crossing.
32.6 Pull off to right into entrance to Stop 1A.

STOP 1. Starting in the Piedmont Lowlands Section of York County, Stop 1 will be broken up into two sections. At the first section, Stop 1A, Dr. Carol deWet will discuss the depositional environment of the carbonate bedrock introducing a new type of depositional facies amid a fantastic backdrop of pinnacled bedrock. At Stop 1B, Dave Hopkins will discuss the quarry operations and the relationship of the local geology to the mining of this valuable product. In addition, we will view exposures of paleokarst where Triassic sediments have infilled caves within the Cambrian Ledger Formation. We will also have a chance to examine a major unconformity where Triassic breccia/fanglomerate overlies Cambrian carbonates.

1A. Discussant: Dr. Carol deWet, Franklin and Marshall College.

Introduction. The Middle Cambrian Ledger Formation is pervasively dolomitized throughout much of its geographic extent. In York Co., Pennsylvania, however, part of the Ledger Formation retains its original calcite mineralogy, revealing an exceptionally well-preserved shelf margin system characterized by algal fabrics and fibrous cavity-filling cements. The lithologies and fabrics in the limestone include dense microbialite with fenestrae, cavities containing *Renalcis* and stromatolites, and intraclast beds. These are interpreted as shallow subtidal, high-energy deposits.

Centimeter-to-meter scale primary cavities are lined with multiple generations of fibrous submarine cement. Petrography of the cements shows radial-fibrous calcite, associated with radiaxial fibrous calcite and fascicular-optic calcite. Evidence for precursor high magnesium calcite mineralogy of the cements consists of abundant microdolomite inclusions, relict high Mg and low Sr values.

Geologic setting. The eastern Laurentian Cambro-Ordovician continental shelf margin bore the brunt of deformation during the closure of the Iapetus Ocean and most Laurentian shelfbreak deposits have either been eroded, stratigraphically displaced, recrystallized and/or dolomitized. This shelf margin succession has been coined "The Great American Bank" by Ginsburg (in Demicco and Mitchell, 1982). Recent discovery of an *in situ*, undolomitized shelf margin deposit from Middle Cambrian "Great American Bank" strata has provided new information on the Laurentian shelfbreak. The Ledger Formation typically consists of light-colored, medium to coarsely crystalline dolomite. Massive and thick-bedded, it contains few preserved sedimentary structures except cross-stratification. Petrographically, much of the formation consists of dolomitized ooid grainstone. Based upon the purity and light color of the dolomite, and the presence of weakly defined shallowing-upward cycles, the Ledger Formation is interpreted as a carbonate shelf deposit (Gohn, 1976), dominated by shelf margin ooid shoals (Taylor and Durika, 1990).

In York County, Pennsylvania, however, part of the Ledger Formation retains its original calcite mineralogy, revealing an exceptionally well-preserved shelf margin system characterized by algal fabrics and fibrous cavity-filling cements. Informally named the "Ledger limestone", this undolomitized area contains karst pinnacles up to 10 m high which provide three-dimensional exposures of the limestone (Figure 4). Borehole core

Figure 4. Pinnacled surface of Ledger limestone, J.E. Baker quarry, near West York, York County, PA. Photo by Helen Delano.

provides another 60 m of limestone. A Middle Cambrian age for the Ledger Formation was suggested by Gohn (1976), based on stratigraphic location, and the lack of diagnostic Early Cambrian fossils (Figure 2).

The Cambrian period documents a transition from stromatolite-dominated organic buildups to Paleozoic metazoan-calcimicrobial reefs and bioherms (Turner et al., 1993). This may reflect a trend in increasing complexity, from primarily algal trapping and binding mechanisms to algal-growth-framework development. Descriptions of Middle Cambrian *in situ* shelf margin biogenic deposits are rare (Evans et al., 1995), therefore, the nature of the Ledger limestone's microbial deposits and cements are important in furthering the understanding of shelf margin evolution and early cementation.

Most of the cavity-filling cement bands within the Ledger limestone consist of fibrous calcite with some of the cement bands consisting of crenulated dark and light calcite, termed

Figure 5. Field photograph illustrating the irregular contact between submarine cavities, filled with light colored cement and sediment, and the surrounding microbialite (dark color).

herringbone calcite (Sumner and Grotzinger, 1996). Sumner and Grotzinger (1996) suggest that herringbone calcite is a growth morphology that reflects distinct geochemical paleoenvironmental conditions, anoxic or anaerobic, early Paleozoic seawater. Herringbone calcite in some of the Ledger limestone's cavities is enigmatic in light of this interpretation because to precipitate the quantity of cement seen in the Ledger limestone high energy currents must have been pumping large volumes of seawater through the microbialite (Marshall, 1983). This scenario suggests that the porewaters would have been in constant exchange with open seawater, and would have been oxygenated. The authors suggest that high levels of decaying organic matter within cavities may resolve this paradox.

Lithologies. The Ledger limestone consists of two carbonate lithologies: (1) microbialite; and (2) packstone-grainstone. The microbialite is composed of algal-bacterial sediment with internal cavities that range in size from a few millimeters to 1.5 meters long by 0.5 meter high. Packstone-grainstone allochems vary in proportion, but consist of oncolites, ooids, peloids, intraclasts, and rare skeletal fragments. Centimeter-high domal stromatolites constitute discreet horizons that may be traced across several pinnacles.

Microbialite. Microbialite is dark gray to black, weakly to massively bedded, with a lumpy and gnarled texture (Figure 5). The rocks are not laminated like stromatolites or algal mats, nor are there discrete outlines of clotted micrite as in thrombolites. The microbialite contains millimeter-to-centimeter long, irregularly shaped cavities that are elongate parallel to bedding and resemble fenestrae. These cavities are filled with fibrous to equant calcite cement. Larger cavities (up to 1.5 meters long) have flat or undulose bases and upper surfaces that widen and narrow in three dimensions. Well-preserved *Renalcis* -like algae rim the insides of the larger cavities, and are overlain by fibrous calcite cement (Figure 6). Fragments of the cavity walls are encased in algal coatings and may be

Figure 6. Field photograph showing a large cavity lined with well-preserved *Renalcis.*-like algae, overlain by internal sediment and fibrous cement (arrow).

cemented to cavity floors. Lateral and vertical contacts are gradational except at cavity boundaries, which are sharply defined. Microbialite is dissected by lenses containing coarse-grained intraclast lags and slumped wall fragments. Petrographically, the microbialite is composed of dense, clotted micritic clumps and uneven layers of micritic spheres and peloids. Stromatolites consist of laterally linked and separate domes of micro-laminated micrite. The stromatolites consistently overlay mm- to cm-scale microbialite and are usually overlain by grainstone horizons, followed by mm- to cm-scale microbialite (Figure 7).

Figure 7. Field photograph of a band of stromatolites underlain by, and overlain with microbialite.

Packstone-Grainstone. Rounded and subrounded clasts of peloidal micrite, ooids, oncoids, and intraclasts form packstone to grainstone horizons within the microbialite facies. The grainstones are commonly, cross-bedded (35-45o) and occur in lenses filled with light colored cement and sediment, which may have coarse-grained pebble lags at the base.

Facies interpretation. Most of the lithologies within the Ledger limestone have an algal (+/- bacterial) origin. The stromatolites and oncolites formed by algally-mediated carbonate precipitation. Algae rim the cm- to m-scale microbialite cavities, and there is no clearly defined transition from those algal forms into the rest of the microbialite. Our interpretation is that the microbialite formed through a process of algal (+/- bacterial) growth and syndepositional cementation. Algae appear to have filled most available ecological niches; from forming a rigid structure, to dwelling in cryptic spaces, to forming stabilized and partially-stabilized colonies of stromatolites and oncolites.

Microbialite. The algal microbialite was deposited in a subtidal, shallow-water setting based upon the absence of vadose cements or other features indicative of subaerial exposure. *Renalcis*-like algae is interpreted as relatively high energy, open framework algae capable of partially constructing a calcified rigid structure (Pratt, 1984). Wright (1994) suggests that microbial buildups are indicative of seas with high nutrient levels, characteristic of up-welling sites. The presence of lenses and collapse features within the microbialite indicates relief off of the sea floor. The large open cavities within this lithology indicate that it must have been at least semirigid to maintain such spaces. Internal marine sediment overlying, and overlain by, cavity cements indicates that marine water circulated through the cavities concurrent with framework development (Barnaby and Read, 1990). The fibrous form of the cements, their association with marine sediment, and lack of dissolution features suggest a syndepositional, submarine origin for the cements and their enclosing matrix. The quantity of cement indicates that porewater circulation was robust and long lasting. The microbialite is interpreted, therefore, as having formed as an undulating, organic plain along a high-energy shelf margin setting. The Ledger limestone's close association with ooid shoals substantiates its interpretation as having formed under subtidal conditions on a high-energy shelf margin.

The microbialite's small cavities resemble fenestrae, and fenestrae are often associated with intertidal environments, but in this case the rock contains no other indications of vadose conditions. The calcite-filled small cavities within the microbialite facies are interpreted as gas voids formed through organic decay, and/or small-scale versions of the meter-scale cavities.

Packstone-Grainstone. Coarse-grained lags and cross-bedding of subaqueous sands is evidence of a high energy, depositional environment. Coarse sands and gravels in channels running seaward between reefs has been recognized in ancient shelf margin sediments, and is a feature commonly observed in modern settings (James, 1983). Oncolites are moderate energy, peritidal to subtidal microbial cryptalgal structures. Their presence indicates photic zone conditions, and the oncolites attest to periodic water agitation to roll the algal ball and initiate further algal growth. In the Ledger rocks, these features, coupled with the lack of subaerial exposure indicators such as meniscus cement, substantiates that these rocks formed in subtidal, high-energy conditions.

The stromatolites indicate a period of substrate stabilization that produced attached algal deposits in contrast to the mobile oncolites. The stromatolites probably reflect shallower water conditions than the microbialite reef that was deposited under completely subtidal conditions. The stromatolites could have formed in either subtidal or intertidal conditions, based on comparison with extant examples.

Cavity-filling cements. Most of the Ledger limestone cement consists of multiple generations of cream-colored to gray, nonferroan, fibrous crystals, which line cavities (Figure 8). Baroque dolomite occludes final porosity in the larger voids. The baroque dolomite is interpreted as a burial precipitate and is not considered in this paper. Stubby to bladed calcite crystals within millimeter-scale cavities are similar to the fibrous cements and are considered a subset of them in this discussion.

The largest cavity found in the field measures 2 m long by 0.5 m high at its widest point. Individual cement bands within it range from 5 mm to 2 cm in thickness. Smaller cavities have thinner cement bands, ranging from 0.5 mm to 6 mm in thickness. Bands composed of herringbone calcite are generally thicker, averaging from 5 to 8 mm thick. The cement distribution is always isopachous around the inside of cavities, and may coat fragments of the wall rock contained within cavities. Individual bands are defined by submillimeter layers or lenses of internal sediment or by color changes reflecting the density of inclusions. Inclusion-rich calcite layers are cloudy, and gray to creamy white. Bands with few inclusions are clear to pale white.

Figure 8. core photograph showing multiple generations of fibrous marine cement (light bands) which precipitated inside of a cavity that was first encrusted with *Renalcis*-like algae (dark material).

The bands consist of elongate or fibrous crystals. Fibrous crystals consist of bundles of crystals radiating perpendicular to the substrate towards the pore center. Radiaxial cements predominate, but there are also fasicular-optic calcite crystals. There are frequent lateral transitions between these types of fibrous calcite, and radial-fibrous crystals are also present. They may be overlain by another generation of fibrous crystals, or stubbier crystals that may have growth orientations at right angles to the crystal below it. Coarse, more equant calcite distinguishes some bands, and may directly overlie internal sediment. Fibrous and herringbone calcite cements occur within the same cavity. Generally, the first pore-lining cement is fibrous, followed by 2-4 bands of fibrous cement, overlain by 1-4 bands of herringbone calcite cement. Herringbone calcite is a recently described calcite morphology consisting of dark and light crenulated bands (Sumner and Grotzinger, 1996). Each band is composed of a row of elongate crystals with their long axes aligned perpendicular to banding in the cement growth direction. Petrographically, the elongate crystals are optically length-slow (Sumner and Grotzinger, 1996). Within the cavity-filling cements of the Ledger limestone, herringbone calcite occurs as 3-cm-thick light gray bands. There are distinguishable boundaries within the bands of the herringbone cement marked by microdolomite and inclusion-rich zones. All of the cements are nonferroan (Dickson, 1966).

Marine Origin. Numerous lines of evidence indicate the origin of the fibrous cements. Ooids have nuclei composed of fragments of fibrous cement. Micritic sediment containing minute marine shell fragments overlies cement bands, and is in turn overlain by fibrous cement. The cements grew symmetrically around the inside of open voids. There are no dissolution vugs present in the matrix. The abundance of large and small syndepositional cavities means that the microbialite had to be at least semirigid during its formation to maintain the abundance of open space. This implies synsedimentary cementation to provide the rigidity.

All of these lines of evidence indicate syndepositional, submarine cementation on a large scale. The cements are interpreted as marine precipitates because such cements are common in modern, high-energy reef settings, but have not been recorded as burial diagenetic precipitates. Fibrous cements must have been present, and then eroded and reworked, to be available as ooid nuclei. Internal sediment between cement generations indicates that sedimentation was coeval with cement precipitation. The even, isopachous coating of fibrous cement around the inside of cavities indicates that the pores were saturated with water at all time. Vuggy dissolution implies that undersaturated fluids penetrated the limestone, but there are no vugs within the Ledger limestone facies, indicating that it was not exposed to undersaturated waters as commonly occurs during subaerial exposure or exposure to fresh meteoric water. If the microbialite was at least partially rigid during its deposition, as implied by the presence of so many cavities, it must have experienced rapid submarine cementation.

Fibrous cements with undulose extinction patterns are termed radiaxial fibrous or fascicular-optic calcite. Radial fibrous calcite lacks the undulose extinction pattern (Kendall, 1985). All three types of fibrous calcite cements are present within the Ledger limestone. Fibrous calcite is commonly interpreted as a marine precipitate (James and Klappa, 1983; Aissaoui, 1985; Kendall, 1985). Fibrous calcite is abundant in platform margin and reef slope limestones (Mazullo et al., 1990). Radiaxial fibrous calcite is interpreted by Sami and James (1996) to be a primary marine cement that may undergo variable amounts of textural and geochemical alteration during diagenesis. Its marine origin is supported by its close association with internal sediment and near-marine isotopic values (Saller, 1986).

Herringbone Calcite The nonferroan nature of the cements, clearly evident by staining of petrographic sections, indicates that calcite precipitated in an oxidizing and/or iron-free environment (Dickson, 1966; Sami and James, 1996). However, the presence of bands of herringbone calcite indicates a change in precipitation environment within the cement cavities. Herringbone calcite cement is interpreted as forming under anoxic conditions (Sumner and Grotzinger, 1996). This indicates the Ledger cavities experienced occasional periods of oxygen depletion, possibly related to sluggish ocean water circulation through specific cavities for relatively short time periods. In conjunction, decaying microbial material would have consumed oxygen faster that it could be replenished by seawater circulation, promoting localized oxygen depletion, further enhancing the growth of herringbone calcite. Herringbone calcite is associated with a Mg-calcite precursor (Sumner and Grotzinger, 1996), supporting with the hypothesis for an original high-Mg calcite mineralogy for the Ledger limestone cements.

Acknowledgments. Acknowledgment is made to the Donors of The Petroleum Research Fund, administered by the American Chemical Society, for support of this research. A Franklin and Marshall College grant for independent research was awarded to S. Gaswirth. Thank you to Baker Refractories for access to the site, and to Dave Hopkins for his field assistance. John Taylor introduced us to these rocks. S. Sylvester provided guidance and assistance with analytical work at Franklin and Marshall College.

34.0   Leave Stop 1A. Return along Baker Road
34.5   Turn right into J.E. Baker Quarry, Stop 1B.

1B. Discussant: David A. Hopkins, Manager, Baker Refractories. Text from Ganis and Hopkins, 1984; Hopkins, 1985.

The quarry. Baker Refractories produces a full range of dolomite refractories, agricultural materials, and mineral fillers from a single quarry located just west of the city of York. The farms on which the current operation is located were first acquired in 1946, and stripping for the main quarry began in mid 1950 (Hopkins, 1985).

   This dolomite deposit was discovered more by accident than good geologic work. A friend of the Baker family was selling some property and asked if the Bakers would be interested. A man was sent out to inspect the property and collect some outcrop samples. Much to everyone's surprise, the samples indicated the potential of a deposit of high quality dolomite, so the property was optioned, drilled and purchased (Hopkins, 1985).

General Geology. The dolomite that has been so critical to the existence of Baker Refractories is known as the Ledger Formation (Figure 2). This formation is named for exposures found near the small hamlet of Ledger in Lancaster County. The dolomite is part of a sequence of Cambrian and lower Ordovician carbonates that underlie what is known as the Conestoga Valley. In York County, the Conestoga Valley is a long, narrow, northeast trending structure that is bounded to the south by lower Cambrian clastics, and to the north by both lower Cambrian clastics and in part by over-lapping Triassic sediments. (Hopkins, 1985).

   The high purity Ledger Formation at this location (Figure 9) is primarily composed of massive-gray to light-gray dolomite with dark-gray mottling. It is typically coarse to medium crystalline and highly fractured. There are also extensive areas of very high purity oolitic dolomite. These oolites commonly concentrate to form dark gray bands that provide the only good indicator for bedding. The lower purity dolomites are generally gray with a light pink to purple tint, and more finely crystalline. Bedding can rarely be noted in the quarry (with the exception of the oolitic bands), but where it is observed, it is normally disrupted by numerous closely spaced faults that show small to moderate offsets. The beds generally dip about 22° to the south. A large number of small-scale, high-angle faults are present and many of these faults contain slickensides that indicate that the last movement along the faults was typically strike slip. No fossils have ever been found at this locality. It is assigned a middle Cambrian age solely on its stratigraphic position. Along the north side of the Conestoga Valley, the Ledger is estimated to be approximately 300 m. (1,000 ft.) thick (Hopkins, 1985).

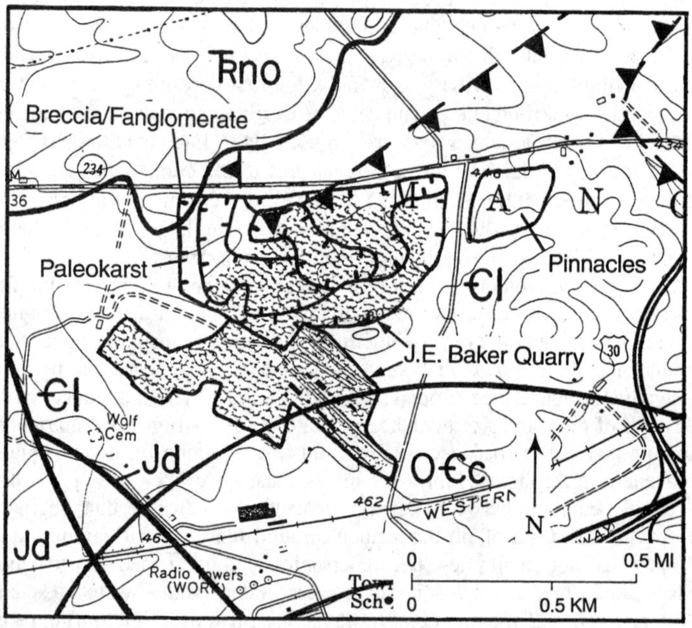

Figure 9. Map showing the J.E. Baker quarry, pinnacles, paleokarst, and Triassic breccia areas. Jd=Diabase, T no =New Oxford, Cl=Ledger, Occ=Conestoga. Geology from Wilshusen (1979b).

Paleokarst and Triassic limestone fanglomerate. A series of Triassic-age caves can be observed along the top bench at the south end of the pit (Figure 9). These paleokarst features are filled with red shale and, in some areas, small to large blocks of dolomite are found within the shale and could indicate roof falls within the caves. Shale in these caves shows bedding that is typically parallel to the cave floor. Core drilling has indicated that the main zone of this shale contamination strikes in a southwesterly direction back from the face and continues for about 150 m. (500 ft.)

Mining operations in the northwest corner of the quarry have recently exposed the contact zone with the overlying Triassic limestone fanglomerate (Figure 10a) along the southern edge of the Gettysburg basin. The clasts here are mostly dolomite, cemented with red shale. They are typically angular, and range in size from small chips to large boulders, all chaotically arranged (Figure 10b). All phenoclasts appear to be locally derived. Evidence suggests that most were probably from the sides of caves, sinks, or scarps. No quartz or quartzite phenoclasts are noted (Ganis and Hopkins, 1984).

Both quartz and limestone fanglomerates occur in numerous locations along the northwest and southeast margins of the Triassic valley in Pennsylvania. Exposures are typically poor, so any information about the nature of the contact zones is of interest in

a.)

b.)

Figure 10.a.) Ledger limestone exposed through a cover of Triassic limestone fanglomerate. b.) Triassic limestone fanglomerate showing angular clasts and range of sizes. Large clast in foreground 0.3 m (1 ft) across. Photos by Helen Delano.

questions of tectonic activity and development of the rift basins. A rock core drilled near the contact about 3 km west-southwest of this location penetrated over 244 m (800 ft) of Triassic rock including 79 m (260 ft) of conglomerate (Cloos and Pettijohn, 1973). Interpretations of this data and that from other locations suggested the basin margin was faulted rather than the simple overlap that had been previously thought. Plate tectonics models led to the view that the Triassic basins here and elsewhere along the central Atlantic margin were rift basins. Recently, new work on some of the limestone fanglomerates (Faill, in press) suggests that there is not sufficient evidence for syndepositional faulting along the southeast margin to support this. The contact here shows Triassic conglomerate deposited directly on pinnacled dolomite. It is possible that the paleokarst surface upon which the Triassic sediments were deposited had greater local relief than has generally been imagined. Perhaps future work at this site will supply evidence to help resolve these and other questions.

Limestone fanglomerate similar to that observed in the quarry occurs at other localities in Pennsylvania. Dissolution of the different sized limestone clasts can produce a very porous unit that can significantly affect groundwater flow and contain vugs of cave proportion. Sinkholes have been reported in areas underlain by Triassic redbeds. In one case near Dillsburg, northwestern York County, the sinkholes formed during the drilling of new water wells. The sinkholes occurred when they had penetrated the limestone fanglomerate. Since the wells were under artesian conditions, it is assumed that the pressure release from the confined fanglomerate affected the annular area around nearby wells that were tapped into the same aquifer and this release of pressure triggered the sinkholes (PAGS Case File 14-1920).

One additional item to note here is the presence of hematite along the contact between the Ledger carbonates and the Triassic breccia. The source of the iron is uncertain. It is possibly a hydrothermal occurrence, or a complex relation between groundwater, iron mobilization, and carbonate dissolution. See Kochanov (1997) for a related account of iron oxide coatings on Ledger pinnacles and Smith (1977) for a discussion of the origins of zinc and lead ores in York County.

The J.E. Baker site is a tale of two carbonates. First, at the pinnacles, the bedrock is a limestone facies and the karst is well developed. Second, in the active quarry, oolitic dolomite exhibits no karst development. It is a classic example of how lithologic variations of the carbonate rock can be a determining factor in the areal extent of the karst topography. Finally, clasts from the underlying carbonate rocks in the Triassic fanglomerate have been dissolved in zones along the unconfomable contact, which produce an interesting variation of karstic bedrock.

34.5	Leave Stop 1. Right from quarry entrance onto Baker Road.
34.9	Turn Left at Traffic light onto Rt. 30 East.
35.1	Bear right on Rt. 30 East (toward Lancaster).
38.0	Exit Right Rt. 74.
38.3	Turn left at end of ramp at traffic light onto Rt. 74 N.
39.0	Contact with Triassic New Oxford Formation. Entering Gettysburg-Newark Lowland Section of Piedmont Province.
41.2	Intersection with Rt. 283 E, continue on Rt. 74.
44.0	enter borough of Dover.
43.4	Junction with route 921, stay on Rt. 74.
44.4	Contact with Triassic Gettysburg Formation.
45.7	Conewago Mountain. Ridge is underlain by red pebbly sandstone and conglomerate of the Conewago Member of the Gettysburg Formation.
46.4	Small sandstone quarry on right.
47.3	Cross Conewago Creek.
47.7	Diabase boulders on right.
48.1	View straight ahead of the north end of South Mountain, Blue Ridge Physiographic Province.
48.3	Turn left at traffic light, staying on Rt. 74.
49.7	Turn Right at stop sign - stay on Rt. 74.
50.2	Sharp left turn, stay on Rt. 74.
51.1	Turn right at stop sign, stay on Rt. 74.
51.5	SLOW sharp curve to left.
52.2	Emu farm on right.
56.3	Stop sign, junction with Rt. 194. Bear right onto Rt. 74. Pass schools, enter Borough of Dillsburg. Rt. 74 is Baltimore Street.
58.0	Turn right at traffic light onto US Rt. 15 North, get in left lane. Diabase outcrop on right. The minerals prehnite, chabazite, heulandite, and calcite have been found in the fractures at this locality.
58.2	Turn left at traffic light onto Rt. 74.
58.8	Northern end of South Mountain on left which is the northern tip of the Blue Ridge Province. Leaving the Piedmont, entering the Great Valley Section of the Ridge and Valley Province.
59.5	Cumberland/York County Line. On a clear day, Blue or First Mountain is visible in distance on right. This marks the beginning of the Appalachian Mountain Section of the Ridge and Valley Province.
60.6	Sharp right turn, cross railroad track, cross Yellow Breeches Creek on Rt. 74.
62.3	Intersection of Old Stone House Road which rides atop a diabase dike.
62.8	Left onto Rt. 174, Boiling Springs Road. Allenberry Playhouse on right.
64.7	Cross Appalachian Trail.
65.0	Boiling Springs Tavern, park in lot.

STOP 2. Stop 2 examines the hydrogeology of the Boiling Springs area in the Great Valley Section. At this site, diabase dikes have played a major role in determining the location of Pennsylvania's seventh largest spring.

Discussant: Albert E. Becher, Hydrogeologist (retired), U.S. Geological Survey. Text from Becher (1991).

Hydrogeology and the Source of the Springs at Boiling Springs. Boiling Springs is one of the three largest springs in the Cumberland Valley and ranks as the seventh largest in the state. Average daily discharge is about 16.5 million gallons based on six instantaneous measurements made between 1944 and 1971 by the U.S. Geological Survey (Flippo, 1974). The water is used solely for recreational purposes in Boiling Springs Lake and the world-renowned trout-fishing waters of Yellow Breeches Creek. Until recently, the sequence of carbonate rock formations that underlie the southern part of the Cumberland Valley was considered to be the source of water for the springs.

Flow from Boiling Springs is through openings in strongly folded limestone of the Elbrook Formation (Figures 2 and 11). The Elbrook Formation is composed of interbedded calcareous shale, argillaceous limestone, and medium-bedded to massive limestone. Rocks typical of the formation are well exposed north of Boiling Springs Tavern in an abandoned quarry highwall behind Aniles Ristorante & Pizzaria (Figure 12). The four Boiling Springs caves are also located here (Stone, 1932; Reich, 1974).

A diabase dike of Jurassic age extends across the valley from the north of Boiling Springs, splits into two branches and encloses the spring area (Figure 11). To the south, on the flank of South Mountain, alluvial, colluvial, terrace, and residual deposits overlie the carbonate rocks to depths up to several hundred feet. Deposits are thickest near the contact between quartzites of South Mountain and the overlying carbonate rocks.

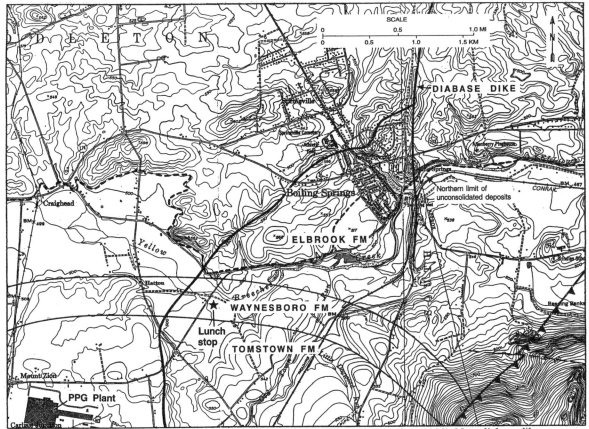

Figure 11. Geologic map of the Boiling Springs and lunch area. Modified from Becher (1991). Note diabase dike.

There are two major areas of groundwater discharge from Boiling springs; one in the walled basin north of the tavern (Figure 13) and the other near the southwestern shore in the northwest corner of the lake. In both areas, the discharges can be seen as boils that rise several inches above the water surface. Head differences between water in the openings and in the basin may be several tens of feet. The water level in the basin is 10 to 15 feet above yellow Breeches Creek.

Characteristics of the spring discharge openings can be seen best in the basin (Figure 13). Here two intersecting linear zones of discharge that parallel local joint directions are visible. Additional flows may be present under the north wall.

Three alternative sources of water were considered for Boiling Springs. The magnitude and degree of fluctuation of the flow, water quality, its seasonal variability, and geologic factors, support South Mountain as the major source. Many stream channels on the flank of South Mountain lose water to the colluvium and some are often dry before reaching Yellow Breeches Creek. It is probable that precipitation and runoff from South Mountain infiltrates the colluvium and moves downward through permeable zones into solution-enlarged openings in the carbonate bedrock. Groundwater then travels under Yellow Breeches Creek and finally discharges, under pressure, through openings at the narrow end of the funnel-like area created by the branching diabase dike.

A drainage area of more than 20 square miles is needed to collect the amount of water discharged by Boiling springs based on the average basin-wide groundwater discharge of 0.81 million gallons per day per square mile. From the groundwater divide, separating drainage to the Conodoguinet and Yellow Breeches

Figure 12. Elbrook Formation exposed in the face of an old quarry. The opening of Boiling Springs Cave # 1 is visible in lower center. Photo by Helen Delano.

459

Creeks, south to the springs, the drainage area is only about 3 square miles. Ample drainage area exists only on the flank of South Mountain.

Temperature of the spring water at the boils in the basin fluctuates only 0.2° C annually (Figure 15) and lags air temperature changes by 4 to 6 months. The temperature of other perennial springs that obtain water from shallow local sources fluctuates with the seasons by as much as 5° C. The specific conductance (electrical measure of dissolved ionic species) of water from Boiling Springs is about half that of ground water in nearby wells. The specific conductance values correspond with well water from the carbonate rocks on the flank of South Mountain. Wells on South Mountain are completed as open casings that penetrate only a few feet into bedrock below the consolidated overburden. The lag in spring water temperature fluctuation suggests a longer residence time; the slight fluctuation suggests deep, well-mixed water. The lower specific conductance means lesser contact with carbonate rock. A South Mountain source fits these interpretations.

Spring and well data suggest that the movement of water under Yellow Breeches Creek occurs along the entire flank of South Mountain. Much of this water is then discharged into the Yellow Breeches but some moves under the apparent groundwater divide into the Conodoguinet Creek drainage basin (Becher and Root, 1981).

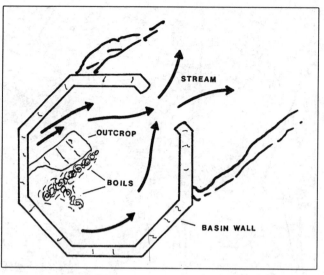

Figure 13. Diagram of walled spring basin behind Boilg Springs Tavern (Becher, 1991).

65.0 Leave Boiling Springs Tavern, turn right out of parking lot, and proceed west on Rt. 174.
65.2 Turn left at stop sign onto Walnut Street.
66.2 Dickinson College experimental well field in field on right
66.9 Cross Yellow Breeches Creek, then turn left into South Middleton Township Park parking lot. Lunch and discussion of Cumberland Valley geology.

Discussant: Helen Delano, Pennsylvania Geological Survey.

South Mountain geology. Although the primary reason for this stop is lunch, there is some relevant geology, both here and along the route we will be taking after we leave the park.

The park is just north of the base of South Mountain (Figure 11), the northernmost extension of the Blue Ridge physiographic province. The backbone of South Mountain is composed of resistant Cambrian sandstones and conglomerates of the Chilhowee Group.

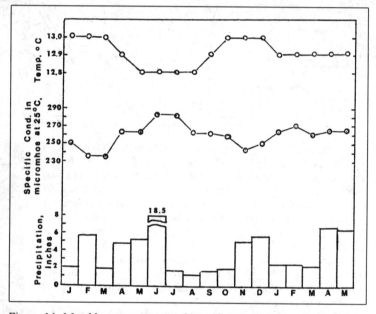

Figure 14. Monthly temperature and specific conductance of water from Boiling Springs and precipitation at Carlisle (Becher, 1991).

The older Catoctin volcanic complex is exposed further to the east of the front and lies in the center of the South Mountain Anticlinorium (Sevon and Potter, 1991). North and west of South Mountain, the Great Valley is underlain by mostly carbonate rocks, Cambrian and Ordovician in age, with a mantle of regolith, colluvium from the mountain, and alluvium (Root, 1978). This setting is an important aspect of the hydrogeologic story at Boiling Springs, and affects the karst geology of this part of the valley.

The bedrock at South Middleton Park is mapped as the Cambrian Waynesboro Formation (Figure 2) and is overlain by colluvium and terrace deposits of undetermined thickness (Root, 1978). The Waynesboro is largely dolomite with limestone interbeds. Near the top of the section, there is a 30 m (100 ft) thick unit of coarse interbedded quartzitic sandstone and gray mudstone. The older Tomstown Formation, with thick massive dolomites and some limestone, siltstone and shaly interbeds, underlies the valley margin up to the quartz sandstones of the Antietam member of the Chilhowee Group. The Tomstown in this area is completely covered by a thick blanket of unconsolidated, unsorted accumulation of quartzitic sediment ranging in size from clay to boulders (Sevon, et al, 1991). Paleosols in and on these surficial deposits exposed in quarries along South Mountain suggests a complex formational history extending at least into pre-Wisconsinan (Pleistocene) time and possibly much further into the Cenozoic (Clark, 1991; Sevon et al., 1991). The colluvium may be more accurately described as coalescing alluvial fans, and relationships between the bedrock surface and overlying deposits suggest that the carbonate bedrock surface has been lowered by dissolution beneath the accumulating cover.

The park is also the site of an on-going study at Dickinson College (Carlisle) to characterize the suspended and dissolved sediment load of the Yellow Breeches Creek. Since 1994, the creek has been monitored and sampled weekly to determine sediment load and total discharge. The average denudation rate for the drainage basin above the bridge was calculated to be 17.0 m/M yr. using the data from the period 1994-1996. (Potter and Niemitz, 1997). Dissolved solids accounted for approximately 80 percent of the total load, with higher percentages in dry years. The drainage basin area of 323.6 square km (125 square mi.) is underlain by approximately two-thirds carbonate bedrock and one-third clastic sedimentary rocks.

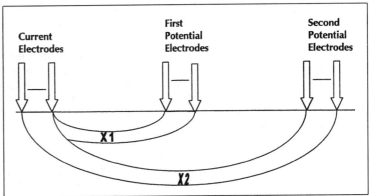

Figure 15. Example of current flow during electrical imaging surveying, where X1 and X2 are successive measurement locations for successive potential electrode pairs.

Immediately after lunch we will drive by sites where the practical implications of this geology were demonstrated. The South Middleton Township water tower is immediately adjacent to the site of a sinkhole that opened during the pump test of a new well. . The 6 m (20 ft) in diameter, 6 m (20 ft) deep sinkhole exposed a thick sequence of gravel, no bedrock was observed. Depth to bedrock was available from the well, however, and was listed as 32.5 m (107 ft). Across the road from the well site is the PPG Industries float glass plant (Figure 6). It is an interesting process where molten glass is extruded onto a bed of molten tin. The glass then cools gradually as it moves down the half-mile length of the plant Test borings, taken when the plant was built, typically showed depths to bedrock ranging from 50 to 75 feet. One boring did not reach bedrock at 230 feet. Well logs indicated the unconsolidated material as primarily gravels overlying silty clays that are interpreted as residuum.

66.10 Leave Park, turn left out of parking lot onto Rt. 174.
68.1   Water tower on right site of 20 ft x 20 ft. sinkhole when new municipal well was developed in 1972.
68.2   PPG Glass factory on left, Land O'Lakes plant on right.
69.5   Turn right at traffic light in Mt. Holly Springs onto Rt. 34 North.
70.5   Cross Yellow Breeches Creek.
74.1   Turn right onto ramp to I-81 North. Continue on I-81.
80.6   The northern extension of the Diabase dike at Boiling Springs makes this ridge. The Appalachian Trail runs along the ridge, and crosses I-81 on a special bridge.
81.4   Crossing contact between carbonate rocks of southern part of the valley and northern Martinsburg Shale.
83.0   Cross Conodoguinet Creek, Blue Mountain on left.
86.2   Bear right onto Exit 19 to Pa Rt. 581.
87.0   The "walls of silence" (sound control walls).
89.5   View to south (right) diabase hills (same as seen between York and Dillsburg).
93.8   I-83 joins Rt. 581. Stay left towards Harrisburg.
94.8   Cross Susquehanna River.
98.0   Harrisburg East Mall is about .5 mi. south of here. The mall is constructed over the Ordovician St. Paul limestone and Martinsburg shale. The shale is thrust over the limestone under the southern part of the mall. Numerous caverns were encountered and much of the foundation is on caissons. Two caverns, Big Pit Cave and Crystal Paradise Cave, were mapped prior to filling. Crystal paradise Cave contained profuse secondary mineral deposits, including fluorescent aragonite (Inners and others, 1978; Reich in Smeltzer, 1964, p. 41).
98.7   Eisenhower interchange, take Rt. 322 straight ahead.
100.2 Belt of carbonate rocks is narrow here, hills on both sides are shales of the Hamburg Klippe.
101.0 Outcrop of Hamburg shale on right
103.4 Cross Swatara Creek, Ordovician Epler Fm. exposed in bank.
104.1 Bear right onto Hummelstown Exit
104.2 Turn right at traffic light onto Middletown Rd.
104.5 Turn right onto Indian Echo Cavern entrance road.
104.8 Rail road crossing, Middletown and Hummelstown Railroad.
104.9 Parking lot., Indian Echo Caverns.

STOP 3. Leaving the Cumberland Valley, we enter the western half of the Lebanon Valley. Stopping at Indian Echo Caverns, a commercial cave, staff from Science Applications International Corporation will demonstrate and discuss a three dimensional electrical imaging technique for characterizing carbonate bedrock. A 7 x 7 3D EI survey will be demonstrated over the caverns. Data from the survey will be provided to the field trip attendees at the site.

Discussants: Jeffrey L. Lieberfinger, P.G., Rick Hoover, Science Applications International Corporation. Text from Leberfinger, Reccelli-Snyder, and Hoover (1998)

Background. Sinkholes are often a major development hazard in areas underlain by carbonate rocks. Road and highway subsidence, building foundation collapse, and dam leakage are a few of the problems associated with sinkholes. To avoid costs caused by structural instability associated with sinkholes, potential sinkhole problems must be identified.

Frequently, in carbonate bedrock systems, borings are drilled without regard to karst limitations and generally do not intersect areas of concern in the subsurface. Poor location of borings can result from inadequate subsurface data, and could misrepresent the subsurface system leading to additional costs for remedial design or additional investigation. Rapid reconnaissance surveys using remote sensing (e.g. aerial photograph evaluation) and surface geophysical techniques integrated with a boring plan are best used to aid in the proper location of test wells to identify subsurface features related to karst development.

In the past, electrical resistivity techniques performed by experienced geophysicists had proven to be effective tools for characterizing the subsurface (Roman, 1951) but certain limitations caused these techniques to be utilized less over the years. First, the technique was very labor intensive. A resistivity crew could range from three to five people. Secondly, interpretation of the data generally took a long time and finally, there was no standardized method for obtaining accurate subsurface representations.

Recently, the development of computer-controlled, multi-electrode, resistivity-survey systems and the development of resistivity-inversion-modeling software (Loke and Barker, 1996) have allowed for more cost-effective electrical imaging (EI) surveys and better representation of the subsurface.

Methods. EI surveys can be used to determine the location of variations in geologic and soil strata, the soil/bedrock interface, bedrock fractures, faults, and caves. The method has also been used effectively to delineate abandoned waste disposal sites and landfill boundaries and to map hydrogeologic and mineral resource boundaries.

EI surveys measure the electrical resistivity of a wide range of materials in the subsurface (Table 1).

Fundamental to all resistivity methods is the concept that current (I) can be impressed into the ground and the effects of this current within the ground can be measured. The effect of potential (V) or differences of potential, ratio of potential differences, or some other parameter that is directly related to these variables is the commonly measured effect of the impressed current. The fundamental theory involved in the different methods used today is based upon Laplace's equation for obtaining the electrical potential and the pattern of current flow about one or more electrodes placed on or in the ground (Van Nostrand and Cook 1966). The principal differences among various methods of EI lie in the number and spacing of the current and potential electrodes, the variable quantity determined, and the manner of presenting the results.

In application, a series of measurements is made between a variety of current electrode pairs and potential electrode pairs. In general, as the distance between the two electrodes increase, the apparent resistivity p_a is measured at greater depths and across increasing volumes of ground as shown in Figure 15.

Table 1. Resistivity values of some selected earth materials.

Material	Resistivity (ohms)
Clay	1-60
Sand, wet to moist	20-200
Shale	1-500
Sandstone	150-450
Porous limestone	100-1,000
Dense Limestone	1,000-1,000,000
Metamorphic Rocks	50-1,000,000
Igneous Rocks	100-1,000,000

EI Data Collection. In this demonstration, the EI data collection uses a Sting/Swift multielectrode system manufactured by Advanced Geoscience, Inc. (AGI) of Austin, Texas. The EI equipment is composed of three primary components: 1) the Sting R1 resistivity meter with data storage capability; 2) the Swift automatic multielectrode switching system, which is an accessory for the Sting; and 3) the Sting/Swift cables which contain fixed cylindrical stainless steel switches that attach to stainless steel electrodes placed into the ground.

The amount of current, potential, and the configuration of electrodes are analyzed to yield an apparent resistivity value between electrodes. The EI system automatically energizes different electrodes to measure apparent resistivities at new horizontal locations and depths. The EI system can be used to determine a three-dimensional (3-D) resistivity model for the subsurface using the data obtained from a 3-D electrical imaging E-SCAN type of survey (Li and Oldenburg, 1992). The electrodes for such a survey are arranged in a rectangular grid. It should first be emphasized that full 3D surveys are not merely a series of 2D surveys, but consist of an entirely different approach.

During preparation for data collection, the operator programs the Sting for the chosen number of current pairs to energize (in electrode spacing measurements) and the maximum separation (in electrode spacing measurements) to measure the potentials. These two numbers determine the total number of measurements to be collected along the electrode spread and the total depth of investigation. The Sting digitally records all of this information for use in data processing and quality assurance.

The pole-pole electrode configuration is commonly used for 3D surveys, such as the E-SCAN method (Li and Oldenburg 1992). In the measurement sequence shown in Figure 16a, each electrode in turn is used as a current electrode and the potentials at all the other electrodes are measured. Note that because of reciprocity, it is only necessary to measure the potentials at the electrodes with a higher index number than the current electrode in Figure 16a. For a 7 by 7 electrode grid, the number of measurements is 1176.. It can be very time consuming to make such a large number of measurements with typical single-channel resistivity meters commonly used for 2D surveys. For example, it could take several hours to make the 1176 measurements for a 7 by 7 survey grid with a standard low frequency earth resistance meter although speed is partly determined by the grid spacing and the magnitude of the measured resistances (Griffiths and Barker, 1993). To reduce the number of measurements required without seriously degrading the quality of the model obtained, an alternative measurement sequence has been tested (Figure 16b). In this proposed "cross-diagonal survey" technique, the potential measurements are only made at the electrodes along the horizontal, vertical and the 45 degree diagonal lines passing through the current electrode. The number of datum points with this arrangement for a 7 by 7 grid is reduced to 476. For large survey grids, it is also common to limit the maximum spacing used in the measurements to about 8 to 10 times the minimum electrode spacing. To map large areas with a limited number of electrodes in a multi-electrode resistivity meter system, the roll-along technique can be used (Dahlin and Bernstone, 1997)

Data Modeling and Interpretation. The apparent resistivity p_a, as measured by the EI system, is the product of a large area of the subsurface responding to the impressed current. Interpretation of apparent resistivity data collected in the field without reduction provides a qualitative product very similar to many electromagnetic (EM) methods. Because the earth is not homogeneous, it is useful to model the resistivities at discrete locations in order to make a more quantified interpretation. Inverse modeling of the data is performed using RES3DINV™ (Loke, 1997) to produce a three-dimensional resistivity model based on the apparent resistivity data.

Final data processing involves the generation of color-enhanced contour maps of the data using a two-dimensional mapping program. EI resistivity models are presented in cross-section or 3D model blocks, with inline distance shown along the horizontal axis, depths, or elevation along the vertical axis. The geoelectrical model presents the electrical stratigraphy (electro-stratigraphy) of the subsurface.

Following the data collection and inversion modeling, the EI electrostratigraphy information is used to interpret the potential gross stratigraphy of the traverses. In general, dry materials have higher resistivity than similar wet materials because moisture increases their ability to conduct electricity.

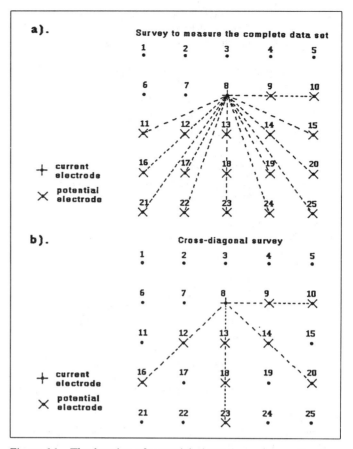

Figure 16. The location of potential electrodes corresponding to a single current electrode in the arrangement used by (a) a survey to measure the complete data set and (b) a cross-diagonal survey.

This resistivity change, if indicated in the observed electrostratigraphy, can represent water table depths. Beneath the water table, silt, clay-free sands, and gravels will have a much higher resistivity than silt or clay under similar moisture condition. This is due to finer grained materials acting as better conductors. In the bedrock, competent rock will have a high resistivity. Saturated fractured or weathered rock would show a much lower resistivity than the competent rock. Very high resistivities can indicate air filled voids.

The identified electric boundaries separating layers of different resistivities may or may not coincide with boundaries separating layers of different lithologic composition. These differences may result from the gradational presentation of the electrostratigraphy. Therefore, the electrostratigraphy can vary from the geologic stratigraphy, and caution should be exercised when reviewing and applying the electrical profiles.

New Methodologies. New advances in EI have allowed for three-dimensional surveys and cross-borehole surveys which will make this technique even more successful for fracture characterization. In areas where very complex subsurface features are present, three-dimensional-survey, data-collection techniques and data-inversion software have recently been developed. Also, the resolution of electrical surveys carried out with electrodes on the ground surface decreases exponentially with depth. One method to obtain reasonably good resolution at depth is by making measurements with the electrodes in boreholes. New down-hole cables have recently been developed to allow for down-hole data collection. These techniques will allow for the use of EI for a much wider range of applications to obtain subsurface information.

A bit about Indian Echo Caverns. The "debbil hole" (Hummelstown Sun, 1892) nicely illustrates solution effects controlled by fractures. The two branches of the cave are developed along orthogonal master joints (Figure 17). The lesser features of the cave also appear to be primarily controlled by jointing. Bedding has only a minor influence. The less soluble dolomite beds are often more eroded than the limestone because, due to their more brittle nature, they are more intensely fractured. Preferential development of solution openings in dolomite beds because of their fractured character is a common feature in the Lebanon Valley (MacLachlan and Root, 1966).

A detailed description of the cave can be found in Smeltzer, (1964).

104.9 Leave Parking lot, retrace route to Middletown Road.
105.4 Turn left onto Middletown Road.

105.6	Turn right onto access road for Rt. 322. Small sinkholes along south bank of access road.
106.1	Traffic light. Proceed straight and merge onto Rt. 322.
106.7	Exit right onto Rt. 322 East toward Ephrata.
107.2	BullFrog Valley Road, site of recurrent sinkholes; Hershey Medical Center on right.
108.6	Traffic Light at Fishburn Road (Truck Rt. 743 S.).
109.1	Traffic light at Pa Rt. 743, Cocoa Avenue.
109.9	Milton Hershey School
112.1	Village of Cambelltown.
112.4	Traffic light, stay straight.
113.2	Traffic light; turn left (north) onto Rt. 117 towards Palmyra.
113.9	Airport on left.
116.8	Rt. 117 becomes Forge Road.
117.3	Sinkhole on left side of road. Note offset curbing and depression.
117.4	Right turn onto Oak Street. Note lack of stormwater drains along the street. This is common throughout Palmyra.
117.5	Stop sign, turn right onto Duke Street. Note ball field/stormwater basin on right. Discharge outlet drains onto Duke Street.
117.7	Left onto Birch Street. Park in lot of Southeast Family Park. Buses follow parking lot around to Duke Street. Turn right onto Duke Street travel 0.4 mi., then park on north side of Cherry Street (along curbing).

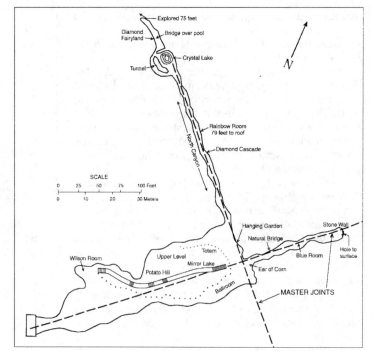

Figure 17. The orientation of Indian Echo Caverns is related to two master joints. From Smeltzer (1964).

STOP 4. The route to Stop 4 is along the western quarter of the Lebanon Valley. The relatively flat karst landscape sets the stage for a walking tour of Palmyra, the "sinkhole capital" of south-central Pennsylvania. Attendees will follow the course of stormwater through a portion of the Borough and see two different outlets, each having its own unique story.

Discussant: William E. Kochanov, Geologist, Pennsylvania Geological Survey. Text from Kochanov (1993, 1995).

Storm water drainage in the Borough of Palmyra: physiographic and geologic setting. The Borough of Palmyra lies within the eastern half of the Great Valley Section of the Ridge and Valley Province of Pennsylvania. Locally called the Lebanon Valley, the northern two-thirds is characterized by moderately dissected uplands underlain by Ordovician shales and sandstones. The southern third is characterized by a gently undulating karstic land surface underlain by Cambro-Ordovician limestones and dolomites. In comparison to the hummocky karst topography found in central Kentucky and Tennessee, the karst landscape in the Lebanon Valley can be described as flat.

Most of the Borough of Palmyra is underlain by the Ordovician Epler Formation (Figures 2 and 18). The Epler is described by Geyer (1970) as being a thick-bedded, strongly laminated, finely crystalline limestone interbedded with medium to thick-bedded, laminated, grayish-yellow-weathering, crystalline dolomite. Nodular chert is present and abundant. In outcrop, the surface of the Epler weathers along planar discontinuities and, in places, can resemble "elephant skin" in texture. The finely crystalline composition and thinly laminated bedding work in conjunction with the attitude of the bedding and joints to produce a pinnacled relief surface. Differential dissolution is more pronounced within the limestone beds whereas the dolomitic beds are more in positive relief.

The primary structural feature in the Lebanon Valley is a nappe system. Local structures are characterized by overturned bedding, folds that are approximately parallel to the regional strike (NE-SW), and joints that are mostly parallel to the strike and dip of bedding. No major faulting has been mapped in the Palmyra area (Geyer, 1970; Kochanov, 1988).

From a karst perspective, the Epler Formation is a major sinkhole producer in the eastern half of the Great Valley. Statistics compiled on the Epler show 676 sinkholes on record covering an area of 325 km^2 (201 mi^2.) (Kochanov, 1993). This distribution of sinkhole occurrences is skewed more towards urban areas simply because more sinkhole records are available when they occur in urban areas as opposed to those that occur in rural areas. Factors that contribute to sinkhole formation, such as stormwater drainage and failure of utility lines, are also more common to the urban setting.

Storm water and karst problems. In a karst area, the natural plumbing system of drains and pipes lies beneath a relatively thin veneer of unconsolidated sediment. It is a dynamic state of coexistence between the sediment, water, and bedrock. In karst areas, the subsurface drains serve as entryways for percolating water so that it can migrate through the fractures in the bedrock to the water table.

The stormwater drainage problem is compounded in karst areas by the fact that land development reduces the surface area available for rainwater to infiltrate naturally into the ground. A typical residential development with quarter-acre lots may reduce the natural ground surface by 25 percent; whereas, a shopping center may reduce it by 100 percent. If storm water gathered over a specific area is collected and directed into an area that normally would not receive it, the concentration of water may unplug existing karst drains resulting in subsidence.

Methods used in managing storm water can vary. Piped systems are typical in established urban settings where curbside drains are used to direct stormwater runoff below ground level into large diameter pipes and then convey the water to a natural or artificial drainageway. Stormwater basins are excavated structures that can either hold stormwater runoff or slowly let it discharge over a period of time at a designated discharge point. Another method is to use existing karst drains as stormwater drains; sinkholes are modified to allow stormwater runoff to more easily enter the karst plumbing system. Baffles and grills are integrated with the drain design to lessen the impact and velocity of storm water and prevent debris from entering the drain.

It makes hydrologic sense to use sinkholes and surface depressions as stormwater drains from a groundwater recharge perspective. However, meteoric water can overwhelm a karst drain, at times, causing the drain to back up and cause localized

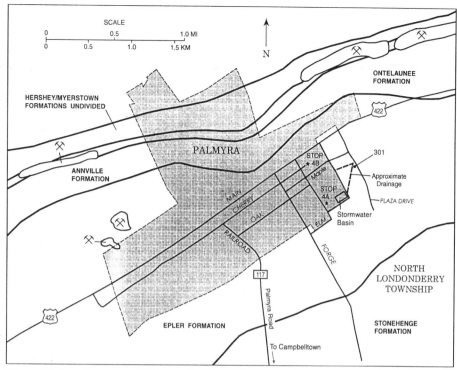

Figure 18. Location map of Palmyra showing carbonate geology, stormwater basin, limestone quarries and the location of Stops 4A and 4B.

flooding. Water entering a karst drain can also carry a wide variety of pollutants and introduce them directly into the carbonate aquifer system. Finally, water is the main triggering mechanism that initiates the piping of surficial material into subsurface voids.

The argument can be made that by not using sinkholes for stormwater drainage, one is interfering with the natural recharge system. This argument begins to fall apart in established urban areas where the founding fathers may not have completely understood the relationships between precipitation, recharge, and the karst plumbing system. Attempts to deal with problems related to the karst system, after the fact, are often complicated by the attitude of residents who demand instant resolution to a problem that has been a long time in the making.

The Borough of Palmyra is essentially a landlocked community. The natural topographic drainageways are located outside the Borough's municipal boundary. If one were to travel the streets of Palmyra, one would note the lack of stormwater drains. Stormwater runoff is directed to the municipal borders via overland flow atop main and side streets, along earthen swales, and through a limited subsurface piping system. Sinkholes have been modified as stormwater drains in Palmyra with limited success.

4A. The Walking Tour, Plaza Drive. As a partial solution to the stormwater management problem in Palmyra, a large, shallow stormwater detention basin was co-constructed with the Southeast Family Park (Figure 18). Completed at the end of summer in 1992, the basin was designed to collect runoff from approximately 110 acres and to help alleviate sinkhole problems in another section of Palmyra. The relatively shallow basin discharges into North Londonderry Township. At the discharge point, water flows along a poorly developed "channel" across private property. The drainage coincides with a fracture trace that is visible on aerial photography from 1963 (USDA, AHN-1DD-194). The drainage continues along the base of a small hill and continues in a northeasterly direction towards Plaza Drive. Another drainage through the backyards of the residential area to the west, travels along the north side of the small hill, and joins with the first drainage. This meeting point has been known to have standing water at times in the recent past. Small sinkholes have also occurred.

During the spring of 1993, a series of sinkholes opened on and adjacent to the property located at 301 Plaza Drive. Sinkholes opened in the back yard, beneath a swimming pool, beneath the house, and in the front yard. Sinkholes had also occurred in front of the house on Plaza Drive (Figure 19). Damage to the house was so extensive that the property was condemned and the house razed (Kochanov, 1995).

The coincidental construction of the stormwater basin and the occurrence of sinkholes resulted in litigation between local residents and the Borough. The argument was that the water draining from the basin was causing the subsidence problems.

Historically, the area has had sinkhole problems prior to the construction of the detention basin. A correlation between the basin and sinkhole occurrences would be difficult to prove. The presence of visible fracture traces from air photo analysis supports the view of structural control of surface and ground water.

The Walking Tour, Stop 4B, Cherry Street. The Cherry Street section has had sinkhole problems dating back to the mid-1960's. Sinkholes have most notably affected the properties at 914 Cherry Street and 34 South Duke Street as well as the utility service and road surface at the east end of Cherry Street.

At the 914 property, the final blow came in 1988 when eight sinkholes opened in roughly a 20-m. (65 feet) in diameter area (Figure 20). The house had severe structural damage with one 5-m. (15-foot) sinkhole located in the basement of the house. Part of the final solution was to move the house and garage away from the stormwater drain. The estimated cost to move the house was approximately $20,000 (Kochanov, 1995).

Sinkholes opened in August of 1993 at the end of the driveway of the South Duke Street property after a period of heavy rain. The main sinkhole was located at the end of the driveway (Figure 20). The Borough repaired the sinkhole but it reopened about a month later. This time, however, the Borough needed access to the Duke Street property in order to make the repairs. In Palmyra, the Borough is responsible for damages from curb to curb. Anything on private property is the property owners' responsibility. So the property owners, in this case, would be responsible for paying whatever damages outside the curb area (i.e. sidewalk, driveway) the Borough caused in the repair of the sinkhole This resulted in the property owner not granting the Borough access to fix the sinkhole

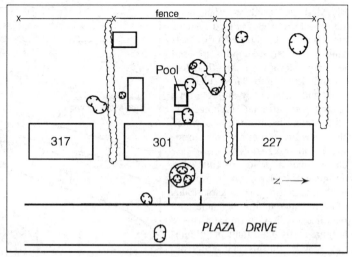

Figure 19. Sinkhole locations with respect to the 301 Plaza Drive property (Kochanov, 1995).

The sinkhole occurrences can be related, in part, to the stormwater drain at the east end of Cherry Street. In this instance a sinkhole has been modified to accept stormwater drainage. It is likely that the capacity for the drain to effectively convey water to the water table is exceeded at times. Stormwater backs up and unplugs nearby karst drains resulting in localized subsidence. What may have been underestimated in both examples was the impact of local limestone quarries. Data compiled by Foose (1953) showed that the lowering of the water table from pumping in nearby quarries affected a large area (Figure 21). Quarries are still active in the Hershey/Palmyra area and pumping rates may vary with seasonal weather changes. This in turn could affect the cone of depression and the stability of the bedrock/soil interface.

In Palmyra, the major focus of the sinkhole problems is on stormwater management. Efforts by the Borough are continuing to establish a long-term solution of a piped-stormwater system. This however is pending on financial assistance.

118.1　Reboard buses. Proceed straight to traffic light; turn left onto Rt. 422 west.
118.2　Macadam patch in east lane. During maintenance to repair a small sinkhole, a large void (approximately 5 m. in diameter and 7 m. deep) was discovered beneath the highway in April of 1995.
118.4　At the corner of Rt. 422 and Green Street is another sinkhole/stormwater drain.
119.7　Main Street sinkhole site on left, house next to VFW. Sinkhole opened in the basement of a two-story apartment building causing structural damage as well as ripping two 1,040 liter (275 gallon) fuel oil tanks from pipes causing some leakage.
122.7　Entering Hershey.
122.9　Hershey Chocolate factory on right. Note "kiss-shaped" light fixtures on Cocoa Avenue.
123.2　Junction with Rt. 743. Turn right (north).
123.3　Cross over bridge, turn left onto Park Boulevard. Hershey Park amusement park on right
124.0　Turn Left into Employee parking lot of Hershey Entertainment, Inc.

STOP 5. An excellent exposure of the Ontelaunee/Epler Formation serves as a backdrop to discuss the engineering geology of Chocolate World. As the final stop in the field trip, attendees will have some free time to go and get some chocolate!

Discussant: William E. Kochanov, Geologist, Pennsylvania Geological Survey. Geologic and engineering portions of the text modified in part from Foose (1953) and (Foose and Humphreville, 1979).

From the permeating aroma that fills the air to the Hershey Kiss streetlights lining Chocolate Avenue, visitors to "the sweetest place on earth" are probably not aware of the distinctive engineering history in the town of Hershey. It is somewhat fitting to begin the story of Hershey's karst geology, with mentioning the town's namesake, Milton

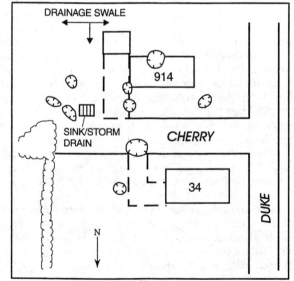

Figure 20. Sinkhole locations in the vicinity east Cherry Street, Palmyra. The stormwater drain is located within a sinkhole (Kochanov, 1995).

466

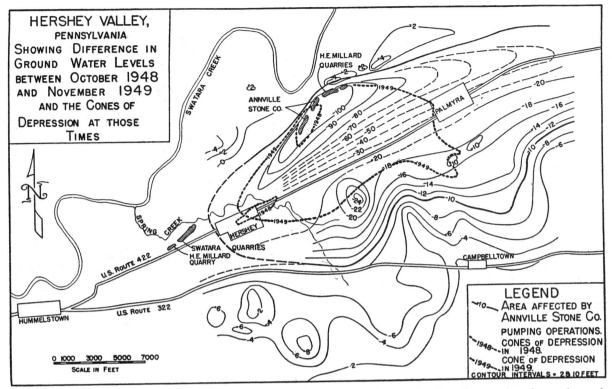

Figure 21. Figure from Foose (1953) showing the extent of a cone of depression resulting from the pumping of groundwater from limestone quarries near Hershey. Note that the zone of influence includes the Borough of Palmyra.

Hershey. In 1903 he selected his site partly based on the presence of good dairy country and a good water supply. He got both in the limestone region of Hershey. Mr. Hershey was also keen on details and it was this habit which made him a sharp observer. One story tells of Mr. Hershey driving along the road and seeing something in a field that he had never seen before. He got out of his car, walked some distance into the field and discovered a new sinkhole. When he got too close to the edge, the ground caved in and Mr. Hershey went with it (Hostetter, 1971).

Geologic setting. Hershey lies within the Great Valley Section of the Ridge and Valley Physiographic Province. It is at the western end of the Lebanon Valley where it is known as the "Hershey Valley." Hershey is underlain by Ordovician carbonates and, as in Palmyra, primarily by the Epler Formation (Figure 22). Extensive deformation of the Cambro-Ordovician section has resulted in sheared bedding surfaces, well-developed joint sets, cleavage, sharply hinged fold axes, thrust and normal faults. The beds generally strike N65°E to N75°E and have overturned dips of 65° 75° south. The carbonates are bounded to the north by the middle Ordovician Martinsburg shales and siltstones and by a major fault zone separating the carbonates from the Triassic sedimentary rocks to the south. (Foose and Humphreville, 1979)

Recent karst development. Detailed geologic mapping conducted during the late 1940's indicated a fairly "normal" karst landscape with minimal subsidence activity. It was during the middle to late 1940's many springs and wells in the Hershey area were affected by groundwater withdrawal from nearby limestone quarries. The dewatering effect created a severe cone of depression over 6 miles long and 3 miles wide (Knight, 1970; Foose 1953). The lowering of the groundwater surface dried up springs and surface streams, and caused a large number of sinkholes to form within the area where groundwater had been lowered by more than 3 m (10 ft). The behavior of groundwater flow and the relationship to sinkhole development during that time period has been previously described by Foose (1953). Conditions were stabilized by an extensive grouting program at the limestone quarries which resulted in the return to basically normal groundwater levels, the return of flowing springs and streams, and the cessation of catastrophic sinkhole formation (Foose and Humphreville, 1979).

Since that time the Hershey Valley has experienced rapid urban growth which in turn has greatly impacted the use of local water resources as well as increased stormwater runoff. This brings to mind the story of Palmyra and its efforts to deal with stormwater management in a karst area. Also, if one examines the data in Foose (1953), an argument can be made for the Borough of Palmyra with regard to sinkhole occurrence and the pumping of groundwater from limestone quarrying operations (Figure 21).

Engineering. Due to the multitude of tourists wishing to tour the chocolate factory, Hershey Foods decided in the early 1970's to construct a separate building to show all aspects of the chocolate-making industry through an educational tour and exhibits. The site was located adjacent to the factory (Figure 23) so that it could be available to the largest number of tourists. The structure was planned to be approximately 91 m (300 ft) square and have relatively light loads, and it would be sited across the middle of a well-drained valley with relief of 3-5 to 5 m (12 to 16 ft) (Foose and Humphreville, 1979).

A geologic investigation of the site was conducted in 1971 by drilling 16 NX holes on roughly 30 m (100 ft) centers. Three additional holes were drilled along the valley axis. Surface examination indicated no evidence of recent sinkhole activity, but there was evidence of undrained surface depressions and incipient sinkholes in the upper part of the local drainage basin several hundred feet to the west (Foose and Humphreville, 1979).

Analysis of the data (Foose and Humphreville, 1979) indicated:
1. Eleven percent of the total footage drilled consisted of cavity. The largest number and size of cavities was along the valley axis. Additionally, it was very likely that drilling did not penetrate all cavities.
2. In an additional eleven percent of the drilling, the bedrock was badly fractured with abundant solution-widened bedding and joint surfaces. It was so bad that the rock would not support caissons or piling.
3. Relief of the bedrock surface ranged over 9 m (30 ft). To

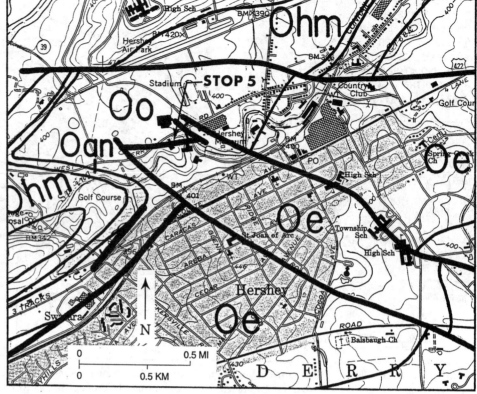

Figure 22. Map showing the general geology of the Hershey area and Stop 5. Modified from Berg and Dodge (1981). Oe = Epler Formation, Oo = Ontelaunee Formation, Oan = Annville Formation, Ohm = Hershey/Meyerstown Formations, undivided. Bold lines indicate faults; thinner lines indicate formational contacts.

achieve the necessary grade level along the north and south sides of the building foundation, some excavation of rock would be required while all the while dealing with the deep and irregular bedrock surface.

Consideration of these geologic observations indicated that additional drilling would be necessary to establish the depth of competent bedrock. This would provide a firm foundation for the structure or develop a means to ensure structural integrity based on spread footings. Important factors that had to be considered were:
1. The relatively light structural loads of the building.
2. The relatively high cost of emplacing caissons.
3. The irregular karst bedrock surface and variations of soil density along the axis of the valley.
4. The requirement that up to as much as 3-3.5 m (10-11 ft) of fill be used in the valley to bring it up to the designed grade of the building.
5. Surface runoff and underground drainage had gradients from west to east directly toward and under the building. Drainages could trigger sinkhole formation.
6. Fluctuation of groundwater on a seasonal or longer-term base that could cause soil migration and induce sinkhole development.

Safety and structural integrity were the primary considerations, but economics were also important Spread footings were more economic than caissons or pilings but the problems just outlined would need to be addressed.

The decision was made to use spread footings that would rest on nearly incompressible, modified 2A crushed stone. Also, an additional 2.5 m (8 ft.) of stone was added as a surcharge above grade level within the area of the building. This was done to create maximum compaction of the underlying overburden and to partially negate the effects of the different soil densities and associated differential compaction above the irregular bedrock surface. It was reasoned that this would "deflate" incipient sinkhole areas by driving the over-burden into near-surface cavities. This approach would provide a thick pad of porous and incompressible stone on which to install the spread footings (Foose and Humphreville, 1979).

Concurrent with the base stabilization program, a drainage intake was designed to capture all surface runoff along the valley axis toward the building and a large diameter pipeline was constructed around the west and south sides of the building to convey the runoff to Spring Creek. All drainage was designed to drain away from the building or into the pipeline. Drain pipe under the stone fill was designed to convey influent water away from the building thus reducing the potential for sinkhole formation. Trees and shrubs were placed in large confined planters to prevent excess water from entering the groundwater system (Foose and Humphreville, 1979).

124.0	Leave Chocolate World, right onto Park Boulevard.
124.6	Traffic light. Turn left onto Rt. 39 east, right lane is best.
126.5	Traffic light, continue straight.
126.8	Rt. 39 ends Junction with Rt. 322 west. Take ramp onto Rt. 322 west towards Harrisburg.
129.0	Cross Swatara Creek.
129.5	Traffic light at Chambers Hill Road.
132.9	Eisenhower Interchange. Take exit to Derry Street, I-83 North.
133.3	Stay left, continue on ramp and merge onto I-83.
134.5	Exit for Union Deposit, continue to traffic light, make a right onto Union Deposit Road.
134.8	Traffic light, turn right onto East Park Drive.
135.1	Turn right into Sheraton parking lot. End of trip.

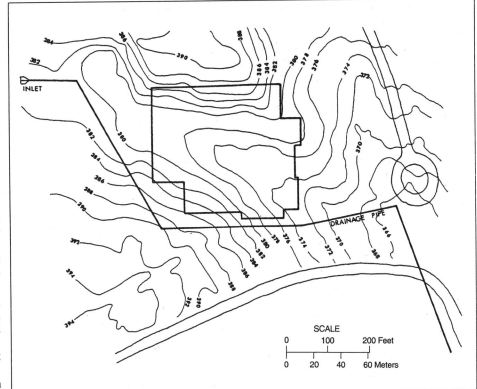

Figure 23. Map showing the original topography of the Chocolate World site (Foose and Humphreville, 1979).

Acknowledgements. The field trip organizer wishes to acknowledge the following individuals, businesses, and municipalities for their help in contributing to the content of the guidebook and completing the many logistical exercises that went into the making of this field trip. Helen Delano, Jim Dolimpio, Bill Sevon, John Barnes, Caron O'Neil, Tina Miles, J. Peter Wilshusen (deceased) PA Geological Survey; David Hopkins, Baker Refractories; Carol deWet, Franklin and Marshall College; Doug Chichester, U.S. Geological Survey; Albert E. Becher; Jeff Lieberfinger and Rick Hoover, Science Applications International Corporation; Boiling Springs Tavern; Sal at Aniles Ristorante & Pizzaria; Sherry Capello, Palmyra Borough; Indian Echo Caverns; Ken Patrick, Hershey Entertainment, Inc; Pat Ludwick. A special thanks to the PA Geological Survey for the time to organize the field trip.

REFERENCES

Aissaoui, D.M., 1985, Botryoidal aragonite and its diagenesis, Sedimentology, v. 32, p. 345-361.

Barnaby R.J. and Read, J.F., March 1990, Carbonate ramp to rimmed shelf evolution: Lower to Middle continental margin, Virginia Appalachians: Geological Society of America Bulletin, v. 102, p. 391-404.

Becher, A.E., 1991, Hydrogeology and the source of Boiling Springs, springs, Cumberland County, Pennsylvania: Field Conference of Pennsylvania Geologists, Inc., 56th Annual Field Conference Guidebook, p. 189-193.

Becher, A.E. and Root, S.I., 1981, Groundwater and geology of the Cumberland Valley, Cumberland County, Pennsylvania: Penn. Geological Survey, 4th ser., Wat. Res. Rpt. 50, 95 p.

Belesky, R.M., Hardy, H.R., Jr. and Strouse, F.F., 1987, Sinkholes in airport pavements: Engineering implications, in Beck, B.F. and Wilson, W.L., eds., Karst Hydrogeology: Engineering and Environmental Applications, Proceedings of the 2nd Multidisciplinary Conference on Sinkholes and the Environmental Impacts of Karst, Orlando, Florida, February 9-11, 1987, A.A. Balkema/Rotterdam/Boston, publisher, p.411-417.

Berg, T.M., Edmunds, W.E., Geyer, A.R., and others, 1980, Geologic map of Pennsylvania: Pennsylvania Geological Survey, 4th Series, Map 1, scale 1:250,000, 3 sheets.

Berg, T.M. and Dodge, C.M., 1981, Atlas of preliminary geologic quadrangle maps of Pennsylvania: Pennsylvania Geological Survey, 4th series, Map 61, 623 maps, scale 1:62,500.

Berg, T.M., McInerney, M.K., Way, J.H. and MacLachlan, D.B., 1983, Stratigraphic correlation chart of Pennsylvania: Pennsylvania Geological Survey, 4th series, General Geology Report 75, 1 sheet.

Berry, M.H., 1986, Sinkhole Investigation: South-Central Pennsylvania, in Dilamarter, R.R., Donkin, J., Ammerman, L., and Williams, P., eds., Proceedings of the Environmental Problems in Karst Terranes and their Solutions Conference, October 28-30, 1986, Bowling Green, Kentucky, National Water Well Association, publisher, p. 325-344.

Chichester, D.C., 1996, Hydrogeology of, and simulation of ground-water flow in, a mantled carbonate-rock system, Cumberland Valley, Pennsylvania: U.S. Geological Survey, Wat. -Res. Inv. Rep. 94-4090, 39 p.

Clark, G. Michael, 1991, South Mountain Geomorphology in, Sevon, W. D, and Potter N., eds. , Geology in the South Mountain Area, Pennsylvania: Guidebook for the 56th Annual Field Conference of Pennsylvania Geologists, p. 55-94.

Cloos, E. and Pettijohn, F.J., 1973, Southern border of the Triassic Basin, west of York, Pennsylvania: Fault or overlap?: Geol. Soc. of Am. Bull., v. 84, p. 523-536.

Dahlin, T. and Bernstone, C., 1997, A roll-along technique for 3D resistivity data acquisition with multi-electrode arrays, Procs. SAGEEP'97 (Symposium on the Application of Geophysics to Engineering and Environmental Problems), Reno, Nevada, March 23-26 1997, vol 2, 927-935.

Demicco, R.V., and Mitchell, R.W., 1982, Facies of the Great American Bank in the central Appalachians, in Lyttle, P.T., ed., Central Appalachian Geology - Field Trip No. 7 Guidebook, NE-SE Geological Society of America Meeting, American Geological Institute, Falls Church, Virginia, p. 171-266.

Demicco, R.V., 1985, Platform and off-platform carbonates of the Upper Cambrian of western Maryland, USA: Sedimentology, v. 32, p. 1-22.

deWet, C.B. and Gaswirth, S.B., 1999, Cambrian fibrous cements within an algal-dominated shelf margin complex, the Ledger Formation, York County, Pennsylvania: in Beck, B.F., ed., Seventh Multidisciplinary Conference on Sinkholes and the Engineering and Environmental Impacts of Karst, Harrisburg, PA, A.A. Balkema, Rotterdam.

Dickson, J.A.D., 1966, Carbonate identification and genesis as revealed by staining: Journal of Sedimentary Petrology, v. 36, p. 491-505.

Evans, K.R., Rowell, A.J., and Rees, M.N., 1995, Sea-level changes and stratigraphy of the Nelson Limestone (Middle Cambrian), Neptune Range, Antarctica, Journal of Sedimentary Research, v. B65, p. 32-43.

Faill, R.T., in press, Carbonate Fanglomerates – Indicators of a non-faulted basin?, Geological Society of America Abstracts with Program for 1999 Northeast Section meeting.

Flippo, H.N., Jr., 1974, Springs of Pennsylvania: Pennsylvania Department of Environmental Resources, Water Resources Bulletin 10, 46 p.

Foose, R.M., 1953, Ground-water behavior in the Hershey Valley, Pennsylvania: Geol. Soc. of Am. Bull., v. 64, p. 623-646.

Foose, R.M. and Humphreville, J.A., 1979, Engineering geological approaches to foundations in the karst terrain of the Hershey Valley: Assoc. of Eng. Geol. Bull., v. XVI, no. 3, p. 355-382.

Ganis, G.R. and Hopkins, D., 1984, Stratigraphy, structural style and economic geology of the York-Hanover Valley: Harrisburg Area Geological Society, 3rd Annual Field Trip, 51 p.

Geyer, A.R., 1970, Geology, mineral resources and environmental geology of the Palmyra quadrangle, Lebanon and Dauphin counties: Penn. Geological Survey, 4th ser., Atlas 157D, 46 p.

Geyer, A.R. and Wilshusen, J.P., 1982, Engineering characteristics of the rocks of Pennsylvania: Pennsylvania Geological Survey, 4th series, Environmental Geology Report 1, 300 p.

Gohn, G.S., 1976, Sedimentology, stratigraphy, and paleogeography of Lower Paleozoic carbonate rocks, Conestoga Valley, southeastern Pennsylvania, Unpublished Ph.D. dissertation, University of Delaware, 315 p.

Griffiths D.H. and Barker R.D. 1993, Two-dimensional resistivity imaging and modeling in areas of complex geology. Jour. of App. Geophysics, v. 29, 211-226.

Hopkins, D.A., 1985, Refractory dolomite production in a geologically complex area: in Glaser, J.D., and Edwards, J., eds., Industrial minerals of the mid-Atlantic states (Proceedings of the twentieth Forum of Industrial Minerals): Maryland Geological Survey, Spec. Pub. No. 2, p. 117-124.

Hostetter, H.H., 1971, The Body, Mind and Soul of Milton Snavely Hershey: p. 25.

Hummelstown Sun, Friday, February 12, 1892, The Debbil Hole; Discovery of the New Cave: v. XXI, no. 13.

Inners, J.D., Wilshusen, J.P., and Branthoover, G.L., 1978, Engineering geology and electric power developments on the lower Susquehanna River, Pennsylvania and Maryland: Fieldtrip guidebook 1978 Annual Meeting, Association of Engineering Geologists, Hershey, PA, Field Trip No. 1, p. 3-110.

James, N.P., 1983, Reef Environment, in Scholle, P.A., Bebout, D.G., and Moore, C.H. eds., Carbonate Depositional Environments, The American Association of Petroleum Geologists, Tulsa, OK, Ch. 8.

James, N.P., and Klappa, C.F., December 1983, Petrogenesis of Early Cambrian reef limestones, Labrador, Canada: Journal of Sedimentary Petrology, v. 53, n. 4, p. 1051-1096.

Kendall, A.C., 1985, Radiaxial fibrous calcite: A reappraisal: SEPM Special Publications n. 36, p. 59-74.

Knight, F.J., 1971, Geologic problems of urban growth in limestone terrain's of Pennsylvania, Association of Engineering Geologists Bulletin, v.8, no. 1, p. 91-101.

Kochanov, W.E., 1988, Sinkholes and karst-related features of Lebanon County, Pennsylvania: Pennsylvania Geological Survey, 4th ser., Open-File Report 8802, five maps, scale 1:24,000.

_____, 1989, Karst mapping and applications to regional land management practices in the Commonwealth of Pennsylvania: in Beck, B.F., ed., Engineering and Environmental Impacts of Sinkholes and Karst, Proceedings of the 3rd Multidisciplinary Conference on Sinkholes and the Engineering and Environmental Impacts of Karst, St. Petersburg Beach, Florida, October 2-4, 1989, A.A. Balkema, Rotterdam, p. 363-368.

_____, 1993, Areal Analysis of karst data in the Great Valley of Pennsylvania: in Beck, B.F., ed., Engineering and Environmental Impacts of Sinkholes and Karst, Proceedings of the 4th Multidisciplinary Conference on Sinkholes and the Engineering and Environmental Impacts of Karst, Panama City, Florida, January 25-27, 1993, A.A. Balkema, Rotterdam, p. 37-41.

_____, 1995, Storm-water management and sinkhole occurrence in the Palmyra area, Lebanon County, Pennsylvania, in Beck, B.F. (ed.), Karst Geohazards, Engineering and Environmental Problems in Karst Terrane, Proceedings of the 5th Multidisciplinary Conference on Sinkholes and the Engineering and Environmental Impacts of Karst, Gatlinburg, Tennessee, 1995, A.A. Balkema, Rotterdam, p. 285-290.

_____, 1997, The mummy's shroud: Pennsylvania Geology, v. 28, no. 1/2, p.2-8.

Kochanov, W.E., and Sullivan, R.M., 1994, Finding Phytosaurs in Pennsylvania: the story of Stahle, Sinclair, and Zions View: Pennsylvania Geology, v. 25, no. 1, p. 3-8.

Lapham, D.M. and Geyer, A.R., 1969, Mineral collecting in Pennsylvania: Pennsylvania Geological Survey, 4th ser., General Geology Report 33, 3rd Edition, 164 p.

Leberfinger, J.L., Reccelli-Snyder, H.L., and Hoover, R.A., 1998, *3D* Electrical Imaging: A method for characterization of bedrock fracture zones, voids, and potential collapse features: Science Applications International Corporation, Middletown, Pennsylvania.

Li Y., and Oldenburg, D.W., 1992, Approximate inverse mappings in DC resistivity problems: Geophysical Jour. Int'l., v. 109, p. 343-362.

Loke, M.H. and Barker R.D., 1996a, Rapid Least-squares inversion of apparent resistivity pseudosections by a quasi-Newton Method: Geophysical Prospecting, v.44, p.131-152.

Loke, M.H., and Barker, R.D., 1996b, Practical techniques for 3D resistivity surveys and data inversion: Geophysical Prospecting, v. 44, p. 499-523.

Loke, M.H., 1997, RES3DINV ver. 2.0 for Window 3.1, 95, and NT, Advanced Geosciences, Inc., 66 p.

MacLachlan, D.B., 1967, Structure and stratigraphy of the limestones and dolomites of Dauphin County, Pennsylvania: Penn. Geological Survey, Bull. G 44, 168 p.

Marshall, J.F., 1983, Submarine cementation in a high-energy platform reef; One Tree Reef, southern Great Barrier Reef, Journal of Sedimentary Petrology, v. 53, p. 1133-1149.

Mazullo, S.J., Bischoff, W.D., and Lobitzer, H., 1990, Diagenesis of radiaxial fibrous calcites in a subunconformity shallow-burial setting: Upper Triassic and Liassic, Northern Calcareous Alps, Austria: Sedimentology, v. 37, p. 407-425.

McGlade, W.G. and Geyer, A.R., 1976, Environmental geology of the Greater Harrisburg metropolitan area, Pennsylvania Geological Survey, 4th Series, Environmental Geology Report 4, 42 p.

Miller, B.L., 1934, Limestones of Pennsylvania: Pennsylvania Geological Survey, 4th Series, Mineral Resources Report 26, 729 p.

Myers, P.B., Jr. and Perlow, M., Jr. (1984), Development, occurrence and triggering mechanisms of sinkholes in the carbonate rocks of the Lehigh Valley, eastern Pennsylvania, in Beck, B.F., ed., Sinkholes: Their geology, engineering and environmental impact, Proceedings of the First Multidisciplinary Conference on Sinkholes, Orlando, Florida, October 15-17, 1984, A.A. Balkema/Rotterdam/Boston, p. 111-115.

O'Neill, B.J., 1964, Atlas of Pennsylvania's mineral resources - Part 1. Limestones and dolomites of Pennsylvania: Pennsylvania Geological Survey, Mineral Resource Report 50, 6 maps, text, tables, scale 1:250,000.

Parizek, R.R., White, W.B., and Langmuir, D., 1971, Hydrogeology and geochemistry of folded and faulted carbonate rocks of the central Appalachian type and related land use problems, Pennsylvania State University, Earth and Mineral Science Experimental Station Circular 82, 181 p.

Parizek, R.R. and White, W.B, 1985, Applications of Quaternary and Tertiary geological factors to environmental problems in central Pennsylvania: in 50th Annual Field Conference of Pennsylvania Geologists Guidebook, State College, Pa., p. 63-119.

Pierce, K.L., 1965, Geomorphic significance of a Cretaceous deposit in the Great Valley of southern Pennsylvania: U.S. Geological Survey, Professional Paper 525-C, p. C

Potter, N. and Niemitz, J. W., 1997, Denudation rate for the Great Valley at Yellow Breeches Creek, Cumberland Co., Pennsylvania: Geological Society of America Abstracts with Programs, vol. 29, #1, p. 73.

Pratt, B.R., 1984, *Epiphyton* and *Renalcis* - Diagenetic microfossils from calcification of coccoid blue-green algae, Journal of Sedimentary Petrology, v. 54, p. 948-971.

Reich, J.R., Jr., 1974, Caves of southeastern Pennsylvania: Penn. Geological Survey, Gen. Geol. Rpt. 65, 120 p.

Roman, Irwin, 1951, Resistivity Reconnaissance in American Society of Testing and Materials: Symposium on Surface and subsurface reconnaissance: American Society of Testing Materials Special Technical Publication 122, p. 171-220.

Root, S.I., 1978, Geology and Mineral Resources of the Carlisle and Mechanicsburg Quadrangles, Cumberland County: Pennsylvania. Pennsylvania Geological Survey, 4th Series, Atlas 138ab, 1 sheet.

Royer, D.W., 1983, Summary groundwater resources of Lebanon County, Pennsylvania: Penn. Geological Survey, Water Resources Report 55, 84 p.

Saller, A.H., 1986, Radiaxial calcite in lower Miocene strata, subsurface Enewetak Atoll, Journal of Sedimentary Petrology, v. 56, p. 743-762.

Sami, T.T., and James, N.P., January 1996, Synsedimentary cements as Paleoproterozoic platform building blocks, Pethei Group, Northwestern Canada: Journal of Sedimentary Research, v. 66, p. 209-222.

Sevon, W.D. and Potter, N., Jr., eds., 1991, Geology in the South Mountain area: Field Conference of Pennsylvania Geologists, Inc., 56th Annual Field Conference of Pennsylvania Geologists Guidebook, 236 p.

Sevon, W. D., 1991, Stop 6 description, in Sevon, W. D, and Potter N., eds., Geology in the South Mountain Area, Pennsylvania: Guidebook for the 56th Annual Field Conference of Pennsylvania Geologists, p. 176-188.

Smeltzer, B.L., 1964, Caves of the southern Cumberland Valley: Nat. Speleo. Soc., Mid-Atlantic Region, Bull. 6.

Smith, R.C., II, 1977, Zinc and lead occurrences in Pennsylvania: Pennsylvania Geological Survey, 4th ser., Mineral Resources Report 72, 318 p.

Smith, R.E, and Riddle, D.J., 1984, Terrain conductivity and its application in site assessment of karstic terrain in central Pennsylvania and Maryland: in Proceedings, Geologic and geotechnical problems in karstic limestones of the northeastern United States, Association of Engineering Geologists/American Society of Civil Engineers Meeting, Frederick, Maryland, May 24-25, 21 p.

Stone, R.W., 1930, Pennsylvania Caves: Pennsylvania Geological Survey 4th ser., Bulletin G 3, 63 p.

Sumner, D.Y., and Grotzinger, J.P., May 1996, Herringbone Calcite: Petrography and environmental significance: Journal of Sedimentary Research, v. 66, n. 3, p. 419-429.

Taylor, J.F., and Durika, N.J., 1990, Lithologies, trilobite faunas, and correlation of the Kinzers Ledger, and Conestoga Formations in the Conestoga Valley, in Scharnberger, C.K., ed., Carbonates, Schists, and Geomorphology in the vicinity of the Lower Reaches of the Susquehanna River, 55th Annual Field Conference of Pennsylvania Geologists Guidebook, p. 136-155.

Trojan, E.J. 1974, The Route 202 sinkhole - A case history: in Crawford, W.A., and Crawford, M.L., eds., Geology of the Piedmont of southeastern Pennsylvania, Guidebook, 39th Annual Field Conference of Pennsylvania Geologists, King of Prussia, Pennsylvania, p. 32-40.

Turner, E.C., Narbonne, G.M., and James, N.P., 1993, Neoproterozoic reef microstructures from the Little Dal Group, northwestern Canada, Geology, v. 21, p. 259-262.

Van Nostrand, R.G. and Cook, K.L., 1966, Interpretation of Resistivity Data, U.S. Geological Survey Prof. Pap. 499.

White, W.B., 1966, Correlation of caves and erosion surfaces in southern Cumberland Valley of Pennsylvania: Nat. Spel. Soc. Bull. 28, p. 92-93.

White, W.B., 1976, Geology and Biology of Pennsylvania Caves, Pennsylvania Geological Survey, 4th ser., General Geology Report 66.

White, E.L., Aron, G., and White, W.B., 1986, The influence of urbanization on sinkhole development in central Pennsylvania: Environmental Geology and Water Sciences, v. 8, no. 1&2, p. 91-97 also in Beck, B.F., ed., Sinkholes: Their geology, engineering and environmental impact, Proceedings of the First Multidisciplinary Conference on Sinkholes, Orlando, Florida, October 15-17, 1984, A.A. Balkema/Rotterdam/Boston, publisher, p.275-282.

White, E.L., and White, W.B. 1979, Quantitative morphology of landforms in carbonate rock basins in the Appalachian Highlands, Geological Society of America Bulletin, v. 90, p. 385-396.

Wilshusen, J.P., 1979a, Geologic hazards in Pennsylvania: Pennsylvania Geological Survey, 4th ser., Educational Series 9, 56 p.

Wilshusen, J.P., 1979b, Environmental geology of the greater York area, York County: Pennsylvania Geological Survey, 4th Series, Environmental Geology Report 6, 4 maps, scale 1:50,000.

Wilshusen, J.P. and Kochanov, W.E., in press, Land Subsidence: Carbonate Terrane: in Shultz, C.H., ed., The Geology of Pennsylvania, Pittsburgh Geological Society and Pennsylvania Geological Survey, Chapter 49, Part A.

Wright, V.P., 1994, Early Carboniferous carbonate systems: an alternative to the Cainozoic paradigm, Sedimentary Geology, v. 93, p. 1-5.

Email addresses

Abdullah, W. – wabdulla@kisr.edu.kw
Aley, T. – oul@tri-lakes.net

Bakalowicz, M. – 1: baka@dstu.univ-montp2.fr
                  2: m.bakalowicz@brgm.fr
Beck, B.F. – pelaor@usit.net
Benson, R.C. – info@technos-inc.com
Buskirk Jr, E.D. – dredb@paonline.com

Campbell, C.W. – wcampbel@hsv.vista-inc.com
Cohen, H.A. – hcohen@sspa.com
Colpitts, R.M. – rcolpits@zeolite.reno.nv.us
Connair, D.P. – cindpc@dames.com
Cooley, T.L. – tlcool0@pop.uky.edu
Cooper, A.H. – 1: a.cooper@bgs.ac.uk
              2: tony.cooperb@btinternet.com
Crawford, N.C. – canda@premiernet.net
Currens, J.C. – currens@kgs.mm.uky.edu

Daly, D. – dalydona@tec.irlgov.ie
Drew, D. – ddrew@tcd.ie
Drumm, E.C. – edrumm@utk.edu
Dunscomb, M.H. – markd@schnabel-eng.com

Engel, S.A. – cinsae@dames.com

Faivre, S. – Sanja.Faivre@public.srce.hr
Fenstermacher, R.F. – rfenster@wlgore.com
Forth, R.A. – r.a.forth@ncl.ac.uk

Gary, M.K. – mgary@mde.state.md.us
George, L.D. – pela@dbtech.net
George, S.E. – segeorge@qstmail.com
Gray, R.E. – r.gray@gaiconsultants.com
Gubbels, T. – tgubbels@eos.hitc.com

Hubbard Jr, D.A. – dhubbard@geology.state.va.us
Hyatt, J.A. – jhyatt@valdosta.edu

Jacobs, P.M. – jacobsp@uwwvax.uww.edu
Jancin, M. – mjancin@chester-engineers.com
Jenkins, S. – tjcl@worldnet.att.net

Kannan, R.C. – 1: rkannan@gate.net
              2: WESLargo@aol.com
Kaufmann, O. – olivier.kaufmann@fpms.ac.be
Kochanov, W.E. – kochanov.william@dcnr.state.pa.us

Lamont-Black, J. – john.lamont-black@ncl.ac.uk
Lane, S.R. – slane@julian.uwo.ca
Lange, A.L. – karstgeo@compuserve.com
Laughland, J.C. – jeffco@intrepid.net
Lolcama, J.L. – jlolcama@sspa.com

Mason, J.M. – masonjm@obg.com
McDonald, R. – robm@terradat.co.uk
Memon, B.A. – pela@dbtech.net
Mortimore, R. – rory.mortimore@btinternet.com
Mumtaz Azmeh, M. – pela@dbtech.net
Murray, B.S. – brian.s.murray@cpmx.saic.com

Nettles, N.S. – nastynet@gte.net
Nyquist, J.E. – nyq@nimbus.temple.edu

Orndorff, R. – rorndorf@usgs.gov

Paukštys, B. – bernardas@iti.lt
Pavelek II, M.D. – mdp@iwl.net
Ponta, G.M. – pela@dbtech.net

Ray, J.A. – ray-j@nrdep.nr.state.ky.us
Redwine, J.C. – jcredwin@southernco.com

Rehwoldt, E. – erehwoldt@schnabel-eng.com
Reiffsteck, Ph. – Philippe.Reiffsteck@lcpc.fr
Reith, C. – creith@schnabel-eng.com
Rice, J.E. – rice.j@adlittle.com
Rodríguez Molina, C. – Suelos@Caribe.net
Roth, M.J.S. – rothm@lafayette.edu
Russill, N. – nick@terradat.co.uk

Šebela, S. – 1: sebela@ns.zrc-sazu.si
           2: izrk@zrc-sazu.si
Smart, C.C. – csmart@julian.uwo.ca
Smith. T.J. – tsmith@bcieng.com
Smoot, J.L. – jsmoot@utk.edu
Stephenson, J.B. – pelaor@usit.net
Strum, S. – stuart-strum@mail.ehnr.state.nc.us
Sullivan, S.M. – sullivansean@mindspring.com

Terry, M.W. – mwterry@haywardbaker.com

Tolmachev, Vl.V. – karst@kis.ru
Tonkin, M.J. – mtonkin@sspa.com
Tucker, R.B. – canda@premiernet.net

Vázquez Castillo, L. – Suelos@Caribe.net
Vesper, D.J. – vesper.d@geosc.psu.edu

Walker, S.E. – SEW@HaleyAldrich.com
Wall, D.B. – david.wall@pca.state.mn.us
Wallace Pitts, M. – pela@dbtech.net
Whitman, D. – whitmand@fiu.edu
Wilkes, H.P. – hapelly@surfsouth.com

Yang, M.Z. – yang@engr.utk.edu
Younger, P.L. – p.l.younger@ncl.ac.uk
Yuhr, L. – info@technos-inc.com

Zhou, W. – pelaor@usit.net

Author index

Abdullah, W.A. 123
Aley, T. 225
Anikeev, A.V. 77

Bakalowicz, M. 331
Balashova, T.A. 423
Becher, A.E. 449
Beck, B.F. 187
Belgeri, J.J. 157
Benson, R.C. 195
Bonniface, J.P. 141
Buskirk Jr, E.D. 263

Cadden, A.W. 149
Campbell, C.W. 383
Cohen, H.A. 51
Colpitts, R.M. 401
Connair, D.P. 287
Cooley, T. 129
Cooper, A.H. 141, 407
Crawford, N.C. 203, 307
Currens, J.C. 85

Daly, D. 267
Davies, R. 243
Delano, H.L. 449
deWet, C. 449
Drew, D. 267
Drumm, E.C. 373
Dunscomb, M.H. 219

Ebaugh, W.F. 315
Engel, S.A. 287
Ericson, W.A. 165

Faivre, S. 25
Fenstermacher, R.F. 351
Fischer, J.A. 359
Fischer, J.J. 359
Forth, R.A. 141

Gabriel, W.J. 323

Gary, M.K. 273
Gaswirth, S.B. 449
George, L.D. 107, 293
George, S. 225
Gray, R.E. 31
Green, Th.S. 107
Gubbels, T. 67

Hahman, W.R. 401
Harrison, R.W. 57
Hoover, R.A. 449
Hopkins, D.A. 449
Hubbard Jr, D.A. 413
Hyatt, J.A. 37

Jacobs, P.M. 37
Jancin, M. 315
Jenkins, S.A. 45

Kannan, R.C. 135
Kaufmann, O. 91
Kochanov, W.E. 449

Lamont-Black, J. 97, 141
Lane, S.R. 301
Lange, A.L. 225, 253
Laughland, J.C. 389
Lewis, M.A. 203
Lieberfinger, J.L. 449
Lolcama, J.L. 51

Mackey, C. 247
Mackey, J.R. 247
Mason, J.M. 323
Matzat, S.L. 179
McCann, M.S. 395
McDonald, R. 243
McDowell, R.C. 57
Memon, B.A. 107, 365
Mihevc, A. 419
Mollah, M.A. 123

Mortimore, R. 97
Mumtaz Azmeh, M. 365
Murray, B.S. 287

Naples III, C.J. 149
Nettles, N.S. 135
Nyquist, J.E. 45, 247

Orndorff, R.C. 57
Ottoson, R.S. 359

Patton, A.F. 107
Paukštys, B. 103
Pavelek II, M.D. 263
Pidyashenko, S.E. 423
Plagnes, V. 331
Ponta, G.M. 293

Quinif, Y. 91

Ray, J.A. 85
Reccelli-Snyder, H.L. 449
Redwine, J.C. 111
Rehwoldt, E. 219
Reiffsteck, Ph. 25
Reith, C.M. 149
Rice, J.E. 351
Rodriguez-Molina, C. 431
Roth, M.J.S. 247
Russill, N. 243

Saunders, J.M. 407
Šebela, S. 57, 419
Siegel, D.I. 323
Siegel, T.C. 157
Slabe, T. 419
Smart, C.C. 301
Smith, T.J. 165
Smoot, J.L. 395
Srinivasa Gowd, S. 339
Stephenson, J.B. 187

Strasz, R. 263
Strum, S. 63
Sudheer, A.S. 339
Sullivan, S.M. 383

Taylor, D.G. 253
Terry, M.W. 157
Tolmachev, V.V. 171, 423
Tonkin, M.J. 51
Tucker, R.B. 307

Turka, R.J. 31

Vazquez-Castillo, L. 431
Vesper, D.J. 345, 351

Walker, S.E. 179
Wall, D.B. 279
Wallace Pitts, M. 365
Weary, D.J. 57
Webster, J.A. 203

Weems, R.E. 57
White, W.B. 3
Whitman, D. 67
Wilkes, H.P. 37
Winter, S.A. 203

Yang, M.Z. 373
Younger, P.L. 141
Yuhr, L. 195

Zhou, W. 187